Communications
in Computer and Information Science   1968

Rationale

The CCIS series is devoted to the publication of proceedings of computer science conferences. Its aim is to efficiently disseminate original research results in informatics in printed and electronic form. While the focus is on publication of peer-reviewed full papers presenting mature work, inclusion of reviewed short papers reporting on work in progress is welcome, too. Besides globally relevant meetings with internationally representative program committees guaranteeing a strict peer-reviewing and paper selection process, conferences run by societies or of high regional or national relevance are also considered for publication.

Topics

The topical scope of CCIS spans the entire spectrum of informatics ranging from foundational topics in the theory of computing to information and communications science and technology and a broad variety of interdisciplinary application fields.

Information for Volume Editors and Authors

Publication in CCIS is free of charge. No royalties are paid, however, we offer registered conference participants temporary free access to the online version of the conference proceedings on SpringerLink (http://link.springer.com) by means of an http referrer from the conference website and/or a number of complimentary printed copies, as specified in the official acceptance email of the event.

CCIS proceedings can be published in time for distribution at conferences or as post-proceedings, and delivered in the form of printed books and/or electronically as USBs and/or e-content licenses for accessing proceedings at SpringerLink. Furthermore, CCIS proceedings are included in the CCIS electronic book series hosted in the SpringerLink digital library at http://link.springer.com/bookseries/7899. Conferences publishing in CCIS are allowed to use Online Conference Service (OCS) for managing the whole proceedings lifecycle (from submission and reviewing to preparing for publication) free of charge.

Publication process

The language of publication is exclusively English. Authors publishing in CCIS have to sign the Springer CCIS copyright transfer form, however, they are free to use their material published in CCIS for substantially changed, more elaborate subsequent publications elsewhere. For the preparation of the camera-ready papers/files, authors have to strictly adhere to the Springer CCIS Authors' Instructions and are strongly encouraged to use the CCIS LaTeX style files or templates.

Abstracting/Indexing

CCIS is abstracted/indexed in DBLP, Google Scholar, EI-Compendex, Mathematical Reviews, SCImago, Scopus. CCIS volumes are also submitted for the inclusion in ISI Proceedings.

How to start

To start the evaluation of your proposal for inclusion in the CCIS series, please send an e-mail to ccis@springer.com.

Biao Luo · Long Cheng · Zheng-Guang Wu ·
Hongyi Li · Chaojie Li

Editors

Neural
Information Processing

30th International Conference, ICONIP 2023
Changsha, China, November 20–23, 2023
Proceedings, Part XIV

 Springer

Editors
Biao Luo ⓘ
School of Automation
Central South University
Changsha, China

Long Cheng ⓘ
Institute of Automation
Chinese Academy of Sciences
Beijing, China

Zheng-Guang Wu ⓘ
Institute of Cyber-Systems and Control
Zhejiang University
Hangzhou, China

Hongyi Li ⓘ
School of Automation
Guangdong University of Technology
Guangzhou, China

Chaojie Li ⓘ
School of Electrical Engineering
and Telecommunications
UNSW Sydney
Sydney, NSW, Australia

ISSN 1865-0929   ISSN 1865-0937 (electronic)
Communications in Computer and Information Science
ISBN 978-981-99-8180-9   ISBN 978-981-99-8181-6 (eBook)
https://doi.org/10.1007/978-981-99-8181-6

This Springer imprint is published by the registered company Springer Nature Singapore Pte Ltd.
The registered company address is: 152 Beach Road, #21-01/04 Gateway East, Singapore 189721, Singapore

Paper in this product is recyclable.

Preface

Welcome to the 30th International Conference on Neural Information Processing (ICONIP2023) of the Asia-Pacific Neural Network Society (APNNS), held in Changsha, China, November 20–23, 2023.

The mission of the Asia-Pacific Neural Network Society is to promote active interactions among researchers, scientists, and industry professionals who are working in neural networks and related fields in the Asia-Pacific region. APNNS has Governing Board Members from 13 countries/regions – Australia, China, Hong Kong, India, Japan, Malaysia, New Zealand, Singapore, South Korea, Qatar, Taiwan, Thailand, and Turkey. The society's flagship annual conference is the International Conference of Neural Information Processing (ICONIP). The ICONIP conference aims to provide a leading international forum for researchers, scientists, and industry professionals who are working in neuroscience, neural networks, deep learning, and related fields to share their new ideas, progress, and achievements.

ICONIP2023 received 1274 papers, of which 394 papers were accepted for publication in Communications in Computer and Information Science (CCIS), representing an acceptance rate of 30.93% and reflecting the increasingly high quality of research in neural networks and related areas. The conference focused on four main areas, i.e., "Theory and Algorithms", "Cognitive Neurosciences", "Human-Centered Computing", and "Applications". All the submissions were rigorously reviewed by the conference Program Committee (PC), comprising 258 PC members, and they ensured that every paper had at least two high-quality single-blind reviews. In fact, 5270 reviews were provided by 2145 reviewers. On average, each paper received 4.14 reviews.

We would like to take this opportunity to thank all the authors for submitting their papers to our conference, and our great appreciation goes to the Program Committee members and the reviewers who devoted their time and effort to our rigorous peer-review process; their insightful reviews and timely feedback ensured the high quality of the papers accepted for publication. We hope you enjoyed the research program at the conference.

October 2023

Biao Luo
Long Cheng
Zheng-Guang Wu
Hongyi Li
Chaojie Li

Organization

Honorary Chair

Weihua Gui   Central South University, China

Advisory Chairs

Jonathan Chan   King Mongkut's University of Technology
Thonburi, Thailand
Zeng-Guang Hou   Chinese Academy of Sciences, China
Nikola Kasabov   Auckland University of Technology, New Zealand
Derong Liu   Southern University of Science and Technology,
China
Seiichi Ozawa   Kobe University, Japan
Kevin Wong   Murdoch University, Australia

General Chairs

Tingwen Huang   Texas A&M University at Qatar, Qatar
Chunhua Yang   Central South University, China

Program Chairs

Biao Luo   Central South University, China
Long Cheng   Chinese Academy of Sciences, China
Zheng-Guang Wu   Zhejiang University, China
Hongyi Li   Guangdong University of Technology, China
Chaojie Li   University of New South Wales, Australia

Technical Chairs

Xing He   Southwest University, China
Keke Huang   Central South University, China
Huaqing Li   Southwest University, China
Qi Zhou   Guangdong University of Technology, China

Local Arrangement Chairs

Wenfeng Hu   Central South University, China
Bei Sun   Central South University, China

Finance Chairs

Fanbiao Li   Central South University, China
Hayaru Shouno   University of Electro-Communications, Japan
Xiaojun Zhou   Central South University, China

Special Session Chairs

Hongjing Liang   University of Electronic Science and Technology,
                                     China
Paul S. Pang   Federation University, Australia
Qiankun Song   Chongqing Jiaotong University, China
Lin Xiao   Hunan Normal University, China

Tutorial Chairs

Min Liu   Hunan University, China
M. Tanveer   Indian Institute of Technology Indore, India
Guanghui Wen   Southeast University, China

Publicity Chairs

Sabri Arik   Istanbul University-Cerrahpaşa, Turkey
Sung-Bae Cho   Yonsei University, South Korea
Maryam Doborjeh   Auckland University of Technology, New Zealand
El-Sayed M. El-Alfy   King Fahd University of Petroleum and Minerals,
                                     Saudi Arabia
Ashish Ghosh   Indian Statistical Institute, India
Chuandong Li   Southwest University, China
Weng Kin Lai   Tunku Abdul Rahman University of
                                     Management & Technology, Malaysia
Chu Kiong Loo   University of Malaya, Malaysia
Qinmin Yang   Zhejiang University, China
Zhigang Zeng   Huazhong University of Science and Technology,
                                     China

Publication Chairs

Zhiwen Chen   Central South University, China
Andrew Chi-Sing Leung   City University of Hong Kong, China
Xin Wang   Southwest University, China
Xiaofeng Yuan   Central South University, China

Secretaries

Yun Feng   Hunan University, China
Bingchuan Wang   Central South University, China

Webmasters

Tianmeng Hu   Central South University, China
Xianzhe Liu   Xiangtan University, China

Program Committee

Rohit Agarwal   UiT The Arctic University of Norway, Norway
Hasin Ahmed   Gauhati University, India
Harith Al-Sahaf   Victoria University of Wellington, New Zealand
Brad Alexander   University of Adelaide, Australia
Mashaan Alshammari   Independent Researcher, Saudi Arabia
Sabri Arik   Istanbul University, Turkey
Ravneet Singh Arora   Block Inc., USA
Zeyar Aung   Khalifa University of Science and Technology,
                                  UAE
Monowar Bhuyan   Umeå University, Sweden
Jingguo Bi   Beijing University of Posts and
                                  Telecommunications, China
Xu Bin   Northwestern Polytechnical University, China
Marcin Blachnik   Silesian University of Technology, Poland
Paul Black   Federation University, Australia
Anoop C. S.   Govt. Engineering College, India
Ning Cai   Beijing University of Posts and
                                  Telecommunications, China
Siripinyo Chantamunee   Walailak University, Thailand
Hangjun Che   City University of Hong Kong, China

Wei-Wei Che	Qingdao University, China
Huabin Chen	Nanchang University, China
Jinpeng Chen	Beijing University of Posts & Telecommunications, China
Ke-Jia Chen	Nanjing University of Posts and Telecommunications, China
Lv Chen	Shandong Normal University, China
Qiuyuan Chen	Tencent Technology, China
Wei-Neng Chen	South China University of Technology, China
Yufei Chen	Tongji University, China
Long Cheng	Institute of Automation, China
Yongli Cheng	Fuzhou University, China
Sung-Bae Cho	Yonsei University, South Korea
Ruikai Cui	Australian National University, Australia
Jianhua Dai	Hunan Normal University, China
Tao Dai	Tsinghua University, China
Yuxin Ding	Harbin Institute of Technology, China
Bo Dong	Xi'an Jiaotong University, China
Shanling Dong	Zhejiang University, China
Sidong Feng	Monash University, Australia
Yuming Feng	Chongqing Three Gorges University, China
Yun Feng	Hunan University, China
Junjie Fu	Southeast University, China
Yanggeng Fu	Fuzhou University, China
Ninnart Fuengfusin	Kyushu Institute of Technology, Japan
Thippa Reddy Gadekallu	VIT University, India
Ruobin Gao	Nanyang Technological University, Singapore
Tom Gedeon	Curtin University, Australia
Kam Meng Goh	Tunku Abdul Rahman University of Management and Technology, Malaysia
Zbigniew Gomolka	University of Rzeszow, Poland
Shengrong Gong	Changshu Institute of Technology, China
Xiaodong Gu	Fudan University, China
Zhihao Gu	Shanghai Jiao Tong University, China
Changlu Guo	Budapest University of Technology and Economics, Hungary
Weixin Han	Northwestern Polytechnical University, China
Xing He	Southwest University, China
Akira Hirose	University of Tokyo, Japan
Yin Hongwei	Huzhou Normal University, China
Md Zakir Hossain	Curtin University, Australia
Zengguang Hou	Chinese Academy of Sciences, China

Lu Hu	Jiangsu University, China
Zeke Zexi Hu	University of Sydney, Australia
He Huang	Soochow University, China
Junjian Huang	Chongqing University of Education, China
Kaizhu Huang	Duke Kunshan University, China
David Iclanzan	Sapientia University, Romania
Radu Tudor Ionescu	University of Bucharest, Romania
Asim Iqbal	Cornell University, USA
Syed Islam	Edith Cowan University, Australia
Kazunori Iwata	Hiroshima City University, Japan
Junkai Ji	Shenzhen University, China
Yi Ji	Soochow University, China
Canghong Jin	Zhejiang University, China
Xiaoyang Kang	Fudan University, China
Mutsumi Kimura	Ryukoku University, Japan
Masahiro Kohjima	NTT, Japan
Damian Kordos	Rzeszow University of Technology, Poland
Marek Kraft	Poznań University of Technology, Poland
Lov Kumar	NIT Kurukshetra, India
Weng Kin Lai	Tunku Abdul Rahman University of Management & Technology, Malaysia
Xinyi Le	Shanghai Jiao Tong University, China
Bin Li	University of Science and Technology of China, China
Hongfei Li	Xinjiang University, China
Houcheng Li	Chinese Academy of Sciences, China
Huaqing Li	Southwest University, China
Jianfeng Li	Southwest University, China
Jun Li	Nanjing Normal University, China
Kan Li	Beijing Institute of Technology, China
Peifeng Li	Soochow University, China
Wenye Li	Chinese University of Hong Kong, China
Xiangyu Li	Beijing Jiaotong University, China
Yantao Li	Chongqing University, China
Yaoman Li	Chinese University of Hong Kong, China
Yinlin Li	Chinese Academy of Sciences, China
Yuan Li	Academy of Military Science, China
Yun Li	Nanjing University of Posts and Telecommunications, China
Zhidong Li	University of Technology Sydney, Australia
Zhixin Li	Guangxi Normal University, China
Zhongyi Li	Beihang University, China

Ziqiang Li	University of Tokyo, Japan
Xianghong Lin	Northwest Normal University, China
Yang Lin	University of Sydney, Australia
Huawen Liu	Zhejiang Normal University, China
Jian-Wei Liu	China University of Petroleum, China
Jun Liu	Chengdu University of Information Technology, China
Junxiu Liu	Guangxi Normal University, China
Tommy Liu	Australian National University, Australia
Wen Liu	Chinese University of Hong Kong, China
Yan Liu	Taikang Insurance Group, China
Yang Liu	Guangdong University of Technology, China
Yaozhong Liu	Australian National University, Australia
Yong Liu	Heilongjiang University, China
Yubao Liu	Sun Yat-sen University, China
Yunlong Liu	Xiamen University, China
Zhe Liu	Jiangsu University, China
Zhen Liu	Chinese Academy of Sciences, China
Zhi-Yong Liu	Chinese Academy of Sciences, China
Ma Lizhuang	Shanghai Jiao Tong University, China
Chu-Kiong Loo	University of Malaya, Malaysia
Vasco Lopes	Universidade da Beira Interior, Portugal
Hongtao Lu	Shanghai Jiao Tong University, China
Wenpeng Lu	Qilu University of Technology, China
Biao Luo	Central South University, China
Ye Luo	Tongji University, China
Jiancheng Lv	Sichuan University, China
Yuezu Lv	Beijing Institute of Technology, China
Huifang Ma	Northwest Normal University, China
Jinwen Ma	Peking University, China
Jyoti Maggu	Thapar Institute of Engineering and Technology Patiala, India
Adnan Mahmood	Macquarie University, Australia
Mufti Mahmud	University of Padova, Italy
Krishanu Maity	Indian Institute of Technology Patna, India
Srimanta Mandal	DA-IICT, India
Wang Manning	Fudan University, China
Piotr Milczarski	Lodz University of Technology, Poland
Malek Mouhoub	University of Regina, Canada
Nankun Mu	Chongqing University, China
Wenlong Ni	Jiangxi Normal University, China
Anupiya Nugaliyadde	Murdoch University, Australia

Toshiaki Omori	Kobe University, Japan
Babatunde Onasanya	University of Ibadan, Nigeria
Manisha Padala	Indian Institute of Science, India
Sarbani Palit	Indian Statistical Institute, India
Paul Pang	Federation University, Australia
Rasmita Panigrahi	Giet University, India
Kitsuchart Pasupa	King Mongkut's Institute of Technology Ladkrabang, Thailand
Dipanjyoti Paul	Ohio State University, USA
IIu Peng	Jiujiang University, China
Kebin Peng	University of Texas at San Antonio, USA
Dawid Połap	Silesian University of Technology, Poland
Zhong Qian	Soochow University, China
Sitian Qin	Harbin Institute of Technology at Weihai, China
Toshimichi Saito	Hosei University, Japan
Fumiaki Saitoh	Chiba Institute of Technology, Japan
Naoyuki Sato	Future University Hakodate, Japan
Chandni Saxena	Chinese University of Hong Kong, China
Jiaxing Shang	Chongqing University, China
Lin Shang	Nanjing University, China
Jie Shao	University of Science and Technology of China, China
Yin Sheng	Huazhong University of Science and Technology, China
Liu Sheng-Lan	Dalian University of Technology, China
Hayaru Shouno	University of Electro-Communications, Japan
Gautam Srivastava	Brandon University, Canada
Jianbo Su	Shanghai Jiao Tong University, China
Jianhua Su	Institute of Automation, China
Xiangdong Su	Inner Mongolia University, China
Daiki Suehiro	Kyushu University, Japan
Basem Suleiman	University of New South Wales, Australia
Ning Sun	Shandong Normal University, China
Shiliang Sun	East China Normal University, China
Chunyu Tan	Anhui University, China
Gouhei Tanaka	University of Tokyo, Japan
Maolin Tang	Queensland University of Technology, Australia
Shu Tian	University of Science and Technology Beijing, China
Shikui Tu	Shanghai Jiao Tong University, China
Nancy Victor	Vellore Institute of Technology, India
Petra Vidnerová	Institute of Computer Science, Czech Republic

Shanchuan Wan	University of Tokyo, Japan
Tao Wan	Beihang University, China
Ying Wan	Southeast University, China
Bangjun Wang	Soochow University, China
Hao Wang	Shanghai University, China
Huamin Wang	Southwest University, China
Hui Wang	Nanchang Institute of Technology, China
Huiwei Wang	Southwest University, China
Jianzong Wang	Ping An Technology, China
Lei Wang	National University of Defense Technology, China
Lin Wang	University of Jinan, China
Shi Lin Wang	Shanghai Jiao Tong University, China
Wei Wang	Shenzhen MSU-BIT University, China
Weiqun Wang	Chinese Academy of Sciences, China
Xiaoyu Wang	Tokyo Institute of Technology, Japan
Xin Wang	Southwest University, China
Xin Wang	Southwest University, China
Yan Wang	Chinese Academy of Sciences, China
Yan Wang	Sichuan University, China
Yonghua Wang	Guangdong University of Technology, China
Yongyu Wang	JD Logistics, China
Zhenhua Wang	Northwest A&F University, China
Zi-Peng Wang	Beijing University of Technology, China
Hongxi Wei	Inner Mongolia University, China
Guanghui Wen	Southeast University, China
Guoguang Wen	Beijing Jiaotong University, China
Ka-Chun Wong	City University of Hong Kong, China
Anna Wróblewska	Warsaw University of Technology, Poland
Fengge Wu	Institute of Software, Chinese Academy of Sciences, China
Ji Wu	Tsinghua University, China
Wei Wu	Inner Mongolia University, China
Yue Wu	Shanghai Jiao Tong University, China
Likun Xia	Capital Normal University, China
Lin Xiao	Hunan Normal University, China
Qiang Xiao	Huazhong University of Science and Technology, China
Hao Xiong	Macquarie University, Australia
Dongpo Xu	Northeast Normal University, China
Hua Xu	Tsinghua University, China
Jianhua Xu	Nanjing Normal University, China

Xinyue Xu	Hong Kong University of Science and Technology, China
Yong Xu	Beijing Institute of Technology, China
Ngo Xuan Bach	Posts and Telecommunications Institute of Technology, Vietnam
Hao Xue	University of New South Wales, Australia
Yang Xujun	Chongqing Jiaotong University, China
Haitian Yang	Chinese Academy of Sciences, China
Jie Yang	Shanghai Jiao Tong University, China
Minghao Yang	Chinese Academy of Sciences, China
Peipei Yang	Chinese Academy of Science, China
Zhiyuan Yang	City University of Hong Kong, China
Wangshu Yao	Soochow University, China
Ming Yin	Guangdong University of Technology, China
Qiang Yu	Tianjin University, China
Wenxin Yu	Southwest University of Science and Technology, China
Yun-Hao Yuan	Yangzhou University, China
Xiaodong Yue	Shanghai University, China
Paweł Zawistowski	Warsaw University of Technology, Poland
Hui Zeng	Southwest University of Science and Technology, China
Wang Zengyunwang	Hunan First Normal University, China
Daren Zha	Institute of Information Engineering, China
Zhi-Hui Zhan	South China University of Technology, China
Baojie Zhang	Chongqing Three Gorges University, China
Canlong Zhang	Guangxi Normal University, China
Guixuan Zhang	Chinese Academy of Science, China
Jianming Zhang	Changsha University of Science and Technology, China
Li Zhang	Soochow University, China
Wei Zhang	Southwest University, China
Wenbing Zhang	Yangzhou University, China
Xiang Zhang	National University of Defense Technology, China
Xiaofang Zhang	Soochow University, China
Xiaowang Zhang	Tianjin University, China
Xinglong Zhang	National University of Defense Technology, China
Dongdong Zhao	Wuhan University of Technology, China
Xiang Zhao	National University of Defense Technology, China
Xu Zhao	Shanghai Jiao Tong University, China

Contents – Part XIV

Applications

Road Meteorological State Recognition in Extreme Weather Based on an Improved Mask-RCNN

Guangyuan Pan[1,2]([✉]), Zhiyuan Bai[1], Liping Fu[2], Lin Zhao[1], and Qingguo Xiao[1]([✉])

[1] Linyi University, Linyi 276000, Shandong, China
garrypan0512@gmail.com, qingguoxiao@126.com
[2] University of Waterloo, Waterloo, ON N2L3G1, Canada

Abstract. Road surface condition (RSC) is an important indicator for road maintenance departments to survey, inspect, clean, and repair roads. The number of traffic accidents can increase dramatically in winter or during seasonal changes when extreme weather often occurs. To achieve real-time and automatic RSC monitoring, this paper first proposes an improved Mask-RCNN model based on Swin Transformer and path aggregation feature pyramid network (PAFPN) as the backbone network. A dynamic head is then adopted as the detection network. Meanwhile, transfer learning is used to reduce training time, and data enhancement and multiscale training are applied to achieve better performance. In the first experiment, a real-world RSC dataset collected from Ministry of Transportation Ontario, Canada is used, and the testing result show that the reidentification accuracy of the proposed model is superior to that of other popular methods, such as traditional Mask-RCNN, RetinaNet, Swin Double head RCNN, and Cascade Swin-RCNN, in terms of recognition accuracy and training speed. Moreover, this paper also designs a second experiment and proved that the proposed model can accurately detect road surface areas when light condition is poor, such as night time in extreme weather.

Keywords: Deep Learning · Road Recognition · Mask-RCNN · Transfer Learning · Swin Transformer

1 Introduction

Road surface condition (RSC) recognition is an extremely important task worldwide, especially for some northern countries that are extremely cold in winter or that often encounter extreme weather such as heavy rain, frost, and snowstorms in emergencies. Therefore, local governments need a financial expenditure to inspect road surface conditions so that they can clear ice and snow in time and send warnings to the local population in real time.

For example, the Ministration of Transportation Ontario (MTO) of Canada adopts two methods to obtain RSC information of winter roads. One is that a contractor is responsible for collecting road information and sharing it with road supervision. To

B. Luo et al. (Eds.): ICONIP 2023, CCIS 1968, pp. 3–15, 2024.
https://doi.org/10.1007/978-981-99-8181-6_1

achieve this, contractors need to manually and ceaselessly check the RSC of each section of the road. Another method is to apply Road Weather Information System (RWIS) stations that collect RSC images using fixed cameras installed on the road [1]. However, the first manual patrol method has a serious time delay and is very labor-consuming, which is a large waste of money and time. While the second method heavily relies on the number of cameras, frame range and resolution, specialized equipment is still necessary to observe the images manually and subjectively. As a result, there is no way to achieve both accuracy and generalization. In recent years, with the rapid development of computer vision, some deep learning-based methods have been proposed for RSC recognition. For example, G. PAN Proposed a learning-transferred deep residual network (ResNet) and an adaptive hybrid attention-based convolutional neural net (AHA-CNN) to recognize snow coverage of highway road surfaces in the snow season and achieved satisfactory classification results [2, 3]. Based on the reviewed papers, most researchers view road surface recognition as an ordinary classification task [4–6]. For example, K. Gui used resonance frequency and optical technology to detect the degree of icing and water accumulation on the road surface [7]. L. Fu RSC classification is performed using traditional machine learning models [8]. However, the result still needs to be improved because the model can be easily affected by redundant information in some pictures. There are two main kinds of redundant information in this problem: The first is sidewalk snow, which confuses the model and lowers recognition accuracy, and the second is information that is evenly distributed in the whole image, such as street lamp light, sunlight, rain and frost, which can make the images blurry and difficult to look at. Therefore, road surface condition recognition is still a severe open question that needs to be solved.

This paper tries to solve the problem by implementing the target detection technology of computer vision. The proposed model is an improved version of Mask-RCNN that can locate the position of the target of interest and exclude other areas as redundant information. The original Mask-RCNN [9] is popular for its two-stage target detection and has a good ability to extract useful feature information for training [10]. Based on this, this study proposes an improved Mask-RCNN that includes transfer learning and data enhancement for RSC real-time monitoring. The innovations of this paper are as follows:

1. To achieve an improved Mask-RCNN, this paper transforms the traditional classification problem into first segmenting the road in the picture, then detecting the road surface state of the road part, and then classifying on this basis.
2. Two road datasets that are specifically tailored for different scenarios are established. The first one is a picture dataset of highways under extreme weather, namely RWIS dataset, the total number of images is 4617. The second dataset is collected from in-vehicle cameras, and the total sample number is 1050. All the pictures from the second dataset feature poor lighting conditions and are taken at night, making them have poor visibility.
3. To reduce the effect of redundant information for the improved Mask-RCNN model, two powerful networks, namely, Swin Transformer and PAFPN, are installed as the backbone in the Mask-RCNN for better feature extraction. Additionally, this model

uses a dynamic head instead of the traditional bounding box head for more accurate snow-covered area detection. The improved Mask-RCNN model is fine-tuned, tested, and compared with the baseline model on the self-built dataset using pretrained weights.

2 Methodology

The method proposed in this paper is mainly divided into two parts. The first part is a feature extraction subnetwork composed of a Swin Transformer and a PAFPN network. The second part is the detection head subnetwork composed of RPN, dynamic head and FCN (fully convolution networks). The first part is used for layer-by-layer feature extraction of the input image, and the second part is used as a classifier and regression network. The specific network structure is shown in Fig. 1.

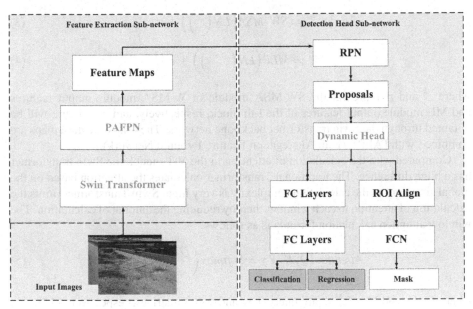

Fig. 1. Improved Mask-RCNN network structure.

2.1 Feature Extraction Network

2.1.1 Swin Transformer

Swin Transformer [11] can convert images into features. The model adopts a hierarchical design based on a CNN to expand the receptive field layer by layer. It contains a total of 4 stages. Each stage will reduce the resolution of the input feature map. When the input starts, the patch embedding cuts the picture into many small blocks and embeds them into the embedding. Each stage consists of patch merging and multiple blocks.

The first part of the model is the patch embedding operation, which consists of two steps: patch partition and linear embedding. In the step, the original input image is cut into small pieces and embedded into vectors. Next, the patch obtained from the first step will be sent to the patch merging. This module performs interval sampling and segmentation on the patch, performs stacking processing, realizes dimensionality reduction, and then performs splicing operations to expand it into a tensor, followed by sending the tensor to the Swin Transformer block, which consists of the Layer Norm (LN) layer, window attention layer (Window MultiHead self-attention, W-MSA), MLP layer and shift window attention layer (SW-MSA), the calculation formula of which is:

$$\hat{z}^l = W\text{-}MSA\left(LN\left(z^{l-1}\right)\right) + z^{l-1} \tag{1}$$

$$z^l = MLP\left(LN\left(\hat{z}^l\right)\right) + \hat{z}^l \tag{2}$$

$$\hat{z}^{l+1} = SW\text{-}MSA\left(LN\left(z^l\right)\right) + z^l \tag{3}$$

$$z^{l+1} = MLP\left(LN\left(\hat{z}^{l+1}\right)\right) + \hat{z}^{l+1} \tag{4}$$

where $\hat{z}^l$ and z^l represent the SW-MSA module or W-MSA module output features and ML module output features of the l-th block, respectively, and the picture will be obtained through the Swin transformer backbone network. The 4-layer feature maps are combined with PAFPN (Path Aggregation Feature Pyramid Network).

Compared with the calculation of attention in the ViT model, the Swin transformer has a huge difference. The traditional Transformer calculates the attention based on the global situation, so the calculation complexity is very high. Swin Transformer limits the calculation of attention to each window, thereby reducing the amount of calculation. The window attention calculation formula is as follows:

$$Attention(Q, K, V) = Softmax\left(\frac{QK^t}{\sqrt{d}} + B\right)V \tag{5}$$

where Q, K, and V $\in R^{M^2,d}$ are query, key and value matrices, respectively [12]. d is the query/key dimension. M^2 is the number of patches in the window, and $\hat{B}$ is the bias matrix. $\hat{B} \in R^{(2M-1)\times(2M-1)}$. The values in B are taken from $\hat{B}$.

2.1.2 Path Aggregation Feature Pyramid Network (PAFPN)

The traditional feature pyramid network (FPN) in a Mask-RCNN has a top-down fusion strategy, which is designed for shallow information on pixels [13] but does not support information interaction between nonadjacent layers, which leads to the lack of multiscale information and the imbalance of semantic information [14]. The improved version utilizes PAFPN, which not only has the same function but also has a bottom-up feature fusion network capable of passing low-level information to higher levels. We make full use of feature fusion, introduce adaptive feature pooling to enrich the extracted

ROI features, introduce fully connected fusion, and obtain more accurate segmentation results by fusing the output of a foreground and background binary classification branch.

Swin Transformer and PAFPN act at the same time to output 5- layer feature maps, and the sizes of the 4-layer feature maps are [56 × 56 × 96], [28 × 28 × 192], [14 × 14 × 384], and [7 × 7 × 768], respectively.

2.2 Detection Head Structure

2.2.1 Dynamic Head

The ROI head pools the proposal generated by the RPN into a 1024-dimensional feature vector and then classifies the feature vector and bounding box regression. The RPN layer judges whether there is a target in the anchors, fine-tunes the prediction frame of the anchors, and then generates candidate frame proposals. The ROI head judges the category of the proposal based on the RPN and fine-tunes the bounding box of the proposal again to generate the final prediction result. This study abandons the traditional Mask-RCNN ROI head and uses the dynamic head instead of the traditional ROI head. Some researchers have shown that regardless of how much the intersection and union ratio threshold is set, as the number of iterations increases, the number of positive samples increases sharply (because the model's ability to classify samples increases as the training progresses, and at this time, if the threshold is increased, higher-quality training samples will be obtained) [15].

The dynamic head contains DLA (dynamic sample allocation). The input image generates candidate regions through the RPN. As increasingly more high-quality samples are generated with the iteration of the training process, the IOU threshold is increased at this time. As the threshold increases, the number of positive samples increases continuously. This process is called DLA, and the specific algorithm is as follows:

$$label = \begin{cases} 1, & if \ maxIoU(b|G) \geq T_{now} \\ 0, & if \ maxIoU(b|G) < T_{now} \end{cases} \tag{6}$$

where T_{now} in (6) represents the current IOU threshold. First, the intersection ratio I of the candidate frame and its matching label frame are calculated. Then, the largest value of KI is selected as the current IOU threshold T_{now}. As the training progresses, T_{now} will increase as I increase. In practice, the KI-th largest IOU value in the batch sample is calculated. Then, every C iteration updates T_{now} with the mean value of the former (because one iteration will generate many batches).

DSL (dynamic smooth L1 loss) is the optimization of the regression loss value in the dynamic head. The specific formulas are shown in Formula (7) and Formula (8).

$$SmoothL1(x|\beta) = \begin{cases} 0.5|x|^2/\beta, \ if \ |x| < \beta \\ x - 0.5\beta, \ otherwise \end{cases} \tag{7}$$

$$DSL(x, \beta_{now}) = \begin{cases} 0.5|x|^2/\beta_{now}, \ if \ |x| < \beta_{now} \\ x - 0.5\beta_{now}, \ otherwise \end{cases} \tag{8}$$

First, the regression loss E of the candidate frame and its matching labeled frame are selected, and then the value with the smallest K is selected as the current value. In practice, the Kth smallest loss value in each batch of samples is calculated, and then is updated with the former intermediate value every C iteration.

The dynamic head can handle the dynamic change process of the candidate frame during the training process of the target detection model very well. It includes two modules, namely, dynamic positive and negative sample allocation and dynamic loss function, which are used for classification and regression branches, respectively. First, to obtain higher-quality candidate boxes in the classification branch, the threshold for screening positive samples is gradually increased, and the setting of the threshold is combined with the overall distribution of candidate boxes. For the regression branch, the form of the regression function is changed to adapt to the change in sample distribution to ensure the contribution of high-quality samples to model training.

3   Experiments and Analysis

The experiments in this paper are all implemented in PyTorch. The code used for this model is open-source code. All experiments were carried out using an i5-10400 CPU, an RTX3060 GPU and the Windows 10 operating system.

This study designed two experiments to verify the effect of the model on road surface recognition. The first experiment utilizes 4617 images from the Ontario RWIS dataset to verify the model's performance in daytime. The second experiment uses pictures collected by onboard cameras installed on snowplows to test the model's performance at night. The total number of photos is 1050.

3.1   Data Collection and Processing

The data used in the experimental part of this research come from Ontario, the most populous province in Canada [16]. The RWIS site uses road cameras placed on both sides of the road to collect real-time photos on the road. The photos are taken from three angles, namely, left, middle, and right, as shown in Table 1.

This study uses the photos taken by the Highway 400 camera as a dataset, three RGB channels, and the image size of the original photo is 480×752 pixels.

First, to ensure that the photos of the dataset under each category are roughly balanced, we mainly manually performed image resolution based on image clarity. The two solid lines of the road in these photos are very clear, and the road photos can be used to train the model to clearly recognize the road surface under better lighting conditions.

1. To unify the complexity of model training, the adjusted picture size is 512×512 pixels.
2. Dataset labeling: As shown in Fig. 2.

A total of 4617 photos were manually labeled and divided into three groups, namely, the training set, verification set, and test set, which contained photos in a ratio of approximately 7:2:1. The training set is used for model training, and the verification set is used for training. The process is used to verify the model, and the test set is used to evaluate and test the model.

Table 1. Road Surface Conditions

Sample Image	Description	5 Class Description
	Between the two solid lines is a dry pavement	Bare road
	There is 30%-50% snow cover between the two solid lines	Partly snow road
	There is frost covering between the two solid lines	Frost road
	Significantly snow cover (more than 90%) between the two solid lines	Fully snow road
	There is a lot of rain cover between the two solid lines	Wet road

(a)   (b)

Fig. 2. Picture annotations, where (a) and (b) represent different road categories, (a) is a partially snow-covered road surface and (b) is a fully snow-covered road surface.

3.1.1 Data Collection and Processing

(1) Data Enhancement

Since there are few data samples in this experiment, data augmentation is used to increase the number of samples to prevent model overfitting and enhance the generalization of the model. The data enhancement methods used include image augment flips color, horizontal flip, perspective, shift scale rotate and image strong augmentation. Part of the enhanced pictures are shown in Fig. 3 [17].

(2) Multiscale Training

The input sizes of the pictures used in this experiment are (2000, 1200), (1333, 800), and (1666, 1000). Whenever the size of the picture is input into the model, a size is randomly selected from the above 3 sizes for training.

3.1.2 Experimental Results

In this experiment, 1 one-stage model and 3 two-stage models are used for training using the organized RWIS dataset. Figure 4 shows the integrated curves of the mAP values of the models for all test set categories after training for 30 epochs. In this study, the time period with the highest mAP value is selected to analyze the model performance on the test set.

To simplify the result table, five categories are named: C1 is bare road, C2 is frost road, C3 is fully snow road, C4 is partly snow road, and C5 is wet road. The indicators in the COCO detection and evaluation standard are used for performance evaluation, including average precision (mAP), average precision with IOU = 0.5 (mAP@0.5), parameters (Params), floating-point operations per second (GFLOPs) and frames per second (FPS). The results compared with other mainstream target detection algorithms are shown in the table below.

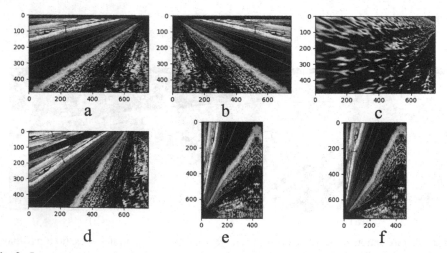

Fig. 3. Image augmentation, where a is the original image, b is the horizontal flip, c is the perspective, d is the shift zoom rotation, e is the image enhancement flip color, and f is the image color enhancement.

It can be seen from Table 2 that the proposed model has the least amount of calculation (GFLOPs) during the training process, but the effect is the best, including the total mAP value and IOU = 0.5. The mAP value and AP value of bare road (C1), frost road (C2) and wet road (C5) are all the highest, among which the total mAP value can reach 82.02, the AP value of bare road can reach 85.9, the AP of frost road can reach 84.2, and the AP value of wet road can reach 87.4. Regarding the category of fully snow road, because the photos of this category have much redundant information, they are extremely complex, and difficult to identify, so the effects of these detection models are not good.

Figure 5 shows the results of the improved model and the RetinaNet, Swin Double head-RCNN, Mask-RCNN and Cascade Swin-RCNN models on the custom test set. A, b, c, d in Fig. 5 are four randomly selected images in the test set, each row of which represents the test results of different models on the same image, from top to bottom model order for the improved Mask-RCNN model, Cascade Swin-RCNN, Swin Double head-RCNN, RetinaNet, and Mask-RCNN.

In Figure a, the original image can be determined as a frost-covered road based on the evaluation criteria. However, only the enhanced Mask-RCNN model in this study accurately identifies the actual road surface, while other models misclassify the image. Figure b illustrates a fully snow-covered road at night. As the RWIS dataset lacks nighttime images, the model must autonomously deduce the road surface type based on training knowledge. Experimental results reveal that the improved Mask-RCNN correctly categorizes a portion of the road surface as snow-covered, unlike the other models which either fail to reason or make incorrect deductions. This underscores the superior learning capacity of the enhanced Mask-RCNN compared to the other four models. Hence, our second experiment aims to enhance the generalization of the improved Mask-RCNN.

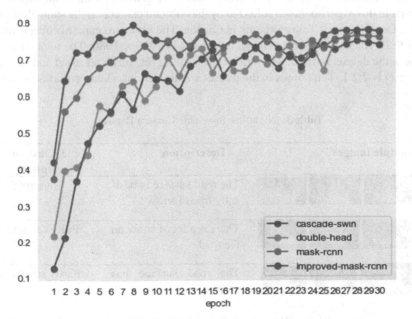

Fig. 4. Model mAP value curve.

Figure c showcases a clear road, with the Improved Mask RCNN accurately and precisely identifying the road, while other models misclassify the image. In Figure d, the original image features a clear road. The improved Mask-RCNN successfully classifies this image, whereas Cascade Swin-RCNN falters in reasoning and fails to recognize the road type. The remaining four models also inaccurately categorize this image. Overall, the enhanced Mask-RCNN exhibits superior and accurate reasoning abilities and higher generalization compared to the other four models.

Table 2. Model comparison.

Model	mAP	mAP@0.5	FPS	Params	GFLOPs	AP of Classes				
						C1	C2	C3	C4	C5
Mask-RCNN [9]	80.24	84.8	24.9	43.77	245.36	85.7	79.1	65.9	86.6	83.9
RetinaNet [18]	79.60	85.4	28.5	**36.19**	206.13	81.0	77.2	**73.9**	84.7	81.2
Cascade RCNN [19]	74.64	86.2	7.0	147.31	646.64	74.8	**84.2**	62.0	**90.1**	62.1
Double head RCNN[20]	80.26	87.4	6.5	77.99	432.55	82.3	83.0	65.5	84.8	85.7
Improved model (ours)	**82.02**	**89.8**	15.4	72.41	**158.35**	**85.9**	**84.2**	65.9	86.7	**87.4**

3.2 Model Testing 2: Snow Detection at Night

3.2.1 Data Collection and Processing

This study designed a second experiment to verify the generalization of the model. The data used in this experiment are collected by the onboard camera on the snowplow of the Ontario Department of Transportation in Canada. The on-board camera collects photos with 3 RGB channels, and the original image size is 1920×1080. The total number of photos in the dataset is 1050. The ratio of the training set, verification set and test set in the dataset is 7:2:1. The photos in the dataset are divided into 3 categories according to

Table 3. Nighttime In-vehicle Camera Dataset

Sample images	Description	3 Class Description
	The road surface is basically free of snow	Bare road
	There is a lot of snow on the road	Fully snow road
	The road surface has some snow	Partly snow road

different road surfaces: bare road, partly snow road and fully snow road. The photos are all collected by snowplows at night. The specific dataset is shown in Table 3.

3.2.2 Experimental Results

This experiment compared the model detection effect of the model proposed in this paper, the Faster-RCNN model and the RetinaNet model. The specific model effect comparison results are shown in Table 4.

Table 4. Model comparison.

Model	mAP@0.5	GFLOPs	Params	FPS
Improved model (ours)	**80.30**	**158.35**	72.40	15.5
Faster-RCNN [21]	70.00	193.79	41.13	28.2
RetinaNet	63.10	206.13	**36.19**	**28.6**

It can be seen from the table that our proposed model has a great advantage in recognition effect compared with other models. And, this recognition effect is still achieved

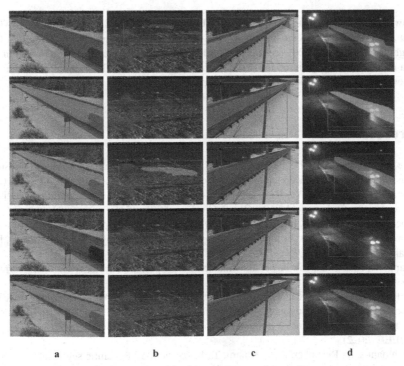

a   b   c   d

Fig. 5. Comparison of model recognition effects

under the condition that the model complexity is not as good as that of the other two models.

4 Conclusion

In this study, to improve the accuracy of road surface recognition under extreme weather, we improved the traditional Mask-RCNN model, during training adopts the methods of multiscale training, data augmentation and transfer learning. In the experimental part, we designed two experiments to verify the effectiveness of our work. The first experiment is to verify that our model can accurately segment the road surface and accurately identify different road surface parts under extreme weather. The results show that the improvement our model is superior to the traditional in identifying different types of road surfaces. Have greatly enhanced the classification effect and positioning effect, and these enhancements are not obtained without increasing the complexity of the model. The second experiment of this study investigates the performance of our proposed model for extreme weather road recognition at night when the lighting conditions are not good. The research has shown that the model is capable of accurately recognizing roads even in challenging conditions.

However, this study still has some limitations. The next step will be to improve the model to improve model detection, and the dataset will be expanded to include different types of extreme weather road surfaces at various resolutions.

Acknowledgements. This research is supported by National Natural Science Foundation of China under grant no. 62103177, Shandong Provincial Natural Science Foundation for Youth of China under grant no. ZR2023QF097 and the National Science and Engineering Research Council of Canada (NSERC) under grant no. 651247. The authors would also like to thank the Ministry of Transportation Ontario Canada for technical support.

References

1. Xie, Q., Kwon, T.J.: Development of a highly transferable urban winter road surface classification model: a deep learning approach. Transp. Res. Rec. **2676**(10), 445–459 (2022)
2. Pan, G., Muresan, M., Yu, R., Fu, L.: Real-time winter road surface condition monitoring using an improved residual CNN. Can. J. Civ. Eng. **48**(9), 1215–1222 (2021)
3. Chen, Q., Pan, G., Zhao, L., Fan, J., Chen, W., Zhang, A.: An adaptive hybrid attention based convolutional neural net for intelligent transportation object recognition. IEEE Trans. Intell. Transp. Syst. **24**, 7791–7801 (2022)
4. Wang, Y., Zhang, D., Liu, Y., et al.: Enhancing transportation systems via deep learning: a survey. Transp. Res. Part C Emerg. Technol. **99**, 144–163 (2019)
5. Ur Rahman, F., Ahmed, Md.T., Amin, Md.R., Nabi, N., Ahamed, Md.S.: A comparative study on road surface state assessment using transfer learning approach. In 2022 13th International Conference on Computing Communication and Networking Technologies (ICCCNT), pp. 1–6. IEEE (2022)
6. Vachmanus, S., Ravankar, A.A., Emaru, T., Kobayashi, Y.: Semantic segmentation for road surface detection in snowy environment. In: 2020 59th Annual Conference of the Society of Instrument and Control Engineers of Japan (SICE), pp. 1381–1386. IEEE (2020)

7. Gui, K., Ye, L., Ge, J., Cheikh, F.A., Huang, L.: Road surface condition detection utilizing resonance frequency and optical technologies. Sens. Actuators A Phys. **297**, 111540 (2019)
8. Linton, M.A., Fu, L.: Connected vehicle solution for winter road surface condition monitoring. Transp. Res. Rec. J. Transp. Res. Board **2551**(1), 62–72 (2016)
9. He, K., Gkioxari, G., Dollár, P., Girshick, R.: Mask R-CNN. In: Proceedings of the IEEE International Conference on Computer Vision, pp. 2961–2969 (2017)
10. Casado-García, Á., Heras, J.: Ensemble methods for object detection. In: ECAI 2020, pp. 2688–2695. IOS Press (2020)
11. Liu, Z., Lin, Y., Cao, Y., et al.: Swin transformer: hierarchical vision transformer using shifted windows. In: Proceedings of the IEEE/CVF International Conference on Computer Vision, pp. 10012–10022 (2021)
12. Vaswani, A., et al.: Attention is all you need. In: Advances in Neural Information Processing Systems, vol. 30 (2017)
13. Liu, S., Qi, L., Qin, H., Shi, J., Jia, J.: Path aggregation network for instance segmentation. In: Proceedings of the IEEE Conference on Computer Vision and Pattern Recognition, pp. 8759–8768 (2018)
14. Lin, T.-Y., et al.: Microsoft COCO: common objects in context. In: Fleet, D., Pajdla, T., Schiele, B., Tuytelaars, T. (eds.) ECCV 2014. LNCS, vol. 8693, pp. 740–755. Springer, Cham (2014). https://doi.org/10.1007/978-3-319-10602-1_48
15. Zhang, H., Chang, H., Ma, B., Wang, N., Chen, X.: Dynamic R-CNN: towards high quality object detection via dynamic training. In: Vedaldi, A., Bischof, H., Brox, T., Frahm, J.-M. (eds.) ECCV 2020. LNCS, vol. 12360, pp. 260–275. Springer, Cham (2020). https://doi.org/10.1007/978-3-030-58555-6_16
16. Statistics Canada: Population and dwelling counts, for census metropolitan areas, 2011 and 2006 censuses. Statistics Canada, Ottawa (2014)
17. Buslaev, A., Iglovikov, V.I., Khvedchenya, E., Parinov, A., Druzhinin, M., Kalinin, A.A.: Albumentations: fast and flexible image augmentations. Information **11**(2), 125 (2020)
18. Lin, T.-Y., Goyal, P., Girshick, R., He, K., Dollár, P.: Focal loss for dense object detection. In: Proceedings of the IEEE International Conference on Computer Vision, pp. 2980–2988 (2017)
19. Cai, Z., Vasconcelos, N.: Cascade R-CNN: delving into high quality object detection. In: Proceedings of the IEEE Conference on Computer Vision and Pattern Recognition, pp. 6154–6162 (2018)
20. Wu, Y., et al.: Rethinking classification and localization for object detection. In: Proceedings of the IEEE/CVF Conference on Computer Vision and Pattern Recognition, pp. 10186–10195 (2020)
21. Ren, S., He, K., Girshick, R., Sun, J.: Faster R-CNN: towards real-time object detection with region proposal networks. In: Advances in Neural Information Processing Systems, vol. 28 (2015)

I-RAFT: Optical Flow Estimation Model Based on Multi-scale Initialization Strategy

Shunpan Liang[1,2]([✉]), Xirui Zhang[1], and Yulei Hou[3]

[1] School of Information Science and Engineering, Yanshan University, Qinhuangdao 066004, Hebei, China
liangshunpan@ysu.edu.cn
[2] School of Information Science and Engineering, Xinjiang University of Science, Korla 841000, Xinjiang, China
[3] School of Mechanical Engineering, Yanshan University, Qinhuangdao 066004, Hebei, China

Abstract. Optical flow estimation is a fundamental task in the field of computer vision, and recent advancements in deep learning networks have led to significant performance improvements. However, existing models that employ recurrent neural networks to update optical flow from an initial value of 0 suffer from issues of instability and slow training. To address this, we propose a simple yet effective optical flow initialization module as part of the optical flow initialization stage, leading to the development of an optical flow estimation model named I-RAFT. Our approach draws inspiration from other successful algorithms in computer vision to tackle the multi-scale problem. By ting initial optical flow values from the 4D cost volume and employing a voting module, we achieve initialization. Importantly, the initialization module can be seamlessly integrated into other optical flow estimation models. Additionally, we introduce a novel multi-scale extraction module for capturing context features. Extensive experiments demonstrate the simplicity and effectiveness of our proposed model, with I-RAFT achieving state-of-the-art performance on the Sintel dataset and the second-best performance on the KITTI dataset. Remarkably, our model achieves these results with a 24.48% reduction in parameters compared to the previous state-of-the-art MatchFlow model. We have made our code publicly available to facilitate further research and development (https://github.com/zhangxirui-1997/I-RAFT).

Keywords: Optical Flow · Multi-scale Feature Extraction · Vision Attention

1 Introduction

Optical flow, by describing the correspondence between pixels in two consecutive frames, reveals the motion of objects in an image. It serves as a foundation for

B. Luo et al. (Eds.): ICONIP 2023, CCIS 1968, pp. 16–29, 2024.
https://doi.org/10.1007/978-981-99-8181-6_2

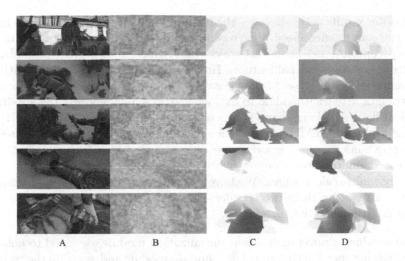

Fig. 1. The model inference visualization results. column A is the first image of the image pair, column B is the calculation result of the initialization module, column C is the final inference result of the model and column D is the real optical flow. column B the optical flow initialization result is visualized in the same way as the optical flow and is appropriate as a starting point for the optical flow update, despite the errors.

more advanced vision tasks, particularly in video-related applications such as action recognition and video super-resolution. Currently, optical flow estimation techniques have been applied in various domains, including autonomous driving, drone positioning, and navigation, visual effects in film-making, and object tracking.

The accuracy of optical flow estimation algorithms based on deep learning has been improving and has far surpassed the traditional algorithms [3,20,24]. In 2015 FlowNet [3] applies deep learning to the optical flow estimation problem first; RAFT [24] proposes to build 4D cost volume to match pixel points, and overcomes the long displacement problem and achieves better results by iterating through recurrent neural network updates. Recently, optical flow estimation models have been introduced locally with attention mechanisms to optimize the feature extraction and optical flow refinement parts, and good results have been obtained [7,10,11,29]. However, there is still much room for improvement in the detailed representation of optical flow and optical flow occlusion.

Many state-of-the-art optical flow models [2,7,29] leverage Gated Recurrent Unit (GRU) to facilitate optical flow updates, leading to improved results. However, the utilization of GRUs introduces several challenges: substantial computational and memory overhead, instability, and poor convergence, thereby creating a performance bottleneck in the models. In most existing models, the optical flow update is initiated from a value of 0, lacking a reasonable initialization strategy. To address this limitation, we introduce a novel multi-scale-based initialization module that enables pre-initialization prior to the optical flow update iteration.

This module significantly alleviates the training difficulty of the model and accelerates the convergence speed, thereby enhancing the overall fitting performance.

Deep convolutional neural networks (CNN) have shown great capability in extracting high-dimensional features. However, several models [2,7,29] with the best results so far have used shallow convolutional neural networks incorporating a lightweight attention module to extract features, but ignoring the specificity of optical flow. All along, many improved optical flow estimation models [1,2,7, 12,18,29] based on RAFT use the same encoder to extract feature information and context information, ignoring the difference between feature information and context information. Context information requires more detailed features, especially multi-scale features. We have designed a more efficient and targeted structure for each of these two encoders.

Our contributions are threefold:

1. We introduce a novel optical flow initialization module, designed to enhance the learning speed of the optical flow update module and expedite the training process. We embedded the optical flow initialization module on RAFT, and the improvement was 3.2% and 13.4% on the Sintel clean and final passes, respectively. We believe that the optical flow initialization module can be well integrated with RAFT and its improved methods and enhance the results.
2. We designed a targeted context encoder and a feature encoder. The context encoder uses a deeper convolutional neural network than before and the multi-scale feature extraction module MSBlock. the feature encoder uses a shallow convolutional neural network and incorporates a lightweight attention module.
3. We have compiled the above improvement ideas and proposed I-RAFT, which achieves the best results at this stage. The improvement is 6.78% and 2.97% on the Sintel clean and final passes, respectively, compared to FlowFormer, and 3.5% on the KITTI test set.

2   Related Work

2.1   Optical Flow Estimation

Optical flow estimation plays a crucial role in computer vision, aiming to determine the motion of each pixel across consecutive video frames. Traditionally, optical flow estimation has been formulated as an optimization problem, where the objective is to maximize the matching value between pairs of pixels in an image. The pioneering work of Horn and Schunck [6] introduced a variational approach to compute optical flow. Although the traditional method, in terms of accuracy and speed, can adequately handle simple scenarios, it often falls short when confronted with realistic and complex scenes.

Models based on deep learning for optical flow estimation can be divided into three generations, and the representative models are FlowNet [3], PWC-Net [20] and RAFT [24]. In 2015 Fischer et al. [3] proposed the model FlowNet, the first application of deep learning to the optical flow domain with end-to-end training, first with spatial compression in the systolic part of the network and then with optical flow refinement in the extended part. Sun et al. [20] proposed the PWC-Net model, which proposes a pyramid structure to learn the features of two images separately, following three simple and well-established principles of traditional methods: feature pyramid processing, warping operation, and cost volume calculation, not only to obtain high accuracy estimation but also to achieve a balance between model size and computational accuracy.

The current mainstream optical flow estimation model is the third generation, and a large part of the latest research results are derived improvements based on the RAFT [24]. In 2020 Teed et al. [24] proposed the RAFT model, which uses two sets of convolutional neural networks to extract image information, including spatial feature information of both images and context information of the first image, constructs a multi-scale 4D cost volume for all pixel pairs in both images based on the two spatial features, and looks up from the correlation matrix by the existing optical flow, and together with the background features. The optical flow is fed into the recurrent neural network GRU to update the optical flow. The emergence of RAFT has heralded a significant breakthrough in the advancement of deep learning-based models for optical flow estimation.

Given the recent surge of attention mechanisms in the field of computer vision, the integration of attentional mechanisms into optical flow estimation tasks has garnered significant interest and attention. Sui et al. [18] proposed that CRAFT introduces a cross-attention mechanism to reduce the error due to large displacement of motion blur. Zhao et al. [29] proposed the GMFlowNet model, which proposes patch-based overlapping concerns to extract more context features to improve the matching quality. Jiang et al. [11] proposed that GMA introduces a global motion aggregation module based on an attention mechanism for finding long-range correlations between pixels in the first image and globally aggregating the corresponding motion features. Huang et al. [7] proposed the FlowFormer model to tokenize 4D cost volume constructed from pixel pairs, encode cost tokens into cost memory using the Alternate Group Transformer (AGT) layer, and decode the cost memory with a cyclic Transformer with dynamic positional cost queries.

Taken together, the optical flow estimation model has clear structural boundaries and clear functions for each part, and with the introduction of the attention mechanism, the model accuracy is greatly improved. Although many models have been improved for specific difficult problems, they all ignore the problems of slow computation, unstable training, and inadequate feature representation in the iterative part of the optical flow update, and we have done research to address these aspects.

2.2 Multi-scale Features in Vision

In the real world, where different objects vary greatly in size, and where objects on an image are affected by being far smaller and closer, multi-scale feature extraction is one of the indispensable methods for solving current computer vision problems. Computer vision problems often require context information and the extraction of large-scale features to obtain coarse information; small-scale features are also required to obtain detailed information for fine-grained tasks to enable higher accuracy calculations. Perceiving features at different scales is important for understanding fine-grained vision tasks, and He et al. [5] proposed the SPP architecture, which proposes a pyramidal structure incorporating multi-scale features.

Image segmentation is a task that requires accurate classification down to the pixel level, acquiring enough detailed features while expanding the field of perception, much like the need for optical flow estimation. The optical flow estimation model needs to differentiate between moving objects and backgrounds and distinguish between different moving objects. Although image segmentation is an intensive pixel-by-pixel prediction task, the semantics of each pixel usually depends on nearby pixels and context information at a distance, so multi-scale feature fusion is also widely used in image segmentation. Tao et al. [23] propose a multi-scale attention mechanism that allows the network to learn how to combine predictions from multiple scales of inference, thus avoiding confusion between different classes and dealing with finer details. Meng et al. [17] proposed an ESeg model that uses BiFPN to fuse multi-scale features without using high resolution and deep convolution, achieving faster speeds and better accuracy.

The most challenging problem in target detection tasks is the problem of scale variation of targets, where the same type of target appears in different shapes and sizes, and may even have some very small, very large, or extreme shapes (e.g. elongated, narrow, and tall, etc.), which makes target localization and identification extremely difficult. Among the existing algorithms proposed for the problem of target size variation, the more effective ones are mainly image pyramids and feature pyramids, which share the idea of using multi-scale features to detect objects of different sizes. The computational complexity of image pyramids is high, and feature pyramids are now commonly used to accomplish target detection using multi-scale features fusion and multi-scale feature prediction paired with, for example, FPN [14], EfficientDet [22], RetinaNet [15], etc.

2.3   Vision Attention Mechanisms

Transformer has been a great success in natural language processing, but thanks to its powerful representation capabilities, researchers have also applied Transformer to various tasks in the field of computer vision [4].

Transformer also has more applications in optical flow estimation models, e.g. FlowFormer [7], GMA [11], etc. Perceiver IO [10] proposes a generic Transformer backbone network that, although requiring a large number of parameters, achieves a high level of optical flow. Although the introduction of attention brings better results in the task of optical flow estimation, the corresponding number of parameters and the computational cost are considerably higher and the training process is more unstable.

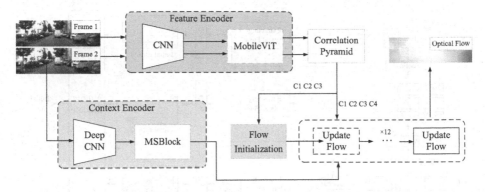

Fig. 2. I-RAFT structure. Extracted using shallow convolutional neural network and MobileViT module, context feature is extracted using deep convolutional neural network and MSBlock, the optical flow is initialized with $\{C1, C2, C3\}$ in 4D cost volume, $\{C1, C2, C3, C4\}$ and context information in 4D cost volume is fed into the optical flow update model, and the optical flow is finally obtained after several iterations.

3 Method

The deep learning model is a black box system with poor interpretability, but the function of the various parts of the optical flow estimation model can be determined. We first identified the main roles of the components in the model: the role of the feature encoder is to obtain a feature of the image which is used for pixel point matching, so the feature matrix should express the features of that pixel point as closely as possible; The role of the 4D cost volume is to calculate the correlation between pairs of pixel points of two images; the role of the context encoder is to extract features from the point of view of moving objects, which can extract information about the moving objects in the image; The optical flow initialization module is used to initialize the optical flow, indicate the rough direction for subsequent optical flow updates, increase the speed of the optical flow update iterations, and reduce the computational effort; the optical flow update iteration part is used to refine the optical flow using the GRU.

The model of I-RAFT is shown in Fig. 2. Given two consecutive frames I_1, I_2 ,the feature encoder extracts feature information F_1, F_2 from I_1, I_2 and calculates the 4D cost volume $\{C1, C2, C3, C4\}$ and the context encoder extracts context information $Cmap$ from I_1.

The optical flow is then initialized based on the correlation matrix and context information, and finally, the update iteration of the optical flow is completed to output the final optical flow. Section 3.1 introduces the optical flow initialization module, Sect. 3.2 introduces the context feature encoder, and Sect. 3.3 briefly introduces the feature encoder.

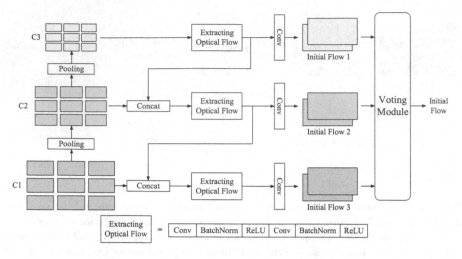

Fig. 3. Diagram of the optical flow initialization structure. According to the idea of the feature pyramid, the red $C1$ represents the first layer, the blue $C2$ represents the second layer and the yellow $C3$ represents the third layer. The optical flow matrix is calculated separately for the third and second layers and passed to the next layer. After the first layer has calculated the optical flow result for that layer, it is sent to the voting module together with the optical flow results of the two previous layers to calculate the final initial optical flow. (Color figure online)

3.1   Optical Flow Initialisation

A common optical flow model will gradually update the iterative optical flow f, usually resulting in an optical flow output $\{f_1, f_2, .., f_i, .., f_N\}$, with f_i representing the optical flow after the i-th update. At each iteration, the update iteration module calculates the updated value Δf of the optical flow and calculates $f_{i+1} = f_i + \Delta f$ according to the update formula. However, in the calculation of the first generation optical flow $f_1 = f_0 + \Delta f$, f_0 being the initial optical flow, which is usually $f_0 = 0$. This creates difficulties in updating the optical flow. We propose an optical flow initialization module based on a multi-scale strategy to improve the quality of the optical flow update iterations.

Our approach is to design a pyramidal structure based on the 4D cost volume to complete the initialization of the optical flow. As shown in Fig. 3, the pathway is first bottom-up, and RAFT pools the last two dimensions of $C1$ at different scales, generating $C2, C3, C4$, so that information on large and small displacements can be obtained. Then there is the top-down pathway and the lateral pathway. Although the higher pyramidal layers are more coarse in their representation of information, they are better at identifying large displacements, and by fusing features from the upper layer of the large field of view with features from this layer, a more stable representation of the results can be obtained, combining both large field of view and fine-grained information. Each layer eventually yields a corresponding initial optical flow. We use a 1×1 convolution layer to extract the information from the 4D cost volume, while the information from the previous layer is upsampled and then convolved for extraction after stitching.

The purpose of the optical flow initialization is to give a rough direction for subsequent iterations of the update and does not require very fine results. $C4$ has an overly large field of view and is irreplaceable for determining extra-long displacements, but the introduction of the pyramid structure makes the layers too high and the parameters larger, but the effect is not significantly enhanced, so we only use $C1, C2, C3$. In each of these three layers, the information from the previous layer is used to generate the initial optical flow for that layer The initial optical flow of the layer. To ensure that our optical flow initialization is stable, we use averaging as the voting method to obtain the final optical flow.

3.2   Context Encoder

There are different objects and backgrounds in the image, and context features are mainly used to allow the model to separate out the individual objects and background layers of the image in order to distinguish the boundaries of moving objects. We used a deep convolutional neural network and MSBlock to complete the context feature extraction. The main objective of the deep convolutional neural network is to extract high latitude features and does not bring in multi-scale features, for this reason, we designed MSBlock to accomplish the fusion of multi-scale features.

First, I_1 is computed by a deep convolutional neural network to obtain the matrix Cmap1. Using a set of small convolutional kernels instead of one large convolutional kernel can improve multi-scale representation at a fine-grained level, so we divide the input features into five pathways to compute a 64-dimensional feature matrix by different pathways. Concatenating these matrices gives a 320-dimensional matrix, which is then compressed to 256 dimensions using 1×1 convolution. In order to extend the field of perception and obtain multi-scale context information, we used atrous convolutions in two of the pathways, which is important for our context feature perception. As the number of convolutions increases, the perceptual field gradually increases and the final output context information incorporates the multi-scale feature information Cmap2.

3.3   Feature Encoder

The role of the feature encoder is to give specific pixel features so that the different pixel features are as orthogonal as possible so that better results can be achieved when performing the relevant volume calculations. Unlike context encoders, there is a greater need to focus on pixel features such as color, texture, light, and dark when encoding features. Attentional mechanisms can take a global view while focusing on each part of the input, allowing for better representational learning. We used a shallow ResNet+mobileViT approach. Although mobileViT is a lightweight attention network, all we needed was a lightweight, efficient attention network for feature extraction to ensure good training results with multiple GRU iterations of optical flow.

Algorithm 1: MSBlock algorithm

    Input : Cmap1
    Output: Cmap2
    `// The first pathway`
1 $x_1^1 \leftarrow MaxPool(Cmap1, ks = 2, s = 1)$
2 $x_1^2 \leftarrow Conv(x_1^1, 256, 64, ks = 1)$
    `// Second access road`
3 $x_2^2 \leftarrow Conv(Cmap1, 256, 64, ks = 3, p = 1)$
4 $x_{temp} \leftarrow x_2^2$
    `// Third to fifth access`
5 **for** i *in [3, 4, 5]* **do**
6      $num \leftarrow (i - 2) * 4$
7      $x_i^1 \leftarrow Concat(Cmap1, x_{temp})$
8      $x_i^2 \leftarrow Conv(x_i^0, 320, 64, ks = 3, p = num, d = num)$
9      $x_{temp} \leftarrow x_i^2$
10 **end**
11 $Cmap2 \leftarrow Concat(x_1^2, x_2^2, x_3^2, x_4^2, x_5^2)$

---

4 Experiments

This section details the experimental results to demonstrate the validity of our constructs. We pre-trained on FlyingChairs [3] and FlyingThings [16], and then fine-tuned on the Sintel and KITTI datasets, respectively. Our method achieves better performance on both clean and final of Sintel and KITTI. Section 4.1 presents the experimental details, Sect. 4.2 presents the experimental results, Sect. 4.3 shows the ablation experimental results, and Sect. 4.4 shows the visualization results.

4.1 Experimental Details

We evaluate using the EPE and F1-All metrics. epe calculates the mean of the Euclidean distance between all predicted optical flow and the true optical flow. f1-all is the percentage of pixels where the stream error is greater than 3 pixels or exceeds 5% of the true length of the optical flow.

    We used PyTorch to build the model, borrowing hyperparameters from RAFT, and the training process used the AdamW optimizer, the OneCycleLR learning rate iterator. Thirty-two optical flow updates were performed at Sintel and 24 optical flow updates at KITTI. We pre-trained the data of FlyingChairs, and FlyingThings combination for 150k iterations respectively. To achieve optimal performance on Sintel, KITTI-2015, and HD1K [13], we further fine-tuned the model with 150k iterations.

Table 1. Experiments on the Sintel and KITTI datasets. * indicates that these methods use a warm start strategy, which relies on previous video frames in the video. "A" denotes the AutoFlow dataset [19]. "C + T" indicates training on the FlyingChairs and FlyingThings datasets only. '+ S + K + H' indicates fine-tuning on a combination of the Sintel, KITTI, and HD1K training sets.

Training Data	Method	Sintel (train)		KITTI-15 (train)		Sintel (test)		KITTI-15 (test)
		Clean	Final	F1-epe	F1-all	Clean	Final	F1-all
A	Perceiver IO [10]	1.81	2.42	4.98	–	–	–	–
	RAFT-A [19]	1.95	2.57	4.23	–	–	–	–
C+T	HD3 [26]	3.84	8.77	13.17	24.0	–	–	–
	LiteFlowNet [8]	2.48	4.04	10.39	28.5	–	–	–
	PWC-Net [20]	2.55	3.93	10.35	33.7	–	–	–
	LiteFlowNet2 [9]	2.24	3.78	8.97	25.9	-	-	-
	S-Flow [27]	1.30	2.59	4.60	15.9	–	–	–
	RAFT [24]	1.43	2.71	5.04	17.4	–	–	–
	FM-RAFT [12]	1.29	2.95	6.80	19.3	–	–	–
	GMA [11]	1.30	2.74	4.69	17.1	–	–	–
	GMFlowNet [29]	1.14	2.71	4.24	15.4	–	–	–
	FlowFormer [7]	**1.00**	2.45	4.09	**14.72**	–	–	–
	MatchFlow [2]	1.03	2.45	**4.08**	15.6	–	–	–
	Ours	1.06	**2.44**	4.10	15.22	–	–	–
C+T+S+K+H	LiteFlowNet2 [9]	(1.30)	(1.62)	(1.47)	(4.8)	3.48	4.69	7.74
	PWC-Net+ [21]	(1.71)	(2.34)	(1.50)	(5.3)	3.45	4.60	7.72
	VCN [25]	(1.66)	(2.24)	(1.16)	(4.1)	2.81	4.40	6.30
	MaskFlowNet [28]	–	–	–	–	2.52	4.17	6.10
	S-Flow [27]	(0.69)	(1.10)	(0.69)	(1.60)	1.50	2.67	4.64
	RAFT [24]	(0.76)	(1.22)	(0.63)	(1.5)	1.94	3.18	5.10
	RAFT* [24]	(0.77)	(1.27)	–	–	1.61	2.86	–
	FM-RAFT [12]	(0.79)	(1.70)	(0.75)	(2.1)	1.72	3.60	6.17
	GMA [11]	–	–	–	–	1.40	2.88	–
	GMA* [11]	(0.62)	(1.06)	(0.57)	(1.2)	1.39	2.48	5.15
	DEQ-RAFT [1]	(0.73)	(1.02)	(0.61)	(1.4)	1.82	3.23	4.91
	GMFlowNet [29]	(0.59)	(0.91)	(0.64)	(1.51)	1.39	2.65	4.79
	FlowFormer [7]	(0.48)	(0.74)	(0.53)	(1.11)	1.18	2.36	4.87
	MatchFlow [2]	(0.49)	(0.78)	(0.55)	(1.1)	1.16	2.37	**4.63**
	Ours	**(0.45)**	**(0.63)**	**(0.51)**	**(0.84)**	1.10	**2.29**	4.70

4.2   Quantitative Experiments

Firstly our initialization module combined with a more sophisticated feature encoder, context encoder, allows for more targeted model effects. The results are shown in Table 1.

Our model was evaluated on the Sintel clean and final passes and achieved EPEs of 1.10 and 2.29 respectively, the current number one. Compared to Flow-Former, the EPE improved by 6.78% and 2.97% respectively, and compared to the current SOTA MatchFlow, the EPE improved by 5.45% and 3.38% respectively.

Table 2. Ablation experiments. Experiments are performed by adding different blocks cumulatively from top to bottom on top of the RAFT.

Experiment	Method	Sintel (train)		KITTI-15 (train)		Params
		Clean	Final	F1-epe	F1-all	
Baseline	RAFT	0.85	1.37	0.75	1.72	5.26M
+ Context Encoder	Deep CNN	0.86	1.15	0.68	1.49	6.17M
	Deep CNN with MSBlock	0.76	1.10	0.65	1.44	7.97M
+ MobilcViT Blocks	MobileViT×2	0.70	1.04	0.61	1.26	8.83M
	MobileViT×4	0.60	0.93	0.54	1.13	9.68M
	MobileViT×6	0.71	0.97	0.51	1.16	10.54M
+ Flow Initialization	–	0.53	0.73	0.53	0.94	11.58M

After the Sintel fine-tuning phase, we further fine-tuned the model on the KITTI2015 training set and evaluated it on the KITTI test set. Our model reached 4.70, ranking second in the KITTI-2015 benchmark. This is a 3.5% improvement compared to FlowFormer, but 0.07 behind MatchFlow. We found an F1-all of 0.84 for I-RAFT in the KITTI training set results, which is a 23.64% improvement over MatchFlow. This data firstly demonstrates that our model can fit the data more accurately, and secondly shows that there is much room for improvement in the generalization ability of the model. We believe that I-RAFT can perform even better when combined with methods to prevent overfitting, but we have not yet completed all of our experiments. Again, overfitting was also seen in the generalization experiments with C+T.

The metrics of I-RAFT were not optimal in the generalization experiments of C+T. When it was evaluated on the training sets of Sintel and KITTI, only one metric ranked first. Comparing all four metrics together, I-RAFT could be ranked third. We analyzed the reasons for this, the parameter learning of the initialization module, which depends on the data distribution of the training, does not give accurate and stable initialization results if it is used for other data sets, to the extent that it negatively affects the optical flow update part.

On balance, I-RAFT's model is effective, with 16.1M and 15.4M parameters for FlowFormer and MatchFlow respectively, while I-RAFT has only 11.6M parameters, which is 27.85% and 24.48% less than the first two models respectively, but better results can be achieved.

4.3   Ablation Experiments

We conducted a series of ablation studies to show the importance and effectiveness of each component of our models. All models in the ablation experiments were pre-trained on the FlyingChairs dataset with batch size set to 12 and learning rate set to 0.0025 for 120k training sessions and then fine-tuned for testing on Sintel and KITTI with batch size set to 8 and learning rate set to 0.0001 for 150k training sessions to obtain the results. The results of the ablation experiments are presented in Table 2.

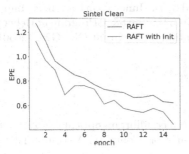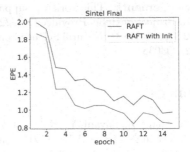

Fig. 4. Controlled experiments on the Sintel dataset before and after RAFT embeds the initialization module. The left panel shows clean pass and the right panel shows final pass.

The above experiments can show that the various modules in the model work. To highlight that the optical flow initialization module is effective, we conducted a second set of generalization tests where we embedded the initialization module into the RAFT model and tested it on the Sintel dataset. The model using the initialization module will fit faster during the training process. At the end of the training, the results for the model without the initialization module invoked are 0.62 and 0.97 for clean and final respectively, while the results with the initialization module introduced are: 0.60 and 0.84, an improvement of 3.2% and 13.4% respectively. A graph of the optimization curves during training Fig. 4 is shown. It can be found that the initialization module can effectively improve the learning ability and fitting speed of the model.

4.4 Visualisation Experiments

In order to see the results of the initialization module more intuitively, we visualized the results of the initialization module, as shown in Fig. 1, where we can see that the optical flow initialization module cannot accurately estimate the exact distance traveled, but can give a fuzzy initialized optical flow, which can better control the direction of the optical flow initialization.

5 Conclusion

We have shown that the optical flow initialization strategy can effectively improve the model learning ability and fitting speed, and the use of a multi-scale fused context encoder can improve the overall prediction ability. Based on these two points we propose a new model I-RAFT and introduce a lightweight attention network. Our model achieved first place in the Sintel dataset and second place in the KITTI dataset. Future work will focus on improving the generalization ability of the model.

Acknowledgements. This work was supported by the Innovation Capability Improvement Plan Project of Hebei Province (No. 22567637H), the S&T Program of Hebei (No. 236Z0302G), and HeBei Natural Science Foundation under Grant (No. G2021203010 & No. F2021203038).

References

1. Bai, S., Geng, Z., Savani, Y., Kolter, J.Z.: Deep equilibrium optical flow estimation. In: Proceedings of the IEEE/CVF Conference on Computer Vision and Pattern Recognition, pp. 620–630 (2022)
2. Dong, Q., Cao, C., Fu, Y.: Rethinking optical flow from geometric matching consistent perspective. arXiv preprint arXiv:2303.08384 (2023)
3. Dosovitskiy, A., et al.: FlowNet: learning optical flow with convolutional networks. In: Proceedings of the IEEE International Conference on Computer Vision, pp. 2758–2766 (2015)
4. Han, K., et al.: A survey on vision transformer. IEEE Trans. Pattern Anal. Mach. Intell. **45**(1), 87–110 (2023). https://doi.org/10.1109/TPAMI.2022.3152247
5. He, K., Zhang, X., Ren, S., Sun, J.: Spatial pyramid pooling in deep convolutional networks for visual recognition. IEEE Trans. Pattern Anal. Mach. Intell. **37**(9), 1904–1916 (2015)
6. Horn, B.K., Schunck, B.G.: Determining optical flow. Artif. Intell. **17**(1–3), 185–203 (1981)
7. Huang, Z., et al.: FlowFormer: a transformer architecture for optical flow. In: Avidan, S., Brostow, G., Cissé, M., Farinella, G.M., Hassner, T. (eds.) Computer Vision-ECCV 2022: 17th European Conference, Tel Aviv, Israel, 23–27 October 2022, Proceedings, Part XVII, pp. 668–685. Springer, Cham (2022). https://doi.org/10.1007/978-3-031-19790-1_40
8. Hui, T.W., Tang, X., Loy, C.C.: LiteFlowNet: a lightweight convolutional neural network for optical flow estimation. In: Proceedings of the IEEE Conference on Computer Vision and Pattern Recognition, pp. 8981–8989 (2018)
9. Hui, T.W., Tang, X., Loy, C.C.: A lightweight optical flow CNN-revisiting data fidelity and regularization. IEEE Trans. Pattern Anal. Mach. Intell. **43**(8), 2555–2569 (2020)
10. Jaegle, A., et al.: Perceiver IO: a general architecture for structured inputs & outputs. arXiv preprint arXiv:2107.14795 (2021)
11. Jiang, S., Campbell, D., Lu, Y., Li, H., Hartley, R.: Learning to estimate hidden motions with global motion aggregation. In: Proceedings of the IEEE/CVF International Conference on Computer Vision, pp. 9772–9781 (2021)
12. Jiang, S., Lu, Y., Li, H., Hartley, R.: Learning optical flow from a few matches. In: Proceedings of the IEEE/CVF Conference on Computer Vision and Pattern Recognition, pp. 16592–16600 (2021)
13. Kondermann, D., et al.: The HCI benchmark suite: stereo and flow ground truth with uncertainties for urban autonomous driving. In: Proceedings of the IEEE Conference on Computer Vision and Pattern Recognition Workshops, pp. 19–28 (2016)
14. Lin, T.Y., Dollár, P., Girshick, R., He, K., Hariharan, B., Belongie, S.: Feature pyramid networks for object detection. In: Proceedings of the IEEE Conference on Computer Vision and Pattern Recognition, pp. 2117–2125 (2017)

15. Lin, T.Y., Goyal, P., Girshick, R., He, K., Dollár, P.: Focal loss for dense object detection. In: Proceedings of the IEEE International Conference on Computer Vision, pp. 2980–2988 (2017)
16. Mayer, N., et al.: A large dataset to train convolutional networks for disparity, optical flow, and scene flow estimation. In: Proceedings of the IEEE Conference on Computer Vision and Pattern Recognition, pp. 4040–4048 (2016)
17. Meng, T., Ghiasi, G., Mahjorian, R., Le, Q.V., Tan, M.: Revisiting multi-scale feature fusion for semantic segmentation. arXiv preprint arXiv:2203.12683 (2022)
18. Sui, X., et al.: Craft: cross-attentional flow transformer for robust optical flow. In: Proceedings of the IEEE/CVF Conference on Computer Vision and Pattern Recognition, pp. 17602–17611 (2022)
19. Sun, D., et al.: AutoFlow: learning a better training set for optical flow. In: Proceedings of the IEEE/CVF Conference on Computer Vision and Pattern Recognition, pp. 10093–10102 (2021)
20. Sun, D., Yang, X., Liu, M.Y., Kautz, J.: PWC-Net: CNNs for optical flow using pyramid, warping, and cost volume. In: Proceedings of the IEEE Conference on Computer Vision and Pattern Recognition, pp. 8934–8943 (2018)
21. Sun, D., Yang, X., Liu, M.Y., Kautz, J.: Models matter, so does training: an empirical study of CNNs for optical flow estimation. IEEE Trans. Pattern Anal. Mach. Intell. 42(6), 1408–1423 (2019)
22. Tan, M., Pang, R., Le, Q.V.: EfficientDet: scalable and efficient object detection. In: Proceedings of the IEEE/CVF Conference on Computer Vision and Pattern Recognition, pp. 10781–10790 (2020)
23. Tao, A., Sapra, K., Catanzaro, B.: Hierarchical multi-scale attention for semantic segmentation. arXiv preprint arXiv:2005.10821 (2020)
24. Teed, Z., Deng, J.: RAFT: recurrent all-pairs field transforms for optical flow. In: Vedaldi, A., Bischof, H., Brox, T., Frahm, J.M. (eds.) Computer Vision-ECCV 2020: 16th European Conference, Glasgow, UK, 23–28 August 2020, Proceedings, Part II 16, pp. 402–419. Springer, Cham (2020). https://doi.org/10.1007/978-3-030-58536-5_24
25. Yang, G., Ramanan, D.: Volumetric correspondence networks for optical flow. In: Advances in Neural Information Processing Systems, vol. 32 (2019)
26. Yin, Z., Darrell, T., Yu, F.: Hierarchical discrete distribution decomposition for match density estimation. In: Proceedings of the IEEE/CVF Conference on Computer Vision and Pattern Recognition, pp. 6044–6053 (2019)
27. Zhang, F., Woodford, O.J., Prisacariu, V.A., Torr, P.H.: Separable flow: learning motion cost volumes for optical flow estimation. In: Proceedings of the IEEE/CVF International Conference on Computer Vision, pp. 10807–10817 (2021)
28. Zhao, S., Sheng, Y., Dong, Y., Chang, E.I., Xu, Y., et al.: MaskFlowNet: asymmetric feature matching with learnable occlusion mask. In: Proceedings of the IEEE/CVF Conference on Computer Vision and Pattern Recognition, pp. 6278–6287 (2020)
29. Zhao, S., Zhao, L., Zhang, Z., Zhou, E., Metaxas, D.: Global matching with overlapping attention for optical flow estimation. In: Proceedings of the IEEE/CVF Conference on Computer Vision and Pattern Recognition, pp. 17592–17601 (2022)

Educational Pattern Guided Self-knowledge Distillation for Siamese Visual Tracking

Quan Zhang⬡ and Xiaowei Zhang(✉)⬡

Qingdao University, Qingdao 266071, Shandong, China
xiaowei19870119@sina.com

Abstract. Existing Siamese-based trackers divide visual tracking into two stages, *i.e.*, feature extraction (backbone subnetwork), and prediction (head subnetwork). However, they mainly implement task-level supervision (classification and regression), barely considering the feature-level supervision in the knowledge learning process, which could result in deficient knowledge interaction among the features of the tracker's targets and background interference during the online tracking process. To solve the issues, this paper proposes an educational pattern-guided self-knowledge distillation methodology by guiding Siamese-based trackers to learn feature knowledge by themselves, which can serve as a generic training protocol to improve any Siamese-based tracker. Our key insight is to utilize two educational self-distillation patterns, *i.e.*, focal self-distillation and discriminative self-distillation, to educate the tracker to possess self-learning ability. The focal self-distillation pattern educates the tracking network to focus on valuable pixels and channels by decoupling the spatial learning and channel learning of target features. The discriminative self-distillation pattern aims at maximizing the discrimination between foreground and background features, ensuring that the trackers are unaffected by background pixels. As one of the first attempts to introduce self-knowledge distillation into the visual tracking field, our method is effective and efficient and has a strong generalization ability, which might be instructive for other research. Codes and data are publicly available.

Keywords: Visual object tracking · Knowledge distillation · Self-knowledge distillation

1 Introduction

Visual tracking is one of the most fundamental tasks in computer vision. In general, trackers estimate the location of a target in every frame of a video, where the information on initial targets is provided beforehand in the first frame. Due to the excellent tracking performance, Siamese-based trackers have attracted more and more attention in the visual tracking field.

As a pioneer Siamese-based tracker, SiamFC [1] adopts a Siamese convolutional network to obtain cross-correlation features to train a similarity metric between the target template and the search region. It consists of two subnetworks, *i.e.*, the backbone subnetwork and the head subnetwork. The backbone

© The Author(s), under exclusive license to Springer Nature Singapore Pte Ltd. 2024
B. Luo et al. (Eds.): ICONIP 2023, CCIS 1968, pp. 30–43, 2024.
https://doi.org/10.1007/978-981-99-8181-6_3

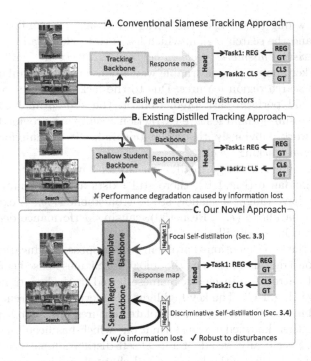

Fig. 1. Methodology comparison between conventional methods (A–B) and ours (C), where "REG" denotes bounding box regression and "CLS" is category classification. Compared to conventional tracking methods (A), which only focus on task-level loss supervision, possibly leading to feature interruption by distractors, and existing distilled tracking methods (B), where the depth gap between student and teacher backbones might cause performance degradation as knowledge is lost in the distillation, our novel approach (C) can enable a Siamese-based tracker to learn feature representations by itself in the feature level and task level simultaneously.

subnetwork encodes the target template and the search region features. The head subnetwork calculates the similarity of the above two features and predicts the tracking results accordingly.

Conventional Siamese-based trackers [10,12] (Fig. 1-A) tend to focus on task-level loss supervision, *i.e.*, the classification task loss and the regression task loss. The rationale can be attributed to the fact that the task-level loss supervision can divide the problem domains (visual tracking task) into two sub-tasks (classification task and regression task) to speed up the training phase. However, though the performance of Siamese-based trackers has improved greatly, more than task-level loss supervision is needed. They could suffer from two limitations:

1) Siamese-based trackers might predict wrong results because of background distractors. At the training stage, the tracker lacks the guidance of feature-level knowledge, *i.e.*, it does not know which part of the features it should learn, and it is likely to learn much false positive knowledge. This error learning can easily introduce the wrong response of the background into the similarity matrix of the template and the search region, which leads to tracking failure.

2) There exists a mismatch problem between the backbone and head. In most cases, the tracker adopts an unlearnable way when matching similarity between templates and search region features. Due to the lack of feature-level guidance, the tracker could perform task prediction on the sub-optimal corresponding graph, *i.e.*, the sub-optimal feature extraction results are provided to the task prediction network. Obviously, this mismatch makes it difficult for the tracker to obtain optimal tracking results.

Though existing knowledge distillation works have provided trackers with features-level guidance (Fig. 1-B), there still exists a fundamental issue: most of the trackers have taught a shallow student backbone (*e.g.*, Alexnet-16) by a normal teacher backbone (*e.g.*, Resnet50), causing performance degradation as knowledge is lost in the distillation.

To address the abovementioned problems, we propose an educational pattern-guided self-knowledge distillation methodology for Siamese-based trackers, which adopt different patterns to teach any Siamese-based tracks how to learn feature-level knowledge (Fig. 1-C). The key highlight is to leverage different educational patterns for different features (*i.e.*, template features and search patches features) to strengthen knowledge transformation, which has been long ignored by existing works. Specifically, we adopt a focal self-distillation pattern for template features to decouple the spatial learning and channel, guiding the tracking network to focus on valuable pixels and channels (see Sect. 3.3). For search region features, we devise a discriminative self-distillation pattern, teaching the tracker to be immune to the background (see Sect. 3.4).

In brief, our contributions can be summarized as follows:

- As the first attempt, we have proposed an educational pattern-guided self-distillation algorithm, teaching Siamese-based trackers how to lean feature-level knowledge upon self-knowledge distillation, which can serve as a generic training protocol to improve any Simamese-based trackers.
- We have devised two different educational patterns for different features. The focal educational pattern could guide Siamese-based trackers in learning knowledge from critical pixels and channels, while the discriminative educational pattern might educates Siamese-based trackers to distinguish foreground knowledge from background knowledge.
- Extensive experiments on five public datasets show the effectiveness of our method. Meanwhile, ablation experiments prove the generalization abilities of our method.

2 Related Work

2.1 Siamese-based Trackers

Benefiting from end-to-end training capabilities and high efficiency, Siamese-based trackers (*e.g.*, [1,5,10,12,15,22]) have recently mainstreamed in the visual tracking community. Since the pioneered application of Siamese network structure in visual tracking does not need to model update, most of the subsequent trackers are optimizing it to achieve more accurate tracking performance. For example, SiamRPN [15] introduces the region proposal subnetwork of object detection into the SiamFC to adapt the scale changes of targets effectively, which formulated visual tracking as a local target-specific anchor-based detection task. SiamRPN++ [14] performs layer-wise and depth-wise feature aggregations more accurately to locate and regress the target bounding boxes. CGACD [6] proposes to learn correlation-guided attention, including pixel-wise correlation-guided spatial attention and channel-wise correlation-guided channel attention in a two-stage corner detection network for estimating tight bounding boxes.

2.2   Self Knowledge Distillation

Self Knowledge distillation means that a teacher model may be found with no new large model and may provide yield information to the student model. Self Knowledge distillation has been proposed by Zhang, Song, Gao, et al. [21]. They use the new deep subnetwork classifier as a teacher to carry out distillation learning on the shallow part of the source network. After the softmax at the end of each classifier, they have established the cross entropy loss between the label and the softmax at the end of each classifier. Then, the KL divergence distillation loss between each classifier's softmax and the deepest classifier's softmax has been established. Finally, L2 distillation loss is established for each classifier's feature and the deepest classifier's feature. In order to establish a parallel feature map distillation, Ji, Shin, Hwang, et al. [13] presents a method of feature self-distillation that can utilize self-knowledge distillation of soft label and feature map distillation.

3   Method

3.1   Overview

Figure 2-A shows the overall method pipeline of the educational pattern-guided self-distillation method. Our tracker has made two technical innovations, including *1) focal self-distillation* and *2) discriminative self-distillation*.

In the network training phase, two inputs, *i.e.*, template and search region, are fed into teacher Siamese-based trackers and student Siamese-based trackers, respectively. Then apply feature-level focal self-distillation (Sect. 3.3) and discriminative self-distillation (Sect. 3.4) between the teacher tracker backbone and the student tracker backbone to focus on valuable pixels and channels and distinguish foreground from background. Next, two task-level loss supervisions are implemented in the prediction results. We will detail these two key components in the following sections.

3.2   Motivations

Motivations

Siamese-based trackers only adopt task-level supervision which might make trackers work in most cases (*e.g.*, simple samples). However, there are some hard samples where the trackers could give wrong predictions, *i.e.*, samples with background distractors. Moreover, failure cases of recent Siamese-based SOTA trackers are more likely to be induced by background-related distractions, which pushes us to rethink whether feature-level supervision can be designed.

Feature-level supervision means performing supervision in template features and search region features. To achieve this, we leverage two different educational self-knowledge distillation patterns to make any Siamese-based trackers learn how to learn features by themselves, where "educational pattern" denotes that the template features and search region features of the trackers are separately implemented by self-knowledge distillation.

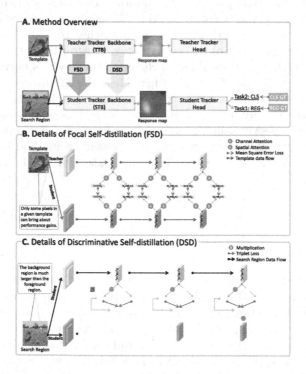

Fig. 2. The architecture and technical details of our proposed method. Our method can serve as a plug-and-play training protocol to improve any Simamese-based trackers (A). During training stages, the focal self-distillation (B) makes Siamese-based trackers focus on the valuable pixel and channels of the template features, while the discriminative self-distillation (C) makes trackers immune to the background and focus on the foreground.

3.3 Focal Self-distillation

Siamese-based trackers have the template and search region features (see Sect. 3.2). We first introduce the self-distillation of the template features $\mathbf{F}_z$. In effect, only some valuable pixels in a given template can bring performance gains, so the tracker should focus on crucial pixels in the learning process of template knowledge. Similarly, the tracker should also pay attention to knowledge learning of critical channels.

To achieve our goals, we propose a focal self-distillation pattern to guide the Siamese-based tracker. The technical details have vividly demonstrated in Fig. 2-B. Our focal self-distillation leverage the classical Siamese-based tracker as a teacher, which massive data have trained. And the student tracker has yet to be trained, but it is the same as the teacher tracker.

As shown in the figure, we transform teacher and student trackers' template features into spatial and channel dimensions, respectively. And then use teachers to guide students learning knowledge. This process can be technically detailed as the following equations:

$$\mathcal{L}_{foc} = \mathcal{L}_C + \mathcal{L}_S \tag{1}$$

$$\mathcal{L}_C = \frac{1}{3N_{CA}} \sum_{k=3}^{5} \sum_{c=1}^{C_z} (A^{ca}(F_{zk_c}^T) - A^{ca}(F_{zk_c}^S))^2 \tag{2}$$

$$\mathcal{L}_S = \frac{1}{3N_{SA}} \sum_{k=3}^{5} \sum_{h=1}^{H_z} \sum_{w=1}^{W_z} (A^{sa}(F_{zk}^T) - A^{sa}(F_{zk}^S))^2, \tag{3}$$

where $\mathcal{L}_{foc}$ denotes the proposed focal self-distillation loss; $\mathcal{L}_C$ and $\mathcal{L}_S$ denote respectively channel self-distillation and spatial self-distillation; C_z denote the channels of the template feature, and $N_{CA} = C_z$; H_z and W_z denote the height and the width of the template features, and $N_{SA} = H_z \times W_z$; T and S denotes the teacher tracker and student tracker, respectively; k denotes the multi-scale features, in this work, $k \in \{3, 4, 5\}$; A_{ca} and A_{sa} denote our feature transfer operation in the channel dimension and the space dimension, which can be technically detailed as the following equations:

$$A^{ca}(F) = C_z \cdot softmax(\frac{1}{H_z W_z} \cdot \sum_{j=1}^{H_z} \sum_{j=1}^{W_z} F_{i,j}) \tag{4}$$

$$A^{sa}(F) = H_z W_z \cdot softmax(\frac{1}{C_z} \cdot \sum_{c=1}^{C_z} F_c), \tag{5}$$

where F denotes inputs; C_z denotes the channels of the inputs (see Eq. 1); (i, j) denotes pixels in the template; H_z and W_z denote the height and the width of the inputs (see Eq. 1); $softmax$ denotes the typical softmax operation.

3.4 Discriminative Self-distillation

In the previous section, we obtain self-distilled template features (see Sect. 3.3). We shall detail how to guide Siamese-based trackers to learn the knowledge of search region features $\mathbf{F}_x$. As shown in Fig. 2-C, the search regions are divided into the foreground region (*i.e.*, the pixels inside the ground-truth bounding box) and the background region (*i.e.*, the pixels outside the ground-truth bounding box). We expect Siamese-based trackers to follow the following two general principles in the process of learning the search region features: *1) The tracker requires a focus on learning about the foreground region. 2) The tracker can autonomously remove knowledge from the background region.*

To ensure compliance with the principles, we propose the discriminative self-distillation pattern. As shown in Fig. 2-C, we leverage the foreground mask to guide the student tracker in learning the foreground features. For background self-distillation, we have two options, *i.e.*, background masks and random masks. The background masks mean we mask the foreground feature, only the background feature as inputs of self-distillation. Obviously, the number of positive pixels is much smaller than the number of negative pixels. Suppose the teacher tracker guides the student by maximizing the difference between the foreground and background features. In that case, the tracker might be inclined to the background region due to the problem of gradient flooding. Meanwhile, even if the number of pixels selected for the background mask is the same as the number of foreground pixels, the knowledge about the background obtained by the tracker is incomplete and far away from the background knowledge. So, the random mask (*i.e.*, randomly mask pixels with the same number of foreground pixels in the search region) is generated to guide the student tracker in staying away from the background features. This process can be technically detailed as the following equations:

$$\mathcal{L}_{dis} = \mathcal{L}_{fore} + \mathcal{L}_{back} \tag{6}$$

$$\mathcal{L}_{fore} = \frac{1}{3N_s} \sum_{k=3}^{5} \sum_{h=1}^{H_s} \sum_{w=1}^{W_s} \sum_{c=1}^{C_s} (M(F_{sk}^T - F_{tk}^S))^2 \tag{7}$$

$$\mathcal{L}_{back} = \frac{1}{3N_s} \sum_{k=3}^{5} \sum_{h=1}^{H_s} \sum_{w=1}^{C_s} \text{Tri}(MF_{sk}^S, MF_{sk}^t, M_{ran}F_{sk}^T), \tag{8}$$

where $\mathcal{L}_{dis}$ denotes the proposed discriminative self-distillation loss; $\mathcal{L}_{fore}$ and $\mathcal{L}_{back}$ denote the foreground self-distillation loss and background self-distillation loss respectively; C_s denotes the channels of the search region feature; H_s and W_s denote the height and the width of the template features, and $N_s = H_s \times W_s \times C_s$; T and S denote respectively the teacher tracker and the student tracker; k denotes the multi-scale features, in this work, $k \in \{3, 4, 5\}$; Tri denotes the typical triplet loss; M_{ran} denotes random mask pixels in the search region and the number of randomly masked pixels is equal to foreground ones, *i.e.*, $W_{M_{ran}} \times H_{M_{ran}} = W_M \times H_M$; M denotes the foreground mask, which can be technically detailed as the following equations:

$$M(i,j) = \begin{cases} 1 & (i,j) \in \text{GT} \\ 0 & \text{otherwise,} \end{cases} \tag{9}$$

where (i,j) denotes pixels; GT denotes ground truth bounding boxes.

3.5 Overall Loss

The overall loss of the proposed method consists of the focal self-distillation loss (see Sect. 3.3), the discriminative self-distillation loss (see Sect. 3.4), and sub-tasks losses (i.e., the classification task loss and the regression task loss). And just like Siamese-based trackers, we leverage the cross-entropy loss for the classification task, i.e., $\mathcal{L}_{\text{CLS}} = CE(\mathbf{CLS}, l_{\text{CLS}})$. $\mathbf{CLS}$ is the classification prediction of Siamese-based trackers and l_{CLS} is the label of the classification task. We use IOU loss to calculate the regression loss $\mathcal{L}_{\text{REG}} = IOU(\mathbf{REG}, l_{\text{REG}})$. $\mathbf{REG}$ is the regression prediction of Siamese-based trackers, and l_{REG} is the label of the regression task. The overall training loss can be technically detailed as the following equations:

$$\mathcal{L} = \mathcal{L}_{\text{CLS}} + \mathcal{L}_{\text{REG}} + \mathcal{L}_{foc} + \mathcal{L}_{dis}, \tag{10}$$

where $\mathcal{L}$ denotes the overall loss of the proposed method; $\mathcal{L}_{\text{CLS}}$ denotes the loss of the classification task; $\mathcal{L}_{\text{REG}}$ denotes the loss of the regression task; $\mathcal{L}_{foc}$ denotes the proposed focal self-distillation loss (see Eq. 1); $\mathcal{L}_{dis}$ denotes the proposed discriminative self-distillation loss (see Eq. 6).

Table 1. Comparison with SOTA trackers on five tracking benchmarks regarding the AUC metric (larger is better). The top three are highlighted in red, green, and blue.

Methods		LaSOT	NFS30	TC128	UAV123	DTB70
SiamFC[1]	2016	.336	-	-	.494	.483
DaSiamRPN[24]	2017	.515	.395	-	.501	.472
SiamDW[22]	2019	.347	.521	.583	.536	.430
SiamRPN++[14]	2019	.495	.502	.573	.613	.597
SiamCAR[11]	2020	.516	.530	.582	.614	.608
Ocean[23]	2020	.526	.553	.585	.621	.631
SiamBAN[4]	2020	.514	.595	.589	.631	.641
CLNet[5]	2020	.499	.547	.564	.633	.650
CGACD[6]	2020	.518	.527	.605	.633	.625
SiamGAT[10]	2021	.539	.543	.585	.633	.604
SiamAPN++[2]	2021	.435	.539	.556	.582	.594
HiFT[2]	2021	.450	.523	.529	.589	.594
ULAST[20]	2022	.474	.451	-	-	-
SiamAPN[8]	2022	.428	.500	.535	.575	.585
CNNInMo[12]	2022	.539	.560	-	.629	.623
Ours		.539	.584	.606	.635	.653

4   Experiment

4.1   Implementation Details

The proposed method is implemented in Python with PyTorch and trained on 2 RTX 2080Ti. For a fair comparison, we follow the same training protocols(*e.g.*, datasets, hyper-parameters, input resolution) as the typical SiamCAR [11] for our method. In training process, we leaverage the Resnet-50 pretrained on the ImageNet [19], in the training

4.2   Datasets and Evaluation Metrics

We evaluate the performance of the proposed method on five benchmarks, including TC128 [17], LaSOT [7], UAV123 [18], NFS30 [9], and DTB70 [16]. As an authoritative evaluation metric, the one-pass evaluation (OPE) metrics, including precision and success rate, are leveraged to assess the tracking performance. Meanwhile, the success plot's area under the curve (AUC) at 20 pixels distance precision threshold is adopted to rank the trackers.

4.3   Comparison With State-of-the-Art Trackers

To prove the validity of our method, we have performed comprehensive experiments with 15 recently published SOTA Siamese-based trackers(including SiamFC [1], DaSiamRPN [24], SiamDW [22], SiamRPN++ [14], SiamBAN [4], SiamCAR [11], Ocean [23], CLNet [5], CGACD [6], SiamGAT [10], CNNInMo [12], HiFT [2], SiamAPN [8], ULAST [20] and SiamAPN++ [3]). The compared results are either reproduced by the released codes or tracking predictions provided by authors.

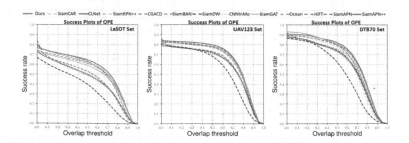

Fig. 3. Quantitative comparisons with SOTA trackers on three benchmarks.

Quantitative Comparisons. As commonly used quantitative evaluation methods, we first evaluate the proposed method using the abovementioned evaluation metrics. The AUC ranking on five benchmarks is reported in Table 1.

Results on LaSOT: As shown in Table 1 (column LaSOT), the proposed method has achieved the best ranking among all compared competitors, where our tracker is equal to CNNInMo and SiamGAT.

Results on UAV123: Table 1 (column UAV123) presents the evaluation results on the UAV123 benchmark. Our method achieves the highest AUC score for all the methods. Compared with the classical SiamCAR, the AUC score of the proposed method has increased by nearly 3% Meanwhile, as shown in Fig. 3, the proposed method significantly outperforms other competitors on the UAV123 benchmark.

Results on DTB70: Table 1 (column DTB70) presents the evaluation result on DTB70. Of all the methods we compared, the proposed method achieves the best AUC score, 0.653. Compared with the classical SiamCAR, our AUC score has increased by nearly 7%. And the quantitative curves is shown in Fig. 3.

Results on NFS30: As shown in Table 1 (column NFS30), the proposed method has achieved the second ranking among all compared competitors. It is worth mentioning that, compared with the classical SiamCAR, our AUC has increased by 10%.

Results on TC128: As shown in Table 1 (column TC128), the proposed method has achieved the best ranking among all compared competitors. It is worth mentioning that, compared with the classical SiamCAR, our AUC has increased by 4%.

4.4 Ablation Studies

Component Evaluations. In order to verify the effectiveness of each component we proposed, as shown in Table 2, we have constructed component evaluation experiments on the DTB70 [16] benchmark.

    1) Effectiveness of the focal self-distillation (see Sect. 3.3). All experiments regarding the effectiveness of focal self-distillation have been highlighted by mark ①. From lines 2 to 7, we have tested two models using channel self-distillation (see Eq. 1) or spatial self-distillation (see Eq. 1) only. As can be seen, compared with the baseline model (line 1), these two models have succeeded in achieving performance gain in the metric. Using single-scale channel self-distillation only (line 2–line 4), the model can increase the AUC score from $0.608 \rightarrow 0.619$. And, using single-scale spatial self-distillation only (line 5–line 7), the model can improve the AUC score from $0.608 \rightarrow 0.618$. The experimental results show that both can bring performance gain, but the effect of growth differs at different scales. See lines 8–10, and we have tested the models using single-scale channel self-distillation and spatial self-distillation simultaneously. As can be seen, compared with the models using only one, these can increase

Table 2. Component evaluations on the DTB70 benchmark.

	FocalSD (Sec. III-C)						DisSD (Sec. III-D)							DTB70
	CSD			SSD			FM			RM			BM	AUC↑
	3	4	5	3	4	5	3	4	5	3	4	5	all	
⓪1	✗	✗	✗	✗	✗	✗	✗	✗	✗	✗	✗	✗	✗	.608
2	✓	✗	✗	✗	✗	✗	✗	✗	✗	✗	✗	✗	✗	.615
3	✗	✓	✗	✗	✗	✗	✗	✗	✗	✗	✗	✗	✗	.616
4	✗	✗	✓	✗	✗	✗	✗	✗	✗	✗	✗	✗	✗	.619
5	✗	✗	✗	✓	✗	✗	✗	✗	✗	✗	✗	✗	✗	.618
6	✗	✗	✗	✗	✓	✗	✗	✗	✗	✗	✗	✗	✗	.615
7	✗	✗	✗	✗	✗	✓	✗	✗	✗	✗	✗	✗	✗	.614
8	✓	✗	✗	✓	✗	✗	✗	✗	✗	✗	✗	✗	✗	.625
9	✓	✗	✗	✗	✓	✗	✗	✗	✗	✗	✗	✗	✗	.628
10	✓	✗	✗	✗	✗	✓	✗	✗	✗	✗	✗	✗	✗	.626
11	✓	✓	✓	✓	✓	✓	✗	✗	✗	✗	✗	✗	✗	.637
12	✓	✓	✓	✓	✓	✓	✓	✗	✗	✗	✗	✗	✗	.640
13	✓	✓	✓	✓	✓	✓	✗	✓	✗	✗	✗	✗	✗	.643
14	✓	✓	✓	✓	✓	✓	✗	✗	✓	✗	✗	✗	✗	.642
15	✓	✓	✓	✓	✓	✓	✗	✗	✗	✓	✗	✗	✗	.639
17	✓	✓	✓	✓	✓	✓	✗	✗	✗	✗	✓	✗	✗	.643
②18	✓	✓	✓	✓	✓	✓	✗	✗	✗	✗	✗	✓	✗	.642
19	✓	✓	✓	✓	✓	✓	✓	✗	✓	✗	✗	✗	✗	.645
20	✓	✓	✓	✓	✓	✓	✓	✗	✗	✓	✗	✗	✗	.649
21	✓	✓	✓	✓	✓	✓	✓	✗	✗	✗	✓	✗	✗	.647
22	✓	✓	✓	✓	✓	✓	✓	✓	✓	✗	✗	✗	✓	.650
23	✓	✓	✓	✓	✓	✓	✓	✓	✓	✓	✓	✓	✗	**.653**

Full Names of the Abbreviations Used in the Table
CSD: Channel Self-distillation   RM: Random Mask
SSD: Spatial Self-distillation   FocalSD: Focal Self-distillation
BM: Background Mask   DisSD: Discriminative Self-
FM: Foreground Mask   distilltion

Primary Target of Each Sub-experiment
⓪ Baseline
① *Verify* the effectiveness of the focal self-kd (Sec. 3.3)
② *Verify* the effectiveness of the discriminative self-kd (Sec. 3.4)

the AUC score from 0.619→0.628. This is enough to show that the combination of the two performs gains at different scales, so we don't need to do more permutations. In line 11, we have tested using simultaneous channel self-distillation and spatial self-distillation at all scales. The AUC score increases to 0.637. The experimental results show that our focal self-distillation can guide Siamese-based trackers to learn template knowledge.

2) Effectiveness of the discriminative self-distillation (see Sect. 3.4). All experiments regarding the effectiveness of discriminative self-distillation have been highlighted by mark ②. Lines 12–18 have reported the results of using a foreground mask (see Eq. 6) or random mask (see Eq. 6) at a single scale only. From lines 12 to 14, we have tested models using only foreground masks at a single scale. Compared to line 11, the AUC score increases from 0.637→0.643. The experimental results show that the foreground mask can bring performance gain. This is undoubtedly true, as it guides the tracker in learning the correct foreground knowledge and inevitably improves performance. And, using a random mask only (line 15–line 18), the model can increase the AUC score from $0.637{\rightarrow}0.643$. In line 23, we have tested using the two mask mechanisms at all scales simultaneously. The AUC score reached the highest score of 0.653. The experimental results show that our discriminative self-distillation can guide a Siamese-based tracker to learn search region knowledge. In line 22, we tested using a background mask to learn the background features of the search region. Compared to line 22, the AUC score decreases from $0.653{\rightarrow}0.650$. Compared to line 11, the AUC score increases from $0.637{\rightarrow}0.650$. The experimental results show that although learning knowledge from the background can bring performance gains, the two mask mechanisms bring different performance improvements. Obviously, the random mask can avoid the gradient flooding problem mentioned earlier, bringing more performance gains.

5 Conclusion

In this paper, we have proposed a generic training protocol to improve any Simamese-based tracker by an educational pattern-guided self-distillation algorithm to teach Siamese-based trackers how to learn feature-level knowledge upon self-knowledge distillation in feature-level supervision during the training stage. The key highlights are our proposed focal self-distillation and discriminative self-distillation to educate the trackers to possess self-learning abilities. Extensive experiments have proved the effectiveness and efficiency of our method and strong generalization ability to improve any Siamese-based trackers. In the near future, we will explore multiple teacher interactions with multiple student trackers to ensure better tracking performance.

References

1. Bertinetto, L., Valmadre, J., Henriques, J.F., Vedaldi, A., Torr, P.H.S.: Fully-convolutional Siamese networks for object tracking. In: Hua, G., Jégou, H. (eds.) ECCV 2016. LNCS, vol. 9914, pp. 850–865. Springer, Cham (2016). https://doi.org/10.1007/978-3-319-48881-3_56
2. Cao, Z., Fu, C., Ye, J., Li, B., Li, Y.: HiFT: hierarchical feature transformer for aerial tracking. In: ICCV, pp. 15437–15446 (2021). https://doi.org/10.1109/ICCV48922.2021.01517
3. Cao, Z., Fu, C., Ye, J., Li, B., Li, Y.: SiamAPN++: Siamese attentional aggregation network for real-time UAV tracking. In: IEEE IROS, pp. 3086–3092 (2021). https://doi.org/10.1109/IROS51168.2021.9636309
4. Chen, Z., Zhong, B., Li, G., Zhang, S., Ji, R.: Siamese box adaptive network for visual tracking. In: CVPR, pp. 6667–6676 (2020). https://doi.org/10.1109/CVPR42600.2020.00670

5. Dong, X., Shen, J., Shao, L., Porikli, F.: CLNet: a compact latent network for fast adjusting Siamese trackers. In: Vedaldi, A., Bischof, H., Brox, T., Frahm, J.-M. (eds.) ECCV 2020. LNCS, vol. 12365, pp. 378–395. Springer, Cham (2020). https://doi.org/10.1007/978-3-030-58565-5_23

6. Du, F., Liu, P., Zhao, W., Tang, X.: Correlation-guided attention for corner detection based visual tracking. In: CVPR, pp. 6835–6844 (2020). https://doi.org/10.1109/CVPR42600.2020.00687

7. Fan, H., Bai, H., Lin, L., Yang, F., Ling, H.: LaSOT: a high-quality large-scale single object tracking benchmark. IJCV **129**, 439–461 (2020)

8. Fu, C., Cao, Z., Li, Y., et al.: Onboard real-time aerial tracking with efficient Siamese anchor proposal network. IEEE TGRS **60**, 1–13 (2022). https://doi.org/10.1109/TGRS.2021.3083880

9. Galoogahi, H.K., Fagg, A., Huang, C., Ramanan, D., Lucey, S.: Need for Speed: a benchmark for higher frame rate object tracking. In: ICCV, pp. 1134–1143 (2017). https://doi.org/10.1109/ICCV.2017.128

10. Guo, D., Shao, Y., Cui, Y., Wang, Z., Zhang, L., Shen, C.: Graph attention tracking. In: CVPR, June 2021

11. Guo, D., Wang, J., Cui, Y., Wang, Z., Chen, S.: SiamCAR: Siamese fully convolutional classification and regression for visual tracking. In: CVPR, pp. 6268–6276 (2020). https://doi.org/10.1109/CVPR42600.2020.00630

12. Guo, M., et al.: Learning target-aware representation for visual tracking via informative interactions (2022)

13. Ji, M., Shin, S., Hwang, S., Park, G., Moon, I.C.: Refine myself by teaching myself: feature refinement via self-knowledge distillation. In: CVPR, pp. 10659–10668 (2021). https://doi.org/10.1109/CVPR46437.2021.01052

14. Li, B., Wu, W., Wang, Q., Zhang, F., Xing, J., Yan, J.: SiamRPN++: evolution of Siamese visual tracking with very deep networks. In: CVPR, pp. 4277–4286 (2019). https://doi.org/10.1109/CVPR.2019.00441

15. Li, B., Yan, J., Wu, W., Zhu, Z., Hu, X.: High performance visual tracking with Siamese region proposal network. In: CVPR, pp. 8971–8980 (2018). https://doi.org/10.1109/CVPR.2018.00935

16. Li, S., Yeung, D.Y.: Visual object tracking for unmanned aerial vehicles: a benchmark and new motion models. In: ICCV, pp. 4140–4146 (2017)

17. Liang, P., Blasch, E., Ling, H.: Encoding color information for visual tracking: algorithms and benchmark. IEEE TIP **24**(12), 5630–5644 (2015). https://doi.org/10.1109/TIP.2015.2482905

18. Mueller, M., Smith, N., Ghanem, B.: A benchmark and simulator for UAV tracking. In: Leibe, B., Matas, J., Sebe, N., Welling, M. (eds.) ECCV 2016. LNCS, vol. 9905, pp. 445–461. Springer, Cham (2016). https://doi.org/10.1007/978-3-319-46448-0_27

19. Russakovsky, O., et al.: ImageNet large scale visual recognition challenge. IJCV **115**, 211–252 (2015)

20. Shen, Q., et al.: Unsupervised learning of accurate Siamese tracking. In: CVPR, pp. 8091–8100 (2022). https://doi.org/10.1109/CVPR52688.2022.00793

21. Zhang, L., Song, J., Gao, A., Chen, J., Bao, C., Ma, K.: Be your own teacher: improve the performance of convolutional neural networks via self distillation. In: ICCV, pp. 3712–3721 (2019). https://doi.org/10.1109/ICCV.2019.00381

22. Zhang, Z., Peng, H.: Deeper and wider Siamese networks for real-time visual tracking. In: CVPR, pp. 4586–4595 (2019). https://doi.org/10.1109/CVPR.2019.00472

23. Zhang, Z., Peng, H., Fu, J., Li, B., Hu, W.: Ocean: object-aware anchor-free tracking. In: Vedaldi, A., Bischof, H., Brox, T., Frahm, J.-M. (eds.) ECCV 2020. LNCS, vol. 12366, pp. 771–787. Springer, Cham (2020). https://doi.org/10.1007/978-3-030-58589-1_46
24. Zhu, Z., Wang, Q., Li, B., Wu, W., Yan, J., Hu, W.: Distractor-aware Siamese networks for visual object tracking. In: Ferrari, V., Hebert, M., Sminchisescu, C., Weiss, Y. (eds.) ECCV 2018. LNCS, vol. 11213, pp. 103–119. Springer, Cham (2018). https://doi.org/10.1007/978-3-030-01240-3_7

LSiF: Log-Gabor Empowered Siamese Federated Learning for Efficient Obscene Image Classification in the Era of Industry 5.0

Sonali Samal[1] , Gautam Srivastava[2]([✉]) , Thippa Reddy Gadekallu[3,4,5] ,
Yu-Dong Zhang[6] , and Bunil Kumar Balabantaray[1]

[1] National Institute of Technology Meghalaya, Shillong, India
{p21cs002,bunil}@nitm.ac.in
[2] Department of Mathematics and Computer Science, Brandon University, Brandon,
Canada
srivastavag@brandonu.ca
[3] Department of Electrical and Computer Engineering, Lebanese American
University, Byblos, Lebanon
thippareddy@ieee.org
[4] College of Information Science and Engineering, Jiaxing University, Jiaxing 314001,
China
[5] Division of Research and Development, Lovely Professional University, Phagwara,
India
[6] School of Computing and Mathematical Sciences, University of Leicester, Leicester
LE1 7RH, UK
yudongzhang@ieee.org

Abstract. The widespread presence of explicit content on social media platforms has far-reaching consequences for individuals, relationships, and society as a whole. It is crucial to tackle this problem by implementing efficient content moderation, educating users, and creating technologies and policies that foster a more secure and wholesome online atmosphere. To address this issue, this research proposes the Log-Gabor Empowered Siamese Federated Learning (LSiF) framework for precise and efficient classification of obscene images in the era of Industry 5.0. The LSiF framework utilizes a Siamese Network with two parallel streams, where log-Gabor input and normal raw input are processed simultaneously. This Siamese architecture leverages shared weights and parameters, enabling effective learning of distinctive features for class differentiation and pattern recognition. The weight-sharing mechanism enhances the model's ability to generalize, increases its robustness, and improves computational efficiency, making it well-suited for resource-constrained and real-time applications. Additionally, federated learning is employed with a client size of three, allowing local model updates on each device. This approach minimizes the need for extensive data transmission to a central server, reducing communication overhead and improving learning efficiency, particularly in environments with limited

B. Luo et al. (Eds.): ICONIP 2023, CCIS 1968, pp. 44–55, 2024.
https://doi.org/10.1007/978-981-99-8181-6_4

bandwidth. The proposed LSiF model demonstrates remarkable performance, achieving an accuracy of 94.30%, precision of 94.00%, recall of 94.26%, and F1-Score of 94.17% with a client size of three.

Keywords: Industry 5.0 · Log-gabor filter · Siamese Network · Federated learning · obscene classification

1   Introduction

Obscene content, encompassing explicit sexual material, graphic violence, and vulgar language has significant impacts on society. Exposure to such content can have immediate and long-term effects, affecting individuals, relationships, and communities [16]. Children and adolescents are particularly vulnerable, as early exposure can hinder their development and understanding. Ultimately, widespread availability of obscene content erodes societal norms and values, favoring instant gratification over meaningful connections. Security challenges, such as web security and the presence of explicit content on social media, have an impact on the integrity and reputation of social media platforms. It is crucial to recognize these impacts and address them to foster a healthier and more respectful society. Industry 5.0, when applied to the context of classifying obscene images and ensuring the safety of women on digital platforms, signifies a transformative leap in technological innovation. Industry 5.0 is a key step toward building a safer and more inclusive online environment in which technology plays a critical role in minimizing dangers and encouraging a respectful and secure online space for all.

Extensive research has been conducted in the field of obscene image classification, aiming to automatically detect and remove inappropriate content from online platforms [1,10,11]. Various techniques have been explored, including skin color exposure, visual perception, and deep learning models trained on explicit images. Initially, skin color exposure was commonly used for nudity classification, but it had limitations due to over-reliance on skin pixels, leading to inconclusive results. The misclassification of short dresses as vulgar content posed challenges and compromised people's dignity. Hence, numerous deep-learning models have been developed for obscenity classification. Additionally, Deep learning methods have emerged as a promising solution to overcome challenges in the detection of obscenity with annotation boxes [13,14]. Nevertheless, the absence of a universally accepted approach for accurately identifying offensive content, particularly with limited data, remains a challenge. It is crucial to develop methods that are less computationally demanding to ensure efficient and practical implementation. The aim is to strike a balance between achieving high accuracy in classifying obscene content while minimizing the computational resources required for the model. This would enable more widespread adoption and usage across various platforms and applications. In [10], the authors utilized Support Vector

Machines (SVM) for obscenity classification, achieving high accuracy in their experiments. However, SVM-based approaches can be computationally expensive and may struggle with complex image features. In recent years, deep learning techniques have gained prominence in the field of obscenity image classification. In [11], the authors employed VGG-16, ResNet-18, -34, and -50 architectures to develop an automated tool capable of classifying obscene content specifically for mobile phones and tablets. The NPDI dataset was utilized for training and testing purposes. Among the different models tested, ResNet-34 demonstrated superior performance, outperforming the other models with an accuracy rate of 75.08%. In [1], the authors utilized another convolutional neural network named VGG-16 model to classify obscene images. The model acquired 93.8% testing accuracy. The classification method is known as Deep One-Class with Attention for Pornography (DOCAPorn) [5] was created to enhance the recognition of pornographic images by incorporating a visual attention mechanism. The aim was to improve the accuracy of the classification process, resulting in an impressive accuracy rate of 98.41%. In [7], the authors introduced a ConvNet pipeline for analyzing frame sequences in videos. The pipeline utilized ResNet-18 for extracting features and employed N-frame feature maps to classify obscenity within the videos. The authors in [13] proposed a detection-based algorithm S3Pooling based bottleneck attention module embedded MobileNetv2 based YOLOv3 which is where obscene images are annotated with a bounding box. In [14], the authors presented a novel approach called attention-enabled pooling embedded swin transformer-based YOLOv3 (ASYv3) for the identification of explicit regions within images. ASYv3 is a modified two-step technique that incorporates attention-based pooling to enhance the detection process.

This paper proposes a Log-Gabor Empowered Siamese Federated Learning for Accurate and Efficient Obscene Image Classification (LSiF). LSiF is specifically designed to address computational demands associated with practical implementation while working with limited datasets. In the proposed framework, a Siamese Network is utilized, where two parallel streams are employed. The first stream takes normal image input, while the second stream incorporates log-Gabor filter-based images. By integrating log-Gabor filters, the model can effectively extract texture analysis, perform feature extraction, and enable object recognition, capturing both intricate and coarse details simultaneously. The log-Gabor filter employs a logarithmic scale for frequencies, enhancing its ability to capture a wider range of frequency information. This combined approach aims to enhance efficiency of obscene image classification tasks. Siamese Networks utilize shared weights and parameters, enabling effective learning of discriminative features for class differentiation. This weight sharing also benefits scenarios with limited labelled data by leveraging relative similarity instead of relying solely on absolute labels. Additionally, this architecture promotes the identification of essential features while disregarding irrelevant attributes, leading to improved generalization and robustness. The shared weights also contribute to computational efficiency, making Siamese Networks suitable for resource-constrained environments. With a client size of 3 in federated learning, model updates are performed locally on each device, reducing the need for transmitting large amounts of data to a central server [8,15]. This minimizes the communication overhead, making the learning process more efficient, especially in bandwidth-constrained environments.

The paper is structured as follows: Sect. 2 outlines our approach for classifying obscene images, while Sect. 3 discusses result analysis. Finally, Sect. 4 concludes the paper by summarizing key points and highlighting research contributions.

2 Methodology

In this section, we present a comprehensive outline of our proposed framework for classifying obscene images, along with a thorough discussion of the dataset employed and the pre-processing technique applied to it. Additionally, we conduct an ablation analysis to examine the proposed model in depth and evaluate its performance.

2.1 Obscene-15k Dataset and Its Pre-processing

A dataset was carefully compiled from multiple sources, encompassing a significant number of images to ensure its richness and diversity. This section presents a thorough explanation of the data-collection process utilized in our study, including detailed information about the size of the dataset and methodologies employed to improve quality of the included images. A collection of more than 15,000 pornographic images was curated from websites where copyright purchase was not necessary. The dataset is titled as the obscene-15k dataset and the detailed specification of the dataset is presented in Table 1.

Table 1. Specifications of the dataset.

Class	Obscene images	Non-Obscene images	Total
Training	5000	5000	10000
Validation	2500	2500	5000
Testing	1500	1500	3000

The dataset has undergone extensive enhancement using various techniques to improve its quality and diversity. Augmentations play a crucial role in the context of obscene image classification for several reasons. First, augmentation introduces diversity into training data by applying transformations such as rotations, translations, flips, zooms, brightness, and contrast changes. This increased variability in data helps the model generalize well to different variations of obscene images. By incorporating these techniques, the dataset is enriched, allowing the model to learn from a more comprehensive range of examples and improving its ability to accurately classify obscene images.

2.2 Theoretical Framework for Log-Gabor Empowered Siamese Federated Learning (LSiF)

This section outlines the proposed LSiF architecture featuring a log-Gabor filter, Siamese Network [4], and federated learning for obscenity classification in images. Figure 1 shows a diagrammatic representation of the proposed architecture.

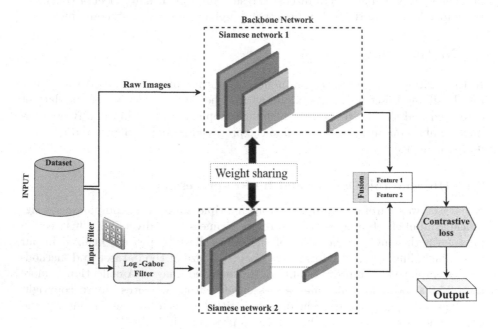

Fig. 1. Proposed Log-Gabor Empowered Siamese Network.

2.3 Siamese Networks

A Siamese Network [6], employed for obscene image classification, operates on a unique structure comprising two identical subnetworks referred to as "Siamese twins". These subnetworks process pairs of images, one explicit and one non-explicit, and learn to measure similarity between them. During training, the network minimizes distance between feature representations of similar pairs and maximizes distance for dissimilar pairs, effectively creating a discrimination boundary between obscene and non-obscene content. This approach is especially valuable in obscene image classification as it enables the network to discern subtle distinctions in explicit content and generalize its understanding to new, unseen explicit material. The Siamese Network's ability to capture fine-grained image differences makes it a robust tool for classifying obscene images, contributing to safer and more secure online environments.

2.4   Log-Gabor Empowered Siamese Network

The Log Gabor filter is applied to each image in the dataset to extract frequency-based features that are relevant for identifying obscene content. The Log Gabor filter highlights specific texture patterns and frequency components that are common in obscene images [17]. By applying the Log Gabor filter as a pre-processing step, the model can focus on relevant frequency-based features, enhancing accuracy of the classification process when dealing with challenging obscene images.

Filter Design. The Log Gabor filter is designed to operate in the logarithmic frequency domain. The filter is defined using the following equation:

$$G(x, y; f_0, \theta) = \exp\left(-\frac{(\log(r/f_0))^2}{2\log(\sigma_r)^2}\right) \cdot \cos\left(2\pi\frac{\log(r/f_0)}{\log(\sigma_o)} + \theta\right) \quad (1)$$

In this equation, $G(x, y; f_0, \theta)$ represents the Log Gabor filter at position (x, y) with parameters f_0 and θ. The term r represents the distance from the center of the filter. The parameters σ_r and σ_o control the spread of the filter in the radial and orientation dimensions, respectively. The term θ denotes the filter orientation.

Filter Application. To apply the Log Gabor filter to an image $I(x, y)$, the obscene/non-obscene image is convolved with the filter in the frequency domain. The procedure involves the following steps:

$$F(u, v) = \mathcal{F}\{I(x, y)\} \quad (2)$$

$$F'(u, v) = F(u, v) \cdot G(u, v; f_0, \theta) \quad (3)$$

$$I'(x, y) = \mathcal{F}^{-1}\{F'(u, v)\} \quad (4)$$

where $F(u, v)$ and $F'(u, v)$ are the Fourier transforms of the image $I(x, y)$ and the filtered image $I'(x, y)$, respectively. $\mathcal{F}$ and $\mathcal{F}^{-1}$ represent the forward and inverse Fourier transforms, respectively. The Log Gabor filter provides a powerful tool for analyzing image frequency content and extracting important features. Its logarithmic frequency representation allows for capturing a wide range of scales, making it suitable for various image processing applications.

2.5   Inclusion of Federated Learning

The incorporation of federated learning forms an essential component of our framework. Our framework incorporates federated learning, utilizing the principles of the federated averaging algorithm, to aggregate locally trained models from separate clients and create a global model [12]. Figure 2 illustrates the proposed Log-Gabor-empowered siamese federated learning for obscene image classification. We designated clients a value of 3 to segment the dataset into three distinct groups The local epochs (LeP) for each dataset are assigned as 20 each.

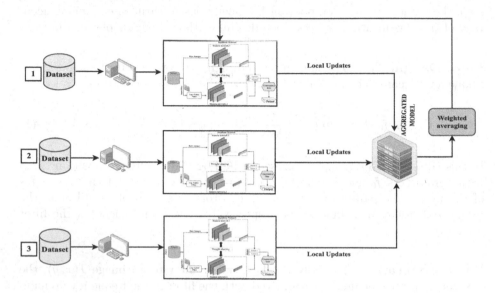

Fig. 2. Inclusion of federated learning in the Proposed Log-Gabor Empowered Siamese Network.

Federated learning is a distributed machine learning approach that enables training models on decentralized data without the need for data sharing. The central server initializes global model parameters, denoted as W_0. A subset of clients (devices or nodes) is selected from the network for participation in the federated learning process. Each client has its local dataset. Selected clients independently train the global model on their local obscene and non-obscene datasets using their local computing resources. The training process typically involves multiple iterations. Let W_i represent the local model parameters at client i, and $f_i(W_i)$ denote the local training objective at client i. After local training, clients send their updated model parameters (weights and gradients) to the central server. The central server aggregates received model updates from clients using a predefined aggregation method. The most common aggregation method is Federated Averaging, where the updates are averaged to compute new global model parameters. Let Aggregate($\cdot$) denote the aggregation function. The updated global model parameters are distributed back to clients. Steps are repeated for a predetermined number of rounds or until convergence criteria are met. In each round, a new subset of clients is selected, and the process of local training, model update, aggregation, and distribution is performed. Mathematically, the federated learning update equation can be expressed as:

$$W_{t+1} = \text{Aggregate}\left(W_t - \eta \sum_i \nabla f_i(W_i)\right) \qquad (5)$$

where W_t represents the global model parameters at iteration t, $\nabla f_i(W_i)$ is the gradient of the local objective function at client i with respect to its local model parameters, and η is the learning rate, respectively. This process of decentralized model training and aggregation allows the global model to be improved while keeping data local and preserving privacy. The iterative nature of federated learning ensures collaboration among clients to achieve a collectively learned model while maintaining data security.

3 Results and Discussion

In this section, the outcome of utilizing the LSiF technique for classifying explicit images is presented, accompanied by a comprehensive examination of the ablation studies. This section also includes a comparison with existing methods in the field, providing an in-depth evaluation of the proposed LSiF method.

3.1 Experimental Setup and Performance Evaluation Metrics

We constructed our model by utilizing the Keras framework, and the training process was carried out on a Quadro RTX 6000 graphics processing unit. The implementation leveraged TensorFlow as the backend framework. Performance measures, such as testing accuracy, precision (PR), recall, and F1-score, provide valuable insights into the effectiveness of a model. Equations 6 to 8 outline the mathematical expressions that define the formulation of the performance evaluation metrics.

$$\text{Accuracy} = \frac{\text{TP} + \text{TN}}{\text{TP} + \text{TN} + \text{FP} + \text{FN}}, \text{Precision} = \frac{\text{TP}}{\text{TP} + \text{FP}}, \qquad (6)$$

$$\text{Recall} = \frac{\text{TP}}{\text{TP} + \text{FN}} \qquad (7)$$

$$\text{F1} - \text{Score} = 2 \times \frac{\text{Precision} \times \text{Recall}}{\text{Precision} + \text{Recall}} \qquad (8)$$

where TP, TN, FP, and FN are true positive, true negative, false positive, and false negative, respectively.

3.2 Measuring Performance: A Comprehensive Assessment

To analyze the classification performance, an evaluation is conducted on the successively evolved models, which include the Siamese Network with Log-Gabor filter (LSi), the proposed Log-Gabor Empowered Siamese Federated Learning

Table 2. Analysis of performance measures on all the evolved models

Methods	Testing Accuracy (%)	Precision (%)	Recall (%)	F1-Score (%)
Siamese Network	85.45	86.62	86.11	85.25
LSi	88.10	88.80	88.00	87.40
LSiF (Client Size = 3) [Proposed]	**98.11**	**98.25**	**98.19**	**98.30**

(LSiF), and the base algorithm Siamese Network, using the Obscene-15k dataset. Table 2 presents an analysis of performance metrics across all the models.

The Siamese Network achieved an accuracy of 85.45%, indicating that 85.45% of the instances were correctly classified by the model. It demonstrated a precision of 86.62% and a recall of 86.11%, which means it effectively identified positive instances and captured a substantial portion of actual positive instances. The F1-Score of 85.25% suggests a balanced performance between precision and recall. Moving on to LSi, it outperformed the Siamese Network in terms of accuracy with a score of 88.10%. It also showed a higher precision rate of 88.80% and a recall rate of 88.00%, indicating an improved ability to correctly identify positive instances and capture a larger proportion of actual positive instances. The proposed LSiF with a client size of 3 demonstrated remarkable performance in all metrics. It achieved an accuracy of 98.11%, indicating a significantly high rate of correct classifications. LSiF also showcased exceptional precision (98.25%) and recall (98.19%) rates, suggesting an excellent ability to identify positive instances and capture almost all actual positive instances. The F1-Score of 98.30% confirms the highly balanced performance between precision and recall. LSiF, with a client size of 3, outperformed both Siamese Network and LSi in accuracy, precision, recall, and F1-Score. It achieved significantly higher scores, indicating superior instance classification. Siamese Network and LSi had respectable performance, with LSi generally surpassing Siamese Network but still falling short of LSiF's exceptional performance.

Table 3. Comparison of the proposed LSiF model on different client size environments with 3 different datasets. FL: Federated Learning, Accuracy: AC, Precision: PR, Recall: RC, F1-Score: F1-S

Dataset	FL (Client Size = 3)				FL (Client Size = 5)			
	AC	PR	RC	F1-S	AC	PR	RC	F1-S
NPDI [2]	87.57	88.00	87.66	88.00	86.00	86.45	86.20	86.30
Pornography-2k [3]	85.00	85.11	85.45	85.47	81.22	81.00	81.19	81.36
Obscene-15k Dataset	98.11	98.25	98.19	98.30	95.56	95.00	95.33	95.23
	FL (Client Size = 7)				FL (Client Size = 10)			
	AC	PR	RC	F1-S	AC	PR	RC	F1-S
NPDI [2]	86.20	86.40	86.33	86.10	84.12	84.57	84.22	84.35
Pornography-2k [3]	80.70	80.55	80.00	80.24	79.27	79.55	79.60	79.19
Obscene-15k Dataset	94.30	94.00	94.26	94.17	93.64	93.33	93.57	93.00

Table 3 provides a comparative analysis of federated learning performance using different datasets: pornography-2k, NPDI, and the new Obscene-15k dataset. It evaluates their effectiveness with varying client sizes. For NPDI and Pornography-2k datasets, increasing client size from 3 to 10 results in declining accuracy, precision, recall, and F1-Score.

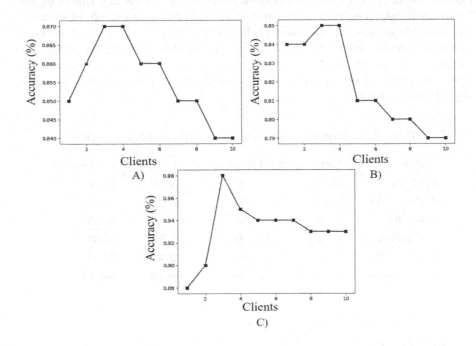

Fig. 3. Trade-off between number of clients and accuracy. A) NPDI, B) Pornography-2k, C) Obscene-15k

This suggests that increasing the number of clients in federated learning has a more noticeable impact on the performance of the model for this particular dataset. In contrast, for the Obscene-15k Dataset, the trend is slightly different. As client size grows, there is a slight dip in accuracy, precision, recall, and F1-Score, but the decline remains modest. This implies the model maintains high performance even with more clients. Overall, this dataset performs better than the others, suggesting it suits federated learning. The accuracy drop with more clients is due to diverse data, making it harder to find a common representation and communication limitations between clients and the server. In summary, the table provides a comparison of the performance of different datasets in federated learning scenarios with varying client sizes. It shows that as the client size increases, there is a tendency for a slight decrease in performance. The outcomes of our second experiment are depicted in Fig. 3, which focuses on examining the relationship between the number of clients and the accuracy of the global model in the context of federated learning.

Comparison of the Proposed LSiF with the State-of-the-Art Methods:-
The proposed LSiF model outperforms compared existing models [1,5,7,10,11] with significantly higher values across all evaluation metrics which is also depicted in Table 4. LSiF achieves an accuracy of 98.11%, precision of 98.25%, recall of 98.19%, and an F1-Score of 98.3%. This indicates that the LSiF model exhibits superior performance compared to the other models in terms of its ability to accurately classify instances and balance precision and recall. In summary, the table illustrates the proposed LSiF model achieving the highest accuracy, precision, recall, and F1-Score among all the models. By implementing these measures, we can effectively address various security challenges [9,13] and mitigate the presence of explicit content, contributing to the creation of a safer digital environment.

Table 4. Performance assessment of the state-of-the-art methods

Models	Accuracy (%)	Precision (%)	Recall (%)	F1-Score (%)
SVM [10]	74.77	75.30	74.20	74.00
ResNet34 [11]	82.20	82.30	80.00	81.95
VGG 16 [1]	85.62	85.76	85.00	85.55
DOCAPorn [5]	91.11	92.00	92.55	92.30
ConvNet [7]	92.35	92.45	92.38	92.00
LSiF [Proposed]	**98.11**	**98.25**	**98.19**	**98.30**

4 Conclusion

This study introduces the Log-Gabor Empowered Siamese Federated Learning (LSiF) framework for classifying obscene images accurately and efficiently. LSiF utilizes a Siamese Network with shared weights to process both log-Gabor and raw input, enhancing feature learning and pattern recognition. This approach improves model generalization, robustness, and computational efficiency, making it suitable for resource-constrained and real-time applications. LSiF also incorporates federated learning with a client size of three, reducing data transmission to a central server and enhancing learning efficiency in low-bandwidth environments. The model achieves impressive performance, with 98.11% accuracy, 98.25% precision, 98.19% recall, and a 98.30% F1-Score.

References

1. Agastya, I.M.A., Setyanto, A., Handayani, D.O.D., et al.: Convolutional neural network for pornographic images classification. In: 2018 Fourth International Conference on Advances in Computing, Communication & Automation (ICACCA), pp. 1–5. IEEE (2018)

2. Avila, S., Valle, E., Araú, A.: NPDI porn dataset, The Institute of Computing at UNICAMP (2018)
3. Avila, S., Thome, N., Cord, M., Valle, E., AraúJo, A.D.A.: Pooling in image representation: the visual codeword point of view. Comput. Vis. Image Underst. **117**(5), 453–465 (2013)
4. Bertinetto, L., Valmadre, J., Henriques, J.F., Vedaldi, A., Torr, P.H.: Fully-convolutional Siamese networks for object tracking. In: Hua, G., Jégou, H. (eds.) Computer Vision–ECCV 2016 Workshops: Amsterdam, The Netherlands, 8–10 and 15–16 October 2016, Proceedings, Part II 14, pp. 850–865. Springer, Cham (2016). https://doi.org/10.1007/978-3-319-48881-3_56
5. Chen, J., Liang, G., He, W., Xu, C., Yang, J., Liu, R.: A pornographic images recognition model based on deep one-class classification with visual attention mechanism. IEEE Access **8**, 122709–122721 (2020)
6. Gao, Z.Y., Xie, H.X., Li, J.F., Liu, S.L.: Spatial-structure Siamese network for plant identification. Int. J. Pattern Recognit Artif Intell. **32**(11), 1850035 (2018)
7. Gautam, N., Vishwakarma, D.K.: Obscenity detection in videos through a sequential convnet pipeline classifier. IEEE Trans. Cogn. Dev. Syst. **15**, 310–318 (2022)
8. Hammoud, A., Otrok, H., Mourad, A., Dziong, Z.: On demand fog federations for horizontal federated learning in IoV. IEEE Trans. Netw. Serv. Manage. **19**(3), 3062–3075 (2022)
9. Hansda, R., Nayak, R., Balabantaray, B.K., Samal, S.: Copy-move image forgery detection using phase adaptive spatio-structured sift algorithm. SN Comput. Sci. **3**, 1–16 (2022)
10. Lin, Y.C., Tseng, H.W., Fuh, C.S.: Pornography detection using support vector machine. In: 16th IPPR Conference on Computer Vision, Graphics and Image Processing, vol. 19, pp. 123–130 (2003)
11. Nurhadiyatna, A., Cahyadi, S., Damatraseta, F., Rianto, Y.: Adult content classification through deep convolution neural network. In: 2017 International Conference on Computer, Control, Informatics and Its Applications, pp. 106–110. IEEE (2017)
12. Rieke, N., et al.: The future of digital health with federated learning. NPJ Digit. Med. **3**(1), 119 (2020)
13. Samal, S., Zhang, Y.D., Gadekallu, T.R., Balabantaray, B.K.: ASYv3: attention-enabled pooling embedded Swin transformer-based YOLOv3 for obscenity detection. Expert. Syst. **40**, e13337 (2023)
14. Samal, S., Zhang, Y.D., Gadekallu, T.R., Nayak, R., Balabantaray, B.K.: SBMYv3: improved MobYOLOv3 a BAM attention-based approach for obscene image and video detection. Expert. Syst. **40**, e13230 (2023)
15. Tran, H.V., Kaddoum, G., Elgala, H., Abou-Rjeily, C., Kaushal, H.: Lightwave power transfer for federated learning-based wireless networks. IEEE Commun. Lett. **24**(7), 1472–1476 (2020)
16. Walters, L.G., DeWitt, C.: Obscenity in the digital age: the re-evaluation of community standards. NEXUS **10**, 59 (2005)
17. Wang, W., Li, J., Huang, F., Feng, H.: Design and implementation of Log-Gabor filter in fingerprint image enhancement. Pattern Recogn. Lett. **29**(3), 301–308 (2008)

Depth Normalized Stable View Synthesis

Xiaodi Wu[1], Zhiqiang Zhang[1(✉)], Wenxin Yu[1(✉)], Shiyu Chen[1], Yufei Gao[2], Peng Chen[3], and Jun Gong[4]

[1] Southwest University of Science and Technology, Sichuan, China
yuwenxin@swust.edu.cn
[2] Hosei University, Tokyo, Japan
[3] Chengdu Hongchengyun Technology Co., Ltd., Sichuan, China
[4] Southwest Automation Research Institute, Chengdu, China

Abstract. Novel view synthesis (NVS) aims to synthesize photo-realistic images depicting a scene by utilizing existing source images. The synthesized images are supposed to be as close as possible to the scene content. We present Deep Normalized Stable View Synthesis (DNSVS), an NVS method for large-scale scenes based on the pipeline of Stable View Synthesis (SVS). SVS combines neural networks with the 3D scene representation obtained from structure-from-motion and multi-view stereo, where the view rays corresponding to each surface point of the scene representation and the source view feature vector together yield a value of each pixel in the target view. However, it weakens geometric information in the refinement stage, resulting in blur and artifacts in novel views. To address this, we propose DNSVS that leverages the depth map to enhance the rendering process via a normalization approach. The proposed method is evaluated on the Tanks and Temples dataset, as well as the FVS dataset. The average Learned Perceptual Image Patch Similarity (LPIPS) of our results is better than state-of-the-art NVS methods by 0.12%, indicating the superiority of our method.

Keywords: Novel View Synthesis · Deep Learning · Normalization

1 Introduction

Novel view synthesis (NVS) has been an exciting problem with applications including mixed reality nd free point-of-view video, allowing users to tour scenes freely without being in them. The task of NVS is to synthesize images depicting a scene under specified new camera poses, using a set of known images. The core objective is that the synthesized images must be as close to the ground truth as possible. Regarding this, recent static scene geometry-based methods

This Research is Supported by National Key Research and Development Program from Ministry of Science and Technology of the PRC (No.2021ZD0110600), Sichuan Science and Technology Program (No.2022ZYD0116), Sichuan Provincial M. C. Integration Office Program, and IEDA Laboratory of SWUST.

B. Luo et al. (Eds.): ICONIP 2023, CCIS 1968, pp. 56–68, 2024.
https://doi.org/10.1007/978-981-99-8181-6_5

[1, 8, 16, 17, 22, 23] can produce images of high quality. However, the rendering process of the above methods merely uses the 3D scene geometry for ray-casting or the query of 3D surface points, which does not fully utilize or even weaken geometric information lying in scene representation. This results in a confusing representation of regions in the target view, such as the translucent state of objects or blurred structure.

To address the above, we propose Depth Normalized Stable View Synthesis (DNSVS), a method based on the pipeline of SVS [17]. DNSVS focuses on the improvement in the refinement process of SVS, utilizing depth information during the refinement stages to enhance the final target view synthesis, which produces clearer novel views. The work most related to ours is SVS [17] and FVS [16]. SVS takes advantage of only the correlation between images and scene representation while FVS does adopt depth information but underperforms SVS due to vastly different designation that fails to utilize all the relevant images.

The key concept of our method is to augment the refinement process with geometric information. Specifically, we use mapped depth values to normalize the target feature by applying element-wise scaling and offsetting operations. Furthermore, we employ the residual structured [5] depth normalization (DN) and U-Net [18] to build paired refinement stages. Given the above, DNSVS can fully utilize geometric information while ensuring a smooth visual effect of the novel views. The proposed model is evaluated on the dataset of Tanks and Temples [9], and FVS [16]. Compared with those state-of-the-art (SOTA) NVS approaches, the produced results achieve significant improvements quantitatively and qualitatively. To summarize, the main contributions of our work are:

- We propose depth normalization (DN) that augments synthesized images with geometric information, which enables rendering clearer novel views.
- A new refinement approach for NVS methods that utilize 3D-level information to enhance 2D novel views. Our DN can be adopted as long as the depth map is available.
- The proposed method has been evaluated on the Tanks and Temples, and the FVS datasets. In general, our approach outperforms the SOTA on LPIPS by 0.12%, and PSNR by 0.03 dB, indicating the superiority of our approach.

2 Related Work

2.1 Novel View Synthesis

People have been working in the field of New Viewpoint Synthesis (NVS) since the early stages, and in recent years, methods combining deep learning have thrived. Flynn [4] et al. first proposed an end-to-end deep learning pipeline based on explicit scene representation, which pioneered the combination of deep learning methods in the field of NVS. This work designed convolutional neural networks that use plane sweep volume (PSV). In recent work based on fixed 3D representations, Riegler and Koltun [16, 17] introduced the concept that warping features from selected source images into the target view and blending them using convolutional networks. Solovev [22] et al. represent the scene with a set of front-parallel semitransparent planes and convert them to deformable layers. Suhuai [23] et al. leverage epipolar geometry to extract patches of each reference view, which can predict the color of a target ray directly.

Our work is most related to [16,17], but we include depth information on a pixel-by-pixel basis in the refinement stage to improve the synthesized images.

Methods based on implicit function representations have also achieved significant results in recent years, with Neural Radiance Fields (NeRF) [11] being a major innovation. NeRF feeds coordinates to multilayer perceptrons (MLP) to learn color and body density, then samples novel views based on volume rendering techniques. After this, the work of Reiser [15] et al. significantly reduced the number of weights of the original NeRF. Kai [25] et al. address a parametrization issue involved in applying NeRF to 360 captures of objects within large-scale, unbounded 3D scenes.

2.2 Normalization in Image Synthesis

Conditional normalization layers are commonly adopted in conditional image synthesis and differ according to various types of input, including the Conditional Batch Normalization (Conditional BatchNorm) [3] and the Adaptive Instance Normalization (AdaIN) [6]. They are first used for style transfer tasks and later for various visual tasks. These conditional normalization layers require external data and typically operate. First, normalize the layer activation to mean zero and deviation unit. Then use the learned affine transformation (learned from external data) to adjust the activation to perform de-normalization.

Based on the concept above, Taesung [13] et al. focus on providing semantic information in the context of regulating normalization activation, using semantic maps at different scales to achieve synthesis from coarse to fine. This work has been a classic method in the field of semantic image synthesis.

There are also other popular normalization layers, which are labeled as unconditional because they do not rely on external data compared to the conditional normalization layers discussed above.

3   Method

3.1   Overview

The general flow of our method is shown in Fig. 1. The input is a randomly captured source image sequence $\{I_n\}_{n=1}^N$. The task to be accomplished by the method is to be able to predict the target image O at any given viewpoint (R_t, t_t) and camera intrinsics K_t.

To start with, we perform preprocessing on the source images. I. We use an ImageNet-pre-trained ResNet18 [5] to encode each I_n to obtain the feature tensor F_n, which adopts upsample using the nearest neighbor interpolation as well as convolution and activation operation. Note this network as ϕ_{enc}, we

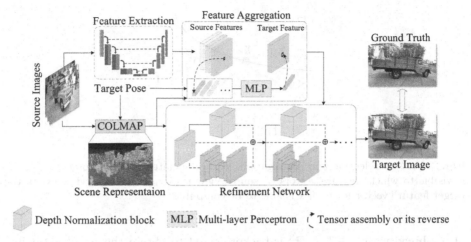

Fig. 1. Overview of Depth Normalized Stable View Synthesis. We extract feature tensors from source images and construct the geometric scene using structure-from-motion, multi-view stereo, and surface reconstruction. Based on the geometric information, the associated feature vectors for each target pixel are aggregated and assembled into a target feature. With our DN-enhanced refinement network, we then render the final novel view.

obtain the encoded feature tensor by $F_n = \phi_{enc}(I_n)$. II. We use COLMAP [19,20] to get the depth maps and point clouds needed to reconstruct the 3D scene representation, as well as to get the camera pose of each source image, including the rotation matrix $\{R_n\}_{n=1}^N$, the camera intrinsics $\{K_n\}_{n=1}^N$ and the translation vectors $\{t_n\}_{n=1}^N$. In this work, the final scene geometric Γ is the surface mesh derived by Delaunay-based 3D surface reconstruction using the point cloud.

After preprocessing, we select source feature vectors for each pixel in the target view O and aggregate its value based on certain rules. Subsection 3.2 gives more details on how feature selection and aggregation work. Note that the source features are selected and aggregated on a pixel-by-pixel basis, to obtain the final photorealistic view, the feature vectors obtained by aggregation are assembled into a tensor and passed into a refinement network to derive the final target view. We will explain how we utilize the depth map to improve the refinement process in Subsect. 3.3.

3.2 Feature Selection and Aggregation

As illustrated in Fig. 2, for a certain point $x \in \Gamma \subset \mathbb{R}^3$, we select only the visible source feature vectors $f_k(x)$ for aggregation. Given a newly specified camera K_t, R_t and t_t, the depth map $D \in \mathbb{R}^{W \times H}$ corresponding to the target view O is computed from the scene geometric Γ. Then the center of each pixel in

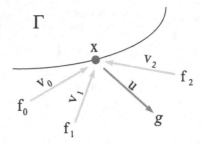

Fig. 2. Feature Selection and Aggregation. With associated (the 3D point $x \in \Gamma \subset \mathbb{R}^3$ is visible to which) feature vectors f_k, source view rays v_k and target view ray u, the target feature vector for each x is able to be computed.

O combined with $x \in \Gamma \subset \mathbb{R}^3$ is backprojected to obtain the set of 3D points corresponding to each pixel $\{x_{w,h}\}_{w,h=1,1}^{W \times H}$, from which the source feature vectors $\{f_k(x)\}_{k=1}^{K}$ visible for each x are derived.

At this point, we apply MLP to the aggregation of each target feature vector corresponding to $\{f_k(x)\}_{k=1}^{K}$ and the associated view rays $\{v_k\}_{k=1}^{k}$, and u. Denote the single target feature vector as g, the network used for aggregation as ϕ_{aggr}, g is computed by

$$g(v, u) = \phi_{aggr}(u, \{(v_k, f_k(x))\}_{k=1}^{K}) \tag{1}$$

where $f_k(x)=F_k(K_k(R_k x + t_t))$ with bilinear interpolation.

Since accuracy is proved in SVS [17], here we use MLP with max pooling for feature aggregation:

$$\phi_{aggr} = \max_{k=1}^{K} MLP([u, v_k, f_k(x)]) \tag{2}$$

where $[\cdot]$ means a concatenation of view rays and source features that are associated with each target feature.

3.3 Refinement with Depth Normalization

Note that the refinement network as ϕ_{ref}, we assemble the feature vectors $\{g_{w,h}\}_{w,h=1,1}^{W \times H}$ derived from the aggregation network ϕ_{aggr} into a tensor $G = \{g_{w,h}\}_{w,h=1,1}^{W \times H}$ and render it to get $O = \phi_{ref}(G)$. Inspired by Taesung [13] et al. which leverages semantic map to enhance image generation, we leverage depth map computed by COLMAP to improve the refinement of the target view.

Depth Normalization (DN). The object of DN is to learn a mapping function that can enhance target features using depth maps, making the target view closer to the scene geometry.

Let the intermediate target feature O' and depth map D be the input with height H and width W. Denote the channel number of O' as C. The value of

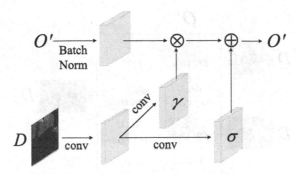

Fig. 3. Depth normalization unit. The learned γ and σ are multiplied and added to the normalized activation element-wise.

output feature map at element-wise $(c \in C, y \in H, w \in W)$ is:

$$\gamma_{c,y,w}\left(D\right) \frac{O'_{c,y,w} - \mu_c}{\beta_c} + \sigma_{c,y,w}\left(D\right) \tag{3}$$

where $\gamma_{c,y,w}\left(D\right)$ and $\sigma_{c,y,w}\left(D\right)$ denote the modulation parameters learned from the depth map D, μ_c and σ_c are the mean and standard deviation of O' at channel c:

$$\mu_c = \frac{1}{HW} \sum_{y,w} O'_{c,y,w}$$

$$\beta_c = \sqrt{\frac{1}{HW} \sum_{y,w} \left(\left(O'_{c,y,w}\right)^2 - \left(\mu_c\right)^2\right)} \tag{4}$$

As Fig. 3 illustrated, we apply a two-layer convolutional network to learn high-dimensional features γ and σ from the depth map D. The learned γ and σ are multiplied and added to the normalized activation element-wise. Specifically, γ performs as a tensor of scaling values, and σ performs as a tensor of bias values.

Given the above, the 3D-level information in the depth map D is thus passed into the 2D target feature map O'. The effect reflected on the target image is that each pixel value is adjusted by the mapped depth value of its corresponding coordinate to more closely match the geometry of the ground truth. Denote $conv(\cdot)$ as convolution operation, the DN can be summarized as:

$$O' = bn(O') \cdot conv_\gamma(D') + conv_\sigma(D')$$
$$D' = ReLU(conv(D)) \tag{5}$$

where O' is intermediate target feature during rendering, and $bn(\cdot)$ denotes batch normalization [7].

Depth Normalization Residual Block. To fully utilize the information in the depth map, we followed the residual structure in [10] with spectral normalization

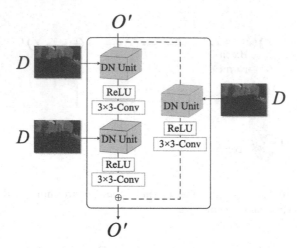

Fig. 4. Depth Normalization Residual Block. O' is the intermediate target feature in refinement process.

[12] applied to each convolution layer. Based on that, we added DN units before each Re LU layer in an end-to-end way (see Fig. 4), meaning the skip connection is also learned in addition to the two stacked DN units.

Comparing with a single DN unit, two stacked DN units can draw depth information more thoroughly. In addition, the skip connection allows the network to better capture feature details and contextual information, improving the expressiveness of the network while mitigating the problem of gradient decay. The effect of each component of the DN residual block will be verified in Subsect. 4.1.

Stage-by-Stage Refinement with DN. To render the final novel view, we kept the sequence consisting of L U-Nets [18] in SVS [17] and combined it with DN residual blocks to build ϕ_{ref} (shown in Fig. 1).

Based on the observations during the experiments, the U-net sequence is able to deliver smooth rendering results. Therefore, to make the target view closer to the scene geometry while taking into account the smoothing effect, we add the output of the DN residual block and U-Net, performing the stage-by-stage rendering. Specifically, a U-Net and a DN residual block are paired to compute an input feature map in parallel, and the respective outputs are added and activated (by Re LU layer) as the input for the next stage. In addition, each pair learns the residual of its input. Recall the input of the refinement network ϕ_{ref} is a feature tensor $G = \{g_{w,h}\}_{w,h=1,1}^{W \times H}$, the stage-by-stage refinement can be described as:

$$\phi_{ref}(G) = \phi_{ref}^{L}(G + \phi_{ref}^{L-1}(G + \ldots)) \tag{6}$$

We will verify the effect of different L values in Subsect. 4.1.

3.4   Loss Function

Inspired by Chen and Koltun [2], the basic idea of the loss function is to match activations in a visual perception network, applying to the synthesized image and separately to the reference image. Therefore, the loss to minimize is:

$$L(O, I_n) = \|O - I_n\|_1 + \sum_l \lambda_l \|\phi_l(O) - \phi_l(I_n)\|_1 \qquad (7)$$

where ϕ is a pre trained VGG-19 [21] network, the hyperparameter λ_l is to adjust the weight of each layer l. For layers ϕ_l ($l \geqslant 1$) we actually adopt 'conv1_2', 'conv2_2','conv3_2', 'conv4_2', and 'conv5_2'. This design allows the evaluation to match lower-layer and higher-layer activations in the perception network ϕ and guides the synthesis network to learn fine details and global layouts.

4   Experiments

We applied PyTorch [14] to build the network. The three networks: ϕ_{enc} for feature extraction, ϕ_{aggr} for feature aggregation, and ϕ_{ref} for refinement are trained end-to-end on a single V100 GPU with a batch size of 1, using the Adam optimizer setting $\beta_1 = 0.9$, $\beta_2 = 0.9999$, learning rate $= 10^{-4}$.

For the fairness of the comparison, all results listed are derived after 600,000 iterations of the entire network without fine-tuning. We use 15 scenes in the Tanks and Temples [9] for training (scenes for evaluation are not included), and use 4 scenes in Tanks and Temples and the FVS [16] dataset for evaluation.

For the Tanks and Temples dataset, we use LPIPS (Learned Perceptual Image Patch Similarit) [26], PSNR (Peak Signal-to-Noise), and SSIM (Structural Similarity) [24] as accuracy metrics. For the FVS dataset, we use LPIPS and SSIM. The following subsections will present ablation studies and comparisons with SOTA NVS methods.

Table 1. Number of Refinement Stages

	Average Accuracy		
$L =$	$\downarrow LPIPS\%$	$\uparrow PSNR$	$\uparrow SSIM$
6	20.21	21.23	0.826
7	**20.19**	21.25	0.826
8	**20.19**	**21.27**	**0.828**
9	**20.19**	**21.27**	0.827

Table 2. Average Accuracy of Various Structures

Datasets		T&T			FVS	
Structures		↓*LPIPS*%	↑*PSNR*	↑*SSIM*	↑*SSIM*	↓*LPIPS*%
A -	Single DN Unit, w/o Residual	18.2	21.15	0.847	0.803	22.04
B	Stacked DN Unit, w/o Residual	18.16	21.21	0.852	0.803	21.91
Full DN Residual Block		**18.11**	**21.27**	**0.856**	**0.806**	**21.86**

4.1 Ablation Study

In Table 1 we verified the refinement network benefits from multiple stages mentioned in Subsect. 3.3. Recall that we denote the number of refinement stages in Eq. (6) as L, we set L from 6 to 9 and record the results respectively. Based on the average accuracy, we adopt $L = 8$ in all later experiments.

(a) Ground Truth   (b) Our full method   (c) Our A   (d) SVS

Fig. 5. Results of ablation study. Both structures of our method convey clearer scene content.

In Table 2 we performed ablation experiments on the DN residual block to verify the efficacy of the components. In structure A, a single DN unit replaced the DN residual block. In structure B, we remove the residual connection in the DN residual block.

The results show that stacked DN units outperform a single unit on most metrics. In addition, the DN residual connection in the full method does bring higher performance than simple stacked DN units. Figure 5 provides visual results of the ablation experiments. The full method delivers images that are closest to the ground truth. More importantly, both our structures ((b) and (c)) convey clearer scene content with respect to geometry.

4.2 Quantitative Results

In Table 3 and Table 4 we compare the quantitative results of our full method with other recent NVS methods. The results of SVS [17] are derived from the published pre-trained models, and the results of the other methods are derived by Riegler and Koltun [16,17] and Jena [8] et al. using the published models or network parameters.

Table 3. Compare with State-to-art on Tanks and Temples dataset

Methods	Accuracy per Scene											
	Truck			M60			Playground			Train		
	↓LPIPS%	↑PSNR	↑SSIM	↓LPIPS%	↑PSNR	↑SSIM	↓LPIPS%	↑PSNR	↑SSIM	↓LPIPS%	↑PSNR	↑SSIM
NeRF [11]	50.74	20.85	0.738	60.89	16.86	0.701	52.19	21.55	0.759	64.64	16.64	0.627
NeRF++ [25]	30.04	22.77	0.814	43.06	18.49	0.747	38.70	22.93	0.806	47.75	17.77	0.681
FVS [16]	13.06	22.93	0.873	30.70	16.83	0.783	19.47	22.28	0.846	24.74	18.09	0.773
SVS [17]	12.41	23.07	0.893	23.70	19.41	0.827	17.38	23.59	**0.876**	19.42	18.42	0.809
NMBG [8]	16.84	**24.03**	0.888	**23.15**	19.54	0.815	22.72	23.59	0.87	24.17	17.78	0.799
Ours	**12.21**	23.22	**0.896**	23.55	**19.59**	**0.835**	**17.29**	**23.64**	0.876	**19.4**	**18.61**	**0.818**

Numbers in bold are the best. On average, our method outperforms the SOTA on LPIPS by 0.12%, and PSNR by 0.03 dB.

Table 4. Compare with State-to-art on FVS dataset

Methods	Accuracy per Scene									
	Bike		Flowers		Pirate		Digger		Sandbox	
	↓LPIPS%	↑SSIM	↓LPIPS%	↑SSIM	↓LPIPS%	↑SSIM	↓LPIPS%	↑SSIM	↓LPIPS%	↑SSIM
NeRF++ [25]	27.01	0.715	30.3	0.816	41.56	0.712	34.69	0.657	23.08	0.852
FVS [16]	27.83	0.592	26.07	0.778	35.89	0.685	23.27	0.668	30.2	0.77
SVS [17]	21.1	0.752	20.66	0.848	29.21	0.76	17.3	0.779	21.55	0.852
Ours	**20.83**	**0.761**	**20.62**	**0.857**	**29.11**	**0.768**	**17.27**	**0.864**	**21.41**	**0.864**

Numbers in bold are the best. On average, our method outperforms the SOTA on LPIPS by 0.12%, and SSIM by 0.01.

On average, the LPIPS of the proposed method outperforms SOTA by 0.12% on both datasets. In addition, the PSNR is 0.03 dB higher on the Tanks and Temple dataset. While improving above, our results still hold the performance on SSIM. This proved that with the DN-enhanced refinement process, our method can synthesize novel views closer to the scene content, delivering images of higher quality.

4.3   Qualitative Results

Figure 6 shows a qualitative comparison with SVS on scenes M60, Train, and Truck. As the image details presented, the proposed DN-enhanced refinement network delivers a sharper and clearer shape of scene content in some regions. These regions might exactly be those neglected by the refinement process in SVS, where 3D-level information in geometric representations is sufficient but weakened by the rendering process. This proved that our depth normalized refinement process can fully utilize geometric information while ensuring a smooth visual effect of the novel views.

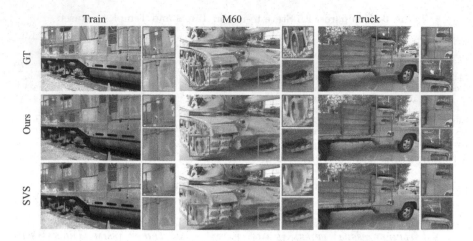

Fig. 6. Qualitative comparison on scenes M60, Train, and Truck

5 Conclusion

This paper proposes the novel view synthesis method DNSVS, which leverages depth information to improve the rendering of views based on the architecture of the SVS method. This also brings a new approach for novel view synthesis methods to utilize 3D-level information to enhance 2D novel views. Benefiting from the depth normalization, DNSVS alleviates the fuzzy phenomenon of SVS. Various experiments show that the proposed method produces promising images and achieves state-of-the-art performance. The LPIPS of our method outperforms state-of-the-art NVS approaches by 0.12%. Other than that, we believe DNSVS can produce images of higher quality with future reconstruction techniques iterated.

References

1. Aliev, K.-A., Sevastopolsky, A., Kolos, M., Ulyanov, D., Lempitsky, V.: Neural point-based graphics. In: Vedaldi, A., Bischof, H., Brox, T., Frahm, J.-M. (eds.) ECCV 2020. LNCS, vol. 12367, pp. 696–712. Springer, Cham (2020). https://doi.org/10.1007/978-3-030-58542-6_42
2. Chen, Q., Koltun, V.: Photographic image synthesis with cascaded refinement networks. In: Proceedings of the IEEE International Conference on Computer Vision, pp. 1511–1520 (2017)
3. Dumoulin, V., Shlens, J., Kudlur, M.: A learned representation for artistic style. arXiv preprint arXiv:1610.07629 (2016)
4. Flynn, J., Neulander, I., Philbin, J., Snavely, N.: Deepstereo: learning to predict new views from the world's imagery. In: Proceedings of the IEEE Conference on Computer Vision and Pattern Recognition, pp. 5515–5524 (2016)
5. He, K., Zhang, X., Ren, S., Sun, J.: Deep residual learning for image recognition. In: Proceedings of the IEEE Conference on Computer Vision and Pattern Recognition, pp. 770–778 (2016)

6. Huang, X., Belongie, S.: Arbitrary style transfer in real-time with adaptive instance normalization. In: Proceedings of the IEEE International Conference on Computer Vision, pp. 1501–1510 (2017)
7. Ioffe, S., Szegedy, C.: Batch normalization: accelerating deep network training by reducing internal covariate shift. In: International Conference on Machine Learning, pp. 448–456. PMLR (2015)
8. Jena, S., Multon, F., Boukhayma, A.: Neural mesh-based graphics. In: Computer Vision–ECCV 2022 Workshops: Tel Aviv, Israel, 23–27 October 2022, Proceedings, Part III. pp. 739–757. Springer (2023). https://doi.org/10.1007/978-3-031-25066-8_45
9. Knapitsch, A., Park, J., Zhou, Q.Y., Koltun, V.: Tanks and temples: benchmarking large-scale scene reconstruction. ACM Trans. Graph. (ToG) **36**(4), 1–13 (2017)
10. Mescheder, L., Geiger, A., Nowozin, S.: Which training methods for gans do actually converge? In: International Conference on Machine Learning, pp. 3481–3490. PMLR (2018)
11. Mildenhall, B., Srinivasan, P.P., Tancik, M., Barron, J.T., Ramamoorthi, R., Ng, R.: Nerf: representing scenes as neural radiance fields for view synthesis. Commun. ACM **65**(1), 99–106 (2021)
12. Miyato, T., Kataoka, T., Koyama, M., Yoshida, Y.: Spectral normalization for generative adversarial networks. arXiv preprint arXiv:1802.05957 (2018)
13. Park, T., Liu, M.Y., Wang, T.C., Zhu, J.Y.: Semantic image synthesis with spatially-adaptive normalization. In: Proceedings of the IEEE/CVF Conference on Computer Vision and Pattern Recognition, pp. 2337–2346 (2019)
14. Paszke, A., et al.: Pytorch: An imperative style, high-performance deep learning library. Adv. Neural Inform. Process. Syst. **32** (2019)
15. Reiser, C., Peng, S., Liao, Y., Geiger, A.: Kilonerf: speeding up neural radiance fields with thousands of tiny mlps. In: Proceedings of the IEEE/CVF International Conference on Computer Vision, pp. 14335–14345 (2021)
16. Riegler, G., Koltun, V.: Free view synthesis. In: Vedaldi, A., Bischof, H., Brox, T., Frahm, J.-M. (eds.) ECCV 2020. LNCS, vol. 12364, pp. 623–640. Springer, Cham (2020). https://doi.org/10.1007/978-3-030-58529-7_37
17. Riegler, G., Koltun, V.: Stable view synthesis. In: Proceedings of the IEEE/CVF Conference on Computer Vision and Pattern Recognition (CVPR), pp. 12216–12225 (June 2021)
18. Ronneberger, O., Fischer, P., Brox, T.: U-Net: convolutional networks for biomedical image segmentation. In: Navab, N., Hornegger, J., Wells, W.M., Frangi, A.F. (eds.) MICCAI 2015. LNCS, vol. 9351, pp. 234–241. Springer, Cham (2015). https://doi.org/10.1007/978-3-319-24574-4_28
19. Schonberger, J.L., Frahm, J.M.: Structure-from-motion revisited. In: Proceedings of the IEEE Conference on Computer Vision and Pattern Recognition, pp. 4104–4113 (2016)
20. Schönberger, J.L., Zheng, E., Frahm, J.-M., Pollefeys, M.: Pixelwise view selection for unstructured multi-view stereo. In: Leibe, B., Matas, J., Sebe, N., Welling, M. (eds.) ECCV 2016. LNCS, vol. 9907, pp. 501–518. Springer, Cham (2016). https://doi.org/10.1007/978-3-319-46487-9_31
21. Simonyan, K., Zisserman, A.: Very deep convolutional networks for large-scale image recognition. arXiv preprint arXiv:1409.1556 (2014)
22. Solovev, P., Khakhulin, T., Korzhenkov, D.: Self-improving multiplane-to-layer images for novel view synthesis. In: Proceedings of the IEEE/CVF Winter Conference on Applications of Computer Vision, pp. 4309–4318 (2023)

23. Suhail, M., Esteves, C., Sigal, L., Makadia, A.: Generalizable patch-based neural rendering. In: European Conference on Computer Vision. Springer (2022). https://doi.org/10.1007/978-3-031-19824-3_10
24. Wang, Z., Bovik, A.C., Sheikh, H.R., Simoncelli, E.P.: Image quality assessment: from error visibility to structural similarity. IEEE Trans. Image Process. **13**(4), 600–612 (2004)
25. Zhang, K., Riegler, G., Snavely, N., Koltun, V.: Nerf++: analyzing and improving neural radiance fields. arXiv preprint arXiv:2010.07492 (2020)
26. Zhang, R., Isola, P., Efros, A.A., Shechtman, E., Wang, O.: The unreasonable effectiveness of deep features as a perceptual metric. In: Proceedings of the IEEE Conference on Computer Vision and Pattern Recognition, pp. 586–595 (2018)

Exploring the Integration of Large Language Models into Automatic Speech Recognition Systems: An Empirical Study

Zeping Min[(⊠)] and Jinbo Wang

Peking University, No. 5 Yiheyuan Road Haidian District, Beijing 100871,
People's Republic of China
zpm@pku.edu.cn, wangjinbo@stu.pku.edu.cn

Abstract. This paper explores the integration of Large Language Models (LLMs) into Automatic Speech Recognition (ASR) systems to improve transcription accuracy. The increasing sophistication of LLMs, with their in-context learning capabilities and instruction-following behavior, has drawn significant attention in the field of Natural Language Processing (NLP). Our primary focus is to investigate the potential of using an LLM's in-context learning capabilities to enhance the performance of ASR systems, which currently face challenges such as ambient noise, speaker accents, and complex linguistic contexts. We designed a study using the Aishell-1 and LibriSpeech datasets, with ChatGPT and GPT-4 serving as benchmarks for LLM capabilities. Unfortunately, our initial experiments did not yield promising results, indicating the complexity of leveraging LLM's in-context learning for ASR applications. Despite further exploration with varied settings and models, the corrected sentences from the LLMs frequently resulted in higher Word Error Rates (WER), demonstrating the limitations of LLMs in speech applications. This paper provides a detailed overview of these experiments, their results, and implications, establishing that using LLMs' in-context learning capabilities to correct potential errors in speech recognition transcriptions is still a challenging task at the current stage.

Keywords: Automatic Speech Recognition · Large Language Models · In-Context Learning

1 Introduction

In today's era of cutting-edge technology, automatic speech recognition (ASR) systems have become an integral part. The advent of end-to-end ASR models, which are based on neural networks [3, 4, 6, 8, 10–13], coupled with the rise of prominent toolkits such as ESPnet [29] and WeNet [32], have spurred the progression of ASR technology. Nevertheless, ASR systems [14, 16, 21, 22, 25, 26, 34] occasionally yield inaccurate transcriptions, which can be attributed to ambient noise, speaker accents, and complex linguistic contexts, thus limiting their effectiveness.

© The Author(s), under exclusive license to Springer Nature Singapore Pte Ltd. 2024
B. Luo et al. (Eds.): ICONIP 2023, CCIS 1968, pp. 69–84, 2024.
https://doi.org/10.1007/978-981-99-8181-6_6

Over the years, considerable emphasis has been placed on integrating a language model [15,30] into the ASR decoding process. Language models have gradually evolved from statistical to neural. Recently, large language models (LLMs) [1,18,19,23,27,33,35] have gained prominence due to their exceptional proficiency in a wide array of NLP tasks. Interestingly, when the parameter scale surpasses certain thresholds, these LLMs not only improve their performance but also exhibit unique features such as in-context learning and instruction following, thereby offering a novel interaction method.

Nevertheless, efforts to harness recent LLMs such as [18,19,27] to improve ASR model performance are still in their early stages. This paper aims to bridge this gap. Our main objective is to investigate the capabilities of LLM's in-context learning to boost ASR performance. Our approach involves giving the LLM a suitably designed instruction, presenting it with the ASR transcriptions, and examining its ability to correct the errors.

We employed the Aishell-1 and LibriSpeech datasets for our experiments and selected well-known LLM benchmarks, such as ChatGPT and GPT-4, which are generally considered superior to other LLMs for their comprehensive capabilities. We concentrated on the potential of GPT-3.5 and GPT-4 to correct possible errors in speech recognition transcriptions.[1] Our initial experiments with the GPT-3.5-16k (GPT-3.5-turbo-16k-0613) model, in a one-shot learning scenario, did not yield lower WER.

Consequently, we undertook further investigation using diverse settings, including variations in the LLM model (GPT-3.5-turbo-4k-0301, GPT-3.5-turbo-4k-0613, GPT-4-0613), modification of instructions, increasing the number of attempts (1, 3, and 5), and varying the number of examples supplied to the model (1, 3, and 5-shot settings). This paper offers a comprehensive review of the experiments, their results, and our observations. Unfortunately, the results indicate that, in the current phase, using the in-context learning capabilities of LLMs to address potential inaccuracies in speech recognition transcriptions is notably challenging and frequently results in a *higher WER*. This could be attributed to the limited proficiency of LLMs in speech transcription.

This study contributes to the field in three ways:

1. **Exploration of LLMs for ASR Improvement:** We explore the potential of large language models (LLMs), particularly focusing on GPT-3.5 and GPT-4, to improve automatic speech recognition (ASR) performance by their in-context learning ability. This is an emerging area of research, and our work contributes to its early development.
2. **Comprehensive Experiments Across Various Settings:** We conduct comprehensive experiments using the Aishell-1 and LibriSpeech datasets and analyze the effect of multiple variables, including different LLM models, alterations in instructions, varying numbers of attempts, and the number of exam-

---

[1] Although we conducted preliminary trials with models like llama, opt, bloom, etc., these models often produced puzzling outputs and rarely yielded anticipated transcription corrections.

ples provided to the model. Our work contributes valuable insights into the capabilities and limitations of LLMs in the context of ASR.

3. **Evaluation of the Performance:** Regrettably, our findings indicate that leveraging the in-context learning ability of LLMs to correct potential errors in speech recognition transcriptions often leads to a higher word error rate (WER). This critical evaluation underscores the current limitations of directly applying LLMs in the field of ASR, thereby identifying an important area for future research and improvement.

2 Related Work

The use of large language models (LLMs) to enhance the performance of automatic speech recognition (ASR) models has been the subject of numerous past studies [5,9,17,24,28,31]. These works have explored various strategies, including distillation methods [9,17] and rescoring methods [5,24,28,31].

In the distillation approach, for instance, [9] employed BERT in the distillation approach to produce soft labels for training ASR models. [17] strived to convey the semantic knowledge that resides within the embedding vectors.

For rescoring methods, [24] adapted BERT to the task of n-best list rescoring. [5] redefined N-best hypothesis reranking as a prediction problem. [31] attempted to train a BERT-based rescoring model with MWER loss. [28] amalgamated LLM rescoring with the Conformer-Transducer model.

However, the majority of these studies have employed earlier LLMs, such as BERT [7]. Given the recent explosive progress in the LLM field, leading to models with significantly more potent NLP abilities, such as ChatGPT, it becomes crucial to investigate their potential to boost ASR performance. Although these newer LLMs have considerably more model parameters, which can pose challenges to traditional distillation and rescoring methods, they also possess a crucial capability, in-context learning, which opens up new avenues for their application.

3 Methodology

Our approach leverages the in-context learning abilities of LLMs. We supply the LLMs with the ASR transcription results and a suitable instruction to potentially correct errors. The process can be formalized as:

$$y = LLM(I, (x_1, y_1), (x_2, y_2), ..., (x_k, y_k), x)$$

where x represents the ASR transcription result, and y is the correct transcription. The pairs $(x_i, y_i)_{i=1}^k$ are the k examples given to the LLM, and I is the instruction provided to the LLM. The prompt is represented by $(I, (x_1, y_1), (x_2, y_2), ..., (x_k, y_k), x)$. The entire process is visually illustrated in Fig. 1.

We conducted thorough experimentation, varying GPT versions, the design of the instruction, and the number of examples k provided to GPT, in order to assess the potential of using Large Language Models (LLMs) to improve Automatic Speech Recognition (ASR) performance. We tested three versions of GPT-3.5, as well as the high-performing GPT-4. We used four carefully crafted instructions and varied the number of examples, where $k = 1, 2, 3$, supplied to the LLM.

Unfortunately, we found that directly applying the in-context learning capabilities of the LLM models for improving ASR transcriptions presents a significant challenge, and often leads to a higher Word Error Rate (WER). We further experimented with multiple attempts at sentence-level corrections. That is, for each transcription sentence x, the LLM generates multiple corrected outputs, and the final corrected result of the transcription sentence x is chosen as the output with the least WER.[2] Regrettably, even with multiple attempts, the corrected output from the LLM still results in a higher WER, further substantiating the challenges associated with directly leveraging the LLM's in-context learning capabilities for enhancing ASR transcriptions.

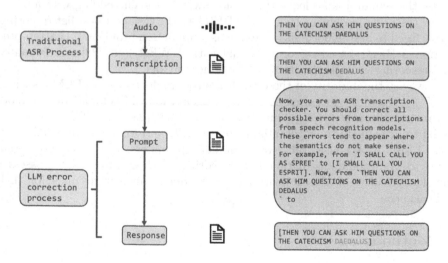

Fig. 1. Overview of the methodology leveraging the in-context learning capability of large language models (LLMs) for potential correction of errors in automatic speech recognition (ASR) transcriptions.

---

[2] Selecting the output with the lowest WER is not practical in real-world scenarios, as we cannot know the actual transcription y. Nonetheless, this technique aids in comprehending the limitations of using LLM's in-context learning capabilities for enhancing ASR transcriptions.

4   Experiments

4.1   Setup

Dataset. For our investigation, we selected two distinct datasets to evaluate the efficacy of utilizing advanced LLMs to improve ASR performance: the Aishell-1 dataset for the Chinese language and the LibriSpeech dataset for the English language. These datasets are greatly appreciated in the ASR research field, serving as standard benchmarks for numerous studies and methodologies.

The Aishell-1 [2] dataset has a total duration of 178 h, with the precision of manual transcriptions exceeding 95%. The dataset is meticulously organized into training, development, and testing subsets.

In contrast, the LibriSpeech [20] dataset comprises approximately 1000 h of English speech sampled at 16kHz. The content is extracted from audiobooks as a part of the LibriVox project. Similar to Aishell-1, LibriSpeech is also partitioned into subsets for training, development, and testing. Furthermore, each subset is classified into two groups based on data quality: clean and other.

ASR Model. To ensure the applicability of our experimental results, we utilized a state-of-the-art hybrid CTC/attention architecture, highly regarded in the field of speech recognition. We employed pretrained weights provided by the Wenet [32] speech community.

The ASR model, trained on the Aishell-1 dataset, includes an encoder set up with a swish activation function, four attention heads, and 2048 linear units. The model employs an 8-kernel CNN module with layer normalization, and normalizes the input layer before activation. The encoder consists of 12 blocks, has an output size of 256, and uses gradient clipping (value $= 5$) to prevent gradient explosions. The model leverages the Adam optimizer with a learning rate of 0.001 and a warm-up learning rate scheduler that escalates the learning rate for the initial 25,000 steps.

The ASR model, trained on the Librispeech dataset, implements a bitransformer decoder and a conformer encoder. The encoder follows the same configuration as that of the Aishell-1 model. The decoder incorporates four attention heads, with a dropout rate of 0.1. The model adheres to the same optimization and learning rate strategies as the Aishell-1 model.

LLM. For the LLM models, considering that ChatGPT and GPT-4 are recognized benchmarks, we inspected three versions from ChatGPT (GPT-3.5-turbo-4k-0301, GPT-3.5-turbo-4k-0613, GPT-3.5-turbo-16k-0613) and GPT-4 (GPT-4-0613). While other LLMs such as Llama, Opt, and Bloom claim to equal or outperform ChatGPT in certain aspects, they generally fall behind ChatGPT, and even more so GPT-4, in terms of overall competency for generic tasks. For the instruction I, we tested four variations, as detailed in Table 1.

Concerning the examples input to the LLM, we assessed 1-shot, 2-shot, and 3-shot scenarios. For the number of attempts, we explored situations with 1-attempt, 3-attempts, and 5-attempts.

Table 1. Instructions for ASR transcription correction.

Instruction ID	Description
Instruction 1	Correct the following transcription from speech recognition
Instruction 2	Now, you are an ASR transcription checker. You should correct all possible errors from transcriptions from speech recognition models. These errors tend to appear where the semantics do not make sense
Instruction 3	I have recently started using a speech recognition model to recognize some speeches. Of course, these recognition results may contain some errors. Now, you are an ASR transcription checker, and I need your help to correct these potential mistakes. You should correct all possible errors from transcriptions from speech recognition models. These errors often occur where the semantics do not make sense and can be categorized into three types: substitution, insertion, and deletion
Instruction 4	I have recently been using a speech recognition model to recognize some speeches. Naturally, these recognition results may contain errors. You are now an ASR transcription checker, and I require your assistance to correct these potential mistakes. Correct all possible errors from transcriptions provided by the speech recognition models. These errors typically appear where the semantics don't make sense and can be divided into three types: substitution, insertion, and deletion. Please use '[]' to enclose your final corrected sentences

Since the LLM output may contain some irrelevant content with ASR transcription, for testing convenience, we devised a method suite to extract transcriptions from LLM output. Specifically, from the prompt perspective, we tell the model to enclosed the corrected transcription in '[ ]', either by presenting the example to the model or directing the model in the instruction. After the LLM generates the text, we initially extract the text within '[ ]' from the LLM output text. In the following step, we eliminate all the punctuation within it. This is because the ground truth transcription provided by the Aishell-1 dataset and Librispeech dataset does not include punctuation.

4.2   Results

In our preliminary experiments, we established a baseline using the GPT-3.5 (GPT-3.5-turbo-16k-0613 version) model. We employed the Instruction 1: *Cor-*

rect the following transcription from speech recognition. We employed a single attempt with one-shot learning. The outcomes of these initial tests, as shown in Table 2, were unsatisfactory.

Table 2. WER (%) results using GPT-3.5-16k-0613 for ASR transcription correction with one-shot learning.

	Aishell-1	LibriSpeech	
		Clean	Other
with LLM	12.36	47.93	51.25
without LLM	4.73	3.35	8.77

```
Prompt: Correct the following transcription from speech recognition. For example,
from `I SHALL CALL YOU AS SPREE` to [I SHALL CALL YOU ESPRIT]. Now, from `STUFFED
INTO YOU HIS BELLY COUNSELLED HIM` to
LLM Output:
[STUFF IT INTO HIS BELLY, COUNSELED HIM]
ASR transcription:
STUFFED {deletion} INTO YOU HIS BELLY COUNSELLED HIM
Extracted LLM output:
STUFF IT INTO {deletion} HIS BELLY COUNSELED HIM
Ground truth:
STUFF IT INTO YOU HIS BELLY COUNSELLED HIM
--------------------------------------------------------------------------
Prompt: Correct the following transcription from speech recognition. For example,
from `I SHALL CALL YOU AS SPREE` to [I SHALL CALL YOU ESPRIT]. Now, from `NUMBER TEN
FRESH NELLIE IS WAITING ON YOU GOOD NIGHT HUSBAND` to
LLM Output:
[NUMBER TEN EXPRESS DELI IS WAITING ON YOU, GOODNIGHT HUSBAND]
ASR transcription:
NUMBER TEN FRESH NELLIE IS WAITING ON YOU GOOD NIGHT HUSBAND
Extracted LLM output:
NUMBER TEN EXPRESS DELI IS WAITING ON YOU GOODNIGHT HUSBAND
Ground truth:
NUMBER TEN FRESH NELLY IS WAITING ON YOU GOOD NIGHT HUSBAND
--------------------------------------------------------------------------
Prompt: Correct the following transcription from speech recognition. For example,
from `I SHALL CALL YOU AS SPREE` to [I SHALL CALL YOU ESPRIT]. Now, from `HELLO
BERTIE ANY GOOD IN YOUR MIND` to
LLM Output:
[HELLO BIRDIE, ANY GOOD IDEA IN YOUR MIND]
ASR transcription:
HELLO BERTIE ANY GOOD IN YOUR MIND
Extracted LLM output:
HELLO BIRDIE ANY GOOD IDEA IN YOUR MIND
Ground truth:
HELLO BERTIE ANY GOOD IN YOUR MIND
```

Fig. 2. Illustrative examples of the LLM's challenges in interpreting and correcting ASR transcriptions from LibriSpeech Clean dataset.

Furthermore, we provide several examples from LibriSpeech Clean in Fig. 2. These instances underscore the challenges the LLM encounters when interpreting and correcting ASR transcriptions, which result in unsatisfactory performance for our task. In the first example, the original transcription read, "STUFFED INTO YOU HIS BELLY COUNSELLED HIM". Yet, the LLM amended "YOU" to "HIS" in the corrected transcription, deviating from the ground truth. In the second instance, the original transcript stated, "NUMBER TEN FRESH NEL-LIE IS WAITING ON YOU GOOD NIGHT HUSBAND;;. However, the LLM altered "FRESH NELLIE" to "EXPRESS DELI", thereby significantly modifying the intended meaning. In the third case, the initial transcription was "HELLO BERTIE ANY GOOD IN YOUR MIND". However, the LLM misinterpreted "BERTIE" as "BIRDIE" and superfluously appended "IDEA" to the corrected transcription.

Results with Different LLM Models. Initially, we varied the LLM models utilized in our experiments, considering three different versions: GPT-3.5-turbo-4k-0301, GPT-3.5-turbo-4k-0613, and GPT-3.5-turbo-16k-0613 models. The results are consolidated in Table 3.

The observations from Table 3 demonstrate that all LLM models have a higher Word Error Rate (WER) than the scenario without the utilization of an LLM. This finding suggests that while LLM models exhibit potential for a broad range of NLP applications, their application for error correction in ASR transcriptions still requires refinement. Notably, the WER for all LLM models in the Aishell-1 dataset is significantly higher than the WER in the scenario without an LLM. A similar pattern is evident in the LibriSpeech dataset, for both clean and other data. Furthermore, the performance of GPT-3.5-turbo-4k-0613 and GPT-3.5-turbo-16k-0613 models is markedly better than that of the GPT-3.5-turbo-4k-0301 model. This disparity could be due to the enhancements in the GPT-3.5 model.

Table 3. WER (%) performance comparison of different LLM models on ASR transcription error correction.

	Aishell-1	LibriSpeech	
		Clean	Other
GPT-3.5-turbo-4k-0301	16.05	57.83	51.20
GPT-3.5-turbo-4k-0613	12.32	47.57	51.10
GPT-3.5-turbo-16k-0613	12.36	47.93	51.25
without LLM	4.73	3.35	8.77

Results with Varying Instructions. Next, we carried out a series of experiments using a variety of instructions. We precisely constructed four different types of instructions, which are displayed in Table 1. These instructions gradually provided more specific guidance for the task. We utilized the GPT-3.5-turbo-16k-0613 model for this purpose. The outcomes for the different instructions are tabulated in Table 4.

Furthermore, we tested varying instructions with two different models, specifically GPT-3.5-turbo-4k-0301 and GPT-3.5-turbo-4k-0613, the results of which are included in Appendix 5. Our findings suggested that supplying detailed instructions to the Language Model (LLM) improves its performance. However, even with extremely detailed instructions, the LLM model does not demonstrate adequate performance in the task of rectifying errors in speech recognition transcriptions. That is to say, the Word Error Rate (WER) escalates after correction.[3]

Table 4. WER (%) comparison for varying instructions with the GPT-3.5-turbo-16k-0613 model.

	Aishell-1	LibriSpeech	
		Clean	Other
Instruction 1	12.36	47.93	51.25
Instruction 2	34.08	48.58	64.60
Instruction 3	22.32	37.21	48.14
Instruction 4	12.22	23.93	17.17
without LLM	4.73	3.35	8.77

Results with Varying Shots. Subsequently, we explored the impact of varying the number of examples given to the model, using 1-shot, 2-shot, and 3-shot configurations. We utilized the GPT-3.5-turbo-16k-0613 model. For instructions, we employed the most detailed Instruction 4, which was proven to yield superior results in Subsect. 4.2. The results are encapsulated in Table 5. Moreover, we also tested Instructions 1, 2, and 3, with the outcomes detailed in Appendix 5. Our observations revealed that providing the model with more examples led to enhanced performance, aligning with findings observed in many NLP tasks involving LLM. However, for the 1-shot, 2-shot, and 3-shot scenarios we experimented with, none of them resulted in a satisfactory WER, indicating an increase in errors post-correction. This is consistent with our previous observation that more progress is needed to harness LLMs effectively for ASR transcription error correction.

---

[3] One might think that more detailed instructions could lead to better performance. This is indeed possible. In fact, we have exhaustively tried a lot of other instructions, but we have not observed a lower WER after corrections made by the LLM.

Table 5. WER (%) comparison for varying shots with Instruction 4 and the GPT-3.5-turbo-16k-0613 model.

	Aishell-1	LibriSpeech	
		Clean	Other
1-shot	12.22	23.93	17.17
2-shot	14.19	23.38	17.68
3-shot	12.71	22.68	17.43
without LLM	4.73	3.35	8.77

Results with Varying Attempts. In the previous subsections, we established that the performance of Language Model Large (LLMs) in correcting errors in Automatic Speech Recognition (ASR) transcriptions is currently unsatisfactory, as corrections generally increase the number of errors. To deepen our understanding of the LLM's limitations in error correction for ASR transcriptions, we conducted further tests allowing the model multiple attempts. Specifically, for each transcription sentence x, the LLM generates multiple corrected outputs, and the final corrected result of the transcription sentence x is chosen as the output with the least Word Error Rate (WER). In practical applications, choosing the output with the lowest WER is not feasible, as the correct transcription y is unknown. Nevertheless, this approach aids in elucidating the constraints of leveraging LLM's in-context learning capabilities for ASR transcription enhancement. We present the results for 1, 3, and 5 attempts in Table 6. We utilized the GPT-3.5-turbo-16k-0613 model. For instructions, we employed the most effective Instruction 4 from Subsect. 4.2. Additionally, in the prompt, we provided the model with three examples, that is, the 3-shot setup. Refer to Table 6 for the experimental results. We discovered that even with up to five trials allowed, and the optimal result taken on a per-sentence basis, the outputs of the LLM still introduce more errors.

Table 6. WER (%) comparison for varying attempts with Instruction 4.

	Aishell-1	LibriSpeech	
		Clean	Other
1 Attempt	12.71	22.68	17.43
3 Attempts	6.81	17.50	12.54
5 Attempts	5.77	15.90	11.49
without LLM	4.73	3.35	8.77

GPT4 Experimentations. We further extended our study to include the latest GPT4 model, currently deemed the most advanced. Due to the high computational demand and RPM restrictions of GPT4, we limited our testing to the LibriSpeech clean test set. We conducted tests using a one-shot setting for the four detailed instructions provided in Table 1. The outcomes are encapsulated in Table 7. Our findings indicated that, despite employing the state-of-the-art GPT4 model, the ASR transcriptions corrected with LLM still yielded a higher number of errors.

Table 7. WER (%) results with the GPT4 model for the LibriSpeech clean test set.

Instruction 1	Instruction 2	Instruction 3	Instruction 4	Without LLM
28.97	23.91	16.76	14.90	3.35

5   Conclusion

This paper has provided an exploratory study on the potential of Large Language Models (LLMs) to rectify errors in Automatic Speech Recognition (ASR) transcriptions. Our research focused on employing renowned LLM benchmarks such as GPT-3.5 and GPT-4, which are known for their extensive capabilities. Our experimental studies included a diverse range of settings, variations in the LLM models, changes in instructions, and a varied number of attempts and examples provided to the model.

Despite these extensive explorations, the results were less than satisfactory. In many instances, sentences corrected by LLMs exhibited increased Word Error Rates (WERs), highlighting the limitations of LLMs in speech applications. Another potential issue is the speed of LLMs; they appear to operate slowly, suggesting that direct integration into ASR systems may not be suitable. This further underscores the challenges of utilizing the in-context learning capabilities of LLMs to enhance ASR transcriptions.

These findings do not imply that the application of LLMs in ASR technology should be dismissed. On the contrary, they suggest that further research and development are required to optimize the use of LLMs in this area. As LLMs continue to evolve, their capabilities might be harnessed more effectively in the future to overcome the challenges identified in this study.

In conclusion, while the use of LLMs for enhancing ASR performance is in its early stages, the potential for improvement exists. This study hopes to inspire further research in this field, with the aim of refining and improving the application of LLMs in ASR technology.

Appendix

Results with Varying Instructions

We conducted experiments with various instructions. Four distinct types of instructions were meticulously designed, as depicted in Table 1. These instructions progressively provided more detailed task directives. The experimental results for the GPT-3.5-turbo-4k-0301 and GPT-3.5-turbo-4k-0613 models, under the conditions of these four instructions, are presented in Table 8 and Table 9, respectively. Our findings suggest that supplying the LLM model with detailed instructions aids in achieving enhanced performance. However, even with highly detailed instructions, the LLM model's performance in the task of correcting speech recognition transcription errors is not satisfactory. That is to say, the Word Error Rate (WER) increases post-correction.

Table 8. WER comparison for varying instructions with the GPT-3.5-turbo-4k-0301 model.

	Aishell-1	LibriSpeech	
		Clean	Other
Instruction 1	16.05	57.83	51.20
Instruction 2	16.81	30.85	36.99
Instruction 3	14.12	24.42	26.19
Instruction 4	14.16	25.56	20.26
without LLM	4.73	3.35	8.77

Table 9. WER comparison for varying instructions with the GPT-3.5-turbo-4k-0613 model.

	Aishell-1	LibriSpeech	
		Clean	Other
Instruction 1	12.32	47.57	51.10
Instruction 2	34.61	48.33	65.06
Instruction 3	23.19	37.05	48.10
Instruction 4	12.13	23.07	17.18
without LLM	4.73	3.35	8.77

Results with Varying Shots

We evaluated the effect of varying the number of examples provided to the model, using 1-shot, 2-shot, and 3-shot configurations. We employed the GPT-3.5-turbo-16k-0613 model for this purpose. Tables 10, 11, and 12 depict the experimental results using Instructions 1, 2, and 3, respectively.

Our findings suggest that providing more examples to the model leads to improved performance. This is consistent with results observed in numerous NLP tasks involving LLM. However, in the 1-shot, 2-shot, and 3-shot scenarios we tested, none yielded a satisfactory WER, indicating an increase in errors after correction. This aligns with our previous observation that additional efforts are required to effectively employ LLMs for ASR transcription error correction.

Table 10. WER comparison for varying shots with Instruction 1 and the GPT-3.5-turbo-16k-0613 model.

	Aishell-1	LibriSpeech	
		Clean	Other
1-shot	12.36	47.93	51.25
2-shot	20.67	70.93	73.32
3-shot	38.49	81.43	76.58
without LLM	4.73	3.35	8.77

Table 11. WER comparison for varying shots with Instruction 2 and the GPT-3.5-turbo-16k-0613 model.

	Aishell-1	LibriSpeech	
		Clean	Other
1-shot	34.08	48.58	64.60
2-shot	45.70	80.04	94.20
3-shot	76.48	80.39	90.79
without LLM	4.73	3.35	8.77

Table 12. WER comparison for varying shots with Instruction 3 and the GPT-3.5-turbo-16k-0613 model.

	Aishell-1	LibriSpeech	
		Clean	Other
1-shot	22.32	37.21	48.14
2-shot	52.52	67.10	72.88
3-shot	86.49	66.78	69.46
without LLM	4.73	3.35	8.77

References

1. Brown, T., et al.: Language models are few-shot learners. Adv. Neural. Inf. Process. Syst. **33**, 1877–1901 (2020)
2. Bu, H., Du, J., Na, X., Wu, B., Zheng, H.: Aishell-1: an open-source mandarin speech corpus and a speech recognition baseline. In: 2017 20th Conference of the Oriental Chapter of the International Coordinating Committee On Speech Databases and Speech I/O Systems and Assessment (O-COCOSDA), pp. 1–5. IEEE (2017)
3. Chan, W., Jaitly, N., Le, Q., Vinyals, O.: Listen, attend and spell: a neural network for large vocabulary conversational speech recognition. In: 2016 IEEE International Conference on Acoustics, Speech and Signal Processing (ICASSP), pp. 4960–4964. IEEE (2016)
4. Chan, W., Jaitly, N., Le, Q.V., Vinyals, O.: Listen, attend and spell. arXiv preprint arXiv:1508.01211 (2015)
5. Chiu, S.H., Chen, B.: Innovative bert-based reranking language models for speech recognition. In: 2021 IEEE Spoken Language Technology Workshop (SLT), pp. 266–271. IEEE (2021)
6. Chorowski, J.K., Bahdanau, D., Serdyuk, D., Cho, K., Bengio, Y.: Attention-based models for speech recognition. In: Advances in Neural Information Processing Systems 28 (2015)
7. Devlin, J., Chang, M.W., Lee, K., Toutanova, K.: Bert: Pre-training of deep bidirectional transformers for language understanding. arXiv preprint arXiv:1810.04805 (2018)
8. Dong, L., Xu, S., Xu, B.: Speech-transformer: a no-recurrence sequence-to-sequence model for speech recognition. In: 2018 IEEE International Conference on Acoustics, Speech and Signal Processing (ICASSP), pp. 5884–5888. IEEE (2018)
9. Futami, H., Inaguma, H., Ueno, S., Mimura, M., Sakai, S., Kawahara, T.: Distilling the knowledge of bert for sequence-to-sequence asr. arXiv preprint arXiv:2008.03822 (2020)
10. Graves, A., Fernández, S., Gomez, F., Schmidhuber, J.: Connectionist temporal classification: labelling unsegmented sequence data with recurrent neural networks. In: Proceedings of the 23rd international conference on Machine learning, pp. 369–376 (2006)
11. Graves, A., Jaitly, N.: Towards end-to-end speech recognition with recurrent neural networks. In: International Conference on Machine Learning, pp. 1764–1772. PMLR (2014)

12. Graves, A., Mohamed, A.r., Hinton, G.: Speech recognition with deep recurrent neural networks. In: 2013 IEEE International Conference on Acoustics, Speech and Signal Processing, pp. 6645–6649. IEEE (2013)
13. Gulati, A., et al.: Conformer: convolution-augmented transformer for speech recognition. arXiv preprint arXiv:2005.08100 (2020)
14. Han, W., et al.: Contextnet: improving convolutional neural networks for automatic speech recognition with global context. arXiv preprint arXiv:2005.03191 (2020)
15. Kannan, A., Wu, Y., Nguyen, P., Sainath, T.N., Chen, Z., Prabhavalkar, R.: An analysis of incorporating an external language model into a sequence-to-sequence model. In: 2018 IEEE International Conference on Acoustics, Speech and Signal Processing (ICASSP), pp. 1–5828. IEEE (2018)
16. Kim, S., Hori, T., Watanabe, S.: Joint ctc-attention based end-to-end speech recognition using multi-task learning. In: 2017 IEEE International Conference on Acoustics, Speech and Signal Processing (ICASSP), pp. 4835–4839. IEEE (2017)
17. Kubo, Y., Karita, S., Bacchiani, M.: Knowledge transfer from large-scale pretrained language models to end-to-end speech recognizers. In: ICASSP 2022–2022 IEEE International Conference on Acoustics, Speech and Signal Processing (ICASSP), pp. 8512–8516. IEEE (2022)
18. OpenAI: Gpt-4 technical report (2023)
19. Ouyang, L., Wu, J., Jiang, X., Almeida, D., Wainwright, C., Mishkin, P., Zhang, C., Agarwal, S., Slama, K., Ray, A., et al.: Training language models to follow instructions with human feedback. Adv. Neural. Inf. Process. Syst. **35**, 27730–27744 (2022)
20. Panayotov, V., Chen, G., Povey, D., Khudanpur, S.: Librispeech: an asr corpus based on public domain audio books. In: 2015 IEEE international conference on acoustics, speech and signal processing (ICASSP), pp. 5206–5210. IEEE (2015)
21. Peng, Y., Dalmia, S., Lane, I., Watanabe, S.: Branchformer: Parallel mlp-attention architectures to capture local and global context for speech recognition and understanding. In: International Conference on Machine Learning, pp. 17627–17643. PMLR (2022)
22. Sainath, T.N., et al.: Two-pass end-to-end speech recognition. arXiv preprint arXiv:1908.10992 (2019)
23. Scao, T.L., et al.: Bloom: A 176b-parameter open-access multilingual language model. arXiv preprint arXiv:2211.05100 (2022)
24. Shin, J., Lee, Y., Jung, K.: Effective sentence scoring method using bert for speech recognition. In: Asian Conference on Machine Learning, pp. 1081–1093. PMLR (2019)
25. Soltau, H., Liao, H., Sak, H.: Neural speech recognizer: acoustic-to-word lstm model for large vocabulary speech recognition. arXiv preprint arXiv:1610.09975 (2016)
26. Tjandra, A., Sakti, S., Nakamura, S.: Listening while speaking: speech chain by deep learning. In: 2017 IEEE Automatic Speech Recognition and Understanding Workshop (ASRU), pp. 301–308. IEEE (2017)
27. Touvron, H., Lavril, T., Izacard, G., Martinet, X., Lachaux, M.A., Lacroix, T., Rozière, B., Goyal, N., Hambro, E., Azhar, F., et al.: Llama: Open and efficient foundation language models. arXiv preprint arXiv:2302.13971 (2023)
28. Udagawa, T., Suzuki, M., Kurata, G., Itoh, N., Saon, G.: Effect and analysis of large-scale language model rescoring on competitive asr systems. arXiv preprint arXiv:2204.00212 (2022)
29. Watanabe, S., et al.: Espnet: end-to-end speech processing toolkit. arXiv preprint arXiv:1804.00015 (2018)

30. Weiran, W., et al.: Improving Rare Word Recognition with LM-aware MWER training. In: Proceedings of Interspeech 2022, pp. 1031–1035 (2022). https://doi.org/10.21437/Interspeech. 2022–10660
31. Xu, L., et al.: Rescorebert: discriminative speech recognition rescoring with bert. In: ICASSP 2022–2022 IEEE International Conference on Acoustics, Speech and Signal Processing (ICASSP), pp. 6117–6121. IEEE (2022)
32. Yao, Z., et al.: Wenet: production oriented streaming and non-streaming end-to-end speech recognition toolkit. arXiv preprint arXiv:2102.01547 (2021)
33. Zeng, A., ct al.: Glm-130b: an open bilingual pre-trained model. arXiv preprint arXiv:2210.02414 (2022)
34. Zhang, B., et al.: Unified streaming and non-streaming two-pass end-to-end model for speech recognition. arXiv preprint arXiv:2012.05481 (2020)
35. Zhang, S., et al.: Opt: Open pre-trained transformer language models. arXiv preprint arXiv:2205.01068 (2022)

Image Inpainting with Semantic U-Transformer

Lingfan Yuan[1,4], Wenxin Yu[1(✉)], Lu Che[1(✉)], Zhiqiang Zhang[2], Shiyu Chen[1],
Lu Liu[1], and Peng Chen[3]

[1] Southwest University of Science and Technology, Sichuan, China
yuwenxin@swust.edu.cn, 790186623@qq.com
[2] Hosei University, Tokyo, Japan
[3] Chengdu Hongchengyun Technology Co., Ltd, Sichuan, China
[4] Instrumentation Technology and Economy Institute, Sichuan,
People's Republic of China

Abstract. With the driving force of powerful convolutional neural networks, image inpainting has made tremendous progress. Recently, transformer has demonstrated its effectiveness in various vision tasks, mainly due to its capacity to model long-term relationships. However, when it comes to image inpainting tasks, the transformer tends to fall short in terms of modeling local information, and interference from damaged regions can pose challenges. To tackle these issues, we introduce a novel Semantic U-shaped Transformer (SUT) in this work. The SUT is designed with spectral transformer blocks in its shallow layers, effectively capturing local information. Conversely, deeper layers utilize BRA transformer blocks to model global information. A key feature of the SUT is its attention mechanism, which employs bi-level routing attention. This approach significantly reduces the interference of damaged regions on overall information, making the SUT more suitable for image inpainting tasks. Experiments on several datasets indicate that the performance of the proposed method outperforms the current state-of-the-art (SOTA) inpainting approaches. In general, the PSNR of our method is on average 0.93 dB higher than SOTA, and the SSIM is higher by 0.026.

Keywords: Image Inpainting · Deep Learning · Transformer · U-Net

1 Introduction

Image inpainting is a challenging computer vision task involving restoring masked images by filling missing regions with meaningful and plausible content. It is widely used in image processing applications, such as face editing, privacy protection, and object removal. Traditional methods for image inpainting

This Research is Supported by National Key Research and Development Program from Ministry of Science and Technology of the PRC (No.2018AAA0101801), (No.2021ZD0110600), Sichuan Science and Technology Program (No.2022ZYD0116), Sichuan Provincial M. C. Integration Office Program, And IEDA Laboratory Of SWUST.

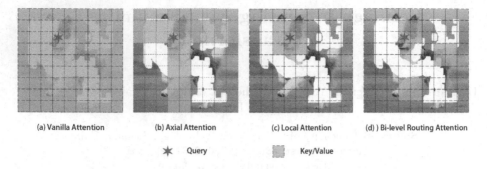

(a) Vanilla Attention   (b) Axial Attention   (c) Local Attention   (d)) Bi-level Routing Attention

✴ Query   Key/Value

Fig. 1. Calculation methods of different types of attention. Bi-level routing attention can effectively reduce the interference caused by masked regions.

rely on borrowing information from known regions to fill in the missing regions. They are limited in cases that involve complex, non-repetitive structures due to the need for high-level semantics understanding. In recent years, CNNs have enabled significant progress in image inpainting by learning from large datasets. Nevertheless, CNNs still face limitations due to their local inductive bias and spatially-invariant kernels, making it difficult to comprehend the global structure and efficiently inpaint large areas of missing content in images.

Transformers originated in natural language processing and have shown significant performance. Unlike the design of CNNs, Transformer-based network structures are naturally good at capturing long-range dependencies in the data by global self-attention. The adaptation of transformers for computer vision was first explored in the Vision Transformer (ViT) [6]. They made a significant contribution by developing a suitable patch-based tokenization method for images, making it possible to use the transformer architecture for image processing. Although self-attention layers in vision transformer models capture long-term dependencies and enable diverse interactions between spatial locations, computation is costly due to their pairwise token affinity across all spatial locations, and they use a substantial amount of memory resources. Recent studies [23] also indicate that transformers are ineffective at capturing high frequencies that mainly convey local information.

The Fourier domain also plays a significant role in extracting frequency-based analysis of image information and has been extensively studied by the research community. It has been demonstrated that Fourier transforms can replace the multi-headed attention layers in transformers, as shown in GFNet [21]. They suggested that this approach captures fine-grained properties of images.

Recent works [25,29] have attempted to use transformers for image inpainting and have achieved remarkable success in improving diversity and large region inpainting quality. However, due to the inherent limitations of transformers, these methods suffer from inadequate modeling of local features, leading to coarse restoration results. More importantly, as shown in Fig. 1, the involvement of the damaged area in attention computation may introduce excessive noise and additional computational complexity, which is not conducive to image inpainting tasks.

To address the issues mentioned above, we propose a single-stage transformer model called the Semantic U-shaped Transformer(SUT), which consists of spectral and Bi-level Routing Attention(BRA) blocks. SUT follows the design of U-shaped architecture, where the encoder and decoder are connected to effectively capture contextual information at different scales. The spectral blocks utilize Fourier Transform to extract appropriate frequency information in the frequency domain, efficiently modeling local information. In deeper layers, we leverage BRA blocks to model global dependencies. By implementing the bi-level routing attention [31], our approach enables each query to prioritize a subset of the key-value pairs that are most semantically relevant. This targeted focus effectively mitigates the interference caused by masked regions, resulting in improved overall performance. Our main contributions are as follows:

- We propose a spectral transformer block for image inpainting, which improves the transformer's capability to capture local information.
- We propose a BRA transformer block for image inpainting, which effectively reduces information interference from damaged regions.
- Experiments on the CelebA-HQ [13] and Places2 [30] datasets demonstrate the superiority of our approach compared to the existing advanced approaches.

2 Related Work

2.1 Image Inpainting

Image inpainting is a long-standing challenge in computer vision, where missing regions in an image need to be filled in seamlessly. Traditional methods have used diffusion-based [1] or patch-based [15] techniques. Although these methods can generate visually persuasive results, their inherent lack of high-level understanding hampers their capacity to yield semantically coherent content.

In recent years, there has been a growing trend in solving the inpainting problem using trainable networks after the Context Encoders [20], which proposed an encoder-decoder framework for this task. MNPS [27] is a method that utilizes context encoders and optimizes both content and texture networks to preserve the structure and details·in the missing regions. GL [10], on the other hand, employs global and local context discriminators to generate consistent results. EC [19] adds the edge images from external or the pre-trained edge generator to complement missing semantics. GC [28] presents the Gated Convolutional that learns a dynamic mask-update mechanism to replace the rigid mask-update. Following the successful integration of transformers into the vision field by DETR [2] and ViT, ICT [25], TFill [29], ZITS [4], all propose effective transformer models for image inpainting. Although these methods successfully model long-range dependencies, local information is disregarded, making it impossible to generate high-quality images using a single-stage transformer model.

2.2 Applications of Fourier Transform in Vision

The Fourier transform has been a crucial tool in digital image processing for decades. With the breakthroughs of CNNs in vision [9], there are a variety of works that start to incorporate Fourier transform in some deep learning method [16] for vision tasks. Some of these works employ discrete Fourier transform to convert the images to the frequency domain and leverage the frequency information to improve the performance in certain tasks, while others utilize the convolution theorem to accelerate the CNNs via fast Fourier transform (FFT). Fast Fourier convolution(FFC) [3] replaces the convolution in CNNs with a Local Fourier Unit and performs convolutions in the frequency domain. Recent advancements have been made, such as the use of fast Fourier convolutions (FFCs) by the LaMa [24], which has shown significant progress in large mask inpainting tasks.

2.3 Efficient Attention Mechanisms

Many research works have emerged to address the computation and memory complexity bottlenecks of vanilla attention. Sparse connection patterns, low-rank approximations, and recurrent operations are among the commonly utilized techniques. With regards to vision transformers, sparse attention has become increasingly popular since the success of Swin Transformer [18]. In Swin Transformer, attention is confined to non-overlapping local windows, and the shift window operation is introduced to facilitate inter-window communication between adjacent windows. To achieve larger and even quasi-global receptive fields while keeping the computation costs reasonable, follow-up works have proposed various handcrafted sparse patterns, such as dilated windows [26] and cross-shaped windows [5]. BiFormer [31] proposes a novel dynamic sparse attention via bi-level routing to enable a more flexible allocation of computations with content awareness. However, its effectiveness in the field of image generation has not yet been explored.

3 Method

Image inpainting aims to generate visually appealing and semantically appropriate content for missing areas by means of sophisticated algorithms and techniques. In this work, we propose a Semantic U-shaped Transformer(SUT) network for image inpainting. In this section, we first describe the overall pipeline and the hierarchical structure of SUT. Then, we provide the details of the Transformer block, which is the basic component of SUT.

3.1 Overall Architecture

As shown in Fig. 2(a), the overall structure of the proposed SUT is a U-shaped hierarchical network with skip-connections between the encoder and the decoder.

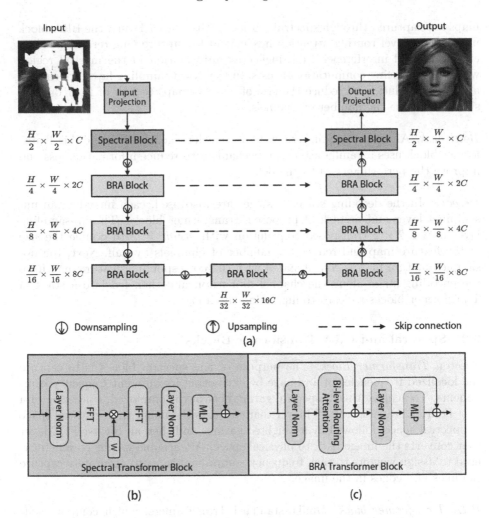

Fig. 2. (a) The structure diagram of our proposed SUT. (b) Details of a spectral Block. (c) Details of a BRA Block.

Specifically, Let $\mathbf{I} \in \mathbb{R}^{H \times W \times 3}$ be an image and $\mathbf{M} \in \{0, 1\}^{H \times W \times 1}$ be the mask denoting whether a region needs to be inpainted (with value 1) or not (with value 0). H and W are the spatial resolutions. The input $\mathbf{I_M} = \mathbf{I} \odot (1 - \mathbf{M})$ is the masked image that contains missing pixels, where $\odot$ is the Hadamard product.

Encoder. SUT firstly applies a 4×4 convolutional layer with *stride* $= 2$ to extract low-level features $\mathbf{X}_0 \in \mathbb{R}^{\frac{H}{2} \times \frac{W}{2} \times C}$. Following the design of U-shaped architectures [22], the feature maps $\mathbf{X}_0$ are processed through K encoder stages, where each stage consists of a stack of proposed Transformer blocks and one down-sampling layer. In the shallow layers of the model, local features in feature maps are captured through spectral blocks. In the deeper layers, the BRA block utilizes a bi-level routing attention mechanism to capture long-term dependencies and avoid interference from ineffective information in the masked region while also reducing computational costs. In the down-sampling layer, we employ a 4 × 4 convolution to reduce the size of feature maps to half of their original size and double the number of channels.

Bottleneck. After the encoding stage, a bottleneck stage is added, and this Transformer block uses a vanilla attention mechanism to reduce information loss and improve the performance of the model.

Decoder. In the decoding stage, K stages are also used, each including an up-sampling layer and a stack of proposed Transformer blocks. The up-sampling layer uses a 2 × 2 transposed convolution with *stride* = 2 to double the size of the feature map and reduce the number of channels by half. Next, the up-sampled feature map is concatenated with the corresponding feature map from the encoding stage along the channel dimension and then fed into a stack of Transformer blocks to learn to inpainting the image.

3.2   Spectral and BRA Transformer Blocks

Spectral Transformer Block. The purpose of the spectral blocks is to capture the localized frequencies of an image by extracting its different frequency components. To achieve this, a spectral gating network is employed, consisting of a Fast Fourier Transform (FFT) layer that converts the image from physical space to spectral space, followed by a weighted gating, and then an Inverse FFT layer that converts the image back to physical space. A learnable weight parameter is used to assign weights to each frequency component so as to accurately capture the lines and edges in the image.

BRA Transformer block. Unlike standard Transformers, which compute self-attention globally across all tokens, our approach is more tailored to image inpainting. By using the routing index matrix $\mathbf{I}^r$, our model can limit attention to the most relevant regions for each query token, reducing the likelihood of noisy interactions with irrelevant features. In detail, Given the input features of the $l-1^{th}$ block $\mathbf{X}_{l-1}$, the encoding process of a BRA Transformer block can be defined as follows:

$$
\begin{aligned}
\mathbf{X}'_l &= \mathrm{BRA}\left(\mathrm{LN}\left(\mathbf{X}_{l-1}\right)\right) + \mathbf{X}_{l-1}, \\
\mathbf{X}_l &= \mathrm{MLP}\left(\mathrm{LN}\left(\mathbf{X}'_l\right)\right) + \mathbf{X}'_l,
\end{aligned}
\tag{1}
$$

where LN denotes the layer normalization and BRA denotes the bi-Level routing attention; We use 3 × 3 depth-wise convolutions to inject positional information into the Transformer.

3.3 Bi-Level Routing Attention

Region Partition and Input Projection. For a given feature map $\mathbf{X} \in \mathbb{R}^{H \times W \times C}$, we divide it into $S \times S$ non-overlapping regions.Specifically, we reshape $\mathbf{X}$ as $\mathbf{X}^r \in \mathbb{R}^{S^2 \times \frac{HW}{S^2} \times C}$. We then derive the query, key, value tensor, $\mathbf{Q}, \mathbf{K}, \mathbf{V} \in \mathbb{R}^{S^2 \times \frac{HW}{S^2} \times C}$, with linear projections:

$$\mathbf{Q} = \mathbf{X}^r \mathbf{W}^q, \quad \mathbf{K} = \mathbf{X}^r \mathbf{W}^k, \quad \mathbf{V} = \mathbf{X}^r \mathbf{W}^v, \tag{2}$$

where $\mathbf{W}^q, \mathbf{W}^k, \mathbf{W}^v \in \mathbb{R}^{C \times C}$ are projection weights for the query, key, and value, respectively.

Region-to-Region Routing with a Directed Graph. Then, we determine which regions need to be involved in the computation by constructing a directed graph. We replace the $\mathbf{Q}^r, \mathbf{K}^r \in \mathbb{R}^{S^2 \times C}$ for each region by using the average values on its corresponding query and key. Next, we obtain the affinity graph between regions by conducting matrix multiplication between $\mathbf{Q}^r$ and the transpose of $\mathbf{K}^r$:

$$\mathbf{A}^r = \mathbf{Q}^r \left(\mathbf{K}^r\right)^T \tag{3}$$

The entries in the adjacency matrix, $\mathbf{A}^r$, quantify the semantic relevance between the two regions. The crucial step that follows is to prune the affinity graph by retaining only the top-k connections for each region. To accomplish this, we obtain a routing index matrix, $\mathbf{I}_r \in \mathbb{N}^{S^2 \times k}$, using the row-wise topk operator.

$$\mathbf{I}^r = \text{topkIndex}\left(\mathbf{A}^r\right) \tag{4}$$

Therefore, the ith row of $\mathbf{I}^r$ comprises the k indices that correspond to the most relevant regions for the ith region.

Token-to-Token Attention. With the utilization of the region-to-region routing index matrix $\mathbf{I}^r$, we are able to implement fine-grained token-to-token attention. For each query token in the region $\mathbf{I}$, it will attend to all key-value pairs residing in the union of k routed regions indexed with $\mathbf{I}^r_{(i,1)}, \mathbf{I}^r_{(i,2)}, \ldots, \mathbf{I}^r_{(i,k)}$. As a result, the masked region will not be involved in attention calculation.

3.4 Loss Functions

Our model is trained on a joint loss, which includes ℓ_1 loss, adversarial loss, perceptual loss [12], and style loss [7].

In particular, we introduce focal frequency loss(FFL) [11], which allows a model to adaptively focus on frequency components that are hard to synthesize by down-weighting the easy ones. The focal frequency loss can be seen as a weighted average of the frequency distance between the real and fake images. Focal frequency loss(FFL) is defined as

$$\mathcal{L}_{ffl} = \frac{1}{HW} \sum_{u=0}^{H-1} \sum_{v=0}^{W-1} w(u,v) \left| F_{gt}(u,v) - F_{out}(u,v) \right|^2, \tag{5}$$

where the image size is $H \times W$, $F_{gt}(u, v)$ is the spatial frequency value at the spectrum coordinate (u, v) of the ground truth image, and $F_{out}(u, v)$ is the spatial frequency value of the output of the network. $w(u, v)$ is the weight for the spatial frequency at (u, v). Our overall loss function is

$$\mathcal{L}_{total} = \mathcal{L}_{\ell_1} + 0.1\mathcal{L}_{adv} + 0.1\mathcal{L}_{perc} + 250\mathcal{L}_{style} + \mathcal{L}_{ffl} \qquad (6)$$

4   Experiments

All of the experiments in this paper are conducted in the dataset of CelebA-HQ [13] and Places2 [30]. The CelebA-HQ dataset is a high-quality version of CelebA that consists of 28,000 train images and 2,000 test images. The Places2 contains 23,7777 training images and 800 test images. The irregular mask dataset used in this paper comes from the work of Liu *et al.* [17]. We compared our model with some popular inpainting methods, including EC [19], CTS [8], and TFill [29], in order to demonstrate its effectiveness. Both our proposed model and the compared models were trained using a single V100 (16G). The number of spectral transformer blocks is 2, the number of BRA transformer blocks is 4, and the number of attention transformer blocks in the bottleneck stage is 8. The *topk* in BRA is set to 1, 4, 16. We used the Adam [14] optimizer to train the model with a learning rate of 10^{-4}.

Table 1. The comparison of PSNR, SSIM, and MAE over the CelebA-HQ and Places2.

Dataset		CelebA-HQ			Places2		
mask ratio		30%–40%	40%–50%	50%–60%	30%–40%	40%–50%	50%–60%
PSNR ↑	EC	24.97	23.11	21.43	24.44	22.87	20.86
	CTS	25.78	23.86	21.04	25.51	23.82	20.72
	TFill	25.78	24.07	21.58	25.56	23.85	21.54
	Ours	**27.61**	**25.72**	**23.01**	**25.65**	**23.98**	**21.91**
SSIM ↑	EC	0.906	0.860	0.766	0.834	0.767	0.658
	CTS	0.916	0.845	0.756	0.864	0.798	0.662
	TFill	0.920	0.884	0.807	0.868	0.799	0.691
	Ours	**0.956**	**0.918**	**0.858**	**0.885**	**0.808**	**0.701**
MAE(%) ↓	EC	3.06	4.23	6.39	3.27	4.25	5.73
	CTS	2.48	4.49	6.62	2.96	3.60	5.57
	TFill	2.56	3.51	5.33	2.86	3.59	5.25
	Ours	**2.13**	**2.92**	**4.43**	**2.83**	**3.54**	**5.08**

4.1 Quantitative Comparison

The Quantitative results in the test dataset of CelebA-HQ and Places2 are shown in Table 1, along with the results produced by popular inpainting methods for comparison. In the case of different ratios of the damaged region, the table demonstrates the inpainting ability of our network, showing that our results are better than other results in PSNR ((peak signal-to-noise ratio), SSIM (structural similarity index) and MAE(mean absolute error). As shown in Table 1, benefits from the semantic-aware long-range modeling and the use of spectral blocks to capture local information, our SUT exhibits superior performance on different datasets, particularly in scenarios with high mask rates where the advantage is more pronounced.

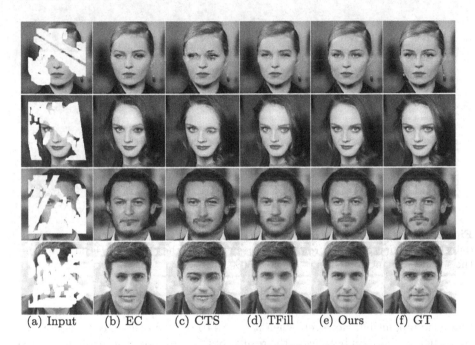

(a) Input   (b) EC   (c) CTS   (d) TFill   (e) Ours   (f) GT

Fig. 3. Qualitative comparisons with EC [19], CTS [8], and TFill [29] on the CelebA-HQ datasets. The first and second rows are the results when the mask ratio is 30%-40%, and the third row and fourth rows are 40%-50%.

On the CelebA-HQ dataset, our SUT exhibits more pronounced advantages, and as the mask rate increases, the extent of improvement in terms of PSNR and SSIM continues to grow. This signifies that our utilization of BRA is more suitable for image inpainting tasks and showcases a stronger resistance to interference from masks.

4.2   Qualitative Comparisons

Figure 3 and Fig. 4 illustrate the visual inpainting results of different methods on the test set of CelebA-HQ and Places2 with mask ratios of 30%-40% and 40%-50%.

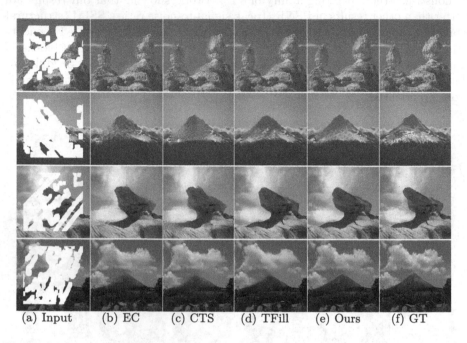

(a) Input   (b) EC   (c) CTS   (d) TFill   (e) Ours   (f) GT

Fig. 4. Qualitative comparisons with EC [19], CTS [8], and TFill [29] on the Places2 datasets. The first and third rows are the results when the mask ratio is 30%-40%, and the second row and fourth rows are 40%-50%.

On the CelebA-HQ dataset, all these methods are capable of performing basic semantic inpainting. However, when it comes to overall coherence and coordination, such as filling in both eyes, the results of CTS are not clear enough, whereas the results of EC and TFill are clear but not entirely reasonable. Our inpainting results in the first row of Fig. 3 demonstrate that the person is looking in the same direction with both eyes, which is a level of detail that was not achieved by the other methods. In the second and third rows, our results demonstrate consistent iris color in the person's eyes and a clearer and more accurate restoration of the mouth. On the Places2 dataset, our SUT has almost no occurrence of artifacts, while other methods have shown extensive artifacts. As shown in Fig. 4, compared to CNN-based methods, our SUT demonstrates superior capabilities in accurately predicting the shapes of rocks and mountains. Furthermore, when compared to Transformer-based methods, our inpainting results exhibit sharper edges.

5 Conclusions

In this paper, we propose SUT, a single-stage image inpainting network with spectral blocks and BRA blocks. The combination of spectral blocks and BRA blocks enables our network to effectively model both global and local information. Furthermore, The SUT can avoid shifts caused by invalid information when calculating attention in inpainting tasks. Various experiments show that the proposed SUT generates promising images and achieves the state-of-the-art performance. Compared with the existing methods, our method improves the PSNR by 0.1-1.83 dB and the SSIM by 0.009-0.051.

References

1. Ballester, C., Bertalmio, M., Caselles, V., Sapiro, G., Verdera, J.: Filling-in by joint interpolation of vector fields and gray levels. IEEE Trans. Image Process. **10**(8), 1200–1211 (2001)
2. Carion, N., Massa, F., Synnaeve, G., Usunier, N., Kirillov, A., Zagoruyko, S.: End-to-end object detection with transformers. In: Vedaldi, A., Bischof, H., Brox, T., Frahm, J.-M. (eds.) ECCV 2020. LNCS, vol. 12346, pp. 213–229. Springer, Cham (2020). https://doi.org/10.1007/978-3-030-58452-8_13
3. Chi, L., Jiang, B., Mu, Y.: Fast fourier convolution. Adv. Neural. Inf. Process. Syst. **33**, 4479–4488 (2020)
4. Dong, Q., Cao, C., Fu, Y.: Incremental transformer structure enhanced image inpainting with masking positional encoding. In: Proceedings of the IEEE/CVF Conference on Computer Vision and Pattern Recognition, pp. 11358–11368 (2022)
5. Dong, X., et al.: Cswin transformer: a general vision transformer backbone with cross-shaped windows. In: Proceedings of the IEEE/CVF Conference on Computer Vision and Pattern Recognition, pp. 12124–12134 (2022)
6. Dosovitskiy, A., et al.: An image is worth 16x16 words: transformers for image recognition at scale. arXiv preprint arXiv:2010.11929 (2020)
7. Gatys, L.A., Ecker, A.S., Bethge, M.: Image style transfer using convolutional neural networks. In: Proceedings of the IEEE Conference on Computer Vision and Pattern Recognition, pp. 2414–2423 (2016)
8. Guo, X., Yang, H., Huang, D.: Image inpainting via conditional texture and structure dual generation. In: Proceedings of the IEEE/CVF International Conference on Computer Vision (ICCV), pp. 14134–14143 (October 2021)
9. He, K., Zhang, X., Ren, S., Sun, J.: Deep residual learning for image recognition. In: Proceedings of the IEEE Conference on Computer Vision and Pattern Recognition, pp. 770–778 (2016)
10. Iizuka, S., Simo-Serra, E., Ishikawa, H.: Globally and locally consistent image completion. ACM Trans. Graph. (ToG) **36**(4), 1–14 (2017)
11. Jiang, L., Dai, B., Wu, W., Loy, C.C.: Focal frequency loss for image reconstruction and synthesis. In: Proceedings of the IEEE/CVF International Conference on Computer Vision, pp. 13919–13929 (2021)
12. Johnson, J., Alahi, A., Fei-Fei, L.: Perceptual losses for real-time style transfer and super-resolution. In: Leibe, B., Matas, J., Sebe, N., Welling, M. (eds.) ECCV 2016. LNCS, vol. 9906, pp. 694–711. Springer, Cham (2016). https://doi.org/10.1007/978-3-319-46475-6_43

13. Karras, T., Aila, T., Laine, S., Lehtinen, J.: Progressive growing of gans for improved quality, stability, and variation. arXiv preprint arXiv:1710.10196 (2017)
14. Kingma, D.P., Ba, J.: Adam: a method for stochastic optimization. arXiv preprint arXiv:1412.6980 (2014)
15. Lee, J.H., Choi, I., Kim, M.H.: Laplacian patch-based image synthesis. In: Proceedings of the IEEE Conference on Computer Vision and Pattern Recognition, pp. 2727–2735 (2016)
16. Li, S., Xue, K., Zhu, B., Ding, C., Gao, X., Wei, D., Wan, T.: Falcon: A fourier transform based approach for fast and secure convolutional neural network predictions. In: Proceedings of the IEEE/CVF Conference on Computer Vision and Pattern Recognition. pp. 8705–8714 (2020)
17. Liu, G., Reda, F.A., Shih, K.J., Wang, T.-C., Tao, A., Catanzaro, B.: Image inpainting for irregular holes using partial convolutions. In: Ferrari, V., Hebert, M., Sminchisescu, C., Weiss, Y. (eds.) ECCV 2018. LNCS, vol. 11215, pp. 89–105. Springer, Cham (2018). https://doi.org/10.1007/978-3-030-01252-6_6
18. Liu, Z., et al.: Swin transformer: hierarchical vision transformer using shifted windows. In: Proceedings of the IEEE/CVF International Conference on Computer Vision, pp. 10012–10022 (2021)
19. Nazeri, K., Ng, E., Joseph, T., Qureshi, F.Z., Ebrahimi, M.: Edgeconnect: generative image inpainting with adversarial edge learning. arXiv preprint arXiv:1901.00212 (2019)
20. Pathak, D., Krahenbuhl, P., Donahue, J., Darrell, T., Efros, A.A.: Context encoders: feature learning by inpainting. In: Proceedings of the IEEE Conference on Computer Vision and Pattern Recognition, pp. 2536–2544 (2016)
21. Rao, Y., Zhao, W., Zhu, Z., Lu, J., Zhou, J.: Global filter networks for image classification. Adv. Neural. Inf. Process. Syst. **34**, 980–993 (2021)
22. Ronneberger, O., Fischer, P., Brox, T.: U-Net: convolutional networks for biomedical image segmentation. In: Navab, N., Hornegger, J., Wells, W.M., Frangi, A.F. (eds.) MICCAI 2015. LNCS, vol. 9351, pp. 234–241. Springer, Cham (2015). https://doi.org/10.1007/978-3-319-24574-4_28
23. Si, C., Yu, W., Zhou, P., Zhou, Y., Wang, X., Yan, S.: Inception transformer. arXiv preprint arXiv:2205.12956 (2022)
24. Suvorov, R., et al.: Resolution-robust large mask inpainting with fourier convolutions. In: Proceedings of the IEEE/CVF Winter Conference on Applications of Computer Vision, pp. 2149–2159 (2022)
25. Wan, Z., Zhang, J., Chen, D., Liao, J.: High-fidelity pluralistic image completion with transformers. In: Proceedings of the IEEE/CVF International Conference on Computer Vision, pp. 4692–4701 (2021)
26. Wang, W., et al.: Crossformer: a versatile vision transformer hinging on cross-scale attention. arXiv preprint arXiv:2108.00154 (2021)
27. Yang, C., Lu, X., Lin, Z., Shechtman, E., Wang, O., Li, H.: High-resolution image inpainting using multi-scale neural patch synthesis. In: Proceedings of the IEEE Conference on Computer Vision and Pattern Recognition, pp. 6721–6729 (2017)
28. Yu, J., Lin, Z., Yang, J., Shen, X., Lu, X., Huang, T.S.: Free-form image inpainting with gated convolution. In: Proceedings of the IEEE/CVF International Conference on Computer Vision, pp. 4471–4480 (2019)
29. Zheng, C., Cham, T.J., Cai, J., Phung, D.: Bridging global context interactions for high-fidelity image completion. In: Proceedings of the IEEE/CVF Conference on Computer Vision and Pattern Recognition, pp. 11512–11522 (2022)

30. Zhou, B., Lapedriza, A., Khosla, A., Oliva, A., Torralba, A.: Places: A 10 million image database for scene recognition. IEEE Trans. Pattern Anal. Mach. Intell. **40**(6), 1452–1464 (2017)
31. Zhu, L., Wang, X., Ke, Z., Zhang, W., Lau, R.W.: Biformer: vision transformer with bi-level routing attention. In: Proceedings of the IEEE/CVF Conference on Computer Vision and Pattern Recognition, pp. 10323–10333 (2023)

Multi-scale Context Aggregation for Video-Based Person Re-Identification

Lei Wu[1], Canlong Zhang[1,2($\boxtimes$)], Zhixin Li[1,2], and Liaojie Hu[3]

[1] Key Lab of Education Blockchain and Intelligent Technology Ministry of Education, Guangxi Normal University, Guilin, China
[2] Guangxi Key Lab of Multi-source Information Mining and Security, Guangxi Normal University, Guilin, China
zcltyp@163.com
[3] The Experimental High School Attached to Beijing Normal University, Beijing, China

Abstract. For video-based person re-identification (Re-ID), effectively aggregating video features is the key to dealing with various complicated situations. Different from previous methods that first extracted spatial features and later aggregated temporal features, we propose a Multi-scale Context Aggregation (MSCA) method in this paper to simultaneously learn spatial-temporal features from videos. Specifically, we design an Attention-aided Feature Pyramid Network (AFPN), which can recurrently aggregate detail and semantic information of multi-scale feature maps from the CNN backbone. To enable the aggregation to focus on more salient regions in the video, we embed a particular Spatial-Channel Attention module (SCA) into each layer of the pyramid. To further enhance the feature representations with temporal information while extracting the spatial features, we design a Temporal Enhancement module (TEM), which can plug into each layer of the backbone network in a plug-and-play manner. Comprehensive experiments on three standard video-based person Re-ID benchmarks demonstrate that our method is competitive with most state-of-the-art methods.

Keywords: Video-based Person Re-identification · Multi-scale Feature Aggregation · Feature Pyramid

1 Introduction

Person Re-Identification (Re-ID) aims to retrieve pedestrian targets with the same identity from multiple non-overlapping cameras, which has a high practical application value in society and industry. Compared with the conventional image-based person Re-ID, video-based person Re-ID can obtain more pedestrian information (e.g., action information and viewpoint information), thus alleviating the negative influence of the common occlusion situation in person Re-ID. Therefore, video-based person Re-ID has begun to receive academic attention and develop rapidly.

B. Luo et al. (Eds.): ICONIP 2023, CCIS 1968, pp. 98–109, 2024.
https://doi.org/10.1007/978-981-99-8181-6_8

Currently, many person Re-ID methods [10,13,17,25] take ResNet [3] as their backbone network to extract features, which can effectively avoid the gradient disappearance and explosion of deep neural networks. However, limited by the size of the receptive field and the pooling operation, some image information is inevitably lost during feature learning. Besides, ResNet focuses more on the local region in the image and lacks modeling the correlation between human parts. These weaknesses limit the person Re-ID ability. To alleviate this problem, the attention mechanism [11,20] has begun to be widely used in video-based person Re-ID and improved the model performance, demonstrating its powerful representation ability by discovering salient regions in images.

The higher-level features of ResNet are abundant in semantics but lacking of details, while the lower-level features have more details but not enough semantics, so previous works [1,16,27] explore the effectiveness of hierarchical features in ResNet for video-based person Re-ID. In fact, features with different levels can complement each other through a specific aggregation. The deeper the layer is, the smaller the scale of its feature map is. Naturally, the feature maps of all layers are stacked together like a feature pyramid, which allows feature aggregation for video-based person Re-ID. However, effectively aggregating features of different layers through the pyramid structure is crucial for dealing with various complicated situations. PANet [14] proposed a bidirectional Feature Pyramid Network (FPN) consisting of a top-down as well as a bottom-up path to aggregate features at each layer of FPN. Similarly, Bi-FPN [9] proposed a nonlinear way to connect high-level and low-level. M2det [24] adopted a block of alternating joint U-shape module to fuse multilevel features. These methods have brought some improvements but require a large number of parameters and computations with complex structure.

In this paper, we innovatively embed the attention mechanism and Gated Recurrent Unit (GRU) into FPN to propose a learning method called Multi-scale Context Aggregation (MSCA), which can recurrently aggregate detail and semantic information of multi-scale feature maps from the backbone by Attention-aided FPN (AFPN), and enable the aggregation to focus on more salient regions in video. Different from previous methods that first extracted spatial features and later aggregated temporal features, our method can aggregate the spatio-temporal features simultaneously due to our proposed Temporal Enhancement Module (TEM). The TEM takes GRU as a primary component, it can be plugged into anywhere in a plug-and-play manner to learn complementary clues in temporal dimension and trained and converged easily due to its fewer parameters.

Overall, the main contributions of this paper are summarized as follows:

- We propose an Attention-aided Feature Pyramid Network (AFPN) to recurrently aggregate the hierarchical features from each layer of the backbone in spatial and channel dimensions; thus, the network can exploit salient clues and avoid some wrong clues during aggregation.

- We propose a Temporal Enhancement Module (TEM) that can be plugged into the backbone in a plug-and-play manner to merge the spatial and temporal information from pedestrian videos.
- Based on AFPN and TEM, we propose a Multi-scale Context Aggregation (MSCA) method for video-based person Re-ID. We conduct extensive experiments on three widely used benchmarks (i.e., MARS, iLIDS-VID and PRID2011), the results demonstrate that our MSCA method is competitive with other state-of-the-art methods, and the ablation studies confirm the effectiveness of AFPN and TEM.

2   Method

2.1   Overview

The overall architecture of our proposed method MSCA is illustrated in Fig. 1. We first adopt ResNet-50 [3] as our backbone and use multi-scale features from Res2, Res3, Res4, and Res5. Then, we plug TEM into the above four stages in a plug-and-play manner for learning temporal information from video features, which makes the features output from each layer contain both spatial and temporal information. We propagate the hierarchical features recurrently in AFPN for context aggregation, combining high-level semantic information with low-level detail information, and the SCA module would be plugged after aggregation to focus on more salient regions for improving the model performance. Finally, we use cross entropy loss and triplet loss as the objective function to optimize the model in the training stage, and features from two aggregate directions are concatenated for testing.

2.2   Attention-Aided Feature Pyramid Network

In video-based person Re-ID, multi-scale information aggregation has been used as one of the main methods to improve performance. The hierarchical features are characterized by insufficient semantic information of low-level features and insufficient detail information of high-level features due to the fact that the ResNet backbone increases the feature dimensions and decreases the feature resolutions across contiguous layers, thus Lin et al. [12] use this intrinsic property to reverse the information aggregation to form FPN based on the backbone by top-down and lateral connection. High-level features with rich semantic information are up-sampled by nearest neighbor interpolation and aggregated with the next layer of features output by lateral connection through element-wise addition, where the lateral connection adopts 1×1 convolution layer for reducing channel dimensions, and then recurrently aggregates swallow features, which take on more detail information and less semantic information. A 3×3 convolution layer is utilized on each aggregated features to generate the final feature map to reduce the aliasing effect of the upsampling operation. Besides, in order to make the aggregated features more discriminative, we introduce a spatial-channel

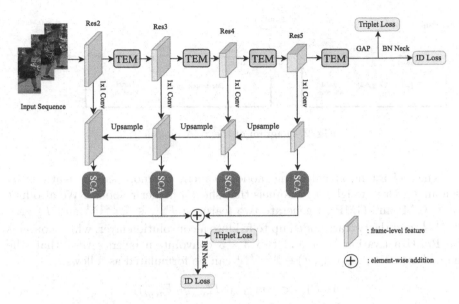

Fig. 1. Illustration of our proposed Multi-scale Context Aggregation. The input frames first fed to the ResNet-50 backbone. For the multi-scale feature maps from each residual block, we propose the AFPN to recurrently aggregate the features and focus on more salient regions by the SCA module. TEM is our proposed temporal enhancement module, which can be plugged into anywhere of the model for learning spatio-temporal information simultaneously. Finally, multiple losses are used to supervise the model in the training stage.

attention (SCA) module in FPN and call it attention-aided FPN (AFPN), the features of each layer after aggregation are fed to the SCA module, which consists of spatial and channel attention. As shown in Fig. 2, two kinds of attention cascade to calculate the spatial attention and channel attention [19], which can be described as follows:

$$F_s = A_s(F) \otimes F \tag{1}$$

$$F_{sc} = A_c(F_s) \otimes F \tag{2}$$

Where $\otimes$ denotes the element-wise multiplication, given the input feature tensor $F \in \mathbb{R}^{C \times H \times W}$, $A_s(\cdot)$ and $A_c(\cdot)$ denote the computation of the spatial and channel attention maps, $F_{sc} \in \mathbb{R}^{C \times H \times W}$ is the final spatial-aided and channel-aided features.

To obtain spatial attention, we adopt two pooling operations GAP and GMP to generate two features: $F_{avg}^s \in \mathbb{R}^{1 \times H \times W}$ and $F_{max}^s \in \mathbb{R}^{1 \times H \times W}$, then the two are concatenated to obtain a two-layer feature descriptor. The feature map is processed using a 7×7 convolution layer and a sigmoid layer to generate a spatial attention map $A_s(F) \in \mathbb{R}^{1 \times H \times W}$, it is calculated as follows:

$$A_s(F) = \sigma(conv_{7 \times 7}([F_{max}^s, F_{avg}^s])) \tag{3}$$

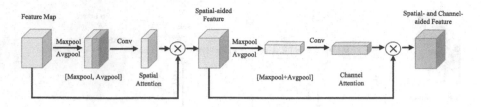

Fig. 2. Diagram of the SCA module

The channel attention helps the model to focus on more salient features by assigning greater weights to channels that show a higher response. We also first adopt GAP and GMP to generate two features: $F_{avg}^c \in \mathbb{R}^{C \times 1 \times 1}$ and $F_{max}^c \in \mathbb{R}^{C \times 1 \times 1}$, and the two are added up to fed into a convolution layer, which contains one ReLU activation layer and two 3×3 convolution layers. After that, the channel attention map $A_c(F) \in \mathbb{R}^{C \times 1 \times 1}$ can be formulated as follows:

$$A_c(F) = \sigma(conv_{3 \times 3}([F_{max}^s + F_{avg}^s])) \tag{4}$$

To eliminate the superimposed effect of distracted information in weighted process, the final video feature is then obtained by the SCA module instead of primitive 3×3 convolution layer.

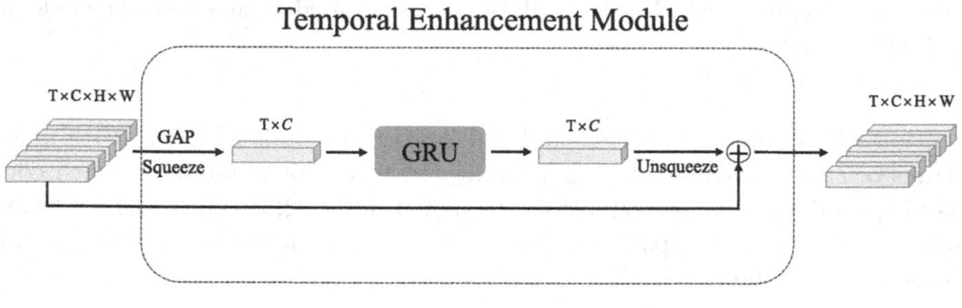

Temporal Enhancement Module

Fig. 3. Detailed structure of our TEM

2.3   Temporal Enhancement Module

Video sequences involve rich temporal information, we design a Temporal Enhancement Module (TEM) based on GRU, and it can be plugged anywhere in the model to capture the temporal clues in the video feature maps in a plug-and-play manner. As shown in Fig. 3, we use GAP to the input feature map $F \in \mathbb{R}^{T \times C \times H \times W}$ and squeeze all the spatial information to obtain temporal vector $F' \in \mathbb{R}^{T \times C}$. In the next, the temporal vector would be processed by GRU and recover the outputted feature map to the same size as the original input, then adopt a skip connection with input to obtain the final temporal enhanced feature $F'' \in \mathbb{R}^{T \times C \times H \times W}$, which incorporates the temporal information in the video and the video feature representation is more discriminative. The whole process of TEM is formulated as follows:

$$F' = Squeeze(GAP(F)) \tag{5}$$

$$F'' = Unsqueeze(GRU(F')) + F \tag{6}$$

2.4 Loss Function

We employ two kinds of losses to jointly supervise the training the model: cross entropy loss and hard triplet loss [4]. We calculated two losses L_{xent} and L_{htri} as follows:

$$L_{xent} = \sum_{i=1}^{N} -q_i \log(p_i) \tag{7}$$

$$L_{htri} = [d_{pos} - d_{neg} + m]_+ \tag{8}$$

Where p_i is the predicted logit of identity i and q_i is the ground-truth label in identification loss, d_{pos} and d_{neg} are respectively defined as the distance of positive sample pairs and negative sample pairs, $[d]_+ = max(\cdot, 0)$ and m is the distance margin, which is set to 0.3 in the training procedure.

In this paper, in order to better supervise the training of model, we adopt output from Res5 in the backbone and output from AFPN in the two directions of feature aggregation. Therefore, the total loss is the summation of the four losses:

$$L_{total} = L_{xent}^{b2t} + L_{htri}^{b2t} + L_{xent}^{t2b} + L_{htri}^{t2b} \tag{9}$$

3 Experiments

3.1 Datasets

MARS [26] dataset is the biggest video-based person re-identification benchmark with 1261 identities and around 20000 video sequences generated by DPM detector and GMMCP tracker. The dataset is captured by six cameras, each identity is captured by at least two cameras and has 13.2 sequences on average. There are 3248 distracter sequences in the dataset, it increases the difficulty of Re-ID.

iLIDS-VID [10] dataset is captured by two cameras in an airport hall. It contains 600 videos from 300 identities. This benchmark is very challenging due to pervasive background clutter, mutual occlusions, and lighting variations.

PRID2011 [5] dataset captures 385 identities by camera A and 749 identities by camera B, but only the first 200 people appear in both cameras.

Table 1. Comparison with State-of-the-Art methods On MARS, iLIDS-VID and PRID2011 datasets.

model	Ref	MARS				iLIDS-VID			PRID2011		
		Rank-1	Rank-5	Rank-20	mAP	Rank-1	Rank-5	Rank-20	Rank-1	Rank-5	Rank-20
ADFD [25]	CVPR2019	87.0	95.4	98.7	78.2	86.3	97.4	99.7	93.9	99.5	100
GLTR [10]	ICCV2019	87.0	95.8	98.2	78.5	86.0	98.0	-	95.5	100	-
MG-RAFA [23]	CVPR2020	88.8	97.0	98.5	85.9	88.6	98.0	99.7	95.9	99.7	100
TCLNet [7]	ECCV2020	88.8	-	-	83.0	86.6	-	-	-	-	-
MGH [21]	CVPR2020	90.0	96.7	98.5	85.8	85.6	97.1	99.5	94.8	99.3	100
SSN3D [8]	AAAI2021	90.1	96.6	98.0	86.2	88.9	97.3	98.8	-	-	-
BiCnet-TKS [6]	CVPR2021	90.2	-	-	-	86.0	-	-	-	-	-
GRL [15]	CVPR2021	91.0	96.7	98.4	84.8	90.4	98.3	99.8	96.2	99.7	100
CTL [13]	CVPR2021	91.4	96.8	98.5	86.7	89.7	97.0	100	-	-	-
GPNet [17]	NN2022	90.2	96.8	98.8	85.1	88.8	98.5	100	96.1	99.8	100
PiT [22]	TII2022	90.2	97.2	-	86.8	92.1	98.9	100	-	-	-
Ours		**91.8**	96.5	98.1	83.2	84.7	94.7	**100**	**96.6**	**100**	**100**

3.2 Evaluation Metrics

We employ the Cumulative Matching Characteristic curve (CMC) and the mean Average Precession (mAP) as evaluation critiria. CMC considers re-ID as a ranking problem and represents the accuracy of the person retrieval with each given query, mAP reflects the true ranking results while multiple ground-truth sequences exist. Conveniently, Rank-1, Rank-5 and Rank-20 are employed to represent the CMC curve.

3.3 Implementation Details

ResNet-50 pre-trained on ImageNet is employed as our backbone, and the input images are all resized to 256×128. We also utilize some commonly data augmentation strategies including random horizontal flipping, random erasing and random cropping. Specifically, in the training stage, we employ a restricted random sampling strategy to randomly sample $T = 8$ frames from each video as input. The ADAM optimizer with an initial learning rate of 5×10^{-5} and a weight decay of 5×10^{-4} for updating the parameters. We train the model for 200 epochs, and the learning rate is reduced by 0.1 per 50 epochs. All the experiments are conducted with Pytorch and a NVIDIA RTX 3090 GPU.

3.4 Comparison with State-of-the-Art Methods

In this section, we compare our proposed method with other state-of-the-art video-based person Re-ID methods on MARS, iLIDS-VID and PRID2011.

Results on MARS. From Table 1, compared with other state-of-the-art methods, the proposed MSCA achieves the best Rank-1 accuracy and competitive Rank-5 and Rank-20 accuracy. According to the results given in CTL baseline [13] and our baseline in Table 2, although both of them adopt ResNet-50 as the backbone, the mAP score of our baseline is 78.3% and that of CTL baseline is 82.7%, thus there is still a large margin in the final mAP result, even though our method performs well. The new and best works for video-based person Re-ID, CTL and PiT [22], the former adopts ResNet-50 as the backbone, but the latter is based on Transformer [2], all of which have utilized some complex modules such as key-points estimator, topology graph learning and hard-to-train transformer. In comparison, our method reaches a best Rank-1 accuracy with effective context aggregation. This demonstrates that our MSCA can aggregate more discriminative information, and the brief feature learning structure also has good generalization performance.

Results on iLIDS-VID and PRID2011. We also conduct several experiments on the two small datasets to demonstrate the advantages and possible flaws of our proposed method over the existing methods as shown in Table 1. We can observe that the result for iLIDS-VID is worse than other state-of-the-art methods, which causes this result is that the video sequences have a large variation of light and the serious occlusions on iLIDS-VID dataset. TEM only considers the temporal correlation but ignores the low-quality video sequences, which will introduce some additional irrelevant information, so as to decreases the model performance. For PRID2011 on the same scale as iLIDS-VID, our method achieves the best Rank-1 accuracy of 96.6%, and outperforms all previous approaches, confirming the superiority of our proposed method.

3.5   Ablation Study

To demonstrate the effectiveness of our proposed methods, we perform ablation studies on MARS dataset and use strong CNN backbone with ResNet-50 as our baseline. The experimental results are reported in Table 2 and Table 3.

Table 2. Ablation analysis of two components on MARS dataset

Model	AFPN	TEM	Rank-1	Rank-5	mAP
Baseline	✗	✗	88.2	95.4	78.3
	✓	✗	89.1	96.1	79.4
	✗	✓	90.7	96.5	82.1
Ours	✓	✓	**91.8**	**96.5**	**83.2**

MSCA: As shown in Table 2, the baseline contains only the ResNet-50 backbone and is supervised by L_{xent}^{b2t} and L_{htri}^{b2t}, the Rank-1 and mAP accuracy of the baseline are 88.2% and 78.3%, respectively. We find that using AFPN to aggregate multi-scale diverse features recurrently, the corresponding performance is

89.1% in Rank-1 and 79.4% in mAP, which is attributed to the complementary information of high-level semantic features and low-level detail features. Moreover, using TEM alone, we can achieve 90.7% in Rank-1 and 82.1% in mAP, the result shows the temporal enhancement module can complement the individual spatial features to learn more temporal information in videos. Eventually, by adding AFPN and TEM to the baseline, the model can further learn more discriminative features with the effective multi-scale context aggregation method.

AFPN: To explore the effectiveness of the SCA module and its position in the FPN, we conduct some experiments under the setting of utilizing the TEM. Table 3 shows the comparison of different plugging positions and the performance gap between our proposed AFPN and vanilla FPN, "w/o SCA" denotes vanilla FPN without SCA module, "1 × 1" denotes plugging the SCA module into the lateral connection and "Up-sample" denotes plugging the SCA module into the Up-sampling process. With the SCA module after propagation, we can find that the Rank-1 accuracy and mAP score are improved by 0.5% and 0.8%, respectively. Meanwhile, depending on the plugging position of the SCA module, the effects are also different. In Table 3, we can observe that plugging the SCA module into the lateral connection and into the downward propagation, they both have different performance degradation compared to plugging the SCA module after propagation.

Table 3. Ablation analysis of different plugging positions on MARS dataset

Position	Rank-1	Rank-5	mAP
w/o SCA	91.3	**96.7**	82.4
1 × 1	91.1	96.1	82.0
Up-sample	91.0	96.6	82.4
Ours	**91.8**	96.5	**83.2**

3.6   Visualization

As shown in Fig. 4, we report examples of different identities with Grad-CAM [18], which is commonly used in computer vision for a visual explanation. To verify the effectiveness of our proposed MSCA, we compare the Grad-CAM visualization of the baseline with our method, and the three example images are selected at intervals of at least 10 frames in three independent sequences of the MARS dataset. The frames contain various conditions such as motion and partial occlusions. As shown in Fig. 4(a), the features extracted from the ordinary frames by our proposed method can capture more information, including torso, legs and accessory that is discriminative to the target person. In Fig. 4(b), our method can focus on the area with more motion compared to the baseline if the subject has significant motion. As shown in Fig. 4(c), the features extracted by our proposed method can capture more body regions without additional bicycle information, which will enhance the representation of the target person. In general, our MSCA can effectively capture more spatio-temporal information and avoid some occlusions to improve the performance.

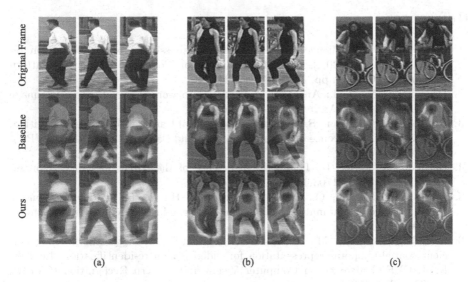

Fig. 4. Visualization of attention maps on different identities of the baseline and our proposed method. (a) Person takes up most of the image. (b) Person with significant motion. (c) Person is half occluded by bicycle

4 Conclusion

In this paper, we propose an innovative multi-scale context aggregation method for video-based person Re-ID. The proposed method can learn more video context information recurrently. AFPN can aggregate the semantic and detail information in multi-scale features, it integrates high-level semantic information into low-level detail information and uses the SCA module to aid the aggregated features to focus on salient regions. Furthermore, we propose a TEM to capture the temporal information among the video frames, and with its plug-and-play property, we can aggregate temporal features while extracting spatial features to enrich the final video feature representations, which is entirely different from previous works. The experimental results on three standard benchmarks demonstrate that our proposed method achieves competitive performance with most state-of-the-art methods.

Acknowledgements. This work is supported by National Natural Science Foundation of China (Nos. 62266009, 62276073, 61966004, 61962007), Guangxi Natural

Science Foundation (Nos. 2019GXNSFDA245018, 2018GXNSFDA281009, 2018GXNS FDA294001), Guangxi Collaborative Innovation Center of Multi-source Information Integration and Intelligent Processing, Innovation Project of Guangxi Graduate Education (YCSW2023187), and Guangxi "Bagui Scholar" Teams for Innovation and Research Project.

References

1. Chen, X., et al.: Salience-guided cascaded suppression network for person re-identification. In: 2020 IEEE/CVF Conference on Computer Vision and Pattern Recognition (CVPR), pp. 3297–3307 (2020)
2. Dosovitskiy, A., et al.: An image is worth 16x16 words: Transformers for image recognition at scale. ArXiv:2010.11929 (2020)
3. He, K., Zhang, X., Ren, S., Sun, J.: Deep residual learning for image recognition. In: 2016 IEEE Conference on Computer Vision and Pattern Recognition (CVPR), pp. 770–778 (2015)
4. Hermans, A., Beyer, L., Leibe, B.: In defense of the triplet loss for person re-identification. arXiv:1703.07737 (2017)
5. Hirzer, M., Beleznai, C., Roth, P.M., Bischof, H.: Person re-identification by descriptive and discriminative classification. In: Scandinavian Conference on Image Analysis (2011)
6. Hou, R., Chang, H., Ma, B., Huang, R., Shan, S.: BiCnet-TKS: learning efficient spatial-temporal representation for video person re-identification. In: 2021 IEEE/CVF Conference on Computer Vision and Pattern Recognition (CVPR), pp. 2014–2023 (2021)
7. Hou, R., Chang, H., Ma, B., Shan, S., Chen, X.: Temporal complementary learning for video person re-identification. In: European Conference on Computer Vision (2020)
8. Jiang, X., Qiao, Y., Yan, J., Li, Q., Zheng, W., Chen, D.: SSN3D: self-separated network to align parts for 3d convolution in video person re-identification. In: AAAI Conference on Artificial Intelligence (2021)
9. Kong, T., Sun, F., bing Huang, W., Liu, H.: Deep feature pyramid reconfiguration for object detection. arxiv:1808.07993 (2018)
10. Li, J., Wang, J., Tian, Q., Gao, W., Zhang, S.: Global-local temporal representations for video person re-identification. In: 2019 IEEE/CVF International Conference on Computer Vision (ICCV), pp. 3957–3966 (2019)
11. Li, S., Bak, S., Carr, P., Wang, X.: Diversity regularized spatiotemporal attention for video-based person re-identification. In: 2018 IEEE/CVF Conference on Computer Vision and Pattern Recognition, pp. 369–378 (2018)
12. Lin, T.Y., Dollár, P., Girshick, R.B., He, K., Hariharan, B., Belongie, S.J.: Feature pyramid networks for object detection. In: 2017 IEEE Conference on Computer Vision and Pattern Recognition (CVPR) pp. 936–944 (2016)
13. Liu, J., Zha, Z., Wu, W., Zheng, K., Sun, Q.: Spatial-temporal correlation and topology learning for person re-identification in videos. In: 2021 IEEE/CVF Conference on Computer Vision and Pattern Recognition (CVPR), pp. 4368–4377 (2021)
14. Liu, S., Qi, L., Qin, H., Shi, J., Jia, J.: Path aggregation network for instance segmentation. In: 2018 IEEE/CVF Conference on Computer Vision and Pattern Recognition, pp. 8759–8768 (2018)

15. Liu, X., Zhang, P., Yu, C., Lu, H., Yang, X.: Watching you: global-guided reciprocal learning for video-based person re-identification. In: 2021 IEEE/CVF Conference on Computer Vision and Pattern Recognition (CVPR), pp. 13329–13338 (2021)
16. Liu, Z., Zhang, L., Yang, Y.: Hierarchical bi-directional feature perception network for person re-identification. In: Proceedings of the 28th ACM International Conference on Multimedia (2020)
17. Pan, H., Chen, Y., He, Z.: Multi-granularity graph pooling for video-based person re-identification. Neural Netw.?: Off. J. Int. Neural Netw. Soc. **160**, 22–33 (2022)
18. Selvaraju, R.R., Das, A., Vedantam, R., Cogswell, M., Parikh, D., Batra, D.: Gradcam: visual explanations from deep networks via gradient-based localization. Int. J. Comput. Vision **128**, 336–359 (2016)
19. Woo, S., Park, J., Lee, J.Y., Kweon, I.S.: Cbam: convolutional block attention module. In: European Conference on Computer Vision (2018)
20. Wu, L., Wang, Y., Gao, J., Li, X.: Where-and-when to look: deep Siamese attention networks for video-based person re-identification. IEEE Trans. Multimed. **21**, 1412–1424 (2018)
21. Yan, Y., et al.: Learning multi-granular hypergraphs for video-based person re-identification. In: 2020 IEEE/CVF Conference on Computer Vision and Pattern Recognition (CVPR), pp. 2896–2905 (2020)
22. Zang, X., Li, G., Gao, W.: Multidirection and multiscale pyramid in transformer for video-based pedestrian retrieval. IEEE Trans. Industr. Inf. **18**(12), 8776–8785 (2022). https://doi.org/10.1109/TII.2022.3151766
23. Zhang, Z., Lan, C., Zeng, W., Chen, Z.: Multi-granularity reference-aided attentive feature aggregation for video-based person re-identification. In: 2020 IEEE/CVF Conference on Computer Vision and Pattern Recognition (CVPR), pp. 10404–10413 (2020). https://doi.org/10.1109/CVPR42600.2020.01042
24. Zhao, Q., et al.: M2det: a single-shot object detector based on multi-level feature pyramid network. In: AAAI Conference on Artificial Intelligence (2018)
25. Zhao, Y., Shen, X., Jin, Z., Lu, H., Hua, X.: Attribute-driven feature disentangling and temporal aggregation for video person re-identification. In: 2019 IEEE/CVF Conference on Computer Vision and Pattern Recognition (CVPR), pp. 4908–4917 (2019)
26. Zheng, L., et al.: Mars: a video benchmark for large-scale person re-identification. In: European Conference on Computer Vision (2016)
27. Zhou, K., Yang, Y., Cavallaro, A., Xiang, T.: Omni-scale feature learning for person re-identification. In: 2019 IEEE/CVF International Conference on Computer Vision (ICCV), pp. 3701–3711 (2019)

Enhancing LSTM and Fusing Articles of Law for Legal Text Summarization

Zhe Chen, Lin Ye, and Hongli Zhang[✉]

Harbin Institute of Technology, Harbin, China
{chen_zhe,hityelin,zhanghongli}@hit.edu.cn

Abstract. The growing number of public legal documents has led to an increased demand for automatic summarization. Considering the well-organized structure of legal documents, extractive methods can be an efficient method for text summarization. Generic text summarisation models extract based on textual semantic information, ignoring the important role of topic information and articles of law in legal text summarization. In addition, the LSTM model fails to capture global topic information and suffers from long-distance information loss when dealing with legal texts that belong to long texts. In this paper, we propose a method for summarization extraction in the legal domain, which is based on enhanced LSTM and aggregated legal article information. The enhanced LSTM is an improvement of the LSTM model by fusing text topic vectors and introducing slot storage units. Topic information is applied to interact with sentences. The slot memory unit is applied to model the long-range relationship between sentences. The enhanced LSTM helps to improve the feature extraction of legal texts. The articles of law after being encoded is applied to the sentence classification to improve the performance of the model for summary extraction. We conduct experiments on the Chinese legal text summarization dataset, the experimental results demonstrate that our proposed method outperforms the baseline methods.

Keywords: Extractive Legal summarization · TS-LSTM · Topic model · Slot memory unit

1 Introduction

Chinese legal texts serve as a public record of proceedings and outcomes of legal cases. They provide transparency by disclosing the rationale, grounds and verdicts of court decisions. These legal texts serve as crucial evidence for the courts in determining and assigning substantive rights and obligations to the parties involved. However, these texts often employ specialized domain knowledge extensively, resulting in lengthy and complex structures that may be unintelligible to the average reader. Moreover, China experienced a notable increase in the online publication of legal texts following the promulgation of the Provisions on the

© The Author(s), under exclusive license to Springer Nature Singapore Pte Ltd. 2024
B. Luo et al. (Eds.): ICONIP 2023, CCIS 1968, pp. 110–124, 2024.
https://doi.org/10.1007/978-981-99-8181-6_9

Online Publication of Legal Texts. Therefore, there is a pressing need to develop efficient methods for Chinese legal text summarization.

Currently, text summarization methods can be categorized into two main types: abstractive and extractive. Abstractive models, employing neural network methods, typically utilize the seq2seq approach. These models encode the entire document, extract textual features, and then generate a complete summary by generating words or phrases one by one. On the other hand, extractive models focus on learning sentence features and directly select important sentences from the original document based on the learned features and aggregate them to form a summary. Abstractive models offer flexibility and reduced redundancy, but the generated summaries often suffer from fluency issues and grammatical inconsistencies. In contrast, extractive models provide readable summaries with higher execution efficiency. Despite the success of neural network-based approaches in text summarization tasks, the task of modeling long-range sentence relationships for summarization remains challenging [12]. To address this issue, hierarchical networks have been commonly employed, treating a document as a sequence of sequences [13]. However, empirical studies have indicated that this approach does not significantly improve summarization performance when modeling inter-sentence relationships [7]. Furthermore, hierarchical methods are slow to train and prone to overfitting [21]. An additional crucial aspect highlighted in summarization studies is the limited utilization of global information in the inductive process, which is crucial for sentence selection. While pre-trained language models have proven to significantly enhance the ability to induce global information by effectively capturing contextual features, they still exhibit limitations in modeling document-level information, particularly in the case of lengthy documents. Latent topics have been utilized in topic modeling [14] to capture the extensive relationships within documents. Notably, the integration of deep learning networks with topic information [5] has demonstrated improved performance.

Few studies have been done on Chinese legal text summarization. There are several challenges in solving the Chinese legal text summarization under the existing text summarization method. Legal texts are usually long, which leads to a loss of information when neural network models extracts textual features and fails to obtain useful information. At present, the memory ability of the most commonly used neural network models in the text field, such as LSTM, is not strong, and their application effect in long text is not good, which also leads to the poor effect of the general text summarization model in the legal field. The semantic similarity between sentences in the text of legal cases is relatively high, and traditional methods calculate the relevance and redundancy of sentences in the text by applying the semantic similarity between sentences when extracting summaries from the text, but these models do not understand the legal domain, and therefore may not be able to learn the deeper legally relevant logical relationships underneath the surface semantics. Intuitively, articles of law can provide additional knowledge for understanding legal cases. The text of a legal case is a long document containing multiple sentences describing the facts of the case and the outcome of the judgement. Some of these sentences are important and should be included in the summary, others are not. Articles of law usually correspond to the most critical information in the case, so incorporating legal articles into the legal text summary extraction algorithm can help improve the performance of the model. In this paper, we propose an approach for extractive summarization of Chinese legal texts. The approach is based on enhanced LSTM (TS-LSTM) and fusing law articles information. The TS-LSTM is based on topic information and slot memory units, which models the relationship between text and topic information, enhances the modeling of long-range sentence relations through slot memory units. Fusing law articles information serves to establish relationships between sentences and legal knowledge information. Position information has been applied in previous text summarization studies, and the proposed approach also encode the position of the text and apply the position encoded information to a sentence classifier. The TS-LSTM method learns richer and deeper semantic feature information through a multi-layer network structure, and the sentence information from the lower layer is passed to the upper layer network. The sentences in the text are encoded with a pre-trained language model. Topic information is used to interact with the input sentence to generate new sentence feature. A slot memory unit is applied to store the feature information $h_1, h_2, ..., h_{t-1}$ before step t and interact with the newly generated information by other structures of the network to generate new feature h_t. The TS-LSTM network consists of a multi-layer structure, with the output h_{l-1} of the $l-1$-th layer as the input to the l-th layer. Law articles information and position encoded information are applied in sentence classifier to improve the ability of the model to select sentences.

Our proposed method has the following advantages: 1) By applying the topic information, the sentence representation can be enriched by the topic information, which can enhance the proposed model extract rich semantic content from the whole text. 2) The use of solt memory unit can alleviate the information loss problem of LSTM for long texts and model long-range inter-sentence relationships for summarization. 3) By enhancing LSTM, the model can effectively extract richer and deeper sentence information. 4) Articles of law is applied in the proposed model, which can help to increase the probability that sentences related to the description of the legal text are included in the summary set and improve the summary extraction capability of the model. We evaluate our model on a Chinese legal text dataset. Experimental results demonstrate the performance of our proposed method exceeds that of the baseline models.

2   Related Work

Topic Modeling for Summarization. Topic modeling is an effective method for capturing document features. [20] introduced a methodology for constructing a document graph that encompasses nodes representing words, sentences, and topics and proposed to train the graph using Markov chain techniques. [7] leveraged topic clusters within the document to generate sub-documents focusing on specific topics. They subsequently applied TextRank, a graph-based ranking algorithm, to generate final summaries from these sub-documents. [19] incorporated both topic-level and word-level attention mechanisms into Convolutional Sequence to Sequence networks for the purpose of abstractive summarization. [25] introduced a method for summarizing multiple documents through the identification and extraction of cross-document subtopics. [16] presented a novel abstractive model that relies solely on convolutional neural networks and is conditioned on the topics of the article. [4] introduced an extractive summarization model based on graph neural networks that incorporates a joint neural topic model to find the latent topics and provides document-level features to facilitate sentence selection.

Extractive Methods for Summarization. Neural networks have made notable progress in the field of extractive summarization. Prior research has primarily treated extractive summarization as a sequence classification problem [6,15,23] or a task of ranking sentences [16]. These strategies capitalize on the capabilities of neural networks to deliver remarkable outcomes in summarization tasks. The utilization of pre-trained language models has significantly enhanced the performance of summarization systems [12,22,24]. Reinforcement learning methods have also been applied to improve the performance of text summarization models [1,17].

3   Proposed Method

This section describes our proposed method. Given a document that consists of sentences, the objective of the proposed method is to learn a sequence of binary labels, where $0, 1$ represents whether the the sentence should be included in summary. Figure 1(a) presents the overview architecture. Our proposed method. generally consists of five parts, which are the 1) Sentence encoder, 2) Topic Model, 3) Position encoder, 4) TS-LSTM network, 5)Law article processed and 6) Sentence classifier. Given the input document, the document encoder learns contextual representations of each sentence with a pre-trained BERT. The purpose of position encoder is to learn the feature of each position. The Neural Topic Model aims to learn the document topic distribution representations. TS-LSTM network applies a multi-layer network structure, where the output of the lower layer is used as the input of the higher layer to generate new feature outputs by combining the topic information of the documents and slot network. After TS-LSTM network, the sentence representations are further sent to a sentence classifier that computes the final labels after fusing the sentence content importance, topic relevance, law articles relevance, redundancy and sentence position important feature.

3.1   Sentence Encoder

Let $\{d_1, d_2, ..., d_m\}$ denote a set of input documents. Each document d_i consists of a collection of sentences $\{s_1, s_2, ..., , s_i, ..., s_n\}$, where a given sentence s_i is formed by several tokens $\{w_{i0}, w_{i1}, ..., w_{iN}\}$. Similar to previous works [4, 22], we employ the BERT model that is a bidirectional transformer encoder pre-trained with a large corpus to encode the input token and generates hidden representations of sentences. Specifically, we first split the text into words, removing stop words and punctuation marks and then recomposing the remaining words into new sentences in the order in which they appeared in the original text. After that, we insert [CLS] and [SEP] tokens at the beginning and end of each sentence, respectively. Then, we put all tokens into BERT encoder and obtain the sentence embedding.

$$h_i = BERT\ ([CLS], w_{i,0}, w_{i,1}, ..., [SEP]) \tag{1}$$

where w_{ij} represents the j-th word of the i-th sentence. h_i represents the sentence embedding of the i-th sentence. After the BERT encoding, we obtain the sentence embedding representations $\{h_1, h_2, ..., h_i, ..., h_N\}$ of document. N is the number of sentences in document.

3.2 Topic Model

We apply the method used in [4] to generate topic embedding of document. NTM (Neural Topic Model) is developed within the framework of Variational Autoencoder [10]. It leverages an encoding-decoding process to learn the latent topic representation.

In the encoder, we employ $\mu = f_\mu(x)$ and $log\sigma = f_\sigma(x)$ to represent the prior parameters used for parameterizing the topic distribution within the decoder networks. The functions f_μ and f_σ are implemented as linear transformations with ReLU activation.

The decoder can be conceptualized as a three-step process for generating the document. Firstly, we utilize Gaussian softmax [13] to sample the topic distribution. Secondly, we learn the probability distribution of predicted words, denoted as $p_w \in R^{|V|}$, through the equation $p_w = softmax(W_\emptyset\theta)$, where $\theta \in R^k$. Here, $W_\emptyset \in R^{|V|*k}$ can be seen as analogous to the topic-word distribution matrix found in LDA-style topic models, and $W_\emptyset^{i,j}$ represents the relevance between the i-th word and the j-th topic. Finally, we generate each word by sampling from p_w in order to reconstruct the input bag-of-words representation x_{bow}. For more detailed information, we refer interested readers to [13].

Given that the intermediate parameters $W_\emptyset$ and θ contain encoded topical information, we leverage them to construct topic representations using the following equation:

$$H^T = f_\emptyset(w_\emptyset^T) \tag{2}$$

$$T_d = \sum_{1 \le i \le k} \theta^{(i)} H_T^{(i)} \tag{3}$$

Here, H_T is denoted a group of topic representations, with a predetermined dimension of d_t, $f_\emptyset$ is a linear transformation with ReLU activation. The value of T_d is obtained as the weighted sum of each topic representation in H_T, aggregating the relevant information.

3.3   Position Encoder

Since position information of sentences is more important in text summarization, and the first or last sentences of a text can usually serve as a summary of the text, position information is often utilized by text summarization models. The goal of this layer is to encode the position information of sentences so that the model can output the probability that a sentence belongs to the summary using sentence position features without knowing the semantic information.

The position feature embedding $P_i^e \in R^d$ of the sentence is randomly initialized. d is the dimension. We used the following formula to calculate the position score of the sentences:

$$score_i^p = WP_i^e \tag{4}$$

Where $W \in R^k$ is the matrix of parameters to be learned.

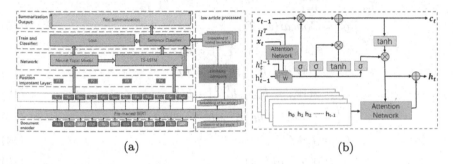

$$\text{(a)} \qquad\qquad\qquad\qquad \text{(b)}$$

Fig. 1. (a) is the overall architecture of our proposed method. (b) is the calculation diagram of the TS-LSTM at step t in l-th layer.

3.4   TS-LSTM Network

This module uses a stacked multi-layer network to compute the features of the text. The calculation process of TS-LSTM at step t of l-th layer is shown in Fig. 1. The network takes as input the sentence embedding x_t of step t, the topic embedding H^T, the output h_t^{l-1} of the previous $l-1$-th layer, the output h_{t-1}^l of the step $t-1$ in l-th layer, and the outputs $h_{1:t-1}^l$ of steps 1 to $t-1$ in l-th layer. We apply the following equation to summarize the structure of the module. $\varPhi$ denotes the calculation of l-th layer at step t.

$$h_t^l = \varPhi(h_t^{l-1}, x_t, H^T) \tag{5}$$

Here, h_t^{l-1} denotes the output of step t at $l-1$ layer, x_t denotes the sentence embedding at step t, and H^T denotes the topic embedding.

Equations [6–19] represent the detailed process of $\varPhi$. In order to fuse the topic vector with the sentence vector and enhance the expressiveness of the sentence vector, we use the attention mechanism to fuse the topic vector with the sentence vector, which is calculated as Equation [6–8]. The calculation process of multi-layer network from lower to higher layers is shown in Equation [9–15]. The Slot network also applies an attention mechanism to fuse the information from the current step with the hidden layer feature information generated from the previous steps to produce a new feature representation. The calculation formula is shown in Equation [16–18], Equation [19] represents the final output h_t of step t in l-th layer.

$$\alpha_{ti} = W_t[x_t, H_i^T] + b_t, i \in \{1, 2, ..., k\} \tag{6}$$

$$\alpha_{ti} = \frac{e^{\alpha_{ti}}}{\sum_{j=1}^{k} e^{\alpha_{tj}}} \tag{7}$$

$$T_t = \sum_{j=1}^{k} \alpha_{tj} H_i^T \tag{8}$$

$$g_t = Concat(x_t, T_t) \tag{9}$$

$$i_t^l = \sigma(W_i[g_t, h_{t-1}^l, h_t^{l-1}] + b_i) \tag{10}$$

$$f_t^l = \sigma(W_f[g_t, h_{t-1}^l, h_t^{l-1}] + b_f) \tag{11}$$

$$o_t^l = \sigma(W_o[g_t, h_{t-1}^l, h_t^{l-1}] + b_o) \tag{12}$$

$$\widetilde{C}_t^l = tanh(W_c[g_t, h_{t-1}^l, h_t^{l-1}] + b_c) \tag{13}$$

$$C_t^l = f_t^l * C_{t-1}^l + i_t^l * \widetilde{C}_t^l \tag{14}$$

$$h_t^l = o_t^l * tanh(C_t^l) \tag{15}$$

$$\beta_{tu}^l = W_s[h_t^l, h_u^l] + b_s, u \in \{1, 2, ..., t-1\} \tag{16}$$

$$\beta_{tu}^l = \frac{e^{\beta_{tu}^l}}{\sum_{i=1}^{t-1} e^{\beta_{ti}^l}} \tag{17}$$

$$Slot_t^l = \sum_{u=1}^{t-1} h_u^l \cdot \beta_{tu}^l \tag{18}$$

$$h_t^l = h_t^l + Slot_t^l \tag{19}$$

Where, $W_g, W_i, W_f, W_o, W_s, W_t$ are the matrix of parameters to be learned, $b_i, b_f, b_o, b_c, b_s, b_t$ are the bias parameters to be learned.

3.5   Law Article Processed

Real legal cases are decided according to the relevant applicable legal provisions. These applicable legal provisions are taken from law books, for example, the Criminal Law of the People's Republic of China. Some legal case texts give the cited law at the end of the text, and some do not mention the cited law. Therefore, we can collect the articles of the law from law books. According to the type of the case text first roughly select the related articles of law, for example, the case text belongs to the civil case text, then we choose the articles of civil law, forming a collection of N law articles $L = (l_1, l_2, ..., l_N)$. The doc2vector method [11] is used to encode the collection of N law articles to generate document document vectors $L^e = (l_1^e, l_2^e, ..., l_N^e)$ of law articles. The doc2vector method also is used to encode legal case texts to generate document document embedding e_d of the legal case texts. We apply the cosine similarity algorithm to calculate the similarity between the legal case texts to be subjected to summary extraction and each article of law, and sort the article of law according to the similarity from the highest to the lowest, we select the top K articles of law from as the relevant articles of law for that legal text. Assuming that the document vector is e_d and the embeddings of articles of law are denoted as $L^e = (l_1^e, l_2^e, ..., l_K^e)$, the method to compute the embedding of the laws related to this document is as follows:

$$\beta_i^l = W_a[e_d, l_i^e] + b_a, u \in \{1, 2, ..., K\} \tag{20}$$

$$\beta_i^l = \frac{e^{\beta_i}}{\sum_{j=1}^{K} e^{\beta_j}} \tag{21}$$

$$e(L) = \sum_{i=1}^{10} \beta_i^l \cdot l_i^e \tag{22}$$

Here W_a and b_a are the parameters to be learnt.

3.6 Sentence Classifier

For the purpose of classification, we take into account various factors such as the relevance of content, relevance to the topic, relevance to the law article, redundancy, and position score. Our approach for classification is based on the method described in [15]. Each sentence is sequentially reviewed, and a logistic layer is utilized to make a binary decision on whether the sentence should be included in the summary. The process is illustrated below:

$$P(y_i = 1|h_i, T_d, s_i) = \sigma(W_c h_i + h_i W_d T_d + h_i W_L e(L) - h_i^T W_r tanh(s_i) + score_i^p) \tag{23}$$

$$s_i = \sum_{j=1}^{i-1} h_i P(y_j = 1|h_j, T_d, s_j) \tag{24}$$

3.7 Training

Since the NTM in the model was trained on our own collected dataset of legal cases and the rest of the model was trained on the summary dataset, We train the NTM and sentence classifier separately. We first train the NTM model and the trained NTM is applied to obtain the topic information of the summarization text. We do not update the parameters of the NTM model while training the rest of the model. In the case of the NTM, the objective function is formulated as the negative evidence lower bound, as demonstrated below [4]:

$$\ell_{NTM} = D_{k\ell}(p(z)||q(z|x)) - E_{q(Z|x)}[p(x|z)] \tag{25}$$

Here, the first term represents the Kullback-Leibler divergence loss, while the second term corresponds to the reconstruction loss. The functions $q(z|x)$ and $p(x|z)$ denote the encoder and decoder networks, respectively.

The binary cross-entropy loss is applied to train and update parameters of the rest of our proposed method. The binary cross-entropy loss can be expressed as follows:

$$\ell_{sc} = \sum_{i=1}^{n}(y_i log(\hat{y}_i) + (1 - y_i)log(1 - \hat{y}_i)) \tag{26}$$

4   Experiment

4.1   Dataset

In this study, we use two legal text datasets, one is legal judgment document crawled from "China Judgements Online", and our topic model is trained on this dataset. Approximately 550,000 legal judgements covering more than 50 allegations were crawled, of which more than 20,000 allegations were distributed as shown in Table 1. The other one is the SFZY dataset, which is a Chinese legal text summarization dataset. The SFZY dataset is a long text dataset for summarization, which is more challenging due to the information loss problem in feature extraction for summarization. The SFZY dataset is derived from the CAIL2020[1]. The dataset contains 13.5K Chinese judgments with a label (0 or 1) for each sentence and a manually summarized summary for each text. We randomly divided the SFZY dataset into training dataset, validation dataset and testing dataset as the ratio of 8:1:1. We perform all the experiments on the partitioned datasets to validate the performance of all the baseline models and our proposed method (Table 2).

4.2   Baselines

We selected the following baseline methods to carry out a comparative of the results achieved with our proposed method. **NN-SE** [3] comprises a hierarchical document encoder and an attention-based content extractor. These components work together to score and select sentences in order to generate extractive summaries. **SummaRuNNer** [15] is an extractive summarization model that utilizes a simple RNN-based sequence classifier. It introduces a novel training mechanism that incorporates abstractive summaries to train the network. **REFRESH** [16] is an extractive summarization technique that utilizes a learning-to-rank framework that takes a global optimization approach by maximizing the ROUGE evaluation metric. **BERTSUM** [12] embeds multiple partitioning tokens into the document to acquire individual sentence representations.

---

[1] http://cail.cipsc.org.cn.

Table 1. Distribution of partial charges for legal judgment document data.

Charge name	Number	Charge name	Number
Larceny	29633	Dangerous driving crime	28197
Intentional injury crime	27869	Traffic accident crime	26795
Fraud	26982	Providing venues for drug users	26810
Defiance and affray crime	25976	Robbery	25617
Casino crime	25109	Disrupting public service crime	25361
Illegally holding drugs crime	24769	Gambling crime	24587
Intentionally destroying possessions	24689	Contract fraud crime	25109
Corruption crime	24766	Official embezzlement crime	25066
Crime of unlawful intrusion intoresidence	24565	Negligence causing death crime	25203

Table 2. The number of sentences of the SFZY dataset.

Text Type	Maximum	Minimum	Average
Document	496	21	57.75
Summarization	69	4	12.69

It holds the distinction of being the inaugural extractive summarization model based on BERT. **HiBERT** [24] transforms the BERT architecture into a hierarchical structure and develops an unsupervised pre-training method tailored for this modified structure. **NeuSum** [28] employs a document encoder to encode each sentence in the document and select the sentences based on their representations, while also considering the previously generated partial summary. **AREDSUM** [2] is an iterative ranking method for extractive summarization. It incorporates redundancy-aware scoring of sentences to be included in a summary, either in conjunction with their salience information or as an additional step. **MATCHSUM** [27] introduces an innovative summary-level framework that focuses on matching the source document with candidate summaries in the semantic space. **DISCOBERT** [22] is an extractive model that introduces a technique to integrate latent topic into a document graph and subsequently refines the sentence representations through a graph encoder. **DeepSumm** [8] leverages the latent information within the document, estimated through topic vectors and sequence networks, to enhance the quality and precision of the generated summary. **HiStruct+** [18] proposed a fresh method for incorporating hierarchical structure information into an extractive summarization model.

4.3   Implementation Details

We utilize the bert-base-ms[2] [26] as the pre-trained BERT version for our document encoder. In order to determine the number of topics, the number of topics we set at settings from 10 to 50 respectively. we set the number of topics to 30

---

[2] https://thunlp.oss-cn-qingdao.aliyuncs.com/bert/ms.zip.

($k = 30$) based on the experimental results of the model on the validation set. The number of sentences we extracted from the text was set to 3, 5, 6, 10 and 15, respectively and the final number determined by the results of the experiment on validation set. The dimension of topic vector is set to 512. For the our proposed method, we employ 3 layers with a hidden size of 256. In the experiment, we set the number of selected articles of law to 10 ($K = 10$). The initial learning rate is set to 2e-3, and the dropout rate is 0.1. The weight and bias parameters in model are initialized randomly and optimized using the Adam optimizer [9].

Table 3. Comparative results on Chinese legal text summarization dataset.

Methods	ROUGE-1	ROUGE-2	ROUGE-L
NN-SE	52.11	30.72	45.65
SummaRuNNer	59.79	39.53	51.55
REFRESH	57.01	35.87	49.21
BERTSUM	59.66	38.36	50.27
HiBERT	58.59	37.97	49.72
NeuSum	58.87	38.13	50.22
AREDSUM	60.19	38.38	51.91
MATCHSUM	60.33	38.92	52.39
DISCOBERT	59.87	38.15	51.55
DeepSumm	61.39	39.75	53.36
HiStruct+	60.62	38.98	52.15
Our method	63.77	40.92	55.31

4.4   Results and Ablation Experiments

On SFZY dataset, our method obtained the highest ROUGE-1 score and ROUGE-2 and ROUGE-L scores comparable to the baseline models as it can be seen from Table 3. Our method achieved comparable or better ROUGE-1 score of 63.77, ROUGE-2 score of 40.92 and ROUGE-L score of 55.31. Our model improves by 3.15, 2.38, 3.9, 3.44, 3.58 on Rouge-1, by 1.94, 1.17, 2.77, 2.0, 2.54 on Rouge-2, and by 3.16, 1.95, 3.76, 2.92, 3.4 on Rouge-L compared to HiStruct+, DeepSumm, DISCOBERT, MATCHSUM, AREDSUM. Our method also exceed other baseline methods such as REFRESH and RNES using reinforcement learning, NeuSum and HSSAS based on sequence networks. Figure 2 shows the values of ROUGE-1,ROUGE-2 and ROUGE-L for different number of sentences extracted from the text to generate the summary.

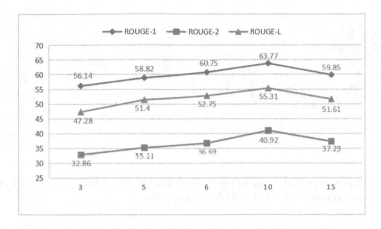

Fig. 2. Experimental results on the number of sentences in the summary.

To verify the effect of topic vector, slot network and law articles on the overall performance of the model, we conduct three ablation experiments. One is that we remove the topic vector from the model. One is that is we remove the slot network from the model. The other is that we remove the law articles vector from the model by modifying Eq. 23 to Eq. 27.

$$P(y_i = 1|h_i, T_d, s_i) = \sigma(W_c h_i + h_i W_d T_d - h_i^T W_r tanh(s_i) + score_i^p) \qquad (27)$$

The experimental results are shown in Fig. 3(a). The **w/o topic vector** denotes the result of removing topic vector from the model. The **w/o Slot** denotes the result of removing slot network from the model. The **w/o law acticles** denotes the result of removing the law articles vector from the model. It can also be concluded from Fig. 3(a) that topic vector, slot network and the law articles vector can improve the performance of the model for text summarization. The topic vector can provide global semantic information to the text summarization model. The slot network can mitigate the information loss of the model in processing long text sequences for text summarization. The law articles vector can provide structural information to the text summarization model.

To verify the effect of our proposed compared with LSTM and multi-layer network, we also conduct two ablation experiments. One is **LSTM** that is a single layer network without topic vector and slot network, the other is **LSTM+Topic+Slot** that is a single layer network with topic vector and slot network. The results of these model compared with our model are shown in Fig. 3(b). As seen in Fig. 3(b), our proposed method outperforms LSTM, indicating that our proposed method can enhance the performance of LSTM networks in legal text summarization tasks and our proposed method outperforms LSTM+Topic+Slot, indicating that our proposed method has the advantage of using a multi-layer network that uses low-level features to generate high-level features.

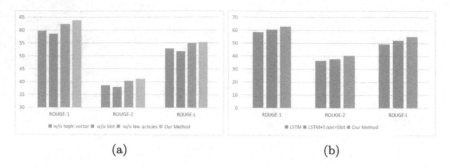

(a)   (b)

Fig. 3. (a) is experimental results of w/o topic vector, w/o Slot, w/o law acticles. (b) is experimental results of LSTM, LSTM+Topic+Slot.

5   Conclusion

In this paper, we propose a method by enhancing LSTM with topic vector and slot unit and fusing law acticles for the extractive summarization of Chinese legal documents. The proposed method captures the structural and semantic features of documents by using the fusion of topic vector, sentence encoding and position information, law acticles information and the slot network is used to mitigate the information loss of LSTM models in processing long text sequences. We apply a pre-trained language model to encode the sentence embeddings of document and law acticles and neural topic model to obtain the topic vector of the document. our proposed method utilizes a multi-layer network to extract rich text feature information. Each layer of the network takes as input the sentence vector, topic vector and the features generated by the lower layer of the network and fuses them into new feature, while capturing long-range textual information using the slot unit. Finally, the sentence classifier is used to calculate content relevance, topic relevance, law acticles relevance, novelty and position scores to extract the sentences and generate summarization. We conducted experiments on the Chinese legal dataset, and the experimental results show that our proposed method outperforms the baseline model.

Acknowledgements. This work is supported by the National Natural Science Foundation of China (NSFC) under Grant 61872111.

References

1. Arumae, K., Liu, F.: Reinforced extractive summarization with question-focused rewards. In: Proceedings of ACL 2018, Melbourne, Australia, July 15-20, 2018, Student Research Workshop (2018)
2. Bi, K., Jha, R., Croft, W.B., Celikyilmaz, A.: AREDSUM: adaptive redundancy-aware iterative sentence ranking for extractive document summarization. In: Proceedings of the 16th Conference of the European Chapter of the Association for Computational Linguistics: Main Volume, EACL, pp. 281–291. Association for Computational Linguistics (2021)

3. Cheng, J., Lapata, M.: Neural summarization by extracting sentences and words. In: Proceedings of the 54th Annual Meeting of the Association for Computational Linguistics, ACL. The Association for Computer Linguistics (2016)
4. Cui, P., Hu, L., Liu, Y.: Enhancing extractive text summarization with topic-aware graph neural networks. In: Proceedings of the 28th International Conference on Computational Linguistics, pp. 5360–5371 (2020)
5. Dieng, A.B., Wang, C., Gao, J., Paisley, J.W.: TOPICRNN: a recurrent neural network with long-range semantic dependency. In: 5th International Conference on Learning Representations, ICLR. OpenReview.net (2017)
6. Dong, Y., Shen, Y., Crawford, E., van Hoof, H., Cheung, J.C.K.: Banditsum: extractive summarization as a contextual bandit. In: Proceedings of the 2018 Conference on Empirical Methods in Natural Language Processing, pp. 3739–3748. Association for Computational Linguistics (2018)
7. Issam, K.A.R., Patel, S., N, S.C.: Topic modeling based extractive text summarization. arxiv:2106.15313 (2021)
8. Joshi, A., Fidalgo, E., Alegre, E., Fernández-Robles, L.: Deepsumm: exploiting topic models and sequence to sequence networks for extractive text summarization. Expert Syst. Appl. **211**, 118442 (2023)
9. Kingma, D.P., Ba, J.: Adam: A method for stochastic optimization. In: 3rd International Conference on Learning Representations, ICLR (2015)
10. Kingma, D.P., Welling, M.: Auto-encoding variational bayes. arXiv.org (2014)
11. Le, Q.V., Mikolov, T.: Distributed representations of sentences and documents. In: Proceedings of the 31th International Conference on Machine Learning, ICML. vol. 32, pp. 1188–1196 (2014)
12. Liu, Y., Lapata, M.: Text summarization with pretrained encoders. In: Proceedings of the 2019 Conference on Empirical Methods in Natural Language Processing and the 9th International Joint Conference on Natural Language Processing, EMNLP-IJCNLP (2019)
13. Miao, Y., Grefenstette, E., Blunsom, P.: Discovering discrete latent topics with neural variational inference. In: Proceedings of the 34th International Conference on Machine Learning, ICML. Proceedings of Machine Learning Research, vol. 70, pp. 2410–2419. PMLR (2017)
14. Mikolov, T., Zweig, G.: Context dependent recurrent neural network language model. In: 2012 IEEE Spoken Language Technology Workshop (SLT), pp. 234–239. IEEE (2012)
15. Nallapati, R., Zhai, F., Zhou, B.: Summarunner: a recurrent neural network based sequence model for extractive summarization of documents. In: Proceedings of the Thirty-First AAAI Conference on Artificial Intelligence, pp. 3075–3081. AAAI Press (2017)
16. Narayan, S., Cohen, S.B., Lapata, M.: Don't give me the details, just the summary! topic-aware convolutional neural networks for extreme summarization. In: Proceedings of the 2018 Conference on Empirical Methods in Natural Language Processing, pp. 1797–1807. Association for Computational Linguistics (2018)
17. Paulus, R., Xiong, C., Socher, R.: A deep reinforced model for abstractive summarization. In: 6th International Conference on Learning Representations, ICLR (2018)
18. Ruan, Q., Ostendorff, M., Rehm, G.: HiStruct+: improving extractive text summarization with hierarchical structure information. In: Findings of the Association for Computational Linguistics: ACL (2022)

19. Wang, L., Yao, J., Tao, Y., Zhong, L., Liu, W., Du, Q.: A reinforced topic-aware convolutional sequence-to-sequence model for abstractive text summarization. In: Proceedings of the Twenty-Seventh International Joint Conference on Artificial Intelligence, IJCAI, pp. 4453–4460. ijcai.org (2018)

20. Wei, Y.: Document summarization method based on heterogeneous graph. In: 9th International Conference on Fuzzy Systems and Knowledge Discovery, FSKD, pp. 1285–1289. IEEE (2012)

21. Xiao, W., Carenini, G.: Extractive summarization of long documents by combining global and local context. In: Proceedings of the 2019 Conference on Empirical Methods in Natural Language Processing and the 9th International Joint Conference on Natural Language Processing, EMNLP-IJCNLP (2019)

22. Xu, J., Gan, Z., Cheng, Y., Liu, J.: Discourse-aware neural extractive model for text summarization. arxiv1910.14142 (2019)

23. Zhang, X., Lapata, M., Wei, F., Zhou, M.: Neural latent extractive document summarization. In: Proceedings of the 2018 Conference on Empirical Methods in Natural Language Processing, pp. 779–784. Association for Computational Linguistics (2018)

24. Zhang, X., Wei, F., Zhou, M.: HIBERT: document level pre-training of hierarchical bidirectional transformers for document summarization. In: Proceedings of the 57th Conference of the Association for Computational Linguistics, ACL (2019)

25. Zheng, X., Sun, A., Li, J., Muthuswamy, K.: Subtopic-driven multi-document summarization. In: Proceedings of the 2019 Conference on Empirical Methods in Natural Language Processing and the 9th International Joint Conference on Natural Language Processing, EMNLP-IJCNLP, pp. 3151–3160 (2019)

26. Zhong, H., Zhang, Z., Liu, Z., Sun, M.: Open Chinese language pre-trained model zoo. Tech. rep. (2019). https://github.com/thunlp/openclap

27. Zhong, M., Liu, P., Chen, Y., Wang, D., Qiu, X., Huang, X.: Extractive summarization as text matching. In: Proceedings of the 58th Annual Meeting of the Association for Computational Linguistics, ACL. pp. 6197–6208. Association for Computational Linguistics (2020)

28. Zhou, Q., Yang, N., Wei, F., Huang, S., Zhou, M., Zhao, T.: Neural document summarization by jointly learning to score and select sentences. In: Proceedings of the 56th Annual Meeting of the Association for Computational Linguistics, ACL, pp. 654–663. Association for Computational Linguistics (2018)

Text Spotting of Electrical Diagram Based on Improved PP-OCRv3

Yuqian Zhao[1], Dongdong Zhang[1]($\boxtimes$), and Chengyu Sun[2]

[1] Department of Computer Science and Technology, Tongji University, Shanghai, China
{2233047,ddzhang}@tongji.edu.cn
[2] Shanghai Key Laboratory of Urban Renewal and Spatial Optimization Technology, Tongji University, Shanghai, China
cy.sun@tongji.edu.cn

Abstract. The text detection and recognition plays an important role in automatic management of electrical diagrams. However, the images of electrical diagrams often have high resolution, and the format of the text in them is also unique and densely distributed. These factors make the general-purpose text spotting models unable to detect and recognize the text effectively. In this paper, we propose a text spotting model based on improved PP-OCRv3 to achieve better performance on text spotting of electrical diagrams. Firstly, a region re-segmentation module based on pixel line clustering is designed to correct detection errors on irregularly shaped text containing vertical and horizontal characters. Secondly, an improved BiFPN module with channel attention and depthwise separable convolution is introduced during text feature extracting to improve the robustness of input images with different scales. Finally, a character re-identification module based on region extension and cutting is added during the text recognition to reduce the adverse effects of simple and dense character on the model. The experimental results show that our model has better performance than the state-of-the-art (SOTA) methods on the electrical diagrams data sets.

Keywords: Electrical diagrams · Optical character recognition · PP-OCRv3 · BiFPN

1 Introduction

It is urgent to automate the management and search of electrical diagrams. As the scale of electrical diagrams expands, traditional manual identification is not only inefficient but also prone to errors. The information in the electrical diagrams is complex. In order to realize the automation process, it is important to detect the text position and recognize the text content in the electrical diagrams.

In recent years, deep learning based text detection and recognition methods have been proposed. These methods hand over the classification features of

B. Luo et al. (Eds.): ICONIP 2023, CCIS 1968, pp. 125–136, 2024.
https://doi.org/10.1007/978-981-99-8181-6_10

electrical diagrams automatically to the networks, thereby improving the adaptability to complex scenarios of electrical diagrams. Laura Jamieson et al. [1] used EAST and LSTM to realize the detection and recognition of text in engineering diagrams. Li et al. [2] used YOLOv5 and PP-OCRv3 [3] to improve the accuracy on electrical cabinet wiring.

However, the forms of text labeling in the electrical diagrams are often diverse, with different shapes, directions, and sizes. The characters involved not only numbers and letters, but also Chinese characters. The accuracy is often poor when directly using the general-purpose OCR model to spotting the text in the electrical diagrams. Figure 1 shows the text detection results of electrical diagrams on PP-OCRv3, where four types of text are prone to errors and omissions in detection and recognition.

To address these limitations and achieve better position detection and content recognition of text in electrical diagrams, we propose a text spotting model based on PP-OCRv3. To improve the detection performance for irregularly shaped text containing both vertical and horizontal characters, a region re-segmentation module based on pixel line clustering is designed. During text feature extraction, we propose an improved BiFPN module with channel attention and depthwise separable convolution to improve the problem of mis-identification caused by text images with different scales. To correct the error of texts with single Chinese character and multi-line numbers, a character re-identification module based on region extension and cutting is introduced, which could improve performance of the model in electrical diagrams. To summarize, our main contributions are as follows:

- We design a region re-segmentation module based on pixel line clustering to solve the text position detection problem caused by irregularly shaped text containing both vertical and horizontal characters.
- We propose an improved BiFPN module with channel attention and depthwise separable convolution for text feature extraction to improve the accuracy of texts with different scales.
- We introduce a charater re-identification module based on region extension and cutting to correct the wrongly text recognition caused by texts with simple Chinese charaters and multi-line numbers.
- Extensive experiments show that our method has better performance than the SOTA methods on the electrical diagrams data sets.

2   Related Work

In the field of OCR technology, some models are proposed to detect the text position. With the introduction of image segmentation, most of the text detection models can achieve the task of detecting arbitrarily shaped text, but the detection capability for dense texts is poor. As a result, PSENet [4] and PAN [5] were proposed to improve the performance of text detection on compact text detection. DBNet [6] introduced a boundary threshold selection strategy to achieve accurate detection results.

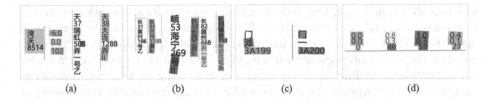

(a)   (b)   (c)   (d)

Fig. 1. Some text instances that are hard to detect and recognize using PP-OCRv3.(a) Irregularly shaped texts that contain both horizontal and vertical characters. (b) Texts with various sizes and shapes. (c) Texts that contain simple Chinese characters, which is easily omitted. (d) Texts that contain dense multi-line numbers.

Other models are proposed to recognize the text content. With the emergence of sequence-to-sequence [7], text recognition models for the extraction and analysis of text sequence features begin to emerge, the most representative of which is CRNN [8]. Sheng et al. [9] used transformer to encode and decode input images, and only use several simple convolutional layers for high-level feature extraction, which verified the effectiveness of transformer in text recognition. Du et al. [10] proposed SVTR, which is based on global and local self-attention mechanisms to obtain large-scale correlation between characters and small-scale correlation between stroke structures, respectively.

In recent years, one-stage text detection and recognition has become the main focus of research in OCR. Most of the one-stage models, like MANGO [11], Mask TextSpotter v3 [12] and PGNet [13], combine the network structure into one overall architecture, which can effectively reduce the inconsistency of detection. However, most of these one-stage models are focused on character-level recognition, which is not applicable to more complex scenes.

PP-OCRv3 is a representative one of the two-stage text spotting models. During text position detection, DBNet is introduced to detect the text regions. And then the text proposals are obtained. During the text feature extraction, cropped text images according to the text proposals are fed into MobileNetv1 to extract features. The feature maps Fs of each text proposals is obtained. During text recognition, a recognition head based on SVTR is proposed to obtain the text recognition result.

However, as a general text spotting model, PP-OCRv3 performs poorly for multi-scale electrical diagrams with irregularly shaped and compact distributed texts. To enhance the performance on text spotting of electrical diagrams, we propose a text spotting model of electrical diagrams based on improved PP-OCRv3.

3   Methodology

In order to solve the problem of text spotting of electrical diagrams, we propose text spotting model of electrical diagrams based on improved PP-OCRv3, which is shown in Fig. 2. Our model includes text position detection, text feature extraction and text recognition. The blue part is the module in the original PP-OCRv3, and the green part is the module we proposed. In order to enhance the detection accuracy for irregularly shaped text, we design a region re-segmentation module based on pixel row clustering in text position detection. In order to improve the robustness for text images of different scales, we introduce an improved BiFPN module in text feature extraction. To correct the detection and recognition errors of simple Chinese characters and multi-line numbers, we add a character re-identification module based on region extension and cutting in text recognition.

Fig. 2. The overall structure of text spotting model of electrical diagrams.

3.1 Text Position Detection Based on DBNet and Region Re-segmentation

The overall framework of the text position detection is shown in Fig. 3. We follow PP-OCRv3 to use DBNet as the detection network, as shown in the green boundary area. Since the electrical diagrams contain irregularly shaped texts that contain both horizontal and vertical characters, the segmentation results of existing methods are not good. Therefore, we design a region re-segmentation module based on pixel line clustering after DBNet. As shown in Fig. 3, the gray part is the wrong segmentation result of the original PP-OCRv3, and the yellow part is the region re-segmentation module we designed.

The specific steps of the region re-segmentation module are as follows.

Step 1: Unusual regions filtering. Calculate the IoUs of each kernel contour and kernel box. The kernel contour is tight and non-rectangular to express the outline of the text segmentation area, while the kernel box is loose and rectangular to express the bounding rectangle of the text segmentation area. $IoU < \theta$ often means that the text kernel has an irregular shape and is regarded as an "unusual region", where θ is a hyper-parameter.

Step 2: Pixel line clustering. Record the width of each pixel line of the unusual contour region. After that, perform line-based clustering on the contour area.

Step 3: Threshold filtering. Set δ as a threshold of the area. Filter the class cluster if the area of it is less than δ. In this way, clusters that are too small to contain text are discarded, while the proper clusters then become the fine-tuned kernels.

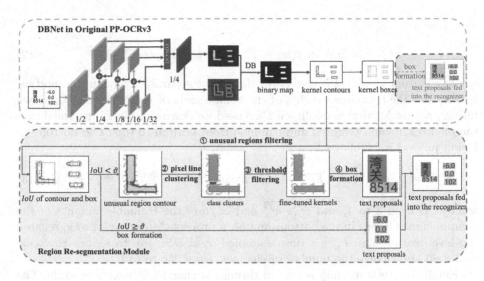

Fig. 3. The structure of text poition detection.

Step 4: Box formation. Generate text proposals according to fine-tuned kernels.

3.2 Text Feature Extraction Based on MobileNetv1 and Improved BiFPN

Our proposed text feature extraction network is designed on the basis of MobileNetv1 of PP-OCRv3. An improved BiFPN module with channel attention and depthwise separable convolution is introduced after MobileNetv1 to extract multi-scale features of texts with different shapes in electrical diagrams. Specifically, a channel attention block is added at the end of BiFPN to enhance information in high-level feature maps and improve the global feature fitting ability. In order to overcome the increase of parameters, depthwise separable convolutions is used to replace all convolutions in the network. Hardswish is uesd as the activation function to further improve the accuracy.

The overall network structure is shown in Fig. 4. The blue area is the structure of the original feature extraction network MobileNetv1 in PP-OCRv3. The feature maps F_1, F_2, F_3, and F_4 are the outputs of different layers of MobileNetv1, and the scales are $64 \times 24 \times 160$, $128 \times 12 \times 160$, $256 \times 6 \times 160$, $512 \times 3 \times 160$. The above feature maps are normalized to 256 channels, which is shown as follows:

$$F_i^{in} = HS(Conv_{1\times1}(F_i)) \tag{1}$$

where i is from 1 to 4, and HS is activation function hardswish. Up-sampling is then performed using nearest neighbor interpolation, and the up-sampled results are weighted and summed from top to bottom. The top-down fusion process is as follows:

$$F_i^{mid} = DSConv(w_i^1 \cdot F_i^{in} + w_i^2 \cdot Upsample(F_{i+1}^{in})) \tag{2}$$

where i is from 1 to 3, $DSConv$ is the depthwise separable convolution, and w_i^1 and w_i^2 are the trainable weights of the components of layer i in the top-down fusion process. Maximum pooling is then used for down-sampling, and the down-sampled results are weighted and summed from bottom to top. The bottom-up fusion process is as follows:

$$F_i^{out} = DSConv(w'_i^1 \cdot F_i^{mid} + w'_i^2 \cdot Downsample(F_{i-1}^{mid}) + w'_i^3 \cdot F_i^{in}) \tag{3}$$

where i is from 2 to 4, and w'_i^1, w'_i^2 and w'_i^3 are the trainable weights of the components of layer i in the bottom-up fusion process. After that, the aggregated top-level feature map F_4^{out} is down-sampled by a $DSConv$ to reduce the scale to 3×80 and change the number of channels from 256 to 512.

Finally, the feature map is passed through a channel attention module. The structure of the channel attention module is shown in Fig. 5. First, adaptive global average pooling is performed. Then feed it into two fully connected layers with a convolution kernel size of 1×1 to achieve non-linearity, and a weight sequence of $512 \times 1 \times 1$ is obtained. This sequence is used to represent the contribution of each channel of the input feature map. Finally, the sequence of weights is multiplied with the input feature map. The obtained result is a feature map improved by the channel attention mechanism. The specific computational procedure is as follows:

$$F_{out} = F_{in} \cdot (sigmoid(Conv_{1\times1}(ReLU(Conv_{1\times1}(AdaptiveAvgPool(F_{in})))))) \tag{4}$$

After down-sampling, we get a sequence feature map F with the scale of $512 \times 1 \times 40$.

3.3 Text Recognition Based on SVTR and Character Re-identification

The structure of text recognition is shown in Fig. 6. We continue to use the recognition head based on SVTR in PP-OCRv3, as shown in the green area. In order to solve the omission of simple Chinese characters and the wrong recognition of multi-line numbers caused by the original model, we design a character re-identification module based on region extension and cutting behind the recognition head, as shown in the blue region in Fig. 6.

The specific steps of the character re-identification module are as follows.

Step 1: Unusual results filtering. Filter the results output by the recognition header. If the recognition result contains only one Chinese character, then go to step 2. If the recognition result only contains numbers and punctuation, and the text proposal corresponding to the recognition result is square, then go to step 3. We consider the rest of the recognition results to be correct.

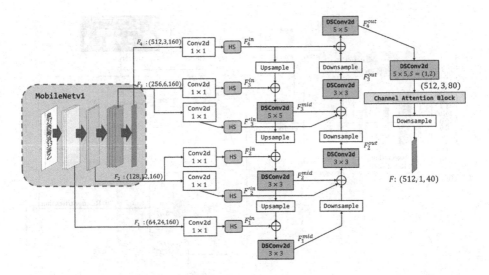

Fig. 4. The framework of text feature extraction network.

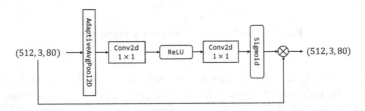

Fig. 5. Channel Attention Block.

Step 2: Vertical extension. We start from the position of the recognized single Chinese character, and extend vertically to the direction of "with text pixels" but "without text proposals" according to the original image and binary map.

Step 3: Horizontal cutting. We cut the text proposals corresponding to the wrong recognition results into two parts horizontally.

Step 4: Re-identification. We feed the corrected text regions in step 2 and step 3 to the text feature extraction and recognition head for re-identification. Finally, we combine the text re-identification results with the previous correct results to get the final recognition result, as shown in the yellow boundary area in Fig. 6.

4 Experiments

4.1 Datasets and Settings

Datasets. When we train the model, we use ICDAR2019-LSVT, ICDAR2017-RCTW and Chinese-Dataset with 127,727 text images as the training set and

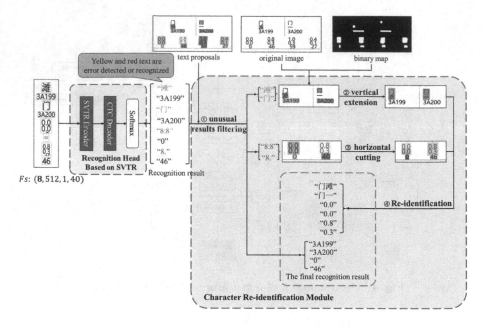

Fig. 6. The framework of text recognition.

ICDAR2015 with 5,230 text images as the validation set for general scene text recognition. To improve the recognition of text in electrical diagrams, we add 23,377 text images of electrical diagrams from State Grid Shanghai Municipal Electric Power Company in the training and validation process.

Settings. During training, we use Adam optimizer with a learning rate of 0.001 and a batch size of 64. We reshape each text instance to 48 × 320 pixels. The threshold θ and δ in text position detection are set to 0.6 and 3.0, respectively. All experiments are performed in PaddlePaddle framework on RTX 3090 GPU. The main results of the experiments including comparison with prior methods and ablation studies, are presented in the following sections.

4.2   Comparison with Prior Methods

Qualitative Evaluation. We show the performance of a one-stage model PGNet [13], PP-OCRv3 and our model in electrical diagrams in Fig. 7. As an efficient one-stage end-to-end text spotting model, the character prediction module of PGNet does not have the sequence encoding ability or the global information acquisition capability. Therefore, PGNet only performs well for recognition of numbers and letters, and cannot achieve recognition of Chinese (as shown in rows 1, 2 and 4 of Fig. 7, it can only recognize numbers and letters effectively). PP-OCRv3 can effectively encode sequence information, and thus has the ability to recognize long Chinese text. However, it is not adaptable to scale changes, resulting in a poor ability to capture texture features in large scale images (as shown in rows 2 and 3 of Fig. 7, with poor performance for long text). In addition, the detection accuracy of PP-OCRv3 for irregular compact text is low due to the lack of targeted fine-tuning operations (as shown in row 4 of Fig. 7, irregularly shaped compact text cannot be detected). Comparatively speaking, our model can be more adapted to the characteristics of electrical diagrams, and has better recognition performance for multi-scale text images.

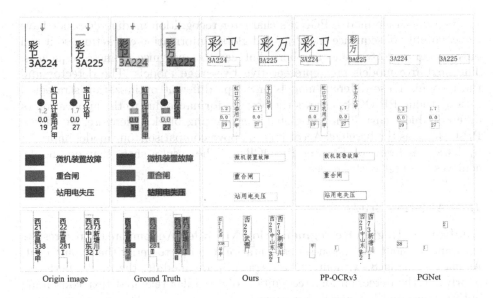

Fig. 7. Examples of results for comparison.

Table 1. Quantitative comparison over different methods.

Method	Detection			Character Accuracy	End-to-End		
	P	R	F1		P	R	F1
PGNet	72.93	70.86	71.88	62.46	55.75	51.39	53.48
EAST+CRNN [8]	83.30	89.52	86.30	68.80	62.12	67.26	64.59
PSENet [4]+NRTR [9]	86.86	86.12	86.98	70.01	68.96	66.80	67.86
ABCNet v2 [14]	89.42	87.24	88.32	72.46	70.74	67.38	69.02
SwinTextSpotter [15]	90.18	88.02	89.05	76.55	73.87	69.82	71.79
PP-OCRv3	86.23	93.43	89.68	80.84	71.89	77.90	74.77
Ours	**90.38**	**95.56**	**92.90**	**85.39**	**78.27**	**82.76**	**80.45**

Quantitative Evaluation. To further compare the accuracy of our model with the SOTA approach, we perform quantitative evaluation on electrical diagram test set, as shown in Table 1.

The column *Detection* shows the accuracy of text position detection. We use P, R and $F1$ as the evaluation indexes of *Detection*. The column *Character Accuracy* indicates the percentage of text images that are correctly recognized during text feature extraction and text recognition. The column $End-to-End$ shows the text detection and recognition accuracy of the whole system.

As a one-stage model, PGNet's character recognition module does not have the capability of sequence encoding and global information extraction, so it can only recognize letters and numbers, but cannot recognize Chinese sequences. The next five models are representative two-stage models. The detector and the recognizer use different models or adopt different optimization strategies. Among them, PP-OCRv3 shows the best performance due to the application of differentiable binarization and global/local mixing block. Therefore, we choose PP-OCRv3 as the baseline. According to above comparison, our model improves by 5.68% in terms of $F1$ over PP-OCRv3. The results show that our model achieves the best performance.

4.3   Ablation Study

Analysis on Region Re-segmentation Module. We compare the test results with and without the region re-segmentation module to verify the impact of it. As shown in Fig. 8, without the re-segmentation module, the model performs poorly for the detection and recognition of irregular text and text containing both vertical and horizontal characters.

Analysis on Improved BiFPN Module. We compare the test results with and without the improved BiFPN module to verify the impact of it, as shown in Fig. 8. The introduction of improved BiFPN enables the recognition branch to focus on the contour information of the characters as well as the texture information, which is more important for text recognition.

Analysis on Character Re-identification Module. We compare the test results with and without the character re-identification module to verify the impact of it. Without the re-identification module, the model has poor location ability for characters with simple structures and numbers with compact positions, which can lead to some missed or incorrect detection, as shown in Fig. 8.

The ablation study results of each module are shown in Table 2. It can be seen that our proposed strategy enables the model to effectively adapt to the actual recognition needs of text in electrical diagrams.

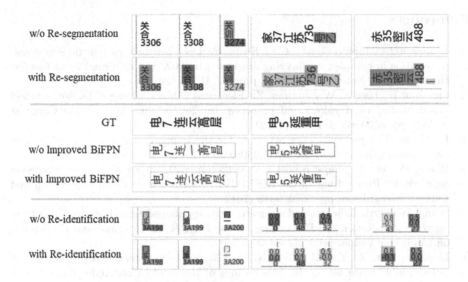

Fig. 8. Ablation study.

Table 2. Ablation study.

Improved BiFPN	Re-segmentation	Re-identification	Detection			Character Accuracy	End-to-End		
			P	R	F1		P	R	F1
✗	✗	✗	86.23	93.43	89.68	80.84	71.89	77.90	74.77
	✗	✗	86.56	93.54	89.91	82.20	74.20	80.17	77.07
		✗	86.56	94.50	90.36	83.09	75.61	80.61	78.03
✓	✓	✓	**90.38**	**95.56**	**92.90**	**85.39**	**78.27**	**82.76**	**80.45**

5 Conclusion and Future Work

In this paper, we propose a text spotting model of electrical diagrams based on improved PP-OCRv3 to enhance the accuracy of text detection and recognition in electrical diagrams. Firstly, we design a region re-segmentation module based on pixel line clustering to help the model better adapt to the irregularly shaped text containing both vertical and horizontal characters. Secondly, we introduce an improved BiFPN module with channel attention and depthwise separable convolution to achieve high-performance recognition of different scales of texts in electrical diagrams. Finally, to correct the mis-identification of simple Chinese characters and compact multi-line numbers, a character re-identification module based on region extension and cutting is used to help the model better adapt to the complex scenarios. Comprehensive experiments show that our model can significantly improve the performance of text spotting and can be effectively applied to the specific text spotting scene of electrical diagrams. For future work, semantic knowledge in electrical diagrams is further fused to improve the overall performance of the model.

References

1. Jamieson, L., Moreno-Garcia, C.F., Elyan, E.: Deep learning for text detection and recognition in complex engineering diagrams. In: 2020 International Joint Conference on Neural Networks (IJCNN), pp. 1–7. IEEE (2020)
2. Shanbin, L., Haoyu, W., Junhao, Z.: Electrical cabinet wiring detection method based on improved yolov5 and pp-ocrv3. In: 2022 China Automation Congress (CAC), pp. 6503 6508. IEEE (2022)
3. Li, C., et al.: Pp-ocrv3: more attempts for the improvement of ultra lightweight OCR system. arXiv preprint arXiv:2206.03001 (2022)
4. Wang, W., et al.: Shape robust text detection with progressive scale expansion network. In: Proceedings of the IEEE/CVF Conference on Computer Vision and Pattern Recognition, pp. 9336–9345 (2019)
5. Wang, W., et al.: Efficient and accurate arbitrary-shaped text detection with pixel aggregation network. In: Proceedings of the IEEE/CVF International Conference on Computer Vision, pp. 8440–8449 (2019)
6. Liao, M., Wan, Z., Yao, C., Chen, K., Bai, X.: Real-time scene text detection with differentiable binarization. In: Proceedings of the AAAI Conference on Artificial Intelligence, vol. 34, pp. 11474–11481 (2020)
7. Sutskever, I., Vinyals, O., Le, Q.V.: Sequence to sequence learning with neural networks. Adv. Neural Inf. Process. Syst. **27** (2014)
8. Shi, B., Bai, X., Yao, C.: An end-to-end trainable neural network for image-based sequence recognition and its application to scene text recognition. IEEE Trans. Pattern Anal. Mach. Intell. **39**(11), 2298–2304 (2016)
9. Sheng, F., Chen, Z., Xu, B.: NRTR: a no-recurrence sequence-to-sequence model for scene text recognition. In: 2019 International Conference on Document Analysis and Recognition (ICDAR), pp. 781–786. IEEE (2019)
10. Du, Y., et al.: SVTR: scene text recognition with a single visual model. arXiv preprint arXiv:2205.00159 (2022)
11. Qiao, L., et al.: Mango: a mask attention guided one-stage scene text spotter. In: Proceedings of the AAAI Conference on Artificial Intelligence, vol. 35, pp. 2467–2476 (2021)
12. Liao, M., Pang, G., Huang, J., Hassner, T., Bai, X.: Mask TextSpotter v3: segmentation proposal network for robust scene text spotting. In: Vedaldi, A., Bischof, H., Brox, T., Frahm, J.-M. (eds.) ECCV 2020. LNCS, vol. 12356, pp. 706–722. Springer, Cham (2020). https://doi.org/10.1007/978-3-030-58621-8_41
13. Wang, P., et al.: Pgnet: real-time arbitrarily-shaped text spotting with point gathering network. In: Proceedings of the AAAI Conference on Artificial Intelligence, vol. 35, pp. 2782–2790 (2021)
14. Liu, Y., et al.: Abcnet v2: adaptive Bezier-curve network for real-time end-to-end text spotting. IEEE Trans. Pattern Anal. Mach. Intell. **44**(11), 8048–8064 (2021)
15. Huang, M., et al.: Swintextspotter: scene text spotting via better synergy between text detection and text recognition. In: Proceedings of the IEEE/CVF Conference on Computer Vision and Pattern Recognition, pp. 4593–4603 (2022)

Topic-Aware Two-Layer Context-Enhanced Model for Chinese Discourse Parsing

Kedong Wang[1,2], Qiang Zhu[1,2], and Fang Kong[1,2(✉)]

[1] Laboratory for Natural Language Processing, Soochow University, Suzhou, China
{kdwang1011,qzhu}@stu.suda.edu.cn
[2] School of Computer Science and Technology, Soochow University, Suzhou, China
kongfang@suda.edu.cn

Abstract. In the past decade, Chinese Discourse Parsing has drawn much attention due to its fundamental role in document-level Natural Language Processing (NLP). In this work, we propose a topic-aware two-layer context-enhanced model based on transition system. Specifically, in one hand, we first adopt a two-layer context-enhanced Chinese discourse parser as a strong baseline, where the Star-Transformer with star topology is employed to enhance the EDU representation. On the other hand, we split the document into multiple sub-topics based on the change of nuclearity of discourse relations. Then we implicitly incorporate topic boundary information via joint learning framework. Experimental results on the Chinese CDTB corpus indicate that, the proposed approach can contribute much to Chinese discourse parsing.

Keywords: Chinese Discourse Parsing · Star-Transformer · Topic Boundary

1 Introduction

According to the Rhetorical Structure Theory [10] (RST), a document can be represented in tree structure called Discourse Rhetorical Structure (DRS). RST constructs each discourse in a sequence of Elementary Discourse Units (EDUs), and further parses the discourse into rhetorical tree. As shown in Fig. 1, where the leaf nodes correspond to EDUs, which combine recursively to form subtrees from the bottom layer. The internal nodes assign nuclearity and rhetorical relations to its child nodes, and the nuclearity includes three categories: *nucleus-satellite* (NS), *satellite-nucleus* (SN) and *nucleus-nucleus* (NN), where the *nucleus* plays a leading role in semantic representation in comparison with the *satellite*.

In the past decade, Chinese Discourse Parsing (CDP) has drawn much attention owing to its fundamental role in document-level Natural Language Processing (NLP). Early researches mainly focused on EDU segmentation, which aims

B. Luo et al. (Eds.): ICONIP 2023, CCIS 1968, pp. 137–148, 2024.
https://doi.org/10.1007/978-981-99-8181-6_11

to identify the correct EDU boundaries, are of great significance for subsequent DRS parsing. Recently, as EDU segmentation work has achieved remarkable performance [6,8], current research of CDP concentrates on DRS parsing, in which discourse constituent tree is first parsed, based on the naked tree to assign the nuclearity and rhetorical relations for discourse rhetorical tree.

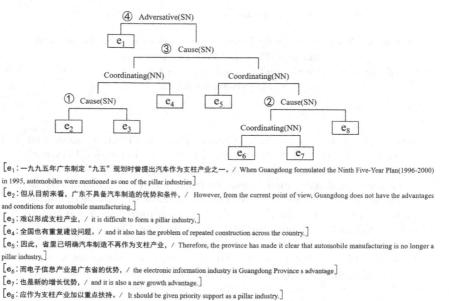

[e_1: 一九九五年广东制定 "九五" 规划时曾提出汽车作为支柱产业之一。 / When Guangdong formulated the Ninth Five-Year Plan(1996-2000) in 1995, automobiles were mentioned as one of the pillar industries]

[e_2: 但从目前来看，广东不具备汽车制造的优势和条件， / However, from the current point of view, Guangdong does not have the advantages and conditions for automobile manufacturing,]

[e_3: 难以形成支柱产业， / it is difficult to form a pillar industry,]

[e_4: 全国也有重复建设问题。 / and it also has the problem of repeated construction across the country.]

[e_5: 因此，省里已明确汽车制造不再作为支柱产业， / Therefore, the province has made it clear that automobile manufacturing is no longer a pillar industry,]

[e_6: 而电子信息产业是广东省的优势， / the electronic information industry is Guangdong Province s advantage]

[e_7: 也是新的增长优势， / and it is also a new growth advantage.]

[e_8: 应作为支柱产业加以重点扶持。 / It should be given priority support as a pillar industry.]

Fig. 1. An example of Chinese Discourse Rhetorical Tree.

After the release of Chinese Connective-driven Discourse Treebank [7] (CDTB), researchers have made certain success in Chinese DRS parsing under neural architecture. Among them, Sun et al. [12] make their first attempt in transition-based system, they construct the DRS in a bottom-up fashion with a sequence of post-order transition actions. Zhang et al. [17] put their sight on a novel top-down model, where the DRS parsing task is regarded as a split point ranking task based on the pointer-network, with the rising attention on pre-trained language models (PLMs), Zhang et al. [16] further enhance the performance based on XLNet [14] in the top-down framework.

The studies mentioned above accomplish different ways to construct DRS. However, on one hand, previous works adopt common encoders such as RNN and CNN on the encoding stage, which is limited for semantic representations. On the other hand, they did not take the multi-granularity information into account, such as sentence boundary information, which proved to be effective in the RST-style English discourse parsing. In this paper, using the transition-based bottom-up framework as backbone model, we consider the bottom-up tree construction as the process of combining sub-topics, in which the EDUs within each sub-topic are combined to form the complete sub-topic and then the whole discourse is constructed by the sub-topics.

2   Related Work

In the field of DRS parsing, a large number of studies were first conducted in English based on RST-DT corpus, including two different types of parsing frameworks in the literature: bottom-up and top-down fashion.

In the former parsing method, including probabilistic CKY-like and shift-reduce parser, where the CKY-like parser uses the dynamic programming algorithm to find the best-scoring tree, and the transition-based parser constructs DT with a series of shift-reduce actions. While the CKY-like parser take $O(n^3)$ time, the transition-based model has the advantage of building the tree in $O(n)$ time with respect to EDU numbers, and can achieve comparable performance compared to dynamic programming methods, which leads to that researchers turned their attention to transition-based approaches. Recently, a novel top-down parsing framework was proposed and achieve state-of-the-art performance both for English and Chinese DRS parsing.

For English RST-style DRS parsing, as mentioned above, the first type of bottom-up parsing method is the probabilistic CKY-like approach [1–3], which relied heavily on manual features. However, due to the high time complexity of CKY-like parsing methods, transition-based models are proposed to replace the original bottom-up methods, where researchers convert DRS parsing into transition sequence prediction task: Wang et al. [13] propose a two-stage parser, they first parse the bare tree structure with nuclearity and then assign relations based on this tree. Yu et al. [15] build a transition-based neural RST parser with implicit syntax features. In recent years, a top-down parsing framework based on a global perspective is proposed, which converts DRS parsing into a split point ranking task: Lin et al. [9] achieve first success in sentence-level discourse parsing, inspired by this work, Zhang et al. [17] propose a unified document-level parsing framework based on pointer-network both for English and Chinese, Kobayashi et al. [4] parse the entire document in three granularity levels and utilize better word embedding for better performance.

For Chinese CDT-style DRS parsing, Kong and Zhou [5] adopt a greedy bottom-up approach to generate the discourse parse trees in a pipeline architecture. Sun et al. [12] propose a shift-reduce parser with different encoders based on Chinese flat text, Zhang et al. [17] first proposed a top-down parsing framework based on pointer-network, and further enhanced with GAN [16]. Previous studies in English RST-style DRS parsing can provide certain assistance to Chinese discourse parsing, however, due to the language difference and the Chinese CDTB corpus is annotated by paragraphs, Chinese discourse parsing still has a long way to go.

3   Model

In this section, we first introduce the basic principles of transition-based bottom-up system. Then we make a detail introduction of **TCCDP** (**T**wo-layer **C**ontext-enhanced **C**hinese **D**iscourse **P**arser). Finally, we describe the proposed topic segmentation algorithm and joint learning framework.

3.1   Transition System

In transition system, DRS parsing is converted into a transition actions prediction task. Specifically, the standard transition action is obtained in post-order traversal of discourse rhetorical tree. When the leaf node is visited export *SHIFT* action, and the internal node returns *REDUCE* action, the meaning of the SHIFT/REDUCE action defined as below:

- SHIFT: Pop the top element of the queue and push it onto the stack.
- REDUCE-Nuc-Rel: Pop the top two elements of the stack, form a subtree and push it onto the stack again. In this action, *Nuc* and *Rel* denote the nuclearity and rhetorical relations for the subtree respectively.

   Similar with Sun et al. [12], we first duplicate a shift-reduce parser using SPINN-based decoder, which consists of the unidirectional Long Short Term Memory (LSTM), called tracking LSTM. At each time step, an auxiliary stack and queue are applied to represent the current transition state. Initially, the queue stores all encoded EDUs in left-to-right order, while the stack is empty. The stack is designed to hold partially generated subtrees and the queue is employed to store unprocessed EDUs, until the queue is empty and the stack contains only one element which represents the whole document.

3.2   Two-Layer Context-Enhanced Chinese Discourse Parser

Traditionally, for a given document $D = \{e_1, e_2, \ldots, e_n\}$, consisting of n EDUs. For ith EDU $e_i = \{w_1, w_2, \ldots, w_m\}$, m means the number of words. We concatenate each word and the POS embedding as inputs to the bi-directional Gated Recurrent Unit (BiGRU), to obtain the contextual representation $H = \{h_1, h_2, \ldots, h_m\}$ for each word based on flat text as Eq. 1. Considering the different importance of each word in EDU encoding, we adopt a self-attention mechanism as Eq. 2, where q is a learnable weight matrix. At last, we concatenate the last state of BiGRU in both directions denoting as $\overrightarrow{h_s}$ and $\overleftarrow{h_s}$, and add it with the weighted summation result as Eq. 3.

$$h_1, h_2, \ldots, h_m = BiGRU(w_1, w_2, \ldots, w_m) \tag{1}$$

$$w_i = \frac{q^T h_i}{\sum q^T h_j} \tag{2}$$

$$e_k = [\overrightarrow{h_s} \oplus \overleftarrow{h_s}] + \sum w_i h_i \tag{3}$$

After obtaining the feature representation $E = \{e_1, e_2, ..., e_m\}$ with intra-EDU encoder, we incorporate an additional inter-EDU GRU encoder to facilitate the EDU-level boundary detection task. We utilize the Star-Transformer as the EDU-level context encoder. For DRS parsing task, EDUs are preferentially combined with adjacent EDUs to form subtrees, so it is of great significance to take the influence of adjacent nodes into account while looking at the overall document.

Star-Transformer applies a topological structure with a relay node and n satellite nodes. Different from the fully-connected Transformer, Star-Transformer eliminates redundant connections instead of a relay node, which is able to incorporate global information that is beneficial to the tree structure construction based on non-adjacent nodes. The state of j-th satellite node represents the feature of the j-th EDU in a document, each satellite node is initially as a vector from the previous intra-EDU encoding layer, and the relay node is initialized to the average of all tokens.

At each time-step, it follows a cyclic updating method, the complete update process is shown in Eq. 4 and 5, in which status update of satellite nodes based on its neighbor nodes, containing three nodes from the previous round e_{t-1}^{i-1}, e_{t-1}^{i} and e_{t-1}^{i+1}, the current processing node e_i and the relay node s^{t-1} from previous round.

$$C_i^t = [e_{t-1}^{i-1}; e_{t-1}^{i}; e_{t-1}^{i+1}; e^i; s^{t-1}] \tag{4}$$

$$e_i^t = MulAtt(e_i^{t-1}, C_i^t, C_i^t) \tag{5}$$

Relay node is updated by all satellite nodes and its status in previous round, we present the update process in Eq. 6

$$s^t = MulAtt(s^{t-1}, [s^{t-1}; E^t], [s^{t-1}; E^t]) \tag{6}$$

3.3   Topic Segmentation Algorithm and Joint Learning Framework

Topic Segmentation Algorithm. Algorithm 1 shows the pseudo-code of proposed method, for a given discourse, it is considered to be a sequence of sentences in the traditional document-level task. However, in this paper, we split the whole document into multiple sub-topics with nuclearity information. Specifically, in the bottom-up construction process, the internal node with NS/SN labels determines the topic boundary of its child nodes. For example, as shown is Fig. 1, the $Cause(SN)$ (denoted with 1) node determines the topic boundary between e_2 and e_3, and the $Adversative(SN)$ (denoted with 4) node determines the topic boundary between e_1 and $e_{2:8}$. Finally, Fig. 2 shows the result of discourse in Fig. 1 by sub-topic segmentation.

Algorithm 1: Topic segmentation

Input: D: Current Discourse Rhetorical Tree
Output: L: Topic Segmentation Labels
1 Divide each document into paragraphs and get D;
2 Initialize L=[ ];
3 **for** *discourse* **do**
4 | Stack S=[ ];
5 | T=[ ];
6 | **if** *root_relation* **then**
7 | | Store all relation nodes with stack S;
8 | | S = level_order_traversal(root_relation);
9 | | **while** *S is not empty* **do**
10 | | | node = S.pop();
11 | | | **if** *node.nuc in [NS, SN]* **then**
12 | | | | lc = left_childs(node);
13 | | | | rc = right_childs(node);
14 | | | | T = set_topic_boundary(lc, rc);
15 | | | **end**
16 | | **end**
17 | **end**
18 | L.append(T)
19 **end**
20 **return** L

Joint Learning Framework. Figure 3 shows the flow chart of the discourse parsing model in this paper. After intra-EDU and inter-EDU encoders to obtain EDU representations, we add an auxiliary task to incorporate boundary information. To be specific, it is an EDU-level sequence labeling task, we employ the I (Inner), E (Edge) annotation rules, since EDUs are treated as the smallest unit in CDP, and the number of EDU within sentence is quite less, only right boundary is built in the sequence labeling task. For given discourse in Fig. 1, we can thus obtain the sequence of sentence boundary labels as $S = [E, I, I, E, I, I, I, E]$, and sub-topic boundary labels as $T = [E, E, I, E, I, I, I, E]$.

3.4 Model Training

For a given discourse tree, which contains n EDUs, it can be constructed with $2n - 1$ transition actions, Eq. 7 defines the Cross-Entropy loss function we used in the training phase. Additionally, we employ L2-norm to prevent over-fitting, and adopt the Adam for the optimization of learnable parameters in the model. The auxiliary task loss function is denoted as L_{sq}, we add up L_{dp} and L_{sq} as the final loss function (Eq. 8)

$$L_{dp} = \frac{1}{2n - 1} \sum_{t=0,\dots,2n-1} -log p_{t+1}^{o(ct)} + \lambda \|\Theta\|_2 s \tag{7}$$

$$L = L_{dp} + L_{sq} \tag{8}$$

[e₁:一九九五年广东制定"九五"规划时曾提出汽车作为支柱产业之一。]｜[e₂:
但从目前来看，广东不具备汽车制造的优势和条件，]｜[e₃:难以形成支柱产
业，][e₄:全国也有重复建设问题。]｜[e₅:因此，省里已明确汽车制造不再作为支
柱产业，][e₆:而电子信息产业是广东省的优势，][e₇:也是新的增长优势，]｜[e₈:
应作为支柱产业加以重点扶持。]
[e₁:When Guangdong formulated the Ninth Five-Year Plan(1996-2000) in 1995, automobiles were
mentioned as one of the pillar industries] ｜ [e₂:However, from the current point of view, Guangdong
does not have the advantages and conditions for automobile manufacturing,] ｜ [e₃:it is difficult to
form a pillar industry,][e₄:and it also has the problem of repeated construction across the country.] ｜
[e₅:Therefore, the province has made it clear that automobile manufacturing is no longer a pillar
industry,][e₆:the electronic information industry is Guangdong Province s advantage][e₇:and it is
also a new growth advantage.] ｜ [e₈:It should be given priority support as a pillar industry.]

Fig. 2. Topic segmentation result

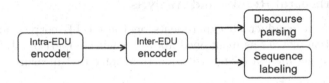

Fig. 3. Model Flow Chart

4   Experiments

4.1   Experimental Settings

Dataset. Following previous studies, we conduct our experiments on the Chinese CDTB corpus [7]. This corpus is annotated under the guidance of the English RST-DT and the PDTB corpus, containing 500 newswire articles in total. Different from the English RST-DT corpus where the whole document is labeled as rhetorical tree structure, the Chinese CDTB corpus is annotated in paragraphs, called Connective-driven Discourse Tree (CDT). Then a total of 2336 valid paragraph trees are included, which further divided into 2002 CDTs for training, 105 CDTs for validation and 229 CDTs for testing respectively.

Evaluation Metrics. To better demonstrate the effectiveness of the proposed scheme in this work, we adopt the strict macro averaged F_1 as previous studies [12,16], including the naked tree structure denoted as *Span*, tree structure with nuclearity label denoted as *Nuc*, tree structure labeled with 16 fine-grained rhetorical relations denoted as *Rel*, and tree structure labeled with both nuclearity and rhetorical relations denoted as *Full* respectively.

Hyper Parameters. For static word representation, we utilize 300D vectors provided by Qiu et al. [11], and we use LTP parser[1] for word segmentation, while the POS embedding is randomly initialized and optimized during the training process. Table 1 shows all hyper-parameters in this work, and all experiments conducted based on Pytorch 0.4.1, and accelerated by NVIDIA RTX 2080ti GPU.

Table 1. Hyper-parameters

Parameter	Value	Parameter	Value
Epoch	20	intra-EDU encoder	256
Batch size	32	Star-Transformer	512
Learning rate	1e-3	tracking LSTM	256

4.2 Experimental Results and Analysis

In this work, we conduct all experiments on the CDTB corpus, and use the same evaluation metrics as mentioned above for fair comparison. Table 2 shows the detailed experimental results with macro averaged F1-measure:

- **Sun et al. [12] (Dup):** Our duplicated transition-based bottom-up parser as [12], where we add two evaluation metrics for rhetorical relations (*Rel*) and DRS construction (*Full*).
- **Baseline (Two-layer):** A shift-reduce parser with two-layer BiGRU encoder, where we adopt an additional inter-EDU BiGRU.
- **TCCDP:** Our context-enhanced model with inter-EDU Star-Transformer encoder.
- **+Joint (Topic Boundary):** This model employs an auxiliary topic boundary detection task to implicitly incorporate topic boundary information

From the results shown in Table 2, our duplicated shift-reduce parser achieves 83.2% on *Span* and 54.4% on *Nuc* with F1-measure in comparison with [12]. Meanwhile, we present two additional evaluation indicators in Table 2, in which it does not perform well on 16 fine-grained relations evaluation (denoted as *Rel*) due to the data sparsity problem. Finally we achieve 46.1% F1-score on DRS construction.

Contribution of Two-Layer Context-Enhancement. As mentioned before, the complete rhetorical tree evaluation metric (*Full*) is on the basis of three components (*Span*, *Nuc* and *Rel*) being fully and correctly identified, which results in lower performance on *Full* metric. Observing the experimental results in Table 2, we can get the following conclusions:

---

[1] http://www.ltp-cloud.com/.

Table 2. Experimental results on CDTB (F1)

Model	Span	Nuc	Rel	Full
Sun et al. [12]	84.0	53.9	–	–
Sun et al. [12] (Dup)	83.2	54.4	48.9	46.1
Baseline (Two-layer)˙	83.5	55.2	49.3	47.1
+Joint (Topic Boundary)	**83.9**	**57.2**	**51.7**	**48.2**
TCCDP	84.0	56.3	49.6	47.6
+Joint (Topic Boundary)	**84.5**	**58.2**	**51.0**	**49.6**

- The proposed two-layer context-enhancement methods can effectively improve the performance of CDP in all four metrics. In comparison with our duplicated parser, it achieves 0.3%, 0.8%, 0.4% and 1.0% improvement respectively, although we just employ the common BiGRU encoder.
- Furthermore, we adopt Star-Transformer as inter-EDU encoder, which makes certain improvements in all four indicators, with 84.0%, 56.3%, 49.6% and 47.6% F1-score respectively. This demonstrates that Star-Transformer encoder with star topology is effective in semantic representation, and can contribute much to Chinese discourse parsing.

Contribution of Topic Boundary Information. In general, we consider the document as being composed of several units that are larger than words, such as sentences, which makes it common to first encode sentences in text-level tasks and then take sentences as the basic unit for document representation. But in the CDP task, the ultimate goal of the encoding stage is to obtain the representation of EDUs. In this paper, we split the document into sub-topics, and implicitly incorporate topic boundary information with a joint learning framework. From the results in Table 2, the following conclusions can be drawn:

- By treating the topic boundary detection task as an EDU-level sequence labeling task, it is able to implicitly incorporate topic information to enhance the EDU representation, and thus improves CDP performance.
- To be specific, observing the experimental results in the fourth row of Table 2, after incorporating topic information, it achieved 0.4%, 2.0%, 2.4%, and 1.1% improvement in the four evaluation metrics on the baseline model, respectively.
- Finally, when we incorporate topic information in the TCCDP, as shown in the last row of the Table 2, which achieves 84.5%, 58.2%, 51.0% and 49.6% for *Span*, *Nuc*, *Rel* and *Full* respectively. We eventually achieve the best F1-score on all evaluation metrics, which further validates the effectiveness of inter-EDU Star-Transformer encoder and topic boundary information.

4.3 Further Experiments on TCCDP

We conduct further experiments on TCCDP, in order to compare with the traditional sentence segmentation. Table 3 shows the relevant statistics in the training set of CDTB, it can be concluded that, topics are the component of document, which is coarser in granularity and larger in scope than sentences.

Table 3. Statistical results of two types of boundary segmentation

Statistical content	Num	Statistical content	Num
Paragraphs	2002	Sentences per paragraph	2.03
Sentences	4058	Topics per paragraph	1.89
Topics	3791	Overlap ratio	72.4%

When we incorporate sentence boundary information, it achieves 0.2% improvement in tree structure, and 1.1%, 1.7% improvement in relations recognition (Nuc and Rel), respectively. Finally it achieves 48.4% F1-score on $Full$, which suggests that sentence boundary information is effective for DRS parsing. By comparing the results shown in Table 4, we can draw the following conclusions:

- Incorporating two different types of boundary information with joint learning has certain improvement on CDP. Obviously, the topic boundary information is more effective compared with sentence segmentation, however, it drops by 0.3% on Rel, which is probably due to the stricter usage of Chinese rhetorical relations both intra- and inter-sentence in Chinese articles.
- In the experiment incorporating both topic and sentence boundary information, which achieves the best performance on $Span$ and Rel. But it drops by 0.4% on $Full$ result from 0.6% decrease on Nuc, while achieving the same F1-score in tree construction ($Span$). Table 3 shows that topic and sentence boundaries overlapped by 72.4%, since topic information is effective for nuclearity recognition, sentence boundaries in non-overlapping parts can weaken the improvement.

Table 4. Further experimental results on TCCDP (F1)

Model	Span	Nuc	Rel	Full
TCCDP	84.0	56.3	49.6	47.6
+Joint (Topic Boundary)	**84.5**	**58.2**	51.0	**49.6**
+Joint (Sentence Boundary)	84.2	57.4	51.3	48.4
+Both	**84.5**	57.6	**51.8**	49.2

The above experimental results indicate that, in the original framework, two-layer context-enhanced encoder can enhance EDU representations in CDP, especially Star-Transformer with star topology acts as a strong encoder. And the incorporation of boundary information with joint learning framework is more effective for DRS parsing in neural architecture.

5   Conclusion

In this paper, we propose a topic-aware context-enhanced model based on transition system, experimental results on the CDTB corpus verify the effectiveness of the proposed scheme. From another perspective, the idea of segmenting discourse into sub-topics with nuclearity information can provide a new approach for DRS parsing in downstream applications, in which we can organize documents by sub-topics instead of sentences. In future work, due to the sparsity of the CDTB corpus, we will explore more effective data augmentation methods.

Acknowledgement. This work was supported by Projects 62276178 under the National Natural Science Foundation of China, the National Key RD Program of China under Grant No. 2020AAA0108600 and the Priority Academic Program Development of Jiangsu Higher Education Institutions.

References

1. Feng, V.W., Hirst, G.: A linear-time bottom-up discourse parser with constraints and post-editing. In: Proceedings of the 52nd Annual Meeting of the Association for Computational Linguistics, ACL 2014, 22–27 June 2014, Baltimore, Volume 1: Long Papers, pp. 511–521. The Association for Computer Linguistics (2014). https://doi.org/10.3115/v1/p14-1048
2. Hernault, H., Prendinger, H., duVerle, D.A., Ishizuka, M.: HILDA: a discourse parser using support vector machine classification. Dialog. Discourse **1**(3), 1–33 (2010). www.dad.uni-bielefeld.de/index.php/dad/article/view/591
3. Joty, S., Carenini, G., Ng, R., Mehdad, Y.: Combining intra-and multi-sentential rhetorical parsing for document-level discourse analysis. In: Proceedings of the 51st Annual Meeting of the Association for Computational Linguistics (Volume 1: Long Papers), pp. 486–496 (2013)
4. Kobayashi, N., Hirao, T., Kamigaito, H., Okumura, M., Nagata, M.: Top-down RST parsing utilizing granularity levels in documents. Proc. AAAI Conf. Artif. Intell. **34**(05), 8099–8106 (2020). https://doi.org/10.1609/aaai.v34i05.6321
5. Kong, F., Zhou, G.: A CDT-styled end-to-end Chinese discourse parser. ACM Trans. Asian Low Resour. Lang. Inf. Process. **16**(4), 26:1–26:17 (2017). https://doi.org/10.1145/3099557
6. Li, J., Sun, A., Joty, S.R.: Segbot: a generic neural text segmentation model with pointer network. In: Lang, J. (ed.) Proceedings of the Twenty-Seventh International Joint Conference on Artificial Intelligence, IJCAI 2018, 13–19 July 2018, Stockholm, pp. 4166–4172. ijcai.org (2018). https://doi.org/10.24963/ijcai.2018/579

7. Li, Y., Feng, W., Jing, S., Fang, K., Zhou, G.: Building Chinese discourse corpus with connective-driven dependency tree structure. In: Conference on Empirical Methods in Natural Language Processing (2014)
8. Lin, X., Joty, S.R., Jwalapuram, P., Bari, S.: A unified linear-time framework for sentence-level discourse parsing. In: Korhonen, A., Traum, D.R., Màrquez, L. (eds.) Proceedings of the 57th Conference of the Association for Computational Linguistics, ACL 2019, Florence, 28 July–2 August 2019, Volume 1: Long Papers, pp. 4190–4200. Association for Computational Linguistics (2019). https://doi.org/10.18653/v1/p19-1410
9. Lin, X., Joty, S.R., Jwalapuram, P., Bari, S.: A unified linear-time framework for sentence-level discourse parsing. arXiv preprint arXiv:1905.05682 (2019)
10. Mann, W.C., Thompson, S.A.: Rhetorical structure theory: toward a functional theory of text organization. Text **8**(3), 243–281 (1988)
11. Qiu, Y., Li, H., Li, S., Jiang, Y., Hu, R., Yang, L.: Revisiting correlations between intrinsic and extrinsic evaluations of word embeddings. In: Sun, M., Liu, T., Wang, X., Liu, Z., Liu, Y. (eds.) CCL/NLP-NABD -2018. LNCS (LNAI), vol. 11221, pp. 209–221. Springer, Cham (2018). https://doi.org/10.1007/978-3-030-01716-3_18
12. Sun, C., Kong, F.: A transition-based framework for Chinese discourse structure parsing. J. Chinese Inf. Process. (2018)
13. Wang, Y., Li, S., Wang, H.: A two-stage parsing method for text-level discourse analysis. In: Barzilay, R., Kan, M. (eds.) Proceedings of the 55th Annual Meeting of the Association for Computational Linguistics, ACL 2017, Vancouver, 30 July–4 August, Volume 2: Short Papers, pp. 184–188. Association for Computational Linguistics (2017). https://doi.org/10.18653/v1/P17-2029
14. Yang, Z., Dai, Z., Yang, Y., Carbonell, J.G., Salakhutdinov, R., Le, Q.V.: Xlnet: generalized autoregressive pretraining for language understanding. In: Wallach, H.M., Larochelle, H., Beygelzimer, A., d'Alché-Buc, F., Fox, E.B., Garnett, R. (eds.) Advances in Neural Information Processing Systems 32: Annual Conference on Neural Information Processing Systems 2019, NeurIPS 2019, 8–14 December 2019, Vancouver, pp. 5754–5764 (2019). www.proceedings.neurips.cc/paper/2019/hash/dc6a7e655d7e5840e66733e9ee67cc69-Abstract.html
15. Yu, N., Zhang, M., Fu, G.: Transition-based neural RST parsing with implicit syntax features. In: Bender, E.M., Derczynski, L., Isabelle, P. (eds.) Proceedings of the 27th International Conference on Computational Linguistics, COLING 2018, Santa Fe, 20–26 August 2018, pp. 559–570. Association for Computational Linguistics (2018). www.aclanthology.org/C18-1047/
16. Zhang, L., Kong, F., Zhou, G.: Adversarial learning for discourse rhetorical structure parsing. In: Zong, C., Xia, F., Li, W., Navigli, R. (eds.) Proceedings of the 59th Annual Meeting of the Association for Computational Linguistics and the 11th International Joint Conference on Natural Language Processing, ACL/IJCNLP 2021, (Volume 1: Long Papers), Virtual Event, 1–6 August 2021, pp. 3946–3957. Association for Computational Linguistics (2021). https://doi.org/10.18653/v1/2021.acl-long.305
17. Zhang, L., Xing, Y., Kong, F., Li, P., Zhou, G.: A top-down neural architecture towards text-level parsing of discourse rhetorical structure. In: Jurafsky, D., Chai, J., Schluter, N., Tetreault, J.R. (eds.) Proceedings of the 58th Annual Meeting of the Association for Computational Linguistics, ACL 2020, Online, 5–10 July 2020, pp. 6386–6395. Association for Computational Linguistics (2020). https://doi.org/10.18653/v1/2020.acl-main.569

A Malicious Code Family Classification Method Based on RGB Images and Lightweight Model

Chenyu Sun[1,2], Dawei Zhao[1,2], Shumian Yang[1,2(✉)], Lijuan Xu[1,2], and Xin Li[1,2]

[1] Key Laboratory of Computing Power Network and Information Security, Ministry of Education, Shandong Computer Science Center (National Supercomputer Center in Jinan), Qilu University of Technology (Shandong Academy of Sciences), Jinan, China
[2] Shandong Provincial Key Laboratory of Computer Networks, Shandong Fundamental Research Center for Computer Science, Jinan, China
yangshm@sdas.org

Abstract. In recent years, malware attacks have been a constant threat to network security, and the problem of how to classify malicious code families quickly and accurately urgently needs to be addressed. Traditional malicious code family classification methods are affected by the proliferation of variants to lead to failure and are no longer adequate for the current stage of research. The visualization method can maximize the malicious code core performance on the image, and the grayscale image has the problem of few and single features. In this paper, we propose a new malicious code visualization method. Specifically, we first convert the original malicious file into a byte file and an asm file using the IDA Pro tool. Secondly, we extract the opcode sequences in the asm file and the byte sequences in the byte file and convert them into a three-channel RGB image by using visualization techniques, which allows for a more comprehensive representation of the features of the malicious sample. Finally, we propose a new neural network architecture, the MobileNetV2 lightweight model combined with Convolutional Block Attention Module (MVCBAM) approach for training and prediction. In addition, we conduct various contrast experiments on the BIG2015 dataset and the Malimg dataset. The Experiments show that the classification accuracy of our proposed model on the two datasets is 99.90% and 99.95%, and the performance of our proposed model was maintained with fewer network parameters than the original MobileNetV2 model and has higher accuracy and faster speed than other advanced methods.

Keywords: Grayscale image · RGB image · Visualization technique · Deep learning · Lightweight model · Malicious code family classification

Supported by organization x.

1   Introduction

With the continuous development of computer technology, malware attacks have also followed, such as ransomware [24], data leakage, and other problems [20] emerge one after another, which have seriously threatened the privacy and security of the country, society, and individuals. According to Kaspersky Lab's 2020 Security Bulletin, between November 2019 and October 2020, 10.18% of user computers were subjected to at least one malware attack. According to data provided by AV-TEST, they found nearly 70 million new malware samples on the Windows platform in 2022. All this shows that malware attacks [19] are still the main way, and it is worth mentioning that the number of malware is increasing, and the number of malware variants [18] is also increasing, so finding ways to effectively classify malicious code and its variants has gradually become the mainstream in the field of cybersecurity.

Traditional static analysis methods [12], such as malicious code classification based on signature and malicious code classification based on heuristic, require a lot of manpower and material resources and require continuous updating of the code base. And through the obfuscation techniques, the characteristics of the malicious code itself are hidden, which can simply bypass the detection tool, which greatly increases the difficulty of classifying malicious code. As the number of variants grows faster and faster, the traditional static analysis method of malicious code has failed. However, by converting malicious code into images [7], using the characteristics of certain similarities in the texture of malicious code images of the same family, the two different scenarios of image processing and malicious code family classification have been successfully unified, and it has been proved to have good results in the field of malicious code family classification. However, most of the current research is based on single-channel grayscale images, which contain a little and single feature, and can not resolve the current malicious code family classification problem [3].

With the influx of artificial intelligence, deep learning has gradually been introduced into the field of malicious code classification [14], compared with traditional machine learning [6] has a deep enough network, can accommodate richer semantic information. And deep learning can save a lot of manpower and material resources to analyze features, and with the continuous increase of data volume, the effect will continue to improve. At the same time, it also has the disadvantages of large calculations and high costs. Aiming at the problem that the currently known classification methods can not make full use of the rich information contained in malicious samples and the inefficiency of deep learning networks. Our contributions in this paper are as follows:

- In order to fully extract the feature information of the malicious samples, a new visualisation method is proposed. First we extract the opcode sequences in the asm file and convert them into two different greyscale images, one is a greyscale image converted using the N-gram method and the other is a Markov image. Then we extract the byte sequence in the byte file and convert the first 256 byte sequences into a greyscale image, and finally we fuse these three greyscale images containing different features into an RGB colour image.

– We devise a framework for malicious family classification which combines MobileNetV2 and Convolutional Block Attention Module (MVCBAM), and compared to other neural network models, our proposed model has fewer bottleneck structures, which reduces the training computation and number of parameters.

– We conduct multifaceted contrast experiments on the BIG2015 dataset and the Maling dataset. The experimental results show that the fused color image is more accurate than the single-channel grayscale image and can correctly classify malicious code families. In addition, compared with the conventional classification model, the MobileNetV2 model with CBAM significantly improved classification efficiency.

The rest of this article is organized as follows: Sect. 2 discusses the related work in the field of classifying malicious code families. Section 3 presents the overall framework of our proposed method and the specific implementation steps for each module. Section 4 shows the configuration and evaluation metrics used for our experiments. Section 5 presents the datasets used in our experiments and the experimental results and analyses we performed. Finally, in Sect. 6, we summarize the full text and plan the direction of our future work.

2   Related Works

In recent years, visualization techniques have been widely used in the field of malicious code family classification. Compared with traditional static analysis methods, visualization methods can retain the feature information of malicious samples intact, and also allow a more intuitive observation of the similarities and differences of each malicious sample, solving the impact caused by obfuscation techniques.

Ni et al. [8] proposed a malicious code classification method using SimHash and CNN. They converted the disassembled malicious code into a grayscale image based on SimHash, and then CNN classified the malicious code family. The method successfully exploited the features provided by the opcode sequences and achieved very high accuracy in malicious code classification. Conti et al. [2] proposed a GEM image which is a combination of a gray-level matrix image, Markov image, and entropy grayscale image, followed by training using an improved CNN architecture. Vasan et al. [16] proposed a malicious code visualization image classification method. The original malicious samples were first converted to color images, followed by the detection and identification of malicious code families using a fine-tuned CNN architecture. Chaganti et al. [1] proposed an efficient neural network model based on EfficientNetB1, which can reduce the computational resource consumption caused during the training and testing of deep learning models. Used three different types of malicious code image representations, the model was evaluated for performance evaluation and effective scaling of network model parameters, but they did not use feature fusion for training and testing. Xue et al. [21] used grayscale images as static features, variable n-gram sequences as dynamic features, used IDF to select meaningful features, and after that used the probabilistic scoring method to connect dynamic and static features, and finally had high classification accuracy.

According to the above content, the existing research usually adopts grayscale processing for malicious binary files, extracts features from grayscale images, and combines neural networks for training and experimental results, but grayscale images are single-channel data images, which contain fewer and single features of malicious code information, which leads to grayscale images in neural networks that are not as intuitive and effective as three-channel data images.

3   Proposed Approach

In this section, we propose a framework for malicious code family classification using lightweight neural network models. The framework consists of a visualization module and an MVCBAM classification module. In the visualization module, we first extract the original opcode sequences in the asm files, and two grayscale images were obtained using two different methods, then extract the byte sequences in the byte files, also generate the grayscale image, and finally fills the three grayscale images into the three RGB channels of the color image to generate a multi-feature fused RGB color images, and then the generated RGB color images are fed into our improved lightweight neural network MVCBAM for classification. Figure 1 depicts the overall framework of our proposed method.

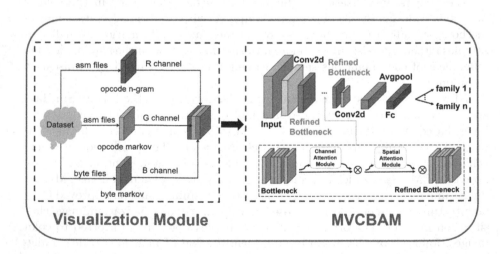

Fig. 1. Overall framework

3.1   Visualization Module

This module extracts different features from malicious samples that have been proven effective in previous work and finally generates a three-channel RGB image in the following steps.

Grayscale Image Based on Opcode Frequency. We use the asm files generated by the disassembly tool IDA Pro disassembly, an assembly instruction is composed of an opcode and multiple operands, our method discards the operands, and only the opcodes is utilized, a single opcode in malware does not have an impact, but the opcode sequence compose of multiple opcodes can be harmful. So we can use the N-gram algorithm to extract some sequences of consecutive opcodes as features, We extract all the opcodes of each malicious asm file. Because in practice N is large that will lead to too large parameter space and will result in dimensional disasters, we choose every three opcodes as a subsequence, after counting the frequency of these subsequences, the top 256 highest subsequence frequencies for each malicious asm files are selected, supplement 0 of less than 256, then fill into the diagonal position of a matrix of size 256×256, finally, a grayscale image is generated, the process is shown in Fig. 2(a).

Markov Image Based on Byte Sequences. In this paper, we extract the byte sequences of each malicious byte file to generate a Markov image. For unification with the image of the first channel, we also generate a matrix initialized to zero and a size of 256×256. In bytes files, we put hexadecimal bytes convert to decimal bytes of 0–255, after the conversion, we set a sliding window of size 2, and in the two consecutive bytes, the first byte is regarded as a row, the second byte is regarded as a column, the corresponding position is added 1, slide to the end of the last byte, generate a two-dimensional transition frequency matrix, after the calculation the ratio of the frequency of each position on the matrix to the sum of the frequencies of each row, because the floating-point type is more complex, we multiply the ratio by 255, become an integer and fill it into the zero matrix, generate a two-dimensional transition probability matrix, the values at each position of the two-dimensional transition probability matrix can be populated as pixel points of a Markov image. The Markov image generation process is shown in Fig. 2(b).

Markov Image Based on Opcode Sequences. The first step is to extract all the opcode sequences in the .text field of all malicious asm files, count the top 255 most frequent opcode sequences, and put the 255 opcode sequences into a one-dimensional array, which is filled with "aaa" at the end of the one-dimensional array, and named a. Then, extract the opcode sequences in the .text field of single malicious asm files and store it in array b, and then determine whether the type of the opcode sequence in array a is the same as the type of the opcode sequence in b, i.e. use the setdiff1d function to find the difference between a

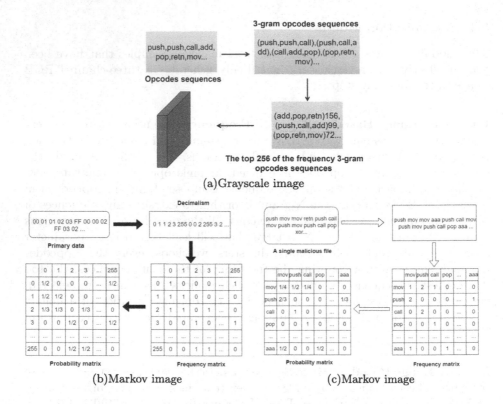

Fig. 2. Three different grayscale images generation process.

and b. If the opcode types of these two arrays are the same, the result of the difference is set to null, otherwise, the unique values in a but not in b are output, and the unique values are set to "aaa" and the output is also a sequence of 256 opcodes of the same type.

Finally, a sliding window of size 2 is set up and the first operand is treated as a row and the second operand as a column, with 1 added to the corresponding position, sliding to the end of the last operand of a single file, after which a two-dimensional transfer probability matrix is generated and finally the same can be visualised as a Markov image. The exact process is shown in Fig. 2(c).

We statistically extract the operand sequences with the most occurrences in the .text field, which represents that these operands are prominent in the overall effect, but extracting the operand sequences in a single malicious asm file and performing the operation of finding the difference is equivalent to extracting the operand sequences with prominent effect in a single malicious asm file, which is the process of feature selection.

Generation of RGB Images. A single feature image has different data sets, leading to different performance problems, to make our method perform well on all datasets, we propose to fuse a single feature grayscale image into a three-channel color image, and the effective information of the color image is three times that of the grayscale image, and we fill the three feature images extracts above into the three channels of RGB to generate a 256 × 256 × 3 image.

3.2 MVCBAM

Since the emergence of Alexnet in 2012, we have begun to widely use deep learning [23] and convolutional neural networks, and these network error rates are getting lower and lower, the depth is getting deeper and deeper, and the number of parameters is increasing, but these are at the cost of the network becoming more bloated and the amount of computation significantly increased, in addition, the attention mechanism [13] has been widely used in neural network models, which can effectively discover key information and increase the weight of key information, so in the case of ensuring accuracy, We propose a method combining MobilNetV2 model and CBAM to classify color images to improve the classification efficiency.

MobileNetV2 Network. The MobileNetV1 network replaces the standard convolution in VGG networks using depth separable convolution, which consists of Depthwise Conv and Pointwise Conv. The MobileNetV2 network [11] is an improvement over the MobileNetV1 network, in summary, two improvements are shown in Fig. 3:

(1) It is found that many of the convolution kernels trained by MobileNetV1 networks using depth separable convolutions are empty, because the ReLU activation function is easy to cause information loss in the face of low-dimensional operations, so MobileNetV2 proposes a Linear activation function to solve the loss of features.

(2) The MobileV2 network combines the residual structure in Resnet and proposes an inverted residual structure. A layer of PW convolution is added before DW convolution, which solves the problem that the number of DW convolution channels cannot be changed and the effect is not good.

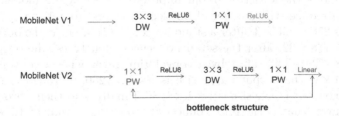

Fig. 3. Improvement of MobileNetV2 Compared to MobileNetV1

CBAM. CBAM (Convolutional Block Attention Module) is a lightweight attention module [22]. It has two modules: the channel attention module is shown in the Fig. 4(a) and the formula is as shown in (1), and the spatial attention module is shown in the Fig. 4(b) and the formula is as shown in (2), with two modules arranged serially, where the channel attention module is by calculating the importance of each channel of the input feature map, if there is a channel contains key information, then focus on it, but will ignore the local information in each channel, The spatial attention module pays different attention to different positions of the input feature map, and does the same processing for each channel, but this ignores the different characteristics of different channels, so in this paper we choose CBAM, which can capture feature information from two dimensions to improve the overall performance.

$$C(M) = \sigma(MLP(MaxPool(M)) + MLP(AvgPool(M))) \qquad (1)$$

where Eq. (1), σ represents the Sigmoid function, MLP represents the shared MLP, and MaxPool and AvgPool represent maximum pooling and average pooling.

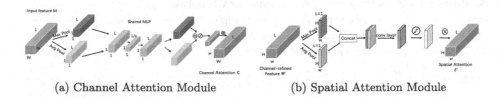

(a) Channel Attention Module   (b) Spatial Attention Module

Fig. 4. Two modules included in the CBAM

$$S'(M) = \sigma(f^{7 \times 7}([MaxPool(M); AvgPool(M)])) \qquad (2)$$

where Eq. (2), σ is the sigmoid function, and $f^{7 \times 7}$ is a convolution kernel of size 7×7.

Table 1 shows the structure of our improved MobileNetV2 model. Our proposed model reduces the number of bottleneck layers from 17 to 7. The input image size is $256 \times 256 \times 3$, after a standard convolution layer, the output image size is $128 \times 128 \times 32$, after the seven bottleneck structures, the output image size is $8 \times 8 \times 320$. Thereafter by a convolution layer with a convolution kernel of size 1×1, the output image size is $8 \times 8 \times 1280$. Through after average pooling, the output image size is $1 \times 1 \times 1280$. Finally, and then, through a convolutional layer, that is the fully connected layer, the output of 1,280 neurons is fully connected to the k neurons in the Softmax, and the data is output after being trained. As shown in Table 1, t is the expansion factor, c is the number of output feature channels, n is the number of iterations of the bottleneck, and s represents the stride, which controls the size of the feature map.

Table 1. Improved MobileNetV2 model structure

Input	Operator	t	c	n	s
$256^2 \times 3$	conv2d	–	32	1	2
$128^2 \times 32$	bottleneck	1	16	1	1
$128^2 \times 16$	bottleneck	6	24	1	2
$64^2 \times 24$	bottleneck	6	32	1	2
$32^2 \times 32$	bottleneck	6	64	1	2
$16^2 \times 64$	bottleneck	6	96	1	1
$16^2 \times 96$	bottleneck	6	160	1	2
$8^2 \times 160$	bottleneck	6	320	1	1
$8^2 \times 320$	1×1	–	1280	1	1
$8^2 \times 1280$	avgpool 8×8	–	–	1	–
$1 \times 1 \times 1280$	conv2d 1×1	–	k	–	–

Based on the improved MobileNetV2 model described above, we add a CBAM module after the last 1×1 convolutional layer of each bottleneck structure, and the shortcut connection is only available when stride=1 and the input and output feature matrices are of the same shape, as shown in Fig. 5(1). The rest of the cases are shown in Fig. 5(2). Which can effectively find the key features in classification detection, increase their weights and effectively reduce information loss. It increases the interpretability of the neural network model while improving the model's performance, which can effectively find the key features in classification detection, increase their weights and effectively reduce information loss. It increases the interpretability of the neural network model while improving the model performance.

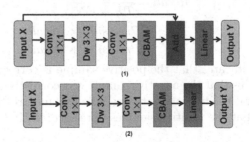

Fig. 5. The bottleneck structure of CBAM is added

Table 2. The BIG2015 dataset

Family Name	Class Id	Number of Samples
Ramnit	1	1541
Lolipop	2	2478
Kelihos_vers	3	2942
Vundo	4	475
Simda	5	42
Tracur	6	751
Kelihos_verl	7	398
Obfuscator	8	1228
Gatak	9	1013

4 Experimental Configuration and Evaluation Metrics

This experiment uses NVIDIA RTX3080 GPU for model training, and the main programming language environment for this experiment is Python3.7. The hyperparameters set in this paper: batch size is 128, training period is 50, learning rate is 0.0001, optimiser is ADAM. To better evaluate our experimental results, we use accuracy, precision, recall, F1-score, and other indicators to conduct statistical analysis of experimental results.

$$Accuracy = \frac{TP + TN}{TP + TN + FP + FN} \tag{3}$$

$$Precision = \frac{TP}{TP + FP} \tag{4}$$

$$Recall = \frac{TP}{TP + FN} \tag{5}$$

$$F1\text{-}score = 2 \times \frac{Precision \times Recall}{Precision + Recall} \tag{6}$$

Where the TP indicates that the positive sample is correctly tested as positive. The TN indicates that the negative sample was correctly tested as negative. The FP indicates that the finger negative sample was incorrectly tested positive. The FN indicates that the positive samples were incorrectly tested negative.

5 Experimental Results and Analysis

5.1 Dataset

As shown in Table 2, the BIG2015 dataset contains 21741 samples from 9 malicious families, of which 10868 samples are labeled training sets and others are unlabeled test sets. Each malicious sample contains two files, bytes files with the

PE header removed in hexadecimal representation, and asm files generated by the disassembly tool IDA Pro containing the machine code of the malicious sample, assembly instructions, etc. In this paper, only the training set with family tags is used for experiments, 80% is used to train the model, and 20% is used to test the model. As shown in Table 3, the Malimg dataset contains 9342 malware images processed in malicious binaries, divided into 25 categories. Again, 80% of the data is used for training and 20% for testing.

Table 3. The Malimg dataset

Family Name	Class Id	Number of Samples	Family Name	Class Id	Number of Samples
Adialer.C	1	122	Lolyda.AA2	14	184
Agent.FYI	2	166	Lolyda.AA3	15	123
Allaple.A	3	2949	Lolyda.AT	16	159
Allaple.L	4	1591	Malex.gen!J	17	136
Alueron.gen!J	5	198	Obfuscator.AD	18	142
Autorun.K	6	106	Rbot!gen	19	158
C2LOP.P	7	200	Skintrim.N	20	80
C2LOP.gen!g	8	146	Swizzor.gen!E	21	128
Dialplatform.B	9	177	Swizzor.gen!I	22	132
Dontovo.A	10	162	VB.AT	23	408
Fakerean	11	381	Wintrim.BX	24	97
Instantaccess	12	431	Yuner.A	25	800
Lolyda.AA1	13	213	Total		9342

5.2   Experiments on the BIG2015 Dataset

Performance Comparison of Single Grayscale Images and RGB Images. To explore the classification performance of single grayscale images and RGB images, we input grayscale images based on opcode frequency, Markov images based on byte sequences, Markov images based on opcode sequences, and fused RGB images into our improved model for classification training. The experimental results are shown in Fig. 6. Through the observation of the graph, using our model for training, the effect of a single grayscale plot varies, but the overall effect is not as good as the fused RGB image, and the classification accuracy can be as high as 99.90% when using RGB image for training, and the F1-score value can also reach 99.5%.

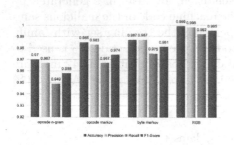

Fig. 6. RGB images experimental results

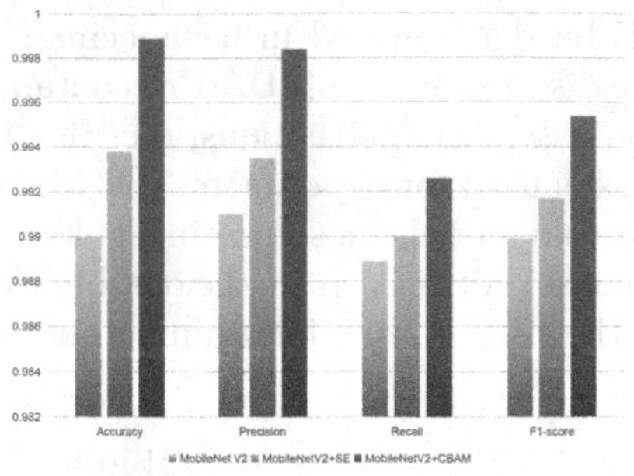

Fig. 7. Effects of different attention mechanisms

Effects of Introducing the Attention Mechanism Module. As can be seen from Fig. 7, the MobileNet V2 neural network model without the attention mechanism performed less on all four metrics than the two models with the attention mechanism, and our CBAM is also higher than the traditional SE attention mechanism method.

Classification Effect of RGB Images in Nine Families. To further prove that the fused RGB images have a good effect, because the number of malicious samples in each family varies greatly, to prove that our model can have good generalization ability in the case of unbalanced data, we use the proposed classification model to train nine families in the dataset, the experimental results are shown in Table 4, it can be seen from the figure that in addition to family 5 and family 7, the F1-score of other families exceeds 98.8%, of which the third family has reached 1. Because it has the smallest number of malicious samples.

Table 4. Nine families classification effects

Class Id	Precision	Recall	F1-score
1	0.9934	0.9837	0.9885
2	0.9979	0.9980	0.9929
3	1	1	1
4	0.9894	0.9852	0.9872
5	0.8	1	0.8888
6	0.9933	0.9867	0.9900
7	0.9873	0.9821	0.9795
8	0.9906	0.9876	0.9890
9	0.9985	0.9918	0.9951

Comparison with the Other Traditional Classification Models. To explore the effectiveness of the classification model proposed in this paper, we compare it with the traditional classification model in terms of the accuracy and loss value of the training set and the test set, the accuracy and loss value change curve with the number of training rounds are shown in Fig. 8(a), (b). From the figure, we can see that compared with other traditional classification models, our model curve has always been stable, whether it is the training set or the test set, it has the highest accuracy and the lowest loss, so the performance of our model is higher than that of other models, and the convergence speed is relatively fast.

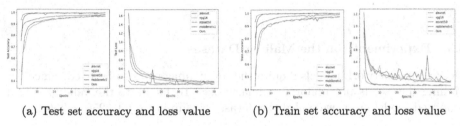

(a) Test set accuracy and loss value   (b) Train set accuracy and loss value

Fig. 8. Comparisions with the traditional classification detection model

Comparison of Parameter Amount, Calculation Amount and Training Speed of Different Classification Models. As can be seen from Table 5, our model is relatively small compared to other models in terms of both parameters and computation, and the time required to train an epoch is greatly reduced.

Performance Comparison with Other Advanced Models. To further verify the effectiveness of our proposed model, we compare other advanced models utilizing the BIG2015 dataset, and the comparison results are shown in Table 6, which shows that our model performs best in terms of accuracy, precision, and F1-score.

Table 5. Results of different classification models

Methods	Params	Flops	Time required for an epoch training
Alexnet	310.08M	14.6M	15.19 s
VGG16	15.44G	70.3M	55.44 s
ResNet50	5.38G	23.53M	32.13 s
MobileNetV1	710.22M	3.22M	17.51 s
Ours	416.69M	2.24M	12.75 s

Table 6. Comparisons with the effects based on the BIG2015

Models	Method	Accuracy	F-score
Jian et al. [3]	RGB+SERLA	0.9831	0.9830
Conti et al. [2]	RGB+Shallow-CNN	0.9856	0.9853
Nataraj et al. [7]	grayscale	0.9782	0.9579
Ni et al. [8]	grayscale+MCSC	0.9886	–
Qiao et al. [10]	RGB+LeNet5	0.9876	–
Pinhero et al. [9]	RGB+VGG3/ResNet50	0.9825	0.9825
Ours	RGB+MVCBAM	0.9990	0.9954

5.3   Experiments on the Malimg Dataset

Most neural network models require the input image size to be fixed. Since the Malimg dataset is not uniform in image size. To facilitate the validity of our proposed model later, we scaled the dataset size uniformly to $256 \times 256 \times 3$.

Comparison with the Other Traditional Classification Models. To explore the effectiveness of the classification detection model proposed in this paper, we compare it with the traditional classification model, the training set accuracy and loss value, and the test set accuracy and loss value change curve with the number of training rounds are shown in Fig. 9(a), (b). From the figure, we can see that compared with other traditional classification models, our proposed model maintains good performance during training and testing, especially with high accuracy compared with lightweight models MobileNetV1 and MobileNetV2.

Table 7. Comparisons with the effects based on the Malimg dataset

Models	Method	Accuracy
Tang et al. [15]	BHMDC	0.9977
Zou et al. [25]	IMCLNet	0.9978
Kumar et al. [4]	DTMIC	0.9892
Sudhakar et al. [5]	MCFT-CNN	0.9918
Xiao et al. [17]	MalFCS	0.9972
Ours	MVCBAM	0.9995

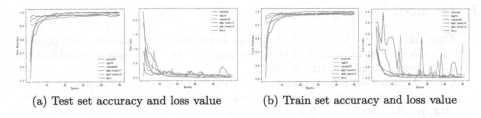

(a) Test set accuracy and loss value   (b) Train set accuracy and loss value

Fig. 9. Comparisions with the traditional classification detection model

Performance Comparison with Other Advanced Models. To further verify the effectiveness of our proposed model, we compare other advanced models utilizing the Malimg dataset, and the comparison results are shown in Table 7, which shows that our model performs best in terms of accuracy.

6 Conclusions and Future Work

In this paper, we propose a malware classification framework based on RGB color images and lightweight neural network models, which visualizes the malicious samples as malware RGB images. Each RGB image is generated from three different grayscale images, this method furthest retains the characteristics of malicious samples. Experiments show that the generated RGB images have excellent classification ability and effectively improved the accuracy of malware classification. In addition, the classification accuracy using our lightweight model is 99.90% and 99.95% in two large benchmark datasets, respectively, and the number of parameters and computations is significantly reduced compared to other methods in the literature.

In future work, we will extract dynamic features, which will help us to cope with the impact of obfuscation techniques. In addition, our proposed method extracts opcode sequences and byte sequences, and in the future, we will try to convert other features into grayscale images to test whether the performance will be improved.

Acknowledgements. This work was supported in part by the Natural Science Foundation of Shandong Province (ZR2021MF132 and ZR2020YQ06), in part by the National Natural Science Foundation of China (62172244), in part by the National Major Program for Technological Innovation 2030-New Generation Artifical Intelligence (2020AAA0107700), in part by the Innovation Ability Pormotion Project for Small and Medium-sized Technology-based Enterprise of Shandong Province (2022TSGC2098), in part by the Pilot Project for Integrated Innovation of Science, Education and Industry of Qilu University of Technology (Shandong Academy of Sciences) (2022JBZ01-01), in part by the Taishan Scholars Program (tsqn202211210), in part of by the Graduate Education and Teaching Reform Research Project of Shandong Province (SDYJG21177), in part by the Education Reform Project of Qilu University of Technology (2021yb63).

References

1. Chaganti, R., Ravi, V., Pham, T.D.: Image-based malware representation approach with efficient net convolutional neural networks for effective malware classification. J. Inf. Secur. Appl. **69**, 103306 (2022)
2. Conti, M., Khandhar, S., Vinod, P.: A few-shot malware classification approach for unknown family recognition using malware feature visualization. Comput. Secur. **122**, 102887 (2022)
3. Jian, Y., Kuang, H., Ren, C., Ma, Z., Wang, H.: A novel framework for image-based malware detection with a deep neural network. Comput. Secur. **109**, 102400 (2021)
4. Kumar, S., Janet, B.: DTMIC: deep transfer learning for malware image classification. J. Inf. Secur. Appl. **64**, 103063 (2022)
5. Kumar, S., et al.: MCFT-CNN: malware classification with fine-tune convolution neural networks using traditional and transfer learning in internet of things. Futur. Gener. Comput. Syst. **125**, 334–351 (2021)
6. Liu, L., Wang, B.S., Yu, B., Zhong, Q.X.: Automatic malware classification and new malware detection using machine learning. Front. Inf. Technol. Electron. Eng. **18**(9), 1336–1347 (2017)
7. Nataraj, L., Karthikeyan, S., Jacob, G., Manjunath, B.S.: Malware images: visualization and automatic classification. In: Proceedings of the 8th International Symposium on Visualization for Cyber Security, pp. 1–7 (2011)
8. Ni, S., Qian, Q., Zhang, R.: Malware identification using visualization images and deep learning. Comput. Secur. **77**, 871–885 (2018)
9. Pinhero, A., et al.: Malware detection employed by visualization and deep neural network. Comput. Secur. **105**, 102247 (2021)
10. Qiao, Y., Jiang, Q., Jiang, Z., Gu, L.: A multi-channel visualization method for malware classification based on deep learning. In: 2019 18th IEEE International Conference on Trust, Security and Privacy in Computing and Communications/13th IEEE International Conference on Big Data Science and Engineering (TrustCom/BigDataSE), pp. 757–762. IEEE (2019)
11. Sandler, M., Howard, A., Zhu, M., Zhmoginov, A., Chen, L.C.: Mobilenetv 2: inverted residuals and linear bottlenecks. In: Proceedings of the IEEE Conference on Computer Vision and Pattern Recognition, pp. 4510–4520 (2018)
12. Sebastio, S., et al.: Optimizing symbolic execution for malware behavior classification. Comput. Secur. **93**, 101775 (2020)
13. Shen, G., Chen, Z., Wang, H., Chen, H., Wang, S.: Feature fusion-based malicious code detection with dual attention mechanism and BILSTM. Comput. Secur. **119**, 102761 (2022)
14. Tang, C., Xu, L., Yang, B., Tang, Y., Zhao, D.: GRU-based interpretable multivariate time series anomaly detection in industrial control system. Comput. Secur. 103094 (2023)
15. Tang, Y., Qi, X., Jing, J., Liu, C., Dong, W.: BHMDC: a byte and hex n-gram based malware detection and classification method. Comput. Secur. **128**, 103118 (2023)
16. Vasan, D., Alazab, M., Wassan, S., Naeem, H., Safaei, B., Zheng, Q.: IMCFN: image-based malware classification using fine-tuned convolutional neural network architecture. Comput. Netw. **171**, 107138 (2020)
17. Xiao, G., Li, J., Chen, Y., Li, K.: MALFCS: an effective malware classification framework with automated feature extraction based on deep convolutional neural networks. J. Parallel Distrib. Comput. **141**, 49–58 (2020)

18. Xu, L., Wang, B., Yang, M., Zhao, D., Han, J.: Multi-mode attack detection and evaluation of abnormal states for industrial control network. J. Comput. Res. Develop. **58**(11), 2333–2349 (2021)

19. Xu, L., Wang, B., Wang, L., Zhao, D., Han, X., Yang, S.: Plc-seiff: a programmable logic controller security incident forensics framework based on automatic construction of security constraints. Comput. Secur. **92**, 101749 (2020)

20. Xu, L., Wang, B., Wu, X., Zhao, D., Zhang, L., Wang, Z.: Detecting semantic attack in SCADA system: a behavioral model based on secondary labeling of states-duration evolution graph. IEEE Trans. Netw. Sci. Eng. **9**(2), 703–715 (2021)

21. Xue, D., Li, J., Lv, T., Wu, W., Wang, J.: Malware classification using probability scoring and machine learning. IEEE Access **7**, 91641–91656 (2019)

22. Yang, N., He, C.: Malaria detection based on resnet+ cbam attention mechanism. In: 2022 3rd International Conference on Information Science, Parallel and Distributed Systems (ISPDS), pp. 271–275. IEEE (2022)

23. Zhao, D., Xiao, G., Wang, Z., Wang, L., Xu, L.: Minimum dominating set of multiplex networks: definition, application, and identification. IEEE Trans. Syst. Man Cybernet. Syst. **51**(12), 7823–7837 (2020)

24. Zhu, J., Jang-Jaccard, J., Singh, A., Welch, I., Harith, A.S., Camtepe, S.: A few-shot meta-learning based Siamese neural network using entropy features for ransomware classification. Comput. Secur. **117**, 102691 (2022)

25. Zou, B., Cao, C., Tao, F., Wang, L.: IMCLNet: a lightweight deep neural network for image-based malware classification. J. Inf. Secur. Appl. **70**, 103313 (2022)

Research on Relation Extraction Based on BERT with Multifaceted Semantics

Meng Dong[1] and Xinhua Zhu[2]([✉])

[1] School of Computer Science and Information Engineering, Guangxi Normal University, Guilin 541004, China
501nlp@stu.gxnu.edu.cn
[2] Guangxi Key Laboratory of Multi-Source Information Mining and Security, Guangxi Normal University, Guilin 541004, China
zxh429@263.net

Abstract. Relation extraction is one of the important tasks in natural language processing, aiming to determine the class of relations to which the entities in a sentence belong. In nowadays, researchers tend to use large-scale corpus to retrain language models for relation extraction, which requires too much relevant resources, besides, when analyzing the corpus, we found that in sentences where entities exist, different entities have different correlations with different parts of the sentence, and this may affect the performance of relation extraction. Based on this, this paper proposes a model based on BERT [1] with multifaceted semantics (BERT-LR) for relation extraction, which learns semantics from multiple aspects centered on entities, and is able to better understand the relationship between entities at different locations and the context and perform relation extraction. First, we make full use of the BERT [1] model that has been pre-trained to provide rich initialization parameters for our model. Second, in order to achieve entity-centered relationship extraction, we propose a BERT [1] and multifaceted semantic relationship extraction model based on BERT [1] with multifaceted semantics consisting of left semantics, right semantics, and global semantics, and use a suitable method to fuse the multifaceted semantics. Third, we found that fixing the embedding layer of the model during the fine-tuning process can achieve better results. Our approach achieves excellent results on the SemEval-2010 Task 8 [2] dataset.

Keywords: BERT-LR · Multifaceted semantics · Embedding layer

1 Introduction

The relationship extraction task refers to identifying and extracting the relationships between entities in a sentence from a given text. The supervised relationship extraction datasets are labeled with the locations of the entities, and

School of Computer Science and Information Engineering, Guangxi Normal University.

this paper focuses on such tasks. For example, *"The <e1> company </e1> fab-ricates plastic <e2> chairs </e2>."* we need to identify the relation Product-Producer(e2, e1). This is a very important task that can be used as an aid for tasks such as building knowledge graphs, text reading comprehension, and information extraction, and has a wide range of applications.

Since the concept of deep learning was first formally introduced by Hinton et al [3] in 2006, it has been gradually applied to relationship extraction by researchers over the years, and various neural network-based methods have been applied to relationship extraction, such as convolutional neural network (CNN) [4–9], recurrent neural network (RNN) [10–15], and attention mechanisms [16,17], all of which have achieved excellent results at that time. In 2017, Transformers [18] was released, which is a model based entirely on attention mechanism, completely abandoning loops and convolutions, followed by BERT [1], which used the mask mechanism as well as next sentence prediction (NSP) as a pre-training task to pre-train the Transformers [18] encoder by self-supervision, and achieved SOTA results in 11 NLP tasks, opening the era of pre-trained models.

Pre-trained models have indeed been applied to relation extraction as well [19–27], with extremely good results. However, as research progresses, more and more researchers tend to inject external knowledge or use large-scale corpus to re-train the models when performing relation extraction, making the methods tend to be complex, which is challenging for smaller labs. Inspired by the paper [28], we found that in the sample drawn from the relationship, the head entity and the tail entity have different correlations with the parts of the sample at different locations, for example, in a sample, the head entity has less correlation with the part of the sample after the tail entity. Based on this, this paper proposes a model (BERT-LR) for relationship extraction based on BERT [1] with multifaceted semantics, we first split the sample into three parts, encoded by BERT [1], and take their sentence encoding (embedding a special first token [CLS] in the setting of BERT [1]) to indicate the left semantics, right semantics, and global semantics, and fuse them to perform the final classification, meanwhile, we fix different embedding layers of BERT [1] for better results when fine-tuning, and the main contributions of this paper are as follows:

(1) Inspired by the paper [28], we propose a model for relation extraction based on BERT [1] with multifaceted semantics

(2) We found that fixing the embedding layer of the model during the fine-tuning process can achieve better results

2 Related Work

In 2014, Zeng [4] et al. used convolutional neural network (CNN) to extract lexical and sentence features without complex preprocessing, making it the most advanced method at that time. In addition to CNN, recurrent neural networks (RNN) have also been applied to relation extraction, for example, Socher [10] et al. used RNN for relation extraction, which used recurrent neural networks to syntactically parse sentences in annotated text and obtained a vector representation of sentences after continuous iteration, effectively considering the syntactic structure of sentences. However, RNN still have shortcomings, and to solve the gradient disappearance and gradient explosion problems, long short-term memory network (LSTM) have also been applied to relation extraction, and Xu [13] et al. and Zhang [15] et al. used LSTM for relation extraction with excellent results. Shen [16] et al. combined convolutional neural networks and attention mechanisms, and were able to extract words that are important to the sentence meaningfully important words.

In 2019, Hennig [19] et al. based the decoder of Transformers [18] on implicit features captured in pre-trained linguistic representations instead of explicit linguistic features required by previous methods, with comparable performance on two relation extraction datasets Tacred [29] and SemEval-2010 Task8 [2]. Joshi [20] et al. improved BERT [1] by proposing a better mask scheme, instead of adding a mask to a random single token, they added a mask to a random pair of neighboring clauses and proposed a new training objective, which enhanced the performance of BERT [1] and achieved excellent results on the relation extraction task. As for Entity-Aware Bert [21], adding an attention layer for predicting structure and entity awareness on top of BERT [1], multiple relations can be extracted in one encoding. Soares [22] et al. added the MTB (Matching The Blanks) task to the pre-training process of BERT [1] and was investigated as a relation encoder to improve the performance of relation extraction. Wu [23] et al. proposed the R-BERT model by focusing on the location of entities in a sentence, using special identifiers to label entities, and extracting vectors of entities for use as classification. KnowBert-W+W [24] proposed a generic approach to embed multiple knowledge bases into large-scale models, thus augmenting the representation of entities with structured, manually curated knowledge. Tian [25] et al. proposed a dependency-driven approach for relation extraction using attention graph convolutional network (A-GCN), and subsequently, Tian [26] et al. realized the syntactic importance of information and trained syntax-induced encoders to improve the relation extraction task. Finally, Li [27] et al. proposed the SPOT model, which learns the representation of entities and relations from the token span and span in the text, respectively, and pre-trained the SPOT model using the knowledge graph extracted from Wikipedia, successfully describing an approach to integrate knowledge information into the linguistic representation model with the most competitive results.

3   Methodology

3.1   Related Introduction

The purpose of relationship extraction is to determine the relationship between the entities contained in a sample, and it is necessary to perform entity identification to mark the entities contained in the sample without labeling the entities, and then perform relationship determination. However, nowadays, relationship extraction data sets are often labeled with the location of the entities, so we often treat the relationship extraction task as a text classification task, and the entities contained in them can be classified into a certain category.

In this paper, the model based on BERT [1] with multifaceted semantics refers to extracting the semantics centered on head entities and tail entities, including left semantics, right semantics and original semantics from the samples by BERT [1] model and integrating them for the final classification, and the specific model structure is shown in Sect. 3.2.

3.2 Model Architecture

Processing Sample. Table 1 shows some samples and Table 2 shows the process of processing the samples

Table 1. Samples

Other	A misty \<e1\> ridge \</e1\> uprises from the \<e2\> surge \</e2\>
Message-Topic(e1,e2)	The Pulitzer Committee issues an official \<e1\> citation \</e1\> explaining the \<e2\> reasons \</e2\> for the award
Message-Topic(e2,e1)	Bob Parks made a similar \<e1\> offer \</e1\> in a \<e2\> phone call \</e2\> made earlier this week

Table 2. Processing Sample

Left sample:	He had chest pains and \<e1\>headaches\</e1\> from \<e2\>mold\</e2\>
Right sample:	\<e1\>Headaches\</e1\> from \<e2\>mold\</e2\> in the bedrooms
Original sample:	He had chest pains and \<e1\>headaches\</e1\> from \<e2\>mold\</e2\> in the bedrooms

For a given sample S, each sample contains a head entity E1 and a tail entity E2. To enable the model to better learn the semantics of the samples from multiple aspects, we split the samples centered on E1 and E2, for a sample "*He had chest pains and \<e1\> headaches \</e1\> from \<e2\> mold \</e2\> in the bedrooms.*", we take "*He had chest pains and \<e1\> headaches \</e1\> from \<e2\> mold \</e2\>*" as the left sample (LS), "*\<e1\> headaches \</e1\> from \<e2\> mold \</e2\> in the bedrooms.*" as the right sample (RS), and the original sample (OS), which are fed into the same BERT [1] model for encoding. In order to enhance the perception of entities, we add entities E1 and E2 as sentence pairs after each sample, and the entities are separated by a special symbol '[SEP]', which forms the second sentence in the form of "*E1[SEP]E2*", the details are as follows:

$$LS = LS + [SEP] + E1[SEP]E2$$

$$RS = RS + [SEP] + E1[SEP]E2$$

$$OS = OS + [SEP] + E1[SEP]E2 \tag{1}$$

The samples are then fed into BERT [1] and we get:

$$H^s = sharedBERT(S) \in R^{n \times d} \tag{2}$$

$sharedBERT(S)$ indicates that the split samples are encoded with shared parameters, H^s represents the hidden state of the last layer of BERT [1], n represents the number of words in the sample, and d represents the dimension of each word.

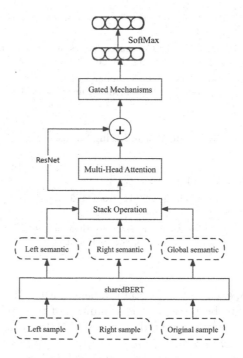

Fig. 1. The Structure of BERT-LR

The Structure of BERT-LR. Figure 1 shows the structure of the model in this paper

For a sample S, after processing, the left sample (LS), the right sample (RS) and the original sample (OS) are obtained. We input the three samples into the same BERT [1] model for encoding, and BERT [1] adds a special '[CLS]' symbol to the first part of the sample when encoding the sample, which can indicate the semantics of the sample for various downstream tasks.

$$H^{ls} - sharedBERT(LS) \in R^d$$

$$H^{rs} = sharedBERT(RS) \in R^d$$

$$H^{os} = sharedBERT(OS) \in R^d \tag{3}$$

When we get the vectors H^{ls}, H^{rs}, H^{os} representing the semantics of the three samples, in order to merge the different semantics so that the model learns more information, we use a multi-headed attention mechanism with the following procedure:

$$H^{concat} = [H^{ls} : H^{rs} : H^{os}] \in R^{3 \times d} \tag{4}$$

$$H_{MHA}^{concat} = MHA(H^{concat}, H^{concat}, H^{concat}) \in R^{3 \times d} \tag{5}$$

[:] stands for the stack operation, which merges the tensor in the specified dimensions, after the stack operation, the dimensions become $R^{3 \times d}$, and Eq. (5) shows that the information in H^{concat} is calculated after the multi-headed attention mechanism [18], where $Q = K = V = H^{concat}$, and the calculation process is as follows:

$$MHA(Q, K, V) = tanh([head_1; head_2; ...; head_h]W^R) \tag{6}$$

$$head_i = Attention(Q_i, K_i, V_i) = Attention(QW^Q, KW^K, VW^V) \tag{7}$$

$$Attention(Q_i, K_i, V_i) = softMax(\frac{Q_i K_i^T}{\sqrt{d_K}})V_i \tag{8}$$

where $Q = K = V = H^{concat} \in R^{3 \times d}$ as the input of the multi-headed attention mechanism, $W^Q \in R^{d \times d_K}$, $W^K \in R^{d \times d_K}$, $W^V \in R^{d \times d_V}$, $W^R \in R^{d \times d}$ are all learnable parameters, $d_K = d_V = \frac{d}{h}$, and h is the number of heads of the multi-headed attention mechanism.

After the multi-headed attention mechanism, we obtain $H_{MHA}^{concat} \in R^{3 \times d}$ as the semantic representation, and to obtain the final semantic representation, we add the residual and gating mechanisms, and the process is as follows:

$$H^{resnet} = LayerNorm(H^{concat} + H_{MHA}^{concat}) \tag{9}$$

$$H'_{resnet} = avePooling(H^{resnet}) \tag{10}$$

$$H^{Final} = H'_{resnet} * (Sigmoid(W_0(H'_{resnet}) + b_0)) \tag{11}$$

$H^{resnet} \in R^{3 \times d}$ as the semantic representation after the residual network, we obtain the $H'_{resnet} \in R^d$ after averaging pooling, the[*]sign represents the element-by-element multiplication, and after the gating mechanism, we obtain the final vector representation $H^{Final} \in R^d$, after which we add the fully connected layer and the softmax layer, represented as follows:

$$h^{'} = W_1(H^{Final}) + b_1$$

$$p = softmax(h^{'}) \tag{12}$$

where $W_0 \in R^{d \times d}$, $W_1 \in R^{L \times d}$, b_0 and b_1 are the bias vectors, L is the number of relations in the dataset, p is the probability of the output, and we use cross entropy as the loss function.

Fixed Embedding Layer. In the process of fine-tuning, we found that fixing the word embedding layer of BERT [1] could achieve better results, so we tried to fix different embedding layers of BERT [1] in order to conduct experiments as follows:

$$bert.embeddings.word\_embeddings.weight.requires\_grad$$

$$bert.embeddings.position\_embeddings.weight.requires\_grad$$

$$bert.embeddings.token\_type\_embeddings.weight.requires\_grad \tag{13}$$

There are three embedding layers in BERT [1], just set each embedding layer to *False* to fix this layer without parameter update, we have tried to fix different embedding layers in our experiments to get good experimental results.

4   Experiments

4.1   Introduction to the Dataset

The SemEval-2010 Task 8 [2] dataset is a classic dataset in the field of relationship extraction, and we use it for our experiments. This dataset contains nine relationship types and one other type. The other type indicates that the relationship type between entities does not belong to any of the above nine relationship types. These nine relationship types are: *Case-Effect, Component-Whole, Content-Container, Entity-Destination, Entity-Origin, Instrument-Agency, Member-Collection, Message-Topic* and *Product-Producer*, however, the dataset has to consider the direction, for example, Entity-Origin (E1, E2) and Entity-Origin (E2, E1) are different, so the dataset should have nineteen relationship types in the final judgment. The dataset contains 10,717 samples, each sentence contains two entities E1 and E2, of which there are 8,000 training samples and 2,717 testing samples, and we evaluate our experiments using the official script of SemEval-2010 Task 8 [2], with the evaluation metric M-F1(macro-averaged F1-scores).

Table 3. Parameters

Parameters	Setting
hidden units in BERT models	768
attention head	8
dropout rate	0.1
learning rate	5e-5/9e-6
epochs	10
maximum sequence length in inputs	150
optimizer	AdamW
batch size	16/32

4.2   Parameters Setting

We first fine-tuned on the basic BERT [1] model to observe the experimental effect on the SemEval-2010 Task 8 [2] dataset, and used it as a reference after obtaining the best results, seeking to reduce the loss during training, and after several experiments, we obtained the parameters as in Table 3, which can achieve the best results.

We have a batch size of 16 for training and 32 for validation, after setting *bert.embeddings.word_embeddings.weight.requires_grad = False* and *bert.embeddings.token_type_embeddings.weight.requires_grad = False* gives the best results, at which point the learning rate for the Bert-base-uncased based model BERT-LR-base is 5e-5 and the learning rate for the Bert-large-uncased based model BERT-LR-large has a learning rate of 9e-6.

4.3   Comparison with Other Methods

We compare our method with a classical variety of methods published on the SemEval-2010 Task 8 [2] dataset, including CNN [4], RNN [10], Bi-LSTM [15], Attention CNN [16], TRE [19], Entity-Aware BERT [21], KnowBert-W+W [24], R-BERT [23], BERTEM+MTB [22], A-GCN+BERT [25], RE-DMP+XLNet [26], as shown in the following Table 4:

Our method is based on the base BERT [1] model as the baseline, with BERT-LR-base representing experiments based on the Bert-base-uncased model and BERT-LR-large representing experiments based on Bert-large-uncased model.

4.4   Ablation Studies

The method proposed in this paper has been strongly validated experimentally, but in order to explore the effects of different factors, we have set up ablation experiments, which are divided into the following two results as shown in Table 5.

From Table 5, we can see that either fixing the embedding layer or using multifaceted semantics can improve the experimental effect more obviously. After

Table 4. Experimental results

Type	Method	M-F1
No-Pretrained	RNN	77.60
	CNN	82.70
	Bi-LSTM	82.70
	Attention CNN	84.30
Pretrained	TRE	87.10
	Entity-Aware BERT	89.00
	KnowBert-W+W	89.10
	R-BERT	89.25
	BERTEM+MTB	89.50
	A-GCN+BERT	89.85
	RE-DMP+XLNet	89.90
Baseline	Bert-base-uncased	88.38
	Bert-large-uncased	89.02
Ours	BERT-LR-base	89.32
	BERT-LR-large	90.00

analysis, we think that when the researchers released the pre-training model, they used super large-scale corpus to train the model, after these trainings, the embedding layer of the model can represent the text more accurately, but we need to use the pre-training model to do the downstream tasks, then during our training, the dataset of these downstream tasks is not big in comparison and are concentrated in a certain domain, which instead may have some adverse effects on the embedding layer of the already trained large model. Besides, the experimental results can also prove that some components in the text do have adverse effects on the model's understanding of the relationships in the entities in the relation extraction, and that it is feasible to carry out a study of multifaceted semantics of relation extraction centered on the entities.

BERT-LR-noF: represents that we use multifaceted semantics, but no fixed embedding layer.

BERT-FE-noLR: represents that we fixed the embedding layer, but did not use multifaceted semantics.

Table 5. Results of Ablation Studies

	Model	M-F1
Base	Bert-base-uncased	88.38
	BERT-LR-noF	89.12
	BERT-FE-noLR	89.13
Large	Bert-large-uncased	89.02
	BERT-LR-noF	89.73
	BERT-FE-noLR	89.52

5 Conclusions

In this paper, we propose a method based on the relationship extraction of BERT [1] with multifaceted semantics, where we obtain different semantic representations by splitting the samples and feeding them into the same model and fuse them well, in addition to which we find that fixing the embedding layer of the pre-trained model achieves better experimental results in the process of fine-tuning. We conduct experiments on the SemEval-2010 Task8 [2] dataset to demonstrate the effectiveness and superiority of our method.

In the field of textual relation extraction, we still need to continue to strengthen the research on pre-trained models, after all, small changes in the model may have an impact on the experiment, and in text, with the growth of sentences and the increase of entities, the difficulty of relation extraction is increased, if the relation extraction is performed in a document, then the model will need to read multiple sentences and understand the context to infer the information and perform relation extraction, these are all challenging work in the future.

Acknowledgements. This work was supported in part by the National Natural Science Foundation of China under the contract number 62062012.

References

1. Devlin, J., Chang, M.-W., Lee, K., Toutanova, K.: BERT: pre-training of deep bidirectional transformers for language understanding. arXiv:1810.04805v2 (2019)
2. Hendrickx, I., et al.: SemEval-2010 task 8: multi-way classification of semantic relations between pairs of nominals. In: International Workshop on Semantic Evaluation (2010)
3. Hinton, G.E., Salakhutdinov, R.: Reducing the dimensionality of data with neural networks. Science **313**(5786), 504–507 (2006). https://doi.org/10.1126/science.1127647
4. Zeng, D., Liu, K., Lai, S., Zhou, G., Zhao, J.: Relation classification via convolutional deep neural network. In: International Conference on Computational Linguistics, pp. 2335–2344 (2014)
5. Liu, C., Sun, W., Chao, W., Che, W.: Convolution neural network for relation extraction. In: Motoda, H., Wu, Z., Cao, L., Zaiane, O., Yao, M., Wang, W. (eds.) ADMA 2013. LNCS (LNAI), vol. 8347, pp. 231–242. Springer, Heidelberg (2013). https://doi.org/10.1007/978-3-642-53917-6_21
6. Collobert, R., Weston, J., Bottou, L., Karlen, M., Kavukcuoglu, K., Kuksa, P.P.: Natural language processing (almost) from scratch. J. Mach. Learn. Res. **12**(76), 2493–2537 (2011). www.pkuksa.org/~pkuksa/publications/papers/jmlr2011nlpfromscratch.pdf
7. Nguyen, T.H., Grishman, R.: Relation Extraction: Perspective from Convolutional Neural Networks (2015). https://doi.org/10.3115/v1/w15-1506
8. Santos, C.N.D., Xiang, B., Zhou, B.: Classifying Relations by Ranking with Convolutional Neural Networks (2015). https://doi.org/10.3115/v1/p15-1061
9. Ye, H., Chao, W., Luo, Z., Li, Z.: Jointly Extracting Relations with Class Ties via Effective Deep Ranking (2017). https://doi.org/10.18653/v1/p17-1166

10. Socher, R., Huval, B., Manning, C.D., Ng, A.Y.: Semantic compositionality through recursive matrix-vector spaces. In: Empirical Methods in Natural Language Processing, pp. 1201–1211 (2012)
11. Lin, Y., Shen, S., Liu, Z., Luan, H., Sun, M.: Neural Relation Extraction with Selective Attention over Instances (2016). https://doi.org/10.18653/v1/p16-1200
12. Lin, C., Miller, T.M., Dligach, D., Amiri, H., Bethard, S., Savova, G.: Self-training improves Recurrent Neural Networks performance for Temporal Relation Extraction (2018). https://doi.org/10.18653/v1/w18-5619
13. Xu, Y., Mou, L., Li, G., Chen, Y., Peng, H., Jin, Z.: Classifying Relations via Long Short Term Memory Networks along Shortest Dependency Paths (2015). https://doi.org/10.18653/v1/d15-1206
14. Zhou, P., et al.: Attention-Based Bidirectional Long Short-Term Memory Networks for Relation Classification (2016). https://doi.org/10.18653/v1/p16-2034
15. Zhang, S., Zheng, D., Hu, X., Yang, M.: Bidirectional long short-term memory networks for relation classification. In: Pacific Asia Conference on Language, Information, and Computation, pp. 73–78 (2015). www.bcmi.sjtu.edu.cn/~paclic29/proceedings/PACLIC29-1009.185.pdf
16. Shen, Y., Huang, X.: Attention-based convolutional neural network for semantic relation extraction. In: International Conference on Computational Linguistics, pp. 2526–2536 (2016). www.aclweb.org/anthology/C16-1238.pdf
17. Lee, J., Seo, S., Choi, Y.: Semantic relation classification via bidirectional LSTM networks with entity-aware attention using latent entity typing. Symmetry 11(6), 785 (2019). https://doi.org/10.3390/sym11060785
18. Vaswani, A., et al.: Attention is all you need. arXiv:1706.03762v5 (2017)
19. Alt, C., Hubner, M.P., Hennig, L.: Improving relation extraction by pre-trained language representations (Cornell University). arXiv preprint arxiv.org/abs/1906.03088 (2019)
20. Joshi, M.S., Chen, D., Liu, Y., Weld, D.S., Zettlemoyer, L., Levy, O.: SpanBERT: improving pre-training by representing and predicting spans. Trans. Assoc. Comput. Linguist. 8, 64–77 (2020). https://doi.org/10.1162/tacl_a_00300
21. Wang, H., et al.: Extracting multiple-relations in one-pass with pre-trained transformers. arXiv (Cornell University) (2019). https://doi.org/10.18653/v1/p19-1132
22. Soares, L., FitzGerald, N., Ling, J., Kwiatkowski, T.: Matching the blanks: distributional similarity for relation learning. In: Meeting of the Association for Computational Linguistics (2019). https://doi.org/10.18653/v1/p19-1279
23. Wu, S., He, Y.: Enriching pre-trained language model with entity information for relation classification. In: Conference on Information and Knowledge Management (2019). https://doi.org/10.1145/3357384.3358119
24. Peters, M.J., et al.: Knowledge Enhanced Contextual Word Representations (2019). https://doi.org/10.18653/v1/d19-1005
25. Tian, Y., Chen, G., Song, Y., Wan, X.: Dependency-driven Relation Extraction with Attentive Graph Convolutional Networks (2021). https://doi.org/10.18653/v1/2021.acl-long.344
26. Tian, Y., Song, Y., Xia, F.: Improving relation extraction through syntax-induced pre-training with dependency masking. In: Findings of the Association for Computational Linguistics (ACL 2022) (2022). https://doi.org/10.18653/v1/2022.findings-acl.147
27. Li, J.: SPOT: knowledge-enhanced language representations for information extraction. In: Proceedings of the 31st ACM International Conference on Information & Knowledge Management (2022)

28. Zhu, X., Zhu, Y., Zhang, L., Chen, Y.: A BERT-based multi-semantic learning model with aspect-aware enhancement for aspect polarity classification. Appl. Intell. **53**(4), 4609–4623 (2022). https://doi.org/10.1007/s10489-022-03702-1
29. Zhang, Y., Zhong, V.W., Chen, D., Angeli, G., Manning, C.D.: Position-aware Attention and Supervised Data Improve Slot Filling (2017). https://doi.org/10.18653/v1/d17-1004

NMPose: Leveraging Normal Maps
for 6D Pose Estimation

Wenhua Liao and Songwei Pei[(✉)]

Beijing University of Posts and Telecommunications, Beijing 100876, China
{wenhua_liao,peisongwei}@bupt.edu.cn

Abstract. Estimating the 6 degrees-of-freedom (6DoF) pose of an object from a single image is an important task in computer vision. Many recent works have addressed it by establishing 2D-3D correspondences and then applying a variant of the PnP algorithm. However, it is extraordinarily difficult to establish accurate 2D-3D correspondences for 6D pose estimation. In this work, we consider 6D pose estimation as a follow-up task to normal estimation so that pose estimation can benefit from the advance of normal estimation. We propose a novel 6D object pose estimation method, in which normal maps rather than 2D-3D correspondences are leveraged as alternative intermediate representations. In this paper, we illustrate the advantages of using normal maps for 6D pose estimation and also demonstrate that the estimated normal maps can be easily embedded into common pose recovery methods. On LINEMOD and LINEMOD-O, our method easily surpasses the baseline method and outperforms or rivals the state-of-the-art correspondence-based methods on common metrics. Code is available at https://github.com/mate-huaboy/NMPose.

Keywords: 6D Pose Estimation · Normal map · Computer Vision

1 Introduction

6D object pose (rotation and translation) estimation, as a fundamental task in computational vision, is dedicated to estimating the rigid body transformation from the object coordinate system to the camera coordinate system, which is widely used in many fields, including robotic manipulation, augmented reality, and autonomous driving. Thanks to the development of deep learning, many recent works have introduced it into 6D pose estimation and achieved impressive results. These methods can be roughly divided into two categories. One is to utilize neural network to directly regress the pose from the image [18]. The other is correspondence-based ones, as shown in Fig. 1, which first establish 2D-3D correspondences and then compute 6D pose using Perspective-n-Point (PnP) algorithm. Nowadays, almost all the best-performing methods belong to the latter, so it has also become the mainstream approach in this field in recent years.

However, wrongly matched outliers and noise in the predicated 2D-3D correspondences seriously affect the robustness and accuracy of the 6D pose [11].

B. Luo et al. (Eds.): ICONIP 2023, CCIS 1968, pp. 178–190, 2024.
https://doi.org/10.1007/978-981-99-8181-6_14

Consequently, the key to improving pose estimation performance lies in establishing better correspondences. This challenge has been the focus of extensive research in recent years, resulting in two main categories of approaches. One is based on sparse 2D-3D correspondences [8,13], in which the correspondences are built by finding the projection of predefined 3D keypoints in image. To mitigate noise, heatmap prediction [7], voting mechanism [8], and other complex, unnatural, and possibly non-differentiable operations are often employed. The other type is dense 2D-3D correspondence-based 6D pose estimation method [5,19]. These methods directly predict the 3D location in the object coordinate system for each pixel in the image, enabling the construction of dense correspondences. Compared to the sparse methods, dense correspondences mitigate the impact of noise on the PnP algorithm, leading to improved results, particularly in complex scenes involving occlusion and truncation.

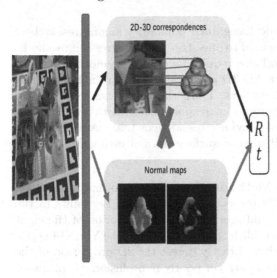

Due to the inherent matching ambiguity of the textureless object surface, noise is inevitable. Constructing more accurate 2D-3D correspondences has become a bottleneck in this field. It is worth noting that the influence of translation and rotation on the object's appearance in the image is reflected differently, some works tend to decouple rotation and translation [5, 18]. In these approaches, translation is directly regressed from the image, while rotation is derived from correspondences. However, it is important to highlight that both rotation and translation can be recovered from correspondences. Therefore, establishing correspondences is not necessary when the objective is solely to recover rotation. This insight motivates us to find an alternative representation closely related to rotation and easier to predict than correspondences.

Fig. 1. overview of our approach. Compared with the traditional method that relies on correspondences, as shown the process with the black arrow in the upper part of the figure, we propose to use the normal maps instead of the correspondences to help estimate 6D pose, as shown the process with the red arrow in the lower part of the figure. (Color figure online)

Based on this idea, we propose using surface normals as intermediate representations to estimate the object's rotation, as illustrated in Fig. 1. Surface normals are independent of translation and effectively capture rotation, aligning with our goal of decoupling rotation and translation. Moreover, predicting surface normals has received significant attention in recent years.

This work makes the following main contributions:

- We propose a correspondence-free method for 6D pose estimation using only surface normals as intermediate features. In the field of 6D pose estimation, corresponding features have become the mainstream approach. However, we demonstrate that normals can serve as a viable alternative option.
- We present a framework that integrates surface normals into various pose recovery algorithms, showcasing the flexibility and scalability of using normals as intermediate representations.
- Through extensive experiments, we demonstrate that the proposed method surpasses or rivals all other state-of-the-art approaches on LINEMOD and LINEMOD-O datasets, showing that using normals as a new option for intermediate representations is promising.

2   Related Work

In recent years, the two-stage method has gained popularity among researchers due to its superior performance compared to one-stage direct regression methods. However, the early two-stage method also required a pose refinement process to improve accuracy, like [1,9], because 2D-3D correspondences are noisy and the PnP is very sensitive to noise and outliers. Thus, the key to achieving better performance lies in establishing improved correspondences. In this section, we will review the significant works of previous researchers that have addressed this problem. Additionally, we will discuss surface normal estimation and its utilization in pose estimation.

Sparse Correspondence-Based Methods. This category of methods [9,13] aims to accurately determine the 2D projection of predefined keypoints in the image. BB8 [9] segments the object and predicts the 2D projection of the eight vertices of the 3D bounding box to establish correspondences. The YOLO-6D [13] adopts the idea of YOLO framework to directly detect the 2D projection of the 3D bounding-box. However, they both perform poorly in occlusion. [7] proposes to predict the 2D projection of 3D key points from the 2D heatmap, and the PVNet [8] adopts a voting-based keypoint positioning strategy, which solved the occlusion problem to a certain extent.

Dense Correspondence-Based Methods. Many recent works focus on establishing pixel-wise correspondences to further enhance the accuracy of 6D pose estimation. CDPN [5] directly predicts dense correspondences. The DPOD [19] predicts UV maps, which are then converted into correspondences. Compared with the correspondences, predicting UV maps is easier, and the restored 3D coordinates lie on the object surface. EPOS [3] proposes to learn a probability distribution of correspondences over object surface fragments. In [16], the entire pose estimation process is differentiable, allowing for the partial learning of correspondences by backpropagating gradients with respect to object pose.

However, noise is inherent in correspondences, and many recent works [10,12, 16] have explored incorporating additional information, such as points, edges, or other types of representations, to improve performance. Although normals could potentially serve as one of these additional information sources, they are not the primary focus of this paper. Our main objective is to explore the potential of normals as a substitute for traditional correspondences.

Surface Normal for Pose Estimation. Some recent works [17,20] in similar fields have recognized the advantages of incorporating normals into pose estimation. They have explored the fusion of normals with other information to achieve more stable and accurate pose estimation results. In contrast, our approach solely utilizes surface normals as an alternative to traditional correspondences, without additional information fusion.

3 Method

The goal of 6D pose estimation is to estimate the rigid body transformation $T = [R|t]$, where R is the rotation matrix, representing the $SO(3)$ transform and t is the translation vector. Considering that t mainly affects the position and size of the object in the image, and R mainly affects the appearance of objects in the image, so like many methods [5,18], we predict them separately. In this work, we use the estimated normals of object in the camera coordinate system and the object coordinate system as intermediate representations instead of the 2D-3D correspondences for predicting the rotation, as illustrated in Fig. 2, the translation can be estimated by translation head. In the following subsections, we demonstrate the advantages of using normals to help estimate 6D pose. In addition, we also demonstrate how to migrate normal features to common pose recovery methods, including interpretable calculation methods (PnP, Umeyama algorithm [14]) and direct regression method by the neural network.

3.1 The Benefits of Using Normals

According to [6], it is proved that the normal is an n-sphere S_n, and a naturally closed geometric manifold defined in $R(n+1)$ space, where n equals 2. As one of the important tasks in computer vision, researchers have studied normal estimation extensively, which means that 6D pose estimation can benefit from the advance of normal estimation. Moreover, since each component of the normal falls within the interval [-1,1], it facilitates learning the pose by providing well-constrained values.

In addition, the normal map directly reflects rotation and is independent of translation. We denote the homogeneous coordinates of the tangent vector of the object surface as $l^o = [l_x, l_y, l_z, 0]^T$, which represents a direction. $T^{4\times4} \in SE(3)$ denotes the rigid body transformation of the object, so it is easy to derive the Eq. (1):

$$l^c = T^{4\times4}l^o = \begin{bmatrix} R^{3\times3} & t \\ 0 & 1 \end{bmatrix} l^o = \begin{bmatrix} R^{3\times3}l^o_{xyz} \\ 0 \end{bmatrix} \quad (1)$$

Here, l^c is the coordinate of the transformed tangent vector, l^o_{xyz} is the x,y,z components of l^o, $R^{3\times3}$ is the rotation of the object, and t is the translation.

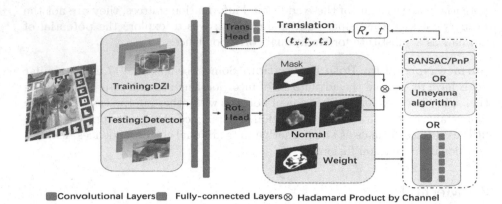

Convolutional Layers ■ **Fully-connected Layers** ⊗ **Hadamard Product by Channel**

Fig. 2. Framework of our work. Given an image, we take Dynamic-Zoom-In on target object (DZI for training, off-the-shelf detections for testing) as input. Then we decouple the prediction of rotation and translation. Concretely, we directly regress the translation, while the rotation is solved by pose recovery methods (RANSAC/PnP, Umeyama Algo. or Network) from the predicted normal maps. RANSAC/PnP and Umeyama Algo. use the normals to calculate the rotation directly; the normal and weight information (the red dashed box in the figure) are fed into the network to regress the rotation. (Color figure online)

Notably, the coordinates of the tangent vector on the object surface before and after transformation are independent of t. For simplicity, we only consider rotation and denote the normal at this point as n^o and the transformation matrix of the normal as M, then:

$$(Mn^o)^T(Rl^o_{xyz}) = 0 \tag{2}$$

so we have:

$$M = ((R^{3\times3})^{-1})^T = R^{3\times3} \tag{3}$$

$$n^c = Rn^o \tag{4}$$

Equation (4) reveals that the relationship between n^o and n^c is unaffected by translation and directly reflects rotation. This aligns with our objective of decoupling rotation and translation.

3.2   Surface Normal Prediction

Normal Angle Error Loss. As shown in Fig. 2, the rotation head outputs the normals of the object in the camera coordinate system and the object coordinate system. We define our loss based on minimizing the angle error between the predicted normal and the ground-truth normal. Specifically, we define the angle error between the predicted normal and the ground-truth normal as $ang_{err}(\hat{n}, \bar{n}) = cos^{-1}(\hat{n}^T\bar{n})$, so our minimized per-pixel surface normal loss is:

$$L_c = \frac{1}{|S|} \sum_{i \in S} ang_{err}(\hat{n}_i^c, \bar{n}_i^c)$$

$$L_o = \frac{1}{|S|} \sum_{i \in S} ang_{err}(\hat{n}_i^o, \bar{n}_i^o)$$

(5)

where the superscripts c and o denote the camera coordinate system and the object coordinate system, respectively. $\hat{\bullet}$ and $\bar{\bullet}$ are the predicted value and the true value respectively. S denotes the object visible surface pixel set, and $|S|$ is the number of elements in the set S.

Cross-Coordinate Loss. Given that n^c and n^o satisfy the constraints of Eq. (4), we enforce cross-coordinate consistency to enhance the stability of normal prediction. We minimize the angular error:

$$f(\bar{R}, \hat{n}^o, \hat{n}^c) = ang_{err}(\bar{R}\hat{n}^o, \hat{n}^c)$$

(6)

Therefore, our cross-coordinate consistency loss is defined as follows:

$$L_{cross} = \frac{1}{|S|} \sum_{i \in S} f(\bar{R}, \hat{n}_i^o, \hat{n}_i^c)$$

(7)

3.3 Rotation Recovery from Normal

In this section, we integrate the normal maps into existing pose recovery algorithms, which have demonstrated their effectiveness. By incorporating the normal maps, we aim to leverage the strengths of these algorithms while replacing the reliance on correspondences.

Explainable Rotation Recovery. We organize all normal vectors $n_i \in S$ into a column-wise matrix.

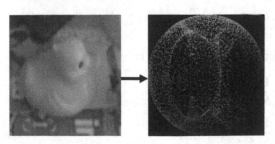

From Eq.(4), we have:

$$N_c^{3 \times |S|} = R N_o^{3 \times |S|}$$

(8)

where:

$$N_c^{3 \times |S|} = \begin{bmatrix} n_1^c & n_2^c & \cdots & n_{|S|}^c \end{bmatrix}$$
$$N_o^{3 \times |S|} = \begin{bmatrix} n_1^o & n_2^o & \cdots & n_{|S|}^o \end{bmatrix}$$

(9)

Fig. 3. View objects in normal space. The duck can be identified by a point cloud consisting of its surface normals. Specifically, the sphere point cloud in the figure consists of the normal coordinates of the duck's visible surface in the camera coordinate system, and its color represents the coordinate of the corresponding normal in the object coordinate system.

From Eq.(8), it can be observed that the R we want to get is actually the least squares solution of the equation, so we can use the Umeyama algorithm [14] to find R directly.

In addition, we treat the surface normals of the object as a point cloud that represents the object's geometry, as shown in Fig. 3, all objects can be viewed as point clouds distributed on a sphere. We define $p = (x, y)$ as the projection point of the normal point cloud in the 2D image. We can calculate p using the following equation:

$$\begin{bmatrix} x \\ y \\ 1 \end{bmatrix} = \frac{1}{n_z^c} K R n^o = \frac{1}{n_z^c} K n^c \qquad (10)$$

Because we already have $\hat{n}^c$ and $\hat{n}^o$, it means that we can use $\hat{n}^c$ to obtain the projection point p, and at the same time, its corresponding 3D coordinate is $\hat{n}^o$. Consequently, we establish the 2D-3D correspondences of the normal point cloud, enabling us to apply various variants of the Perspective-n-Point (PnP) algorithms conveniently. In our experiments, we chose the widely used RANSAC-PnP method.

Direct Rotation Regression by Neural Network. We can directly use the decoupled rotation loss in [16] as our loss:

$$L_R = \underset{x \in M}{avg} \|\bar{R}x - \hat{R}x\|_1 \qquad (11)$$

where M represents the 3D model.

Considering that the quality of normal estimation may vary in different regions, the network may need to allocate different attention to different regions. Inspired by the attention mechanism [15], we introduce weighted angle error. The essence of our problem is to find an optimal rotation R that can minimize the accumulated weighted angle error:

$$\underset{R,w}{argmin} \sum_{i \in S} w_i f(R, \hat{n}_i^o, \hat{n}_i^c) \qquad (12)$$

We expect R in Eq.(12) to converge to the true value $\bar{R}$, leading to:

$$\underset{R,w}{min} \sum_{i \in S} w_i f(R, \hat{n}_i^o, \hat{n}_i^c) = \underset{w}{min} \sum_{i \in S} w_i f(\bar{R}, \hat{n}_i^o, \hat{n}_i^c) \qquad (13)$$

In summary, we obtain the loss for rotation L_R^{naive} and weight loss L_w^{naive} as follows:

$$L_R^{naive} = \frac{1}{|S|} \sum_{i \in S} w_i f(R, \hat{n}_i^o, \hat{n}_i^c)$$

$$L_w^{naive} = \frac{1}{|S|} \sum_{i \in S} w_i f(\bar{R}, \hat{n}_i^o, \hat{n}_i^c) \qquad (14)$$

$$s.t. \sum_{i \in S} w_i = 1$$

It can be seen from Eq.(14) that the larger the angle error $f(\bar{R}, \hat{n}_i^o, \hat{n}_i^c)$ at the surface i, the corresponding weight should be smaller. To facilitate the learning process, the normalization constraint is removed. In an ideal scenario, $\bar{w}_i$ describes the confidence of rotation estimation at i, i.e. $P(\bar{R}|\hat{n}_i^o, \hat{n}_i^c)$, defined as:

$$\bar{w}_i = exp(-f(\bar{R}, \hat{n}_i^o, \hat{n}_i^c)) \tag{15}$$

Hence, our improved weight loss term L_w^{imp} is defined as:

$$L_w^{imp} = \frac{1}{|S|} \sum_{i \in S} \|\hat{w} - \bar{w}\|_1 \tag{16}$$

The improved weighted rotation loss is:

$$L_R^{imp} = \frac{1}{|S|} \sum_{i \in S} \bar{w}_i f(R, \hat{n}_i^o, \hat{n}_i^c) \tag{17}$$

4   Experiments

4.1   Experimental Setup

Implementation Details. For comparison with existing correspondence-based methods, we utilize CDPN's [5] network as our baseline. In order to be suitable for our method, we make a slight modification to the network of CDPN, and we refer to the modified network as Normal-based Disentangled Pose Network (NDPN), as shown in Fig. 2. Specifically, we directly replace the correspondences with the normal features in this article, and the size of the features remains unchanged at 64×64. At the same time, we use the RGB pixel coordinates, which are used as the input of RANSAC/PnP in CDPN, together with RGB as the input of the backbone network. The visible mask loss and translation loss terms are the same as in CDPN. For the direct rotation regression method, we add a channel behind the rotation head to represent the weight of each pixel due to the introduction of the weight mechanism. The rotation regression network consists of four convolutional layers and three fully connected layers. We use allocentric representation [4,16] and add loss L_{cross} after 10% and loss L_R after 70% of the total training epochs to stabilize the network regression. To ensure a fair comparison between the baseline and our method, as done in [5], during training we dynamically zoom in on the target object as input, during evaluation we use the detection results from Faster-RCNN. Our network is trained using the Ranger optimizer, which uses a cosine schedule at 72% of the training epoch, and batch size is set to 24.

Datasets and Metrics. We conduct our experiments on the LINEMOD [2] (LM) and the Occlusion [1] (LM-O) dataset. LM serves as the de facto standard benchmark for 6D pose estimation of textureless objects in cluttered scenes. It includes 13 sequences, each sequence has about 1.2k images. Following the data split configuration in [5], we used 15% of LM as our training set and 85% as our test set. LM-O is a subset of images with severe occlusions in the LM. For LM-O, like [16], we employed publicly available data rendered with a physics-based renderer (pbr) for training. We render ground-truth normal maps for each data using PyTorch3D. The synthetic data and renderer can be obtained from publicly available code.

We use two commonly used metrics for evaluating 6D pose estimation: ADD(-S) and n cm $n°$. ADD(-S) measures the average deviation of the transformed model points and considers the estimated pose correct if the deviation is below a certain percentage of the object diameter (e.g., ADD-0.1d means the deviation is less than 10% of the diameter). n cm $n°$ measures whether the translation error is lower than n cm and the rotation error is lower than $n°$.

4.2   Ablation Study on LM

We conduct ablation experiments on the LM. Table 1 shows that NDPN is significantly higher than the baseline (B0, B1, B2 vs. A0) on all metrics. In fact, NDPN even achieves slightly better performance than GDR-Net [16], which introduces additional guidance information (B2 vs. A1). While our method may be slightly inferior to GDR-Net in terms of translation, we significantly outperform GDR-Net in rotation, resulting in better overall performance in the comprehensive metrics.

Table 1. Ablation Study on LM. Ablation of Cross-coordinate loss, weight mechanism, and the parametrization of translation. (-) denotes unavailable results.

Row	Method	ADD(-S)		5°, 5 cm	2°	2 cm
		0.02d	0.1d			
A0	CDPN [5]	-	89.86	94.31	-	92.81
A1	GDR-Net [16]	35.50	93.70	-	63.20	95.50
B0	NDPN+Umeyama(**Ours**)	42.02	94.37	97.98	73.90	95.26
B1	NDPN+PnP(**Ours**)	40.44	94.25	96.05	63.18	95.25
B2	NDPN+Net(**Ours**)	39.59	94.50	97.69	73.42	95.41
C0	B0: w/o L_{cross}	41.37	94.18	97.93	70.04	95.65
C1	B1: w/o L_{cross}	37.87	94.09	93.34	53.06	95.65
C2	B2: w/o L_{cross}	40.16	94.03	97.49	72.81	95.48
D0	B2: w/o weight mechanism	38.59	93.82	97.51	71.39	94.77
D1	B2:$L_R^{imp},L_w^{imp} \longrightarrow L_R^{naive},L_w^{naive}$	36.12	93.03	96.75	63.45	94.02

Effectiveness of Cross-Coordinate Loss. As described in Sect. 3.2, the Cross-coordinate loss L_{cross} is employed to stabilize the normal prediction, which in turn contributes to stabilizing the rotation estimation. Table 1 illustrates the effectiveness of L_{cross}. In group B (B0, B1, B2) experiments, we use the full normal loss including L_c, L_o, and L_{cross}, while in group C (C0, C1, C2), we remove L_{cross}. A comparison of the experimental results (B0vs.C0, B1vs.C1, B2vs.C2) reveals that the utilization of L_{cross} improves rotation estimation. Specifically, the models with L_{cross} achieved higher accuracy in rotation estimation, with improvements under 2°(73.90% vs. 70.04%, 63.18% vs. 53.06%, 73.42% vs. 72.81%).

Effectiveness of Weighted Angle Error. We conducted experiments where we removed the Weighted Angle Error component from the loss function (D0) and compared it with models using the naive Weighted Rotation Loss and Weighted Loss (L_R^{naive} and L_w^{naive}) (D1). A comparison of the results (B2 vs. D0, B2 vs. D1) revealed that the utilization of the improved weighted losses (L_R^{imp} and L_w^{imp}) led to a significant enhancement in the accuracy of rotation estimation.

4.3 Comparison with State-of-the-Arts

Comparison with Baseline. We conduct experiments on the LM. We train an NDPN for all objects for 200 epochs without any data augmentation. Note in CDPN [5], due to the difficulty of correspondence prediction, it has to use a complex training strategy involving 3 different stages, while we only need one-time end-to-end training.

Table 2. Comparison with baseline on metric ADD(-S). * denotes symmetric objects.

Object	CDPN [5]	+Umeyama	+PnP	+Net	Mean
ape	64.38	81.81	81.52	82.48	81.94
benchvise	97.77	97.96	97.77	98.45	98.06
camera	91.67	93.73	93.63	94.22	93.86
can	95.87	96.95	96.85	97.74	97.18
cat	83.83	93.91	93.81	93.51	93.74
driller	96.23	97.72	97.62	97.32	97.55
duck	66.76	83.38	83.29	84.23	83.63
eggbox*	99.72	99.62	99.62	99.62	99.62
glue*	99.61	99.42	99.42	99.32	99.39
holepuncher	85.82	91.53	91.34	91.25	91.37
iron	97.85	97.34	97.24	94.14	96.24
lamp	97.89	99.62	99.62	99.33	99.52
phone	90.75	93.77	93.58	93.96	93.77
Average	89.86	94.37	94.25	94.50	94.37

Table 2 presents a comparison of the results achieved by NDPN in combination with other commonly used pose recovery methods, including Umeyama algorithm, RANSAC/PnP, and direct regression by network, against the baseline method CDPN. Our method outperforms CDPN by a large margin in terms of the average ADD(-S) metric (94.37, 94.25, 94.50 vs. 89.86), indicating superior performance overall. Notably, we achieve impressive accuracy on small objects such as "ape" and "duck", improving the accuracy from 64.38 to 81.94 and 66.76 to 83.63, respectively, representing an increase of approximately 20%.

Comparison with Other Works. Table 3 compares the results of our method compared with other SOTA methods based on correspondence. Although correspondence-based methods have evolved over the years, we can easily outperform them without refinement. Furthermore, our method even outperforms or competes closely with methods that incorporate additional guidance information.

Table 3. Comparison with state-of-the-arts. We report the Average Recall (%) of ADD(-S). † denotes that these methods add additional information in addition to correspondences. Bold indicates the highest result, and underline indicates the second highest result.

	Methods	LM dataset	LM-O dataset
Correspondence-based	YOLO-6D [13]	56.0	6.4
	PVNet [8]	86.3	40.8
	DPOD [19]	82.3	47.3
	GDR-Net† [16]	93.7	47.4
	HybridPose† [12]	91.3	47.5
	INVNet† [10]	**94.5**	49.6
NDPN(Ours)	+Umeyama	<u>94.4</u>	**50.9**
	+PnP	94.3	<u>50.8</u>
	+Net	**94.5**	50.3

4.4   Runtime Analysis

We conducted runtime experiments on a desktop system equipped with a 3.60GHz CPU and an NVIDIA 3090 GPU. For a 640×480 image containing 8 objects in the LM-O dataset, we measured the runtime of our method using the Umeyama algorithm, PnP, and a direct rotation regression network. The average runtime for each method per image was approximately 17ms, 19ms, and 16ms, respectively.

5   Conclusion and Discussion

In this work, we have introduced the use of normal maps as an alternative to correspondences for estimating 6D poses from RGB images. We have applied normal maps in conjunction with the Umeyama algorithm, PnP, and direct rotation regression network, achieving promising results.

However, it is important to note that this paper does not fully explore the potential of normal maps in certain scenarios where obtaining depth maps can be challenging, but acquiring normal maps is relatively more feasible. Exploring these applications and their implications will be a focus of our future research.

Acknowledgement. This work was supported in part by National Natural Science Foundation of China (NSFC) under Grant No.61772061.

References

1. Brachmann, E., Krull, A., Michel, F., Gumhold, S., Shotton, J., Rother, C.: Learning 6D object pose estimation using 3D object coordinates. In: Fleet, D., Pajdla, T., Schiele, B., Tuytelaars, T. (eds.) ECCV 2014. LNCS, vol. 8690, pp. 536–551. Springer, Cham (2014). https://doi.org/10.1007/978-3-319-10605-2_35

2. Hinterstoisser, S., et al.: Model based training, detection and pose estimation of texture-less 3D objects in heavily cluttered scenes. In: Lee, K.M., Matsushita, Y., Rehg, J.M., Hu, Z. (eds.) ACCV 2012. LNCS, vol. 7724, pp. 548–562. Springer, Heidelberg (2013). https://doi.org/10.1007/978-3-642-37331-2_42

3. Hodan, T., Barath, D., Matas, J.: EPOS: estimating 6D pose of objects with symmetries. In: Proceedings of the IEEE/CVF Conference on Computer Vision and Pattern Recognition, pp. 11703–11712 (2020)

4. Kundu, A., Li, Y., Rehg, J.M.: 3D-RCNN: instance-level 3D object reconstruction via render-and-compare. In: Proceedings of the IEEE Conference on Computer Vision and Pattern Recognition, pp. 3559–3568 (2018)

5. Li, Z., Wang, G., Ji, X.: CDPN: coordinates-based disentangled pose network for real-time RGB-based 6-DoF object pose estimation. In: Proceedings of the IEEE/CVF International Conference on Computer Vision, pp. 7678–7687 (2019)

6. Liao, S., Gavves, E., Snoek, C.G.: Spherical regression: learning viewpoints, surface normals and 3D rotations on n-spheres. In: Proceedings of the IEEE/CVF Conference on Computer Vision and Pattern Recognition, pp. 9759–9767 (2019)

7. Oberweger, M., Rad, M., Lepetit, V.: Making deep heatmaps robust to partial occlusions for 3D object pose estimation. In: Proceedings of the European Conference on Computer Vision (ECCV), pp. 119–134 (2018)

8. Peng, S., Liu, Y., Huang, Q., Zhou, X., Bao, H.: PVNet: pixel-wise voting network for 6dof pose estimation. In: Proceedings of the IEEE/CVF Conference on Computer Vision and Pattern Recognition, pp. 4561–4570 (2019)

9. Rad, M., Lepetit, V.: BB8: A scalable, accurate, robust to partial occlusion method for predicting the 3D poses of challenging objects without using depth. In: Proceedings of the IEEE International Conference on Computer Vision, pp. 3828–3836 (2017)

10. Sang, H., et al.: Trace-level invisible enhanced network for 6D pose estimation. In: 2022 IEEE International Conference on Multimedia and Expo (ICME), pp. 1–6. IEEE (2022)

11. Sheffer, R., Wiesel, A.: Pnp-Net: a hybrid perspective-n-point network (2020)

12. Song, C., Song, J., Huang, Q.: HybridPose: 6D object pose estimation under hybrid representations (2020)

13. Tekin, B., Sinha, S.N., Fua, P.: Real-time seamless single shot 6D object pose prediction. In: Proceedings of the IEEE Conference on Computer Vision and Pattern Recognition, pp. 292–301 (2018)

14. Umeyama, S.: Least-squares estimation of transformation parameters between two point patterns. IEEE Trans. Pattern Anal. Mach. Intell. **13**(04), 376–380 (1991)

15. Vaswani, A., et al.: Attention is all you need. In: Advances in Neural Information Processing Systems 30 (2017)

16. Wang, G., Manhardt, F., Tombari, F., Ji, X.: GDR-Net: geometry-guided direct regression network for monocular 6D object pose estimation. In: Proceedings of the IEEE/CVF Conference on Computer Vision and Pattern Recognition, pp. 16611–16621 (2021)

17. Wang, Y., Jiang, X., Fujita, H., Fang, Z., Qiu, X., Chen, J.: EFN6D: an efficient RGB-D fusion network for 6D pose estimation. J. Ambient Intell. Hum. Comput., 1–14 (2022)
18. Xiang, Y., Schmidt, T., Narayanan, V., Fox, D.: PoseCNN: a convolutional neural network for 6D object pose estimation in cluttered scenes. arXiv preprint arXiv:1711.00199 (2017)
19. Zakharov, S., Shugurov, I., Ilic, S.: DPOD: 6D pose object detector and refiner. In: Proceedings of the IEEE/CVF International Conference on Computer Vision, pp. 1941–1950 (2019)
20. Zhang, H., Opipari, A., Chen, X., Zhu, J., Yu, Z., Jenkins, O.C.: TransNet: category-level transparent object pose estimation. In: Computer Vision-ECCV 2022 Workshops: Tel Aviv, Israel, October 23–27, 2022, Proceedings, Part VIII, pp. 148–164. Springer, Cham (2023). https://doi.org/10.1007/978-3-031-25085-9_9

BFTracker: A One-Shot Baseline Model with Fusion Similarity Algorithm Towards Real-Time Multi-object Tracking

Fuxiao He, Qiang Chen, and Guoqiang Xiao$^{(\boxtimes)}$

College of Computer and Information Science, Southwest University, Chongqin, China
{kylekyle,cq0907}@email.swu.edu.cn, gqxiao@swu.edu.cn

Abstract. The key problem of real-time tracking is to achieve the balance between the tracking accuracy and real-time inference. Currently, the one-shot tracker, which joins multiple tasks into a single network, achieves trade-offs between tracking speed and accuracy. Different from previous trackers' practice of exchanging more extra calculation cost for tracking accuracy, a new one-shot baseline that is simple and efficient is proposed by us. We continue to discuss the conflict under the tracking paradigm of joint detection and re-identification tasks and try to explore the source of this conflict and also devote to alleviate the feature conflict in the one-shot model. On the other hand, we propose a fusion similarity algorithm to focus on handling tracking challenges of sudden appearance changes and potential Kalman Filter prediction errors. On the MOT17 test set, the algorithm proposed can reduce the ID switches by 42.2% compared with the initial algorithm. On the MOT16 test set, the proposed BFTracker improves the ID F1 Score (i.e. IDF1) by 2.8% compared with the most popular one-shot tracker. In particular, Proposed baseline is quite simple and runs at 31 FPS on a single GPU.

Keywords: One-Shot · Multi-Object Tracking · Real-Time Inference

1 Introduction

1.1 The One-Shot Baseline Model

One-shot methods refer to jointly learning multiple tasks in a single network. The proposed one-shot baseline can integrate re-ID task and detection task into a unified network to output results of two tasks, and balance the conflict between the joint re-ID and detection tasks to achieve a good performance of MOT speed and accuracy. Departed one-shot trackers always use DLA-34 [1] as backbone for joint learning (e.g., CenterTrack, FairMOT, and LMOT [2–4]). DLA-34, which upgrades every convolutional layer in upsampling stages to deformable convolutional layer (DCN [5]) and appends multi-layer feature fusion based on ResNet-34, as backbone can fully fuse features from all layers, but introduces the massive computation cost. LMOT [4] adopts the backbone of simplified DLA-34 to decrease the cost, but this damages its efficient feature extraction and fusion, resulting in deteriorated the final MOT performance. Based above mentioned, a new simple one-shot baseline is proposed shown in Fig. 1.

B. Luo et al. (Eds.): ICONIP 2023, CCIS 1968, pp. 191–204, 2024.
https://doi.org/10.1007/978-981-99-8181-6_15

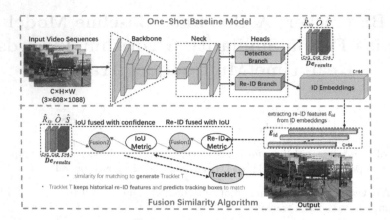

Fig. 1. Proposed baseline with fusion similarity algorithm is named as BFTracker for short. Heads output ID embeddings and detection results, which are used for data association. The "C" means the channel number.

1.2 Feature Conflict

JDE [6] first proposes the one-shot model, which makes the re-ID and detection tasks share the same backbone and neck structure in a single network and share the same features. FairMOT [3] systematically explores the phenomenon of feature conflicts under joint two tasks due to the essential differences in the respectively required features of the re-ID and detection tasks, and it's solution is to use multi-layer feature fusion structures, such as FPN [7] and DLA-34 [1], to fuse the features of all layers, which shows multi-layer feature fusion can alleviate the phenomenon of feature conflict by letting each task respectively extract own required features from the fused features. However, DLA-34's each upsampling operation will follow a operation of multi-layer feature fusion, which exactly brings another problem that not all layers fusion features are required for our final tracking task. In this paper, We will further verify the existence of feature conflict in the one-shot model and try to alleviate the feature conflict.

1.3 Fusion Similarity Algorithm

Data association process mainly calculates the similarity between the current frame's detection boxes and existing tracklets and leverages different strategies to match them and finish assigning the tracklets' identities to matched detections. The flowchart of the data association is illustrated in Fig. 2. On the other hand, we further propose a fusion similarity algorithm to deal with tracking challenges shown in frame t2 of Fig. 2 (b). Our algorithm will focus more on the stability of tracklet's ID (e.g., IDsw [8] and IDF1 [9]), which means that the predicted tracklet will have as few ID switches (i.e. IDsw) as possible. We believe that the basic reason of ID switch is that detections and tracklets are matched by mistake due to above tracking challenges and we will focus on handling the above challenges of all.

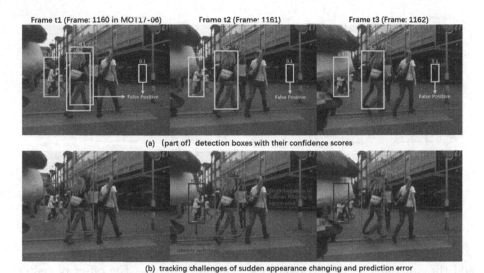

(a) (part of) detection boxes with their confidence scores

(b) tracking challenges of sudden appearance changing and prediction error

Fig. 2. Flowchart of data association and generation of tracklet. (a) introduces part of detection boxes. (b) introduces tracklets generated by motion or appearance similarity for matching which brings the challenges for tracking. Boxes of different colors represent different identities. The predicted tracking box (predicted by Kalman Filter) of the tracklet in the new frame is noted as a dashed box.

2   Method

In method, as shown in Fig. 1, we will first elaborate the detailed structure of the proposed one-shot baseline. Moreover, we will explain the feature conflict phenomenon introduced by the stride shown in Fig. 3. At last, we will introduce the background and process of the proposed algorithm applied in data association shown in Fig. 4.

2.1   The One-Shot Baseline

Backbone. We choose backbones from YOLO series more lightweight detectors to improve the real-time performance. Therefor, we choose YOLOv7's [10] backbone as the feature extraction network and also decrease the computing-cost of backbone network. Following JDE [6] and FairMOT [3], the size of video frames is resized to 608 (H) *1088 (W) as input of the baseline.

Stride in Neck. Shown in Fig. 3(a)-3(d), the output stride means the final scale of the output feature map, while we name the feature map scale that performs the last feature fusion in neck as fusion stride, which introduces the feature conflict phenomenon. Denote that the feature map after the final chose neck has the shape of 152 (H) *272 (W) (output stride = 4) and we finally build the neck network based on BiFPN [11], illustrated as Fig. 3(e).

Detection Branch. We adopt the anchor-free detector CenterNet's [1] detection heads as the detection branch of our one-shot baseline shown in Fig. 1. We

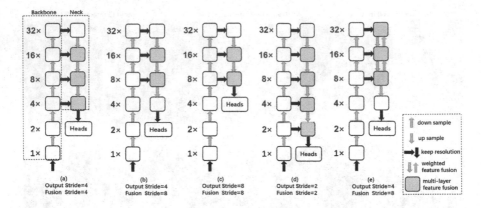

Fig. 3. Stride in neck network.

add three parallel heads to predict the heatmap response, the offset of center, and the bounding box size. The construction of the each head is of a convolution of 3×3, SiLu( ) , and a convolution of 1×1.

Heatmap Head. $\hat{R}_{xy}$ is the estimated heatmap response value in the heatmap at (x, y) location. When $\hat{R}_{xy}$ equals to 1, it is relative to a keypoint of detected center location of object. Heatmap head is to estimate the center response $\hat{R}_{xy} \in [0,1]^{1 \times \frac{H}{4} \times \frac{W}{4}}$ of the object. Heatmap head is trained through a pixel-wise logistic regression with focal loss [12] L_H, as defined in Eq.(1). Further more, R_{xy} is the relative keypoint's real response value at the location of object center. Moreover, hyper-parameters of focal loss are noted as α, β and we set $\alpha=2$ and $\beta=4$ in our experiments. Note that N is detected objects' number in the input image.

$$L_H = -\frac{1}{N} \sum_{xy} \begin{cases} \left(1 - \hat{R}_{xy}\right)^{\alpha} \log\left(\hat{R}_{xy}\right), & R_{xy} = 1; \\ \left(1 - R_{xy}\right)^{\beta} \left(\hat{R}_{xy}\right)^{\alpha} \log\left(1 - \hat{R}_{xy}\right) & \text{otherwise,} \end{cases} \tag{1}$$

Offset Head. To mitigate the quantization errors introduced from output stride, we estimate a local continuous offset $\hat{O} \in R^{2 \times \frac{H}{4} \times \frac{W}{4}}$ for each center point by each pixel. The offset is trained with an L1 Loss computed as L_O, where the predicted offset is noted as $\hat{O}_i$ at center location of the object and O_i is the ground-truth offset.

$$L_O = \frac{\sum_{n=1}^{N} |\hat{O}_i - O_i|}{N} \tag{2}$$

Bounding Box Head. Different from CenterNet's [11] regression approach, we do not predict the size of detection bounding box directly, but estimate $\hat{S} \in R^{4 \times \frac{H}{4} \times \frac{W}{4}}$ of two coordinates from the bounding box at lower right and upper left corners, and then compute the box size (width and height) as $\hat{S}_i = (\hat{W}_i, \hat{H}_i)$ from the regressed coordinates. We use L1 loss to compute L_B as Eq. (3), where we give ground-truth boxes to compute S_i.

$$L_B = \frac{\sum_{n=1}^{N} |\hat{S}_i - S_i|}{N} \tag{3}$$

Re-ID Branch. It is noted that we represent the object appearance feature through the object center location's extracted re-ID feature following FairMOT [3]. The re-ID head outputs ID embeddings $\in R^{64 \times \frac{H}{4} \times \frac{W}{4}}$ and the re-ID feature E_{id} of the object can obtain from ID embeddings at the object center location. Re-ID features are learnt with a classification task and all instance objects with the same identity are regarded as the same class. Extracted re-ID feature vector E_{id} then will be through a fully connected layer and a softmax operation to map it into a class distribution vector $\mathbf{V} = \{p(\mathbf{k}), \mathbf{k} \in [1, K]\}$. Further more, K is the number of identities in the training data and the one-hot representation of the ground-truth class label is noted as L(k). Then we calculate the loss of the re-ID branch is

$$L_R = -\sum_{n=1}^{N} \sum_{k=1}^{K} L(k) \log(p(k)) \tag{4}$$

Loss for the One-Shot Baseline. Re-ID and detection tasks are joint trained and weighted parameters λ (i.e. $\lambda_H = \lambda_O = \lambda_R = 1$, $\lambda_B = 0.5$) are assigned to regulate each branch's loss in the one-shot model. Loss of the entire baseline is

$$L_{\text{Baseline}} = \lambda_H \cdot L_H + \lambda_O \cdot L_O + \lambda_B \cdot L_B + \lambda_R \cdot L_R \tag{5}$$

2.2   Online Data Association

Our initial association combines the processes of the two-stage hierarchical matching strategy proposed by MOTDT [13] and the classification matching strategy based on detection boxes proposed by ByteTrack [14] as shown in Fig. 4. Specifically, in the first association stage, we first compute the appearance similarity between extracted high confidence detections' re-ID features of the current frame and all the tracklets' historical re-ID features and match them. In second association stage, we then compute the motion similarity between the low confidence detections and the unmatched tracklets' predicted tracking boxes and finish the matching. However, previous association algorithms [13,14] only adopt a single similarity to associate, which is limited to handle tracking challenges that are more likely leading to mismatches between detections and tracklets. We will analyze the background of the tracking challenges and propose corresponding strategies later.

Sudden Appearance Changes. In the large flow of people, there would be some objects with similar appearances. Moreover, when objects are occluded, it would cause sudden appearance changes. In frame t2 of Fig. 2 (b), it is very possible that the extracted appearance feature changes suddenly compared with frame t1, and this causes the object of ID=6 identity switching. The above mentioned would cause uncertainty and unreliability to the first association stage's matching results by re-ID metric.

Kalman Filter Prediction Errors. Following ByteTrack's [14] classification matching strategy, all low confidence detection boxes are reserved, and in the second association stage, we use low confidence detections to associate with the

Fusion Similarity Algorithm Applied in the Data Association	
Input: video sequences V; object detections Det; detection score threshold T_{high}=0.4; extracted object features E_{id}	

1 Initialize Tracklet T: T ← ∅	/* the first association stage */
2 for frame f_k in V do	17 Use D_{high} to match all the tracklets T by fusion similarity1
/* **ByteTrack**'s detections classification strategy */	18 D_{remain} ← unmatched object detection boxes from D_{high}
3 D_k ← Det (f_k)	19 T_{remain} ← unmatched tracklets from T
4 D_{high} ← ∅	/* the second association stage */
5 D_{low} ← ∅	20 Use D_{low} to match T_{remain} by fusion similarity2
6 for d in D_k do	/* save the remaining unmatched tracklets after the
7 if d.score >= T_{high} then	second association for 30 frames in case they reappear in the future*/
8 D_{high} ← D_{high} U {d}	21 $T_{re-remain}$ ← remaining unmatched tracklets from T_{remain}
9 end if	22 Delete unmatched D_{low} after the second association
10 else	/* Initialize new tracklets */
11 D_{low} ← D_{low} U {d}	23 for d in D_{remain} do
12 end	24 T ← T U {d}
13 end for	25 end for
/* **MOTDT**'s two-stage association strategy */	26 end for
14 For t in T do	27 return T
15 t ← KalmanFilter(t)	
16 end for	

Output: Tracklet T of the input video

Fig. 4. Fusion similarity algorithm applied in the data association.

unmatched tracklets. When the motion similarities (calculated by IoU distance) between the tracklet's predicted tracking box by Kalman Filter and the current frame's low confidence detection boxes are all close, the background detection (False Positive) may be mismatched with the predicted tracking box, as shown in frame t1 of Fig. 2 (b), which will lead to more aggravated prediction errors of Kalman Filter in the next frame t2. Based on the above, the second association stage's matching results by IoU metric would become unreliable.

2.3   Fusion Similarity Algorithm

Proposed fusion similarity algorithm applied in the data association as shown in Fig. 4. In order to eliminate mismatches that candidate tracklets have close cosine distances with a certain detected object but are far away from the certain detection, our algorithm is to use motion similarity (calculated by GIoU distance D_g) to integrate appearance similarity (calculated by cosine distance D_c). In the first association stage, the fusion *similarity1* distance D_{Sim1} is as

$$D_{Sim1} = mD_c + nD_g \tag{6}$$

where we set weighted parameters m and n respectively as 0.65 and 0.35. Based on D_{Sim1}, we adopt Hungarian algorithm [15] to finish the matching with the matching threshold τ_1 set as 0.45.

In order to reduce mismatches with FP (False Positive) detections, we combine the confidence score Det_{score} and motion similarity (calculated by IoU distance D_i) to match and we choose to match the higher confidence detections. In the second association stage, the fusion *similarity2* distance D_{Sim2} is as

$$D_{Sim2} = 1 - [Det_{score} \times (1 - D_i)] \tag{7}$$

where we adopt Hungarian algorithm to finish the matching based on D_{Sim2} with the matching threshold τ_2 set as 0.85.

3 Experiments

3.1 Datasets and Metrics

We train our model on the mixed data set mixed from seven public data sets: MOT17 [8], CityPerson [16], CUHK-SYSU [17], CalTech [18], ETH [19], PRW [20] and CrowdHuman [21]. MOT16 [8], and MOT17 test sets of two benchmarks are used to evaluate our approach. CLEAR metrics [22], including IDF1 [9], MOTA [8] and currently popular metric HOTA [23] are adopt to measure the accuracy of tracking and we also adopt IDsw [8] and IDF1 metrics which can reflect the change of identity to evaluate the tracklet's ID stability.

3.2 Details of Implementation

We load the pretrained weights of YOLOv7 [10] trained on coco dataset to initialize our approach since the proposed one-shot baseline's backbone is selected as YOLOv7's backbone. Following FairMOT [3] and JDE [6], Adam optimizer is selected and used to train the one-shot baseline for 30 epochs. The learning rate of the first 20 epochs is 5e-4 and of the last 10 epochs is 5e-5. We train our approach for nearly 30 h on a single GPU Tesla V100.

3.3 Experiments for Ablation

All results from ablation experiments are measured on the MOT16 test set to validate the conclusions which we focus on.

Backbone. We rebuild YOLOv5s [24] detector's backbone to obtain the backbone of YOLOv8 by replacing each C3 convolutional model in YOLOv5s with C2f. The backbone set as YOLOv7, shown in Table 1, achieves the best tracking accuracy in IDF1 and MOTA. For tracking speed, YOLOv5s achieves the best real-time performance. DLA-34 as backbone performs not as good as YOLOv7 dose in measuring both MOT accuracy and speed (IDF1, MOTA and FPS), which shows that blind using of the larger network is not necessary to achieve the best results. The more lightweight YOLO series detectors' backbones of the one-shot models perform better in both MOT accuracy and speed.

Feature Fusion. Multi-layer feature fusion (MLFF) is critical to the paradigm of joint re-ID and detection. DLA-34, which adds the MLFF structure based on ResNet34_DCN, can achieve great advantages in the accuracy of MOT as shown in Table 1. By fusing the features of different layers as the output feature map, we can mediate the conflict between the features required by the re-ID and detection tasks to a certain extent. As for our one-shot model, we first choose the most common FPN [7] as the basis of our neck network. However, the weighted feature fusion method (BiFPN) proposed by EfficientDet [11] achieves repeated

Table 1. Comparison of combinations evaluated on the MOT16 test set. For fair comparison, MOTDT [13] association algorithm is adopt to all models to associate.

Combinations			Feature Fusion		Metrics		
Backbone	Neck	MLFF	BFF	IDF1↑	MOTA↑	FPS↑
ResNet34_DCN [5]	–	✗	✗	64.8	68.5	>30
DLA-34 [1]	–	✓	✗	71.9	73.2	<25
YOLOv5 [24]	FPN	✓	✗	68.8	69.3	**>35**
YOLOv5	BiFPN	✓	✓	69.0	69.6	>30
YOLOv7 [10]	FPN	✓	✗	73.4	74.0	>30
YOLOv7	BiFPN	✓	✓	73.6	74.4	>30
YOLOv8 [24]	FPN	✓	✗	69.6	70.6	>30
YOLOv8	BiFPN	✓	✓	69.7	70.8	>30

bidirectional fusion between features and can integrate more abundant feature information, which is faster and more repeated than another bidirectional feature fusion (BFF) method of PANet [25]. Based on the YOLOv5s, YOLOv7 and YOLOv8 as backbones of one-shot models, the initial feature fusion network is replaced from FPN to BiFPN. Shown in Table 1, BiFPN performs better in MOT accuracy than FPN dose due to its bidirectional feature fusion (BFF) structure. **Feature Conflict.** FairMOT [3] uses combinations of stride values at 2, 4 and 8 for ablation study of exploring the best value for the one-shot model's performance and the conclusion is that it can achieve the best MOT results when the (output) stride equals to 4. We follow those stride settings to further explore the phenomenon of feature conflict. Its [3] solution is to use DLA-34 as a multi-layer feature fusion network to integrate features from different layers to alleviate the feature conflict. However, as mentioned in *Introduction*, blindly fusing the features of all layers is not the best choice. Based on our base model (backbone=YOLOv7, neck=FPN), as shown in Fig. 3, we further validate different sources of feature conflicts. We name the feature map scale that performs the last feature fusion from top to bottom in FPN as "fusion stride". We further elaborate the existence of feature conflict with the concept of "fusion stride", and explore preferences of re-ID and detection tasks for separately required features.

As shown in Table 2, the TPR (True Positive Rate at false accept rate 0.1) measures the re-ID features' quality, while AP (Average Precision) reflects results of detection tasks. When the output stride is 4, the tracking accuracy of fusion stride of 4 (MOTA=73.6) is lower than that of fusion stride of 8 (MOTA=74.0). Comparing sub indicators of the two cases, we find that when the fusion stride equals to 4, IDsw and TPR are deteriorated significantly (+11.1% and -0.7 respectively). Different from the re-ID indicators performance, when the fusion stride equals to 4, the AP that measures the detection performance is the best. To further elaborate, when the fusion stride is 8, we continue to fuse the fea-

Table 2. The impact of different "fusion stride" and "output stride" for re-ID and detection tasks measured on MOT16 test set, embedding and detection validation sets from the divided MOT17 data set as same as FairMOT's practice. MOTDT [13] association algorithm is adopt to all models to associate.

Fusion Stride	Output Stride	MOT16 MOTA↑	MOT16 IDsw↓	Emb TPR↑	Det AP↑
8	8	71.5	1211	98.3	83.2
8	4	74.0	1169	96.3	86.4
4	4	73.6	1299	95.6	86.6
2	2	72.1	1203	69.3	85.6

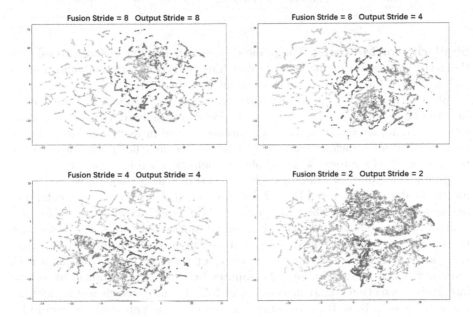

Fig. 5. Clustering analysis of re-ID features extracted by all persons in the test set using t-SNE. Different colors represent different people identities.

tures of the shallower layer, that is, when the fusion stride is 4, and the re-ID performance of our base model is significantly damaged, while the detection performance is slightly improved. This directly proves that there are feature conflicts between the features separately required by the re-ID and detection tasks. It also further implies that the fusion of more shallow features may be beneficial to the detection task but significantly damage the re-ID performance. In particular, our baseline model's neck network adopts the structure of the fusion stride of 8 and the output stride of 4, as shown in Fig. 3(e).

Re-ID features extracted by different models are visualized in Fig. 5. It is obvious that different identities' extracted features are mixed when fusion stride = 2. In contrast, they are well separated of the fusion stride that equals to 8.

Table 3. Comparison of the similarity for matching.

Similarity/Type	Similarity1	Similarity2	IDF1↑	MOTA↑	IDsw↓
Single	IoU	IoU	73.5	74.6	906
Single	Re-ID	IoU	74.5	74.7	890
Fusion	Fusion1	IoU	75.3	74.9	688
Fusion	Fusion1	Fusion2	75.6	75.0	676

Similarity for Matching. The proposed one-shot baseline is adopt to validate the performances of the similarity for matching in tracking accuracy and tracklet stability when faces with complex scenarios on MOT16 test set.

Shown in Fig. 4, metrics *similarity1* and *similarity2* respectively participate in matching of two-stage data association. When the similarity type is single for each matching, shown in Table 3, appearance as the primary similarity metric performs better in tracking accuracy (+1 IDF1) because the appearance descriptor restores the long-term occluded object' identity when it reappears.

To further explore advantages of the fusion similarity for matching, we remain metric similarity2 computed by IoU distance, when metric similarity1 computes by the fusion distance *fusion1* by combining GIoU distance and cosine distance, which would decrease mismatches that candidate tracklets have close cosine distances with a certain detected object but are far away from the certain detection, so the tracklet's stability of the one-shot baseline has greatly improved. Moreover, when metric similarity2 computes by the fusion distance *fusion2* by integrating detection confidence score and IoU distance, which would decrease mismatches with False Positive detections and alleviate potential Kalman Filter prediction errors, the stability of the one-shot baseline is further improved.

3.4 Compared with SOTA MOT Methods

The proposed approaches are extensively evaluated on the MOT16 and MOT17 testing set with abundant evaluation metrics, and later we compare the proposed methods with SOTA MOT models and algorithms.

One-Shot Baselines. Our final baseline model adopts the paradigm of joint re-ID and detection, shown in Fig. 1. The backbone of the proposed baseline is selected as YOLOv7's backbone, and we design the structure of feature fusion network shown in Fig. 3(e). Proposed baseline model's final tracking results is listed in Table 4. Particularly, our baseline model outperforms FairMOT [3] and

JDE [6], which both belong to joint re-ID and detection for tracking. It is worth noting that in order to reflect the advantages of our baseline model, following FairMOT, MOTDT [13] association algorithm is adopt to the proposed baseline to ensure fair comparison. In particular, our baseline model performs very well in terms of speed, reaching an average of 31 FPS on MOT17 test set.

Table 4. The listed comparison results are from original papers of FairMOT and JDE. We test on the same GPU (Tesla V100) to get FPS indicators.

Dataset	Baseline	IDF1↑	MOTA↑	HOTA↑	FPS↑
MOT16	JDE [6]	55.8	64.4	–	20.8
	Ours	**73.6**	**74.4**	**60.1**	**30.8**
MOT17	FairMOT [3]	72.3	**73.7**	59.3	22.6
	Ours	**73.0**	73.6	**59.7**	**31.0**

Table 5. Main-stream association algorithms measured on MOT17 test set.

Association	IDF1↑	MOTA↑	IDsw↓
DeepSORT_2 [26]	72.6	73.4	3805
MOTDT [13]	73.0	73.6	3681
ByteTrack [14]	73.9	74.3	2757
Ours	**74.9**	**74.5**	**2127**

Association Algorithms. We choose the proposed one-shot baseline model and keep on training with main-stream association algorithms, and then evaluate on the MOT17 test set. Listed in Table 5, the proposed fusion similarity algorithm further improves the stability of tracklet in IDsw and IDF1 and tracking accuracy (MOTA, IDF1) of the tracker compared with the most advanced association algorithms. Specifically, compared with the initial association algorithm MOTDT [13] of the baseline model, our association algorithm has a significant improvement in tracklet's ID stability (+1.9 IDF1, −42.2% IDsw) and the final tracking accuracy (MOTA) increases by 0.9%. Further more, compared with the SOTA association algorithm ByteTrack [14], IDF1 increases by 1% and IDsw also greatly decreases by 22.9%.

Table 6. Comparison of the related SOTA one-shot trackers. Note that we obtain listed BFTracker final results directly from the official evaluation by MOT Challenge. We fill the not provided values of original papers by "–".

Dataset	Metrics	Related SOTA One-Shot Trackers				Ours
		CSTrack [27]	FairMOT [3]	OMC [29]	CorrTracker [28]	BFTracker
MOT16	IDF1↑	71.8	72.8	74.1	74.3	**75.6**
	MOTA↑	70.7	74.9	76.4	**76.6**	75.0
	IDsw↓	1071	1074	–	979	**676**
	HOTA↑	–	–	–	61.0	**61.1**
	FPS↑	15.8	25.9	12.8	14.8	**29.7**
MOT17	IDF1↑	71.6	72.3	73.8	73.6	**74.9**
	MOTA↑	70.6	73.7	76.3	**76.5**	74.5
	IDsw↓	3465	3303	–	3369	**2127**
	HOTA↑	–	59.3	–	**60.7**	60.7
	FPS↑	15.8	25.9	12.8	14.8	**29.9**

One-Shot Trackers. BFTracker (proposed baseline model with our proposed algorithm termed as BFTracker) is widely compared with related SOTA one-shot (private detection) trackers evaluated on the MOT16 and MOT17 test set. All the other related one-shot trackers' results shown in the table are from original papers. As shown in Table 6, FairMOT [3] and CSTrack [27] are both classic MOT models of joint re-ID and detection. Furthermore, CorrTracker [28] and OMC [29] respectively extend the work based on the former to achieve better performance. In particular, BFTracker achieves the best results in MOT accuracy (HOTA, IDF1), stability (IDF1, IDsw) and real-time performance (FPS).

4   Conclusion and Future Works

In conclude, we first propose a new simple one-shot baseline of joint re-ID and detection tasks for real-time MOT. Moreover, we further explore the feature conflict phenomenon in the one-shot model and devote to alleviate the feature conflict. On the other hand, we analyze the background of the tracking challenges and further propose the algorithm in combining different cues to appropriately handle sudden appearance changes and alleviate Kalman Filter prediction errors.

For future works, more expansion work similar to attention mechanism can be carried out based on our baseline. We provide the remaining problem of our baseline that the method of predicting objects based on center points can affect the quality of the two major tasks of re-ID and detection. For example, when the center point prediction is not accurate, the detection box based on regression of that point will also deviate more from GT, and the re-ID features extracted based on that point will also become meaningless. We believe that relying on attention mechanism can improve the object localization capability and make our baseline more accurate and complete in predicting object centers, and further improve tracking performance.

References

1. Zhou, X., et al.: Objects as points. In: arXiv:1904.07850 (2019)
2. Zhou, X., Koltun, V., Krähenbühl, P.: Tracking objects as points. In: Vedaldi, A., Bischof, H., Brox, T., Frahm, J.-M. (eds.) ECCV 2020. LNCS, vol. 12349, pp. 474–490. Springer, Cham (2020). https://doi.org/10.1007/978-3-030-58548-8_28
3. Zhang, Y., Wang, C., Wang, X., Zeng, W., Liu, W.: FairMOT: on the fairness of detection and re-identification in multiple object tracking. Int. J. Comput. Vis. **129**(11), 3069–3087 (2021). https://doi.org/10.1007/s11263-021-01513-4
4. Mostafa, R., et al.: LMOT: efficient light-weight detection and tracking in crowds. IEEE Access **10**, 83085–83095 (2022)
5. Dai, J., et al.: Deformable convolutional networks. In: International Conference on Computer Vision (ICCV), pp. 764–773 (2017)
6. Wang, Z., Zheng, L., Liu, Y., Li, Y., Wang, S.: Towards real-time multi-object tracking. In: Vedaldi, A., Bischof, H., Brox, T., Frahm, J.-M. (eds.) ECCV 2020. LNCS, vol. 12356, pp. 107–122. Springer, Cham (2020). https://doi.org/10.1007/978-3-030-58621-8_7
7. Lin, T., et al.: Feature pyramid networks for object detection. In: Computer Vision and Pattern Recognition (CVPR), pp. 2117–2125 (2017)
8. Milan, A., et al.: MOT16: a benchmark for multi-object tracking. In: arXiv: 1603.00831 (2016)
9. Dendorfer, P., et al.: MOTChallenge: a benchmark for single-camera multiple target tracking. In: International Journal of Computer Vision (IJCV) 129.4, pp. 845–881 (2021)
10. Wang, C., et al.: YOLOv7: trainable bag-of-freebies sets new state-of-the-art for real-time object detectors. In: arXiv: 2207.02696 (2022)
11. Tan, M., et al.: EfficientDet: scalable and efficient object detection. In: Computer Vision and Pattern Recognition (CVPR), pp. 10781–10790 (2020)
12. Lin, T, et al.: Focal loss for dense object detection. In: International Conference on Computer Vision (ICCV), pp. 2980–2988 (2017)
13. Chen, L., et al.: Real-time multiple people tracking with deeply learned candidate selection and person re-identification. In: International Conference on Multimedia and Expo (ICME), pp. 1–6 (2018)
14. Zhang, Y., et al.: ByteTrack: multi-object tracking by associating every detection box. In: European Conference on Computer Vision (ECCV), pp. 1–21. Springer, Cham (2022). https://doi.org/10.1007/978-3-031-20047-2_1
15. Kuhn, H.: The Hungarian method for the assignment problem. In: Naval research logistics quarterly 2.1-2, pp. 83–97 (1955)
16. Zhang, S., et al.: CityPersons: a diverse dataset for pedestrian detection. In: Computer Vision and Pattern Recognition (CVPR), pp. 3213–3221 (2017)
17. Xiao, T., et al.: Joint detection and identification feature learning for person search. In: Computer Vision and Pattern Recognition (CVPR), pp. 3415–3424 (2017)
18. Dollár, P., et al.: Pedestrian detection: a benchmark. In: Computer Vision and Pattern Recognition (CVPR), pp. 304–311 (2009)
19. Ess, A., et al.: A mobile vision system for robust multi-person tracking. In: Computer Vision and Pattern Recognition (CVPR), pp. 1–8 (2008)
20. Zheng, L., et al.: Person re-identification in the wild. In: Computer Vision and Pattern Recognition (CVPR), pp. 1367–1376 (2017)
21. Shao, S., et al.: CrowdHuman: a benchmark for detecting human in a crowd. In: arXiv: 1805.00123 (2018)

22. Bernardin, K., et al.: Evaluating multiple object tracking performance: the clear mot metrics. In: EURASIP J., 1–10 (2008)
23. Luiten, J., et al.: HOTA: a higher order metric for evaluating multi-object tracking. Int. J. Comput. Vis. (IJCV) 129.2, 548–578 (2021)
24. Glenn, J.: www.github.com/ultralytics
25. Liu, S., et al.: Path aggregation network for instance segmentation. In: Computer Vision and Pattern Recognition (CVPR), pp. 8759–8768 (2018)
26. Wojke, N., et al.: Simple online and realtime tracking with a deep association metric. In: International Conference on Image Processing (ICIP), pp. 3645–3649 (2017)
27. Liang, C., et al.: Rethinking the competition between detection and Re-ID in Multi-object tracking. In: arXiv:2010.12138 (2020)
28. Wang, Q., et al.: Multiple object tracking with correlation learning. In: Computer Vision and Pattern Recognition (CVPR), pp. 3876–3886 (2021)
29. Liang, C., et al.: One more check: making "fake background" be tracked again. In: Association for the Advance of Artificial Intelligence (AAAI), vol. 36, No. 2, pp. 1546–1554 (2022)

Violence-MFAS: Audio-Visual Violence Detection Using Multimodal Fusion Architecture Search

Dan Si[1], Qing Ye[1], Jindi Lv[1], Yuhao Zhou[1], and Jiancheng Lv[2(✉)]

[1] College of Computer Science, Sichuan University, Chengdu 610065,
People's Republic of China
{sidan_starry,lvjindi}@stu.scu.edu.cn, yeqing@scu.edu.cn
[2] Engineering Research Center of Machine Learning and Industry Intelligence,
Ministry of Education, Chengdu 610065, People's Republic of China
lvjiancheng@scu.edu.cn

Abstract. Audio-visual fusion methods are widely employed to tackle violence detection tasks, since they can effectively integrate the complementary information from both modalities to significantly improve accuracy. However, the design of high-quality multimodal fusion networks is highly dependent on expert experience and substantial efforts. To alleviate this formidable challenge, we propose a novel method named Violence-MFAS, which can automatically design promising multimodal fusion architectures for violence detection tasks using multimodal fusion architecture search (MFAS). To further enable the model to focus on important information, we elaborately design a new search space. Specifically, multilayer neural networks based on attention mechanisms are meticulously constructed to grasp intricate spatio-temporal relationships and extract comprehensive multimodal representation. Finally, extensive experiments are conducted on the commonly used large-scale and multi-scene audio-visual XD-Violence dataset. The promising results demonstrate that our method outperforms the state-of-the-art methods with less parameters.

Keywords: Violence Detection · Multimodal Fusion Architecture Search · Attention Mechanism · Deep Neural Networks

1 Introduction

Recently, deep neural networks (DNNs) have achieved promising success in violence detection due to their powerful feature representation and generalization capabilities in the complex real-world scenario [3,10,11,15,16,20,22]. According to the type of inputs, the existing DNN methods can be broadly categorized into unimodal [3,11,15,16] and multimodal methods [10,20,22]. Specifically, unimodal methods utilize visual information to detect violence, concentrating only

This work is supported by the Key Program of National Science Foundation of China (Grant No. 61836006).

on visual elements such as abuse and fighting. On the other hand, multimodal methods employ both audio and visual information to capture violent features, where audio elements such as screams and gunshots are also highly concerned. Among these methods, the multimodal method achieves a surprisingly competitive performance, since it can efficiently exploit and aggregate the complementary information from multi-source data.

In dealing with multimodal tasks, the architecture design of multimodal fusion networks plays a key role, which involves the effective exploitation of latent information from multiple perspectives. However, the manual construction of fusion networks relies heavily on time-consuming experiments and has no guarantee of optimality. To address this issue, neural architecture search (NAS) is proposed to automate the extremely tedious process of discovering the well-performing network architecture for a specific task, and it is widely used for architecture design in various fields [2,9,12,13,23]. MFAS [12], as a typical method of NAS, focuses on the automatic search for multimodal feature combinations and achieves promising performance. Specifically, given that the features extracted from the hidden layers of each modality contain various insights, MFAS is intended to automatically find the optimal fusion architecture among different layers of the involved modalities.

Inspired by this, we propose a new violence detection method, called Violence-MFAS, to explore high-performance multimodal fusion architectures with MFAS. To the best of our knowledge, this is the first work using NAS in the field of violence detection. Additionally, several studies [1,18] show that spatio-temporal information in videos is extremely important for video classification, and such instructive prior knowledge contributes to enhancing the performance of the searched architectures. Therefore, we design a new search space, where attention mechanisms are applied to focus on vital violent clues in complex scenes and suppress redundant information using spatio-temporal information. The main contributions are summarized below:

- To explore the optimal multimodal fusion network for audio-visual violence detection, we propose an effective method called Violence-MFAS, which is the first work of NAS in violence detection tasks.
- To further improve the performance of fusion architectures, we specifically design a novel search space, where the attention mechanisms are applied to the multilayer neural networks.
- To illustrate the efficiency of our method, we conduct extensive experiments on XD-Violence, the largest audio-visual violence dataset. The results show that our method performs better than the state-of-the-art methods with a lightweight architecture.

2 Related Work

2.1 Violence Detection

Video violence detection with DNN refers to the automatic recognition of the time window of violent behaviors, which can be classified into unimodal and multimodal approaches. The former predominantly employ visual feature extraction techniques to detect instances of violence within videos [3,11,15,16]. Some methods utilized temporal networks, exemplified by Hanson et al. [3] employing bidirectional long short-term memory (LSTM), and Peixoto et al. [11] embracing C3D and CNN-LSTM, to extract key insights embedded in the video. To eliminate irrelevant information and enhance the accuracy of detection, multiple instance learning is widely used in this process [15,16]. Besides, Sultani et al. [15] proposed the inclusion of sparsity and time smoothness constraints in the loss function, and Tian et al. [16] introduced robust temporal feature magnitude learning (RTFM), which significantly enhanced the discernment of positive examples.

However, these methods ignore the importance of audio. And the combined usage of audio and visual modalities [10,20,22] harnesses the power of cross-modal interaction, promoting the extraction of complementary information and reinforcing the reliability of the detection process. Wu et al. [20] introduced a multimodal violence dataset named XD-Violence, accompanied by the proposal of a holistic and localized network, but it only simply uses an early concatenation of the visual and audio features. In contrast, Pang et al. [10] focused on the design of fusion architecture, presenting three modules that effectively integrated and combined modal information. Yu et al. [22] employed distillation techniques to narrow the semantic gap and reduce data noise resulting from cross-modality interactions, albeit at the expense of additional hyperparameters.

Note that the fusion architectures of the aforementioned multimodal methods exhibit two types of issues. Either they usually employed a single combination of late (or early) features failing to capitalize on the rich information exchange between modalities, or they relied heavily on manual construction which is an exhausting process. Therefore, in this work, our focus lies on the exploration of an automated methodology that endeavors to design the optimal fusion architecture for the violence detection task.

2.2   Neural Architecture Search

Neural architecture search (NAS) is a technique that automatically discovers the high-performance architectures of the DNNs for a specific task [23]. Early NAS works mainly based on reinforcement learning (RL) [23,24] or evolutionary algorithms (EA) [13,21]. In RL-based methods, a controller agent is employed to generate a sequence of actions that define the neural architecture. Subsequently, the architecture is trained and evaluated on a validation set, and the resulting validation error is utilized as a reward to update the controller. In EA-based methods, the neural architecture is encoded as a string, and the search process involves performing random recombinations and mutations of these strings. Then every "parent" architecture specified by the string is trained and evaluated, afterwards the most promising architectures are selected to generate "children". These methods demonstrated superior classification accuracy compared to manually constructed architectures. However, their utilization necessitates substantial computational resources, rendering them unaffordable. To alleviate this problem, a sequential model-based optimization (SMBO) [6] strategy was introduced. For example, PNAS [8] based on SMBO progressively grows the complexity of neural network architectures by iteratively adding operators, allowing for the exploration of diverse network architectures while maintaining computational efficiency.

The automatic ability of NAS in designing high-performing architectures has led to its rapid application in various fields [2,9], which also has sparked the idea of utilizing NAS for multimodal tasks. The pioneering work is MFAS [12], which employs the SMBO technique to sequentially explore optimal multimodal fusion architectures. By treating multimodal classification as a combinatorial search problem, MFAS offers a novel approach to address this task efficiently. Given the impressive capabilities of MFAS, we consider applying MFAS to address the aforementioned challenges of violence detection.

3   Method

In this section, the detail of the proposed Violence-MFAS will be introduced, which consists of two parts: 1) a search space including multilayer neural networks based on attention mechanisms of involved modalities and the fusion architecture, and 2) a search algorithm based on MFAS.

3.1   Search Space

An illustration of the search space is shown in Fig. 1, where the upper part represents the visual multilayer neural network and the lower part represents the audio multilayer neural network. x_a^m indicates the embedding of the audio network layer m and x_v^n indicates the embedding of the visual network layer n. The hidden embeddings x_a^m and x_v^n (the output of I3D/Vggish, FC-1, TAM, SAM, CMAM) of multilayer neural networks serve as one part of the search space. The first fusion layer in fusion architecture only contains two inputs: the output of a selected embedding for each modality x_a^m and x_v^n. Other fusion layer contains three inputs: x_a^m, x_v^n and the output of the previous fusion layer h_{l-1}. They are fused using a concatenation operation and the process of the fusion layer $l(l > 1)$ is represented as

$$h_l = \sigma_p \left(W_l \begin{bmatrix} x_a^m \\ x_v^n \\ h_{l-1} \end{bmatrix} \right), \tag{1}$$

where (x_a^m, x_v^n, σ_p) is the search target for fusion layer l, respectively denoting which hidden feature from the audio modality, which hidden feature from the visual modality, and which activation function is utilized. h_{l-1} is the output of the previous fusion layer and W_l is the fusion layer weights that need to be trained. Also, $m \in \{1, \cdots, M\}$, $n \in \{1, \cdots, N\}$, $p \in \{1, \cdots, P\}$. Specifically, two activation functions, ReLU and Sigmoid, are used.

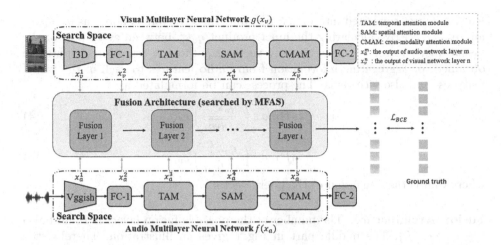

Fig. 1. The overview of our method. x_v and x_a respectively denote the input of the visual and audio network. The upper part in green represents the visual network $g(x_v)$ with three attention modules. The lower part in blue represents the audio network $f(x_a)$ with three attention modules. The middle part in orange represents the fusion architecture that needs to be searched by MFAS. The outputs of different layers from the audio and visual network (i.e. x_a^m and x_v^n) are used in the search space. (Color figure online)

Multilayer Neural Networks. Both audio and visual multilayer neural networks mainly consist of three attention mechanisms: the temporal attention module (TAM), the spatial attention module (SAM), and the cross-modality attention module (CMAM). TAM is a dynamic selection mechanism that determines when to pay attention. SAM is an adaptive spatial region selection mechanism that determines where to concentrate on. And CMAM is designed to effectively capture complementary information.

Given an input feature $x \in R^{T \times S}$, TAM and SAM aim to sequentially infer a 1D attention map $W_T \in R^T$ and $W_S \in R^S$ respectively from the temporal and spatial dimensions, then the attention map is multiplied to the input feature x for refinement as follows:

$$W = \sigma(Conv1D\left(GMP(x)\right)), \tag{2}$$

$$x = W \odot x, \tag{3}$$

where GMP denotes global max pooling, $Conv1D$ denotes a convolution operation, σ denotes the sigmoid function, $\odot$ denotes element-wise multiplication, and W denotes the attention map W_T or W_S. Note that global average pooling is not utilized, because we aim to highlight the specific segments in the video that exhibit a higher propensity for containing elements of violence, rather than the average tonal characteristics of the entire video.

To enrich the information of every single modality and obtain more representative features, CMAM is applied. According to [17], each visual and audio feature is first converted into query, key, and value vectors denoted as q_v, k_v, v_v and q_a, k_a, v_a. By calculating the inner product $q_v k_a^T$ between every pair of the visual query vector q_v and audio key vector k_a, the raw attention weights are obtained for aggregating information from audio features to each of the visual features, and also vice versa. The process can be formulated as:

$$x_v = softmax\left(\frac{q_v k_a^T}{\sqrt{dim}}\right) v_a, \tag{4}$$

$$x_a = softmax\left(\frac{q_a k_v^T}{\sqrt{dim}}\right) v_v, \tag{5}$$

where dim is the dimension of the query vector.

Fusion Architecture. The number of the possible fusion layers is L, and also $l \in \{1, \cdots, L\}$. The middle part in Fig. 1 gives an illustration. Thereby, the combination of fusion architectures is formed by five possible features for each modality ($M = N = 5$), two possible activation functions ($P = 2$), along with the number of fusion layers ($L = 4$) in our designed search space, which results in $(M \times N \times P)^L$ (i.e., 6.25×10^6) possible fusion architectures. However, it is extremely laborious and impractical to explore all the fusion ways in the large search space. To address this challenge, an sequential search algorithm is introduced, which aims to automatically search for the optimal fusion architecture.

3.2 Search Algorithm

Algorithm 1 shows the whole search algorithm. First, audio and visual multilayer neural networks are trained before the search and serve as off-the-shelf multilayer feature extractors during the search (i.e., the weights of the multilayer feature extractors are frozen in the subsequent search process). When exploring possible fusion architectures, a progressive search method is used, which starts from the simplest fusion architecture and then expands each architecture to more complex ones. The more layers a fusion architecture has, the more complex it is. Specifically, start by constructing all possible fusion architectures $\mathcal{B}_1$ with minimal complexity (i.e., $L = 1$, each fusion architecture contains only one fusion layer), then train and evaluate all the simplest architectures. The process is shown in Alg.1 steps from 3 to 7. $\mathcal{M}$ represents the set of sampled architectures and $\mathcal{A}$ represents the accuracy respectively. Subsequently, expand each architecture successively to the more complex hierarchy by adding all possible fusion architectures in $\mathcal{B}_1$. Since the thorough exploration of all possibilities is unrealistic, a predictor π is used to evaluate all possible architectures and select the k most promising architectures to train. The performance of the trained architectures is fed back to the predictor for refinement, and the updated predictor guides the next complex level of the search process. Steps 8 to 14 show the process of progressive exploration. The whole search iteration is executed E_{search} times. After obtaining the top-K fusion architectures, the K networks assembled from f, g, and the fusion architectures are constructed. We trained them and selected the best one as the customized model for the specific task.

Algorithm 1: Violence-MFAS algorithm

Input: multimodal dataset D, visual network f, audio network g, the simplest
      fusion architecture set $\mathcal{B}_1$
Output: the optimal multimodal fusion architecture $\mathcal{M}$

1   $f, g = $ **train-CNN** (f, g, D)// Obtain off-the-shelf multilayer feature
   extractors f and g as the search space
2   **for** e in $1 : E_{search}$ **do**
3 | $\mathcal{S}_1 = \mathcal{B}_1$ //Begin with $L = 1$
4 | $\mathcal{M}_1 = $ **build-fusion-net** $(f, g, \mathcal{S}_1)$ // Build fusion networks
5 | $\mathcal{A}_1 = $ **train-eval-net** $(\mathcal{M}_1, D)$// Train and evaluate
6 | $\mathcal{M} = \mathcal{M} \cup \mathcal{S}_1$, $\mathcal{A} = \mathcal{A} \cup \mathcal{A}_1$// Merge the sampled architectures
7 | $\pi = $ **update-predictor** $(\mathcal{S}_1, \mathcal{A}_1)$
8 | **for** l in $2 : L$ **do**
9 | | $\mathcal{S}'_l = $ **add-one-layer** $(\mathcal{S}_l, \mathcal{B}_1)$ // Unfold one more fusion layer
10 | | $\mathcal{S}_l = $ **sampleK** $(\mathcal{S}'_l, \pi, K)$// Sample K architectures with a predictor π
11 | | $\mathcal{M}_l = $ **build-fusion-net** $(f, g, \mathcal{S}_l)$ // Build fusion networks
12 | | $\mathcal{A}_l = $ **train-eval-net** $(\mathcal{M}_l, D)$ // Train and evaluate
13 | | $\pi = $ **update-predictor** $(\mathcal{S}_l, \mathcal{A}_l)$
14 | | $\mathcal{M} = \mathcal{M} \cup \mathcal{S}_l$, $\mathcal{A} = \mathcal{A} \cup \mathcal{A}_l$// Merge the sampled architectures
15 $\mathcal{M} = $ **topK** $(\mathcal{M}, \mathcal{A}, K)$//Select top-$K$ architectures
16 **return** $\mathcal{M}$

---

4   Experiments

4.1   Dataset and Evaluation Metric

XD-Violence [20] is the only publicly available audio-visual violence dataset
so far. It is also the largest multi-scene violence dataset, which contains 4754
untrimmed videos, including 2405 violent videos and 2349 non-violent videos
with a total duration of 217 h. Six common types (i.e. abuse, car accident, explosion, fighting, riot, and shooting) collected from YouTube movies and in-the-wild
videos are covered. Following previous work [10,20,22], the dataset is divided
into a training set with 3954 videos and a test set with 800 videos. The test set
comprises 500 violent videos and 300 non-violent videos. The frame-level average
precision (AP) is used as the evaluation metric rather than AUC which is more
sensitive to class-imbalanced data.

4.2   Experimental Setup

As the prior works do, we adopt features of each modality as inputs in our
method. More specifically, we use the visual features extracted from I3D network
[1] pretrained on the Kinetics-400 dataset and the audio features extracted from
VGGish network [5] pretrained on a large YouTube dataset. The kernel size
of the 1D-convolution layer in spatial attention is 7 with padding 3, while an
adaptive kernel is used in temporal attention. To obtain the visual and audio

off-the-shelf multi-layer feature extractor, a contrastive learning loss, the same as in [22], is used along with the binary cross entropy during the training process of the multilayer neural networks. E_{search} is set 3. During the search process and training of the searched fusion architecture, only the binary cross-entropy is used. After obtaining the optimal fusion architecture from MFAS, a novel visual-audio model is built. We train the model for 10 epochs with a batch size of 128. For network optimization, Adam is used with an initial learning rate 1e-5 with a cosine annealing algorithm during the training stage. All experiments are performed on a single 1080Ti GPU with PyTorch implementation.

Table 1. Comparison with the state-of-the-art methods on the XD-Violence dataset. A and V represent audio and visual respectively.

Manner	Method	Modality	AP(%)	Param(M)
traditional methods	SVM baseline	V	50.78	/
	OCSVM [14]	V	27.25	/
DNN methods	Hasan et al. [4]	V	30.77	/
	Sultani et al. [15]	V	73.20	/
	Wu et al. [19]	V	75.90	/
	RTFM [16]	V	77.81	12.067
	Li et al. [7]	V	78.28	/
	Wu et al. [20]	A + V	78.64	0.843
	Pang et al. [10]	A + V	81.69	1.876
	Yu et al. [22]	A + V	83.40	0.678
	Violence-MFAS(Ours)	A + V	**84.08** ↑	**0.456** ↓

Table 2. The top 3 found architectures on the XD-Violence dataset according to the validation accuracy during the search. (e.g., the first found architecture shown in line 1 has four fusion layers, where (3, 3, 1) respectively indicates (x_a^m, x_v^n, σ_p). Specifically, (3, 3, 1) denotes that fuse the hidden embedding x_a^3 and x_v^3 and use the first kind of activation function (i.e. ReLU) in the first layer of the fusion architecture using the Eq. 1, while (4, 2, 2) denotes that fuse the hidden embedding x_a^4 and x_v^2 and use the second kind of activation function (i.e. Sigmoid) in the second layer of the fusion architecture. (5, 5, 1) and (3, 5, 2) respectively denote (x_a^m, x_v^n, σ_p) of the third and fourth layer in the fusion architecture.)

#	Fusion Architecture	Fusion Param(M)	AP(%)
1	[(3, 3, 1), (4, 2, 2), (5, 5, 1), (3, 5, 2)]	0.017	**84.08**
2	[(3, 3, 1), (1, 4, 1), (5, 5, 1), (3, 5, 2)]	0.017	83.93
3	[(4, 5, 2), (1, 4, 1), (5, 5, 1) (3, 5, 2)]	0.017	83.90

4.3 Comparison with State-of-the-Arts

Table 1 shows the frame-level AP results on the XD-Violence dataset. Violence-MFAS is superior to all current methods and guarantees a lightweight network in pursuit of high accuracy. The methods in the first two rows are ineffective in violence detection tasks with traditional methods. Compared with the DNN methods only using visual information, at least 5.3% improvement is achieved, suggesting that audio information provides significant complementary features in violence detection tasks. In addition, our method achieves a 0.68% gain over the state-of-the-art method which uses an artificially designed multimodal fusion network. Specifically, the fusion architecture designed through automatic search can efficiently leverage the inherent complementarity among multiple modalities, ingeniously constructing an optimal fusion architecture than the model designed manually. These promising results indicate Violence-MFAS is feasible and advantageous for audio-visual violence detection. Table 2 reports the accuracy of the top three fusion architectures during the search and Fig. 2 gives a vivid illustration of the best one fusion architecture (i.e. Table 2 #1), in which multiple hidden layers of information are efficiently involved in the fusion architecture.

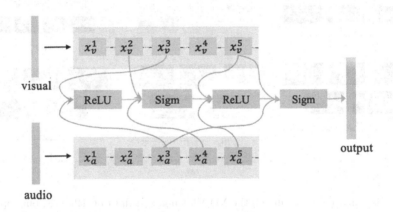

Fig. 2. The DNN architecture of the found fusion module.

4.4 Ablation Studies

To further demonstrate the effectiveness of our method, we conduct ablation experiments shown in Table 3. First, we reimplement the state-of-the-art method [22] discarding its self-distillation module as the vanilla network, which achieves 81.92% without the SAM, TAM module, and the MFAS method. Then, MFAS is added to the vanilla implementation for searching the optimal fusion architecture, and the output of the searched fusion architecture is used for prediction. As the result suggests, MFAS effectively enhances the AP by 0.65%. When further adding TAM and SAM to the multilayer feature extractor, the search space is expanded, and the searched fusion architecture achieves the best result.

This indicates that the outputs of different attention modules in multilayer neural networks contain more representative features, thus allowing a better fusion architecture to be searched with an effective search space.

Table 3. Ablation studies of our proposed method.

#	Multimodal	Attention mechanism	MFAS	AP(%)
1	✔	✘	✘	81.92
2	✔	✘	✔	82.57
3	✔	✔	✘	83.32
4	✔	✔	✔	**84.08**

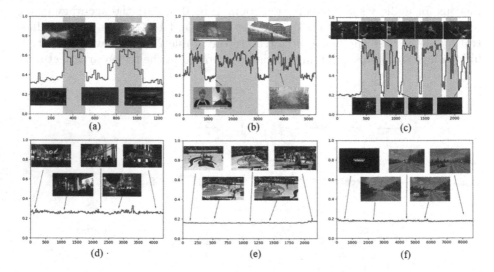

Fig. 3. Visualization of results on the XD-Violence test dataset. Blue regions represent the temporal ground-truths of violent events. (Color figure online)

4.5   Visualizations of Prediction

Figure 3 provides several visualizations of the violence detection results, where (a)(b)(c) are violent videos and (d)(e)(f) are non-violent videos. The architecture pinpoints periods of time when violence occurs and gives much lower scores for non-violent snippets through its discerning capabilities. Note that the normal clip between two violent clips is distinguished successfully. While for non-violent videos, the violent scores consistently remain at a low level. The promising outcomes demonstrate that Violence-MFAS enables the model to grasp the disparity between normal events and violent events.

5   Conclusion

We propose a novel method, Violence-MFAS, for violence detection by finding the optimal multimodal fusion architecture with MFAS (i.e., an efficient automatic fusion architecture search approach for multimodal tasks). To the best of our knowledge, Violence-MFAS is the first work of NAS in violence detection tasks, which can effectively capture complementary information embedded in audio and visual. To further enhance the performance of fusion architectures, a new search space is designed that includes multilayer neural networks based on attention mechanisms. The searched lightweight architecture of the Violence-MFAS validates its remarkable superiority over the state-of-the-art methods through extensive experiments on the largest violence dataset. Considering more fusion strategies to expand search space is a direction for future work.

References

1. Carreira, J., Zisserman, A.: Quo Vadis, action recognition? a new model and the kinetics dataset. In: proceedings of the IEEE Conference on Computer Vision and Pattern Recognition, pp. 6299–6308 (2017)
2. Du, X., et al.: SpineNet: learning scale-permuted backbone for recognition and localization. In: Proceedings of the IEEE/CVF Conference on Computer Vision and Pattern Recognition, pp. 11592–11601 (2020)
3. Hanson, A., Pnvr, K., Krishnagopal, S., Davis, L.: Bidirectional convolutional LSTM for the detection of violence in videos. In: Proceedings of the European Conference on Computer Vision (ECCV) Workshops, pp. 0–0 (2018)
4. Hasan, M., Choi, J., Neumann, J., Roy-Chowdhury, A.K., Davis, L.S.: Learning temporal regularity in video sequences. In: Proceedings of the IEEE Conference on Computer Vision and Pattern Recognition, pp. 733–742 (2016)
5. Hershey, S., et al.: CNN architectures for large-scale audio classification. In: ICASSP 2017–2017 IEEE International Conference on Acoustics, Speech and Signal Processing (ICASSP), pp. 131–135. IEEE (2017)
6. Hutter, F., Hoos, H.H., Leyton-Brown, K.: Sequential model-based optimization for general algorithm configuration. In: Coello, C.A.C. (ed.) LION 2011. LNCS, vol. 6683, pp. 507–523. Springer, Heidelberg (2011). https://doi.org/10.1007/978-3-642-25566-3_40
7. Li, S., Liu, F., Jiao, L.: Self-training multi-sequence learning with transformer for weakly supervised video anomaly detection. In: Proceedings of the AAAI Conference on Artificial Intelligence, vol. 36, pp. 1395–1403 (2022)
8. Liu, C., et al.: Progressive neural architecture search. In: Proceedings of the European Conference on Computer Vision (ECCV), pp. 19–34 (2018)
9. Lv, J., Ye, Q., Sun, Y., Zhao, J., Lv, J.: Heart-Darts: classification of heartbeats using differentiable architecture search. In: 2021 International Joint Conference on Neural Networks (IJCNN), pp. 1–8. IEEE (2021)
10. Pang, W.F., He, Q.H., Hu, Y.j., Li, Y.X.: Violence detection in videos based on fusing visual and audio information. In: ICASSP 2021–2021 IEEE International Conference on Acoustics, Speech and Signal Processing (ICASSP), pp. 2260–2264. IEEE (2021)

11. Peixoto, B., Lavi, B., Martin, J.P.P., Avila, S., Dias, Z., Rocha, A.: Toward subjective violence detection in videos. In: ICASSP 2019–2019 IEEE International Conference on Acoustics, Speech and Signal Processing (ICASSP), pp. 8276–8280. IEEE (2019)

12. Pérez-Rúa, J.M., Vielzeuf, V., Pateux, S., Baccouche, M., Jurie, F.: MFAS: multimodal fusion architecture search. In: Proceedings of the IEEE/CVF Conference on Computer Vision and Pattern Recognition, pp. 6966–6975 (2019)

13. Real, E., et al.: Large-scale evolution of image classifiers. In: International Conference on Machine Learning, pp. 2902–2911. PMLR (2017)

14. Schölkopf, B., Williamson, R.C., Smola, A., Shawe-Taylor, J., Platt, J.: Support vector method for novelty detection. In: Advances in Neural Information Processing Systems, vol. 12 (1999)

15. Sultani, W., Chen, C., Shah, M.: Real-world anomaly detection in surveillance videos. In: Proceedings of the IEEE Conference on Computer Vision and Pattern Recognition, pp. 6479–6488 (2018)

16. Tian, Y., Pang, G., Chen, Y., Singh, R., Verjans, J.W., Carneiro, G.: Weakly-supervised video anomaly detection with robust temporal feature magnitude learning. In: Proceedings of the IEEE/CVF International Conference on Computer Vision, pp. 4975–4986 (2021)

17. Vaswani, A., et al.: Attention is all you need. In: Advances in Neural Information Processing Systems, vol. 30 (2017)

18. Wang, L., et al.: Temporal segment networks: towards good practices for deep action recognition. In: Leibe, B., Matas, J., Sebe, N., Welling, M. (eds.) ECCV 2016. LNCS, vol. 9912, pp. 20–36. Springer, Cham (2016). https://doi.org/10.1007/978-3-319-46484-8_2

19. Wu, P., Liu, J.: Learning causal temporal relation and feature discrimination for anomaly detection. IEEE Trans. Image Process. **30**, 3513–3527 (2021)

20. Wu, P., et al.: Not only look, but also listen: Learning multimodal violence detection under weak supervision. In: Vedaldi, A., Bischof, H., Brox, T., Frahm, JM. (eds.) Computer Vision – ECCV 2020. ECCV 2020. Lecture Notes in Computer Science, vol. 12375, pp. 322–339. Springer, Cham (2020). https://doi.org/10.1007/978-3-030-58577-8_20

21. Ye, Q., Sun, Y., Zhang, J., Lv, J.: A distributed framework for EA-based NAS. IEEE Trans. Parallel Distrib. Syst. **32**(7), 1753–1764 (2020)

22. Yu, J., Liu, J., Cheng, Y., Feng, R., Zhang, Y.: Modality-aware contrastive instance learning with self-distillation for weakly-supervised audio-visual violence detection. In: Proceedings of the 30th ACM International Conference on Multimedia, pp. 6278–6287 (2022)

23. Zoph, B., Le, Q.V.: Neural architecture search with reinforcement learning. arXiv preprint arXiv:1611.01578 (2016)

24. Zoph, B., Vasudevan, V., Shlens, J., Le, Q.V.: Learning transferable architectures for scalable image recognition. In: Proceedings of the IEEE Conference on Computer Vision and Pattern Recognition, pp. 8697–8710 (2018)

DeepLink: Triplet Embedding and Spatio-Temporal Dynamics Learning of Link Representations for Travel Time Estimation

Jiezhang Li[1] , Yue-Jiao Gong[1]($\boxtimes$) , Ting Huang[2] , and Wei-Neng Chen[1]

[1] School of Computer Science and Engineering, South China University of Technology, Guangzhou, China
gongyuejiao@gmail.com
[2] Guangzhou Institute of Technology, Xidian University, Guangzhou, China

Abstract. Estimating the time of arrival is a crucial task in intelligent transportation systems. The task poses challenges due to the dynamic nature and complex spatio-temporal dependencies of traffic networks. Existing studies have primarily focused on learning the dependencies between adjacent links on a route, often overlooking a deeper understanding of the links within the traffic network. To address this limitation, we propose DeepLink, a novel approach for travel time estimation that leverages a comprehensive understanding of the spatio-temporal dynamics of road segments from different perspectives. DeepLink introduces triplet embedding, enabling the learning of both the topology and potential semantics of the traffic network, leading to an improved understanding of links' static information. Then, a spatio-temporal dynamic representation learning module integrates the triplet embedding and real-time information, which effectively models the dynamic traffic conditions. Additionally, a local-global attention mechanism captures both the local dependencies of adjacent road segments and the global information of the entire route. Extensive experiments conducted on a large-scale real-world dataset demonstrate the superior performance of DeepLink compared to state-of-the-art methods.

Keywords: Estimating time of arrival · Spatio-temporal dynamics learning · Graph positional embedding

1 Introduction

The task of estimating arrival time (ETA) plays an important role in intelligent transportation systems. It is widely used in route planning, online scheduling, map navigation, etc [7,8,10,17]. ETA can significantly benefit organizations, such as DiDi, Uber, and Google Maps, in terms of enhancing their route planning and pricing services, since accurate estimation can improve the quality of these services and customer satisfaction. Although considerable research has been done

© The Author(s), under exclusive license to Springer Nature Singapore Pte Ltd. 2024
B. Luo et al. (Eds.): ICONIP 2023, CCIS 1968, pp. 217–229, 2024.
https://doi.org/10.1007/978-981-99-8181-6_17

on travel time estimation [6,15,16,19], accurate prediction is still a challenge. This is because the travel time is influenced by a variety of factors such as traffic conditions, weather, departure times, travel routes, and driver behaviour.

Earlier studies simply predicted arrival times based on historical average data [3], without considering real-time traffic information, leading to unsatisfactory results. Benefiting from the development of GPS and data mining technology, deep learning methods [13,15,16] model the spatio-temporal characteristics of routes and achieve considerable results. However, these early attempts learn the dependencies of links on the route without considering the influence of other links on the traffic network. To address the limitation, recent methods [6,19] formulate the traffic network as a graph and introduce graph neural networks to capture the spatio-temporal correlations of the traffic network.

However, unlike the traffic flow prediction task [8,9,18], the graph size in the ETA task is much larger, reaching hundreds of thousands level. Therefore, it is impossible to model the full graph information, and it is generally necessary to sample the subgraphs based on the routes. The problem with graph subsampling is the absence of canonical positional information of links, which is crucial for graph representation learning of traffic networks. In addition to static positional information, historical routes can be valuable for learning graph representations of links. The correlation between links is stronger if they have a high frequency of co-occurrence in historical routes. This also reflects the driving preferences of drivers, who prefer a certain adjacent link when passing an intersection. We refer to this relationship as potential semantics. Existing studies do not take these important factors into account, resulting in limited performance in learning representation of links, and consequently, in estimating the travel time.

To address the issues, we propose the DeepLink model, which efficiently learns the spatio-temporal dynamic representation of links and applies it to travel time estimation. Our approach includes the following key components.

- First, we introduce triplet embedding to comprehensively represent the static information of links in the traffic graph. This involves three positional embeddings: LaplacianPE, RandomWalkPE, and SemanticPE. LaplacianPE and RandomWalkPE effectively capture the topology of the traffic network, while SemanticPE learns the potential semantics among links based on historical routes.
- Second, we design a network that captures spatio-temporal dependencies between links in real time. By integrating triplet embedding and real-time information, we employ a graph attention network to model dynamic traffic conditions. Furthermore, we utilize a local-global fusion attention mechanism to capture both the local dependencies of adjacent links and the global information of all links on the route.
- Third, after a "deep" understanding of the link properties, a multi-task learning approach is deployed to accomplish our prediction task. Our model simultaneously predicts the entire travel time and the traffic condition status of each link on the route, improving the robustness of the prediction.

The main contributions of this paper are summarized as: 1) introduction of triplet embedding for a comprehensive representation of links, incorporating both network topology and potential semantics from historical routes; 2) design of a spatio-temporal dynamic representation learning network that captures inherent spatio-temporal relationships among links; 3) application of multi-task learning to improve the robustness of traffic time prediction; and 4) evaluation of our model on a large-scale real-world dataset, demonstrating its superiority over state-of-the-art methods.

2   Problem Definition

In this section, we introduce the basic definitions and the problem statement.

Definition 1. (Traffic Network): We define the traffic network as a graph $G^t = \{V, E, X^t\}$, where V represents the set of links and E represents the set of edges. In this paper, the terms "link" and "road segment" are used interchangeably for convenience. Unlike a general graph, the traffic network has the time attribute. G^t represents the traffic network information at timestamp t, where X^t represents the dynamic traffic information (link time and traffic status) of V.

Definition 2. (Route): A route $R = \{v_1, v_2...v_{N_r}\}$ is a sequence of adjacent links, where N_r is the number of links on the route (usually in dozens to hundreds).

Problem Statement: Given a query $Q^t = \{G^t, R, A^t\}$, where A^t denotes external attributes (such as distance, weekID, timeID, weather, driverID), the goal of ETA is to estimate the travel time from the complex traffic data.

3   Methodology

In this section, we present the architecture and modules of our proposed model. As shown in Fig. 1, attribute feature learning embeds external attributes into a low-dimensional feature representation. The triplet embedding module uses three embedding to learn the topology and the potential semantics of the traffic network. STDRL focuses on learning the spatio-temporal dynamic representation of links from both the traffic network and the query route. Finally, the travel time of the whole route and the traffic congestion status of each link are simultaneously predicted in the multi-task learning module.

3.1   Attribute Feature Learning

External attributes are important for traffic prediction tasks. The inclusion of timeID and weekID enables the model to capture temporal periodicity within the traffic data. By incorporating driverID, the model can learn about the driving behavior preferences of individual drivers. Then, weatherID helps the model

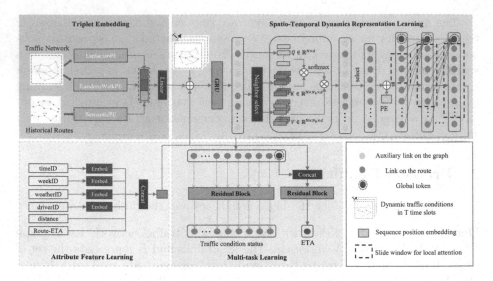

Fig. 1. The framework of DeepLink.

understand the impact of weather variability. To incorporate these categorical values, we embed them into a low-dimensional feature representation. We also introduce normalized distance and Route-ETA as features. Here, Route-ETA denotes the sum of the average travel time of links at the departure time extracted from historical data. These features are concatenated with the continuous and embedded features, creating an attribute feature representation:

$$X_{AF} = \text{Concat}[timeID, weekID...RouteETA] \tag{1}$$

3.2 Triplet Embedding

We introduce the Triplet Embedding approach, which combines Laplacian Eigenvector, RandomWalk, and Semantic positional embedding techniques. These embeddings offer unique perspectives on capturing positional information and semantic relationships among links in the traffic network.

Laplacian Eigenvector Positional Embedding Most GNN methods learn information about structural nodes with constant node positions, thus capturing dependencies between individual nodes. However, when dealing with graphs that have changing node positions, such as subgraphs of traffic networks that vary based on routes, simple GNN methods may not achieve optimal performance. Inspired by [4], we use the Laplacian eigenvectors [1] of the graph as the information representation of the link:

$$L = I - D^{-1/2}AD^{-1/2} = U^{T}\Lambda U \tag{2}$$

where I is the identity matrix, A is the adjacency matrix, D is the degree matrix, $\Lambda = diag(\lambda_0, \lambda_1, \ldots, \lambda_{N-1})$ is the eigenvalue matrix and U is the eigenvector matrix. To obtain the k-dimensional embedding of link l_i, we choose the k ($k < N$) smallest non-trivial eigenvectors. This process results in the graph Laplacian eigenvector positional embedding, denoted as PE_{lap}.

RandomWalk Positional Embedding Random Walk [12] offers an alternative approach for representing the positional information of links in the traffic network. In our study, we employ the method in [5] as a compensatory positional embedding of the links. For a given link i, we use it as the starting point for the random walk. The Random Walk positional embedding, denoted as PE_{rw}, is defined using k steps of the random walk as follows:

$$PE_{rw}^i = [\mathrm{RW}_{ii}, \mathrm{RW}_{ii}^2, \cdots, \mathrm{RW}_{ii}^k] \in \mathbb{R}^k \tag{3}$$

where RW is the random walk operator. The usage of the random walk matrix is simplified by considering only the landing probability of link i to itself, denoted as RW_{ii}, resulting in a low-complexity implementation.

Semantic Positional Embedding Treating routes as text, the co-occurrence frequency of links in historical routes reflects the potential semantics between links. For this purpose, we treat the link sequences in the historical routes as texts and use the skip-gram method [11] to learn the vector representation of links. Specifically, we first initialize the vector of each link and then predict the probability of occurrence of different neighboring links. This is solved by maximizing the likelihood of occurrence of the real neighboring links. Formally, the objective can be formulated as follows:

$$\arg\min_{\theta} -\frac{1}{T} \sum_{t=1}^{T} \sum_{-w \leq j \leq w} \log P(l_{t+j} \mid l_t, \theta) \tag{4}$$

where θ is the parameter to be learned, t denotes the position of the link in the sequence, w denotes the window size of sequences. By employing the skip-gram method, we obtain the link vector representation denoted as PE_{sem}, which is based on the semantic information extracted from historical routes.

3.3   Spatio-Temporal Dynamic Representation Learning

In this module, we focus on learning the representation of links in both the context of the traffic network and individual routes. Directly modeling the entire traffic network can be challenging due to its large scale. So we collect subgraphs based on routes and learn the link information within subgraphs.

Dynamics Representation Learning on the Subgraph. For each link on the subgraph, its dynamic traffic conditions are represented as a sequence $X^t = [x_i^t]_{t=1}^T$, where T denotes the last time slot before the departure time. The x_i^t comprises the average travel time and congestion status at time slot t, as well as the external attribute features. We then append the dynamic traffic conditions to positional embeddings and obtain the input $[\tilde{x}_i^t]_{t=1}^T$:

$$\tilde{X} = \sigma_{\text{ReLU}}(X^t + PE_{lap} + PE_{rw} + PE_{sem}) \tag{5}$$

We employ the Gated Recurrent Unit (GRU) to model the temporal attribute representation of links. GRU [2] has proven to be effective in learning from time sequences, addressing the gradient vanishing problem encountered by traditional RNNs and offering a simpler structure compared to LSTM. Given a sequence $[\tilde{x}_i^t]_{t=1}^T$, the output of GRU can be expressed as:

$$H = \text{GRU}([\tilde{x}_i^t]_{t=1}^T) \tag{6}$$

For each link i, we collect its neighboring links from the subgraph and obtain a matrix $H_{\text{nei}}^i \in \mathbb{R}^{N_k \times d}$, where N_k denotes the maximum number of neighbours of links and d denotes the dimensionality of the hidden state. Then we introduce attention mechanism to capture the spatio-temporal dependencies between links. The calculation process can be expressed as follows:

$$Q_g = HW_Q \in R^{N \times d}$$
$$K_g = H_{\text{nei}}W_K \in R^{N \times N_k \times d}$$
$$V_g = H_{\text{nei}}W_V \in R^{N \times N_k \times d} \tag{7}$$
$$H_g = \text{softmax}\left(\frac{Q_g K_g^T}{\sqrt{d}}\right) V_g$$

where W_Q, W_K and W_V are trainable parameters, and N denotes the number of links on the subgraph.

In addition to the attention mechanism, we incorporate a feedforward neural network (FFN) and layer normalization to further enhance the link representation. This results in the final representation H_g of the links.

Dynamics Representation Learning on the Route. To capture the dependencies between links within a route, we employ a self-attention network for link sequences. In a route, links are often influenced by other links within a certain range, such as the propagation of traffic congestion. We hence utilize local attention to learn the influence of adjacent links within the route. Also, similar to the task of text classification, we introduce a global token that aggregates the information from the entire route.

For a given route, the hidden states of the link sequence, denoted as $H_r \in \mathbb{R}^{N_r \times d}$, can be extracted from the previously calculated link representation $H_g \in \mathbb{R}^{N \times d}$, where $N_r < N$ represents the number of links in the route.

After combining with the sequence-based positional embedding [14], H_r is first projected to three matrices: query $Q_r \in \mathbb{R}^{N_r \times d}$, key $K_r \in \mathbb{R}^{N_r \times d}$ and value $V_r \in \mathbb{R}^{N_r \times d}$.

Then, for link i, the attention score and local representation are calculated as

$$
\text{attn}_i = \begin{cases} \frac{Q_r^i K_r^T}{\sqrt{d}}, & |j - k| \leq \omega \\ -\infty, & \text{otherwise} \end{cases} \tag{8}
$$

$$
H_{\text{local}}^i = \text{softmax}\left(\text{attn}_i\right) V_r
$$

where ω is the window size of local attention.

The global representation of the entire route is calculated as:

$$
H_{global} = \text{softmax}(\frac{Q_r K_r^T}{\sqrt{d}}) V_r \tag{9}
$$

In this way, we obtain the local and global spatio-temporal features $H_{local} \in \mathbb{R}^{N_r \times d}$ and $H_{global} \in \mathbb{R}^d$.

3.4   Multi-Task Learning

We design a multi-task learning module to predict both the travel time and traffic congestion status of the link. We concatenate H_{global} and X_{AF}, and pass them through a residual block to obtain the estimated travel time of the route $\hat{Y}_{eta}$. The loss function is expressed as:

$$
L_{eta} = |\hat{Y}_{eta} - Y_{eta}| \tag{10}
$$

where Y_{eta} is the ground-truth travel time.

The traffic congestion status can be classified into four categories: unknown, unblocked, slow, and jam. Simultaneously predicting the traffic congestion status helps in learning the local representation of the link. We feed H_{local} through a residual network to predict the traffic congestion status. The auxiliary loss is defined as:

$$
L_{aux} = \frac{1}{N_r} \sum_{i=1}^{N_r} CE\left(s_i, \hat{s}_i\right) \tag{11}
$$

$CE(\cdot)$ indicates the cross-entropy function, s_i and $\hat{s}_i$ are the ground-truth and predicted traffic congestion status of link i respectively.

Combining the learning of the two different tasks, we obtain the overall loss function:

$$
L = \lambda_1 L_{eta} + \lambda_2 L_{aux} \tag{12}
$$

where λ_1 and λ_2 are hyper-parameters that control the importance of each task in the overall objective function.

4   Experiments

In this section, We evaluated the performance of DeepLink on a large-scale dataset. Ablation experiments and sensitivity analysis of key parameters are also performed to further analyze our method.

4.1   Dataset

We evaluate our model on the Shenzhen dataset provided by DiDi Chuxing. The Shenzhen dataset contains 8,651,005 trajectories from August 1 to August 31, 2020. Table 1 shows some descriptions and the basic statistics of the dataset. Figure 1 describes the travel time and distance distribution on the probability density functions of the dataset.

Table 1. Statistics of Shenzhen Dataset.

Split	Route number	Average time	Collection date
Training set	6,136,545	13.85	8.1–8.23 (2020)
Validation set	356,105	13.91	8.24 (2020)
Test set	2,157,953	14.03	8.25–8.31 (2020)

4.2   Experimental Settings

DeepLink is trained for over 10 epochs until convergence with a batch size of 128. We train the model on the training set and select the model that performs best on the validation set, and finally evaluate our results on the test set. The model is trained using an AdamW optimizer with an initial learning rate of 1e-4 and a weight decay of 0.01. In attribute feature learning, $timeID$, $weekID$, $driverID$ and $WeatherID$ are embedded in the 3-, 8-, 17- and 2-dimensional spaces, respectively. k in the laplacian eigenvector positional embedding is set to 64 and k in the randomWalk positional embedding is set to 16. $\lambda 1$ and $\lambda 2$ in the multi-task learning module are set to 1 and 1e2 respectively. In the representation learning of a link, the dimensional of the hidden state is 64. All experiments are implemented with Pythoch and server with an NVIDIA RTX3090Ti GPU.

4.3   Evaluation Metrics

We use three evaluation metrics commonly used in ETA, including the mean absolute error (MAE), the absolute percentage error (MAPE), and the root mean square error (RMSE). Their definitions are as follows:
    Mean Absolute Error:

$$MAE = \frac{1}{M} \sum_{i=1}^{M} \left| y_i - \hat{y}_i \right| \tag{13}$$

Mean Absolute Percentage Error:

$$MAPE = \frac{1}{M} \sum_{i=1}^{M} \left| \frac{y_i - \hat{y}_i}{y_m} \right| \tag{14}$$

Root Mean Square Error:

$$RMSE = \sqrt{\frac{1}{M} \sum_{i=1}^{M} (y_i - \hat{y}_i)^2} \tag{15}$$

Where y_i and $\hat{y}_i$ are the ground truth and the prediction of travel time.

Table 2. Comparisons with State-of-the-Art Methods

Model	MAPE(%)	MAE(s)	RMSE(s)
Route-ETA	16.13	135.46	203.54
MlpTTE	18.06	131.26	206.10
conSTGAT	17.03	128.61	185.68
DeepTTE	13.07	102.06	157.31
WDR	12.55	101.14	154.67
DeepLink	**12.15**	**94.62**	**141.16**

4.4 Baselines

We compare our model with the following baselines:

- Route-ETA estimates the arrival time by adding up the real-time average time of each link in the departure time while taking into account the delay time of the intersection.
- MlpTTE is used as a common baseline for deep learning models and is learned through the perceptron mechanism. We use a 5-layer perceptron with ReLu activation to estimate travel times. MlpTTE has the same input as DeepLink, with features embedded in the same dimension. Since MlpTTE cannot handle variable-length sequences, we uniformly sample each route to a fixed length of 128.
- ConSTGAT is a spatio-temporal graph neural network that adopts a graph attention mechanism to obtain the joint relations of spatial and temporal information and captures contextual information based on convolution over local windows.
- DeepTTE is one of the most representative studies based on deep learning in the ETA task. DeepTTE uses geo-convolution networks and LSTM to learn the spatio-temporal dependencies of trajectory data and introduces a multi-task learning approach to predict both entire travel times and local travel times.

– WDR combines wide, deep, and recurrent components to handle the sparse
  features, dense features, and link sequence features respectively, and achieves
  very competitive prediction accuracy.

4.5 Experimental Results

Table 2 shows the results of our model compared to other baselines. Route-ETA
achieves the worst performance indicating that using average times is not sufficient. MlpTTE can not handle variable-length sequences and does not effectively learn the spatio-temporal information. DeepTTE and WDR use the RNN
model to capture spatio-temporal dependencies between links and achieve better
results than the above models. ConSTGAT uses a graph attention mechanism to
learn neighborhood information, but ignores the potential semantic association
of links. Compared with other models, our model shows better performance on
all metrics, illustrating the effectiveness of the proposed approach.

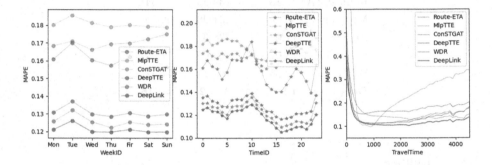

Fig. 2. MAPE changes against the time.

Table 3. Ablation experiments of key modules

Model	MAPE(%)	MAE(s)	RMSE(s)
no LaplacianPE	12.21	94.84	141.35
no RandomWalkPE	12.39	94.96	141.30
no SemanticPE	12.37	94.98	141.26
no STDRLG	12.38	96.15	142.67
no STDRLR	12.22	95.09	141.78
no Auxiliary	12.24	95.01	141.87
DeepLink	**12.15**	**94.62**	**141.16**

To investigate the temporal properties of our model, we also analyze the performance under different time dimensions (week, departure time and travel time).
Figure 2 shows that our model achieves the best performance. Note that, since
MlpTTE cannot handle variable-length sequences, the performance of MlpTTE
get worse when travel time grows.

Table 4. Parameter analysis of different window sizes of local attention

Model	MAPE(%)	MAE(s)	RMSE(s)
Decreasing from 16 to 4	12.21	94.65	**141.12**
Fixed to 8	12.26	94.81	141.50
Increasing from 4 to 16	**12.15**	**94.62**	141.16

4.6 Ablation Study and Parameter Analysis

To examine the effectiveness of the model, we provide ablation experiments for key modules in DeepLink. Table 3 presents the ablation experimental results. We found a significant decrease in performance after removing each module, indicating the effectiveness of key modules. It can be observed that the performance of the model deteriorates after removing each positional embedding. This is because LaplacianPE and RandomWalkPE can effectively extract the topology of the traffic network and SemanticPE can learn the potential semantics between links. The MAE drops from 94.62 to 96.15 and 95.09 after removing the STDRLG and STDRLR, indicating that learning the dynamic representation of links from the graph and route respectively helps to improve the performance.

In addition, we take a parametric analysis of window sizes ω. Table 4 shows the impact of different window sizes per layer configuration. Increasing the window size layer by layer achieved the best results in terms of MAE and MAPE metrics. Decreasing the window size layer by layer is the next best, and fixing the window size gives a worse result. This shows that setting different window sizes per layer is effective in learning local information from different receptive fields.

5 Conclusion

In this paper, we propose a novel travel time estimation method, denoted as DeepLink, to learn the deep understanding of link spatio-temporal dynamics through different perspectives. DeepLink introduces triplet embedding to understand the static information of links. LaplacianPE and RandomWalkPE capture the topology of the traffic network, while SemanticPE learns the potential semantics based on historical routes. In addition, we design a spatio-temporal dynamic representation module to model the real-time traffic information jointly with the triplet embedding. We learn the spatio-temporal dynamic representation of links based on the subgraph and then introduce a local-global attention mechanism to capture the local dependencies of adjacent links and the global information of the entire route. Extensive experiments conducted on a large-scale real-world dataset show that DeepLink achieves state-of-the-art performance.

Acknowledgments. This work was supported in part by the National Natural Science Foundation of China under Grant 62276100, in part by the Guangdong Natural Science Funds for Distinguished Young Scholars under Grant 2022B1515020049, in part by the Guangdong Regional Joint Funds for Basic and Applied Research under Grant 2021B1515120078, and in part by the TCL Young Scholars Program.

References

1. Belkin, M., Niyogi, P.: Laplacian eigenmaps and spectral techniques for embedding and clustering. In: Advances in Neural Information Processing Systems 14 (2001)
2. Chung, J., Gulcehre, C., Cho, K., Bengio, Y.: Empirical evaluation of gated recurrent neural networks on sequence modeling. arXiv preprint arXiv:1412.3555 (2014)
3. De Fabritiis, C., Ragona, R., Valenti, G.: Traffic estimation and prediction based on real time floating car data. In: 2008 11th International IEEE Conference on Intelligent Transportation Systems, pp. 197–203. IEEE (2008)
4. Dwivedi, V.P., Bresson, X.: A generalization of transformer networks to graphs. arXiv preprint arXiv:2012.09699 (2020)
5. Dwivedi, V.P., Luu, A.T., Laurent, T., Bengio, Y., Bresson, X.: Graph neural networks with learnable structural and positional representations. arXiv preprint arXiv:2110.07875 (2021)
6. Fang, X., Huang, J., Wang, F., Zeng, L., Liang, H., Wang, H.: ConSTGAT: contextual spatial-temporal graph attention network for travel time estimation at Baidu maps. In: Proceedings of the 26th ACM SIGKDD International Conference on Knowledge Discovery & Data Mining, pp. 2697–2705 (2020)
7. Geng, X., et al.: Spatiotemporal multi-graph convolution network for ride-hailing demand forecasting. In: Proceedings of the AAAI Conference on Artificial Intelligence, vol. 33, pp. 3656–3663 (2019)
8. Huo, G., Zhang, Y., Wang, B., Gao, J., Hu, Y., Yin, B.: Hierarchical spatiotemporal graph convolutional networks and transformer network for traffic flow forecasting. IEEE Trans. Intell. Transp. Syst. (2023)
9. Ji, J., Wang, J., Jiang, Z., Jiang, J., Zhang, H.: STDEN: towards physics-guided neural networks for traffic flow prediction. In: Proceedings of the AAAI Conference on Artificial Intelligence, vol. 36, pp. 4048–4056 (2022)
10. Li, K., Chen, L., Shang, S., Kalnis, P., Yao, B.: Traffic congestion alleviation over dynamic road networks: continuous optimal route combination for trip query streams. In: International Joint Conferences on Artificial Intelligence Organization (2021)
11. Mikolov, T., Chen, K., Corrado, G., Dean, J.: Efficient estimation of word representations in vector space. arXiv preprint arXiv:1301.3781 (2013)
12. Perozzi, B., Al-Rfou, R., Skiena, S.: DeepWalk: online learning of social representations. In: Proceedings of the 20th ACM SIGKDD International Conference on Knowledge Discovery and Data Mining, pp. 701–710 (2014)
13. Sun, Y., et al.: CoDriver ETA: combine driver information in estimated time of arrival by driving style learning auxiliary task. IEEE Trans. Intell. Transp. Syst. **23**(5), 4037–4048 (2020)
14. Vaswani, A., et al.: Attention is all you need. In: Advances in Neural Information Processing Systems 30 (2017)

15. Wang, D., Zhang, J., Cao, W., Li, J., Zheng, Y.: When will you arrive? Estimating travel time based on deep neural networks. In: Proceedings of the AAAI Conference on Artificial Intelligence, vol. 32 (2018)
16. Wang, Z., Fu, K., Ye, J.: Learning to estimate the travel time. In: Proceedings of the 24th ACM SIGKDD International Conference on Knowledge Discovery & Data Mining, pp. 858–866 (2018)
17. Yuan, N.J., Zheng, Y., Zhang, L., Xie, X.: T-Finder: a recommender system for finding passengers and vacant taxis. IEEE Trans. Knowl. Data Eng. **25**(10), 2390–2403 (2012)
18. Zheng, C., Fan, X., Wang, C., Qi, J.: GMAN: a graph multi-attention network for traffic prediction. In: Proceedings of the AAAI Conference on Artificial Intelligence, vol. 34, pp. 1234–1241 (2020)
19. Zhou, W., et al.: Travel time distribution estimation by learning representations over temporal attributed graphs. IEEE Trans. Intell. Transp. Syst. (2023)

Reversible Data Hiding in Encrypted Images Based on Image Reprocessing and Polymorphic Compression

Yicheng Zou, Yaling Zhang, Chao Wang, Tao Zhang, and Yu Zhang[✉]

College of Computer and Information Science, Southwest University,
Chongqing 400715, China
zhangyu@swu.edu.cn

Abstract. With the rapid development of cloud computing and privacy protection, Reversible Data Hiding in Encrypted Images (RDHEI) has attracted increasing attention, since it can achieve covert data transmission and lossless image recovery. To realize reversible data hiding with high embedding capacity, a new RDHEI method is proposed in this paper. First, we introduce the Image Reprocessing and Polymorphic Compression (IRPC) scheme, which can classify the images and then vacate enough room for embedding. After that, an improved RDHEI method combined with the IRPC scheme and a chaotic encryption algorithm is presented. In this method, the content owner uses the IRPC scheme to reserve embeddable rooms in the original image and then utilizes a six-dimensional chaotic encryption system to encrypt the reserved image into an encrypted image. After receiving the encrypted image, the data hider can embed additional data into it to obtain the marked encrypted image. According to the different keys the receiver has, the embedded data or the original image can be extracted or recovered from the marked encrypted image without error. Extensive experimental results show that the average Embedding Rate (ER) of our proposed method on the datasets BOSSbase, BOWS-2, and UCID is higher than that of the baseline method by 0.1bpp. At the same time, the security performance of the image is also improved.

Keywords: Reversible Data Hiding · The Encrypted Image · Polymorphic Image Processing · Bit Plane Compression · Cloud Computing

1 Introduction

As an effective information protection technology, Reversible Data Hiding (RDH) makes use of the redundant room in plaintext images to embed additional data [9]. Meanwhile, the additional data embedded in encrypted images can be completely extracted and the image can be correctly restored by the receiver [17]. Due to its crypticity and reversibility, RDH is widely used in sensitive fields such as

B. Luo et al. (Eds.): ICONIP 2023, CCIS 1968, pp. 230–243, 2024.
https://doi.org/10.1007/978-981-99-8181-6_18

medical [19], military [16], and law. The concept of data hiding was first proposed in Barton's [2] invention patent, and then various classic RDH methods gradually emerged. These RDH methods are roughly divided into three categories: (1) Lossless compression [3,4,6] vacates the embeddable room by compressing the feature set of the original image. (2) Difference expansion [1,18] embeds additional data by modifying the prediction error. (3) Histogram shifting [12,15,20] utilizes the histogram features of the original image to reserve the room.

Nowadays, with the popularization of cloud storage and cloud computing, people tend to transfer data to the cloud for processing and storage. While technology brings convenience, data security issues have become increasingly serious, such as corporate information leakage and personal portrait infringement. More and more people hope to encrypt images before transmitting them to the cloud, but traditional RDH methods cannot meet the demand. Therefore, the emergence of Reversible Data Hiding in Encrypted Image (RDHEI) plays an essential role in the security of both the embedded data and the original image. RDHEI now has several applications in various fields such as copyright protection, integrity authentication, privacy protection, and secret information transmission. Although the existing RDHEI methods have achieved good performance, there exist great potentials in the utilization rate of the redundant room and the embedding capacity of the images.

In this paper, a high embedding capacity RDHEI method based on the Image Reprocessing and Polymorphic Compression (IRPC) mechanism is proposed. As a room reservation scheme, IPRC mechanism ensures that the structural characteristics of different regions in the image can be effectively used to reserve enough room for embedding additional data. In the image encryption of RDHEI, a six-dimensional neural network chaotic system is introduced to encrypt the image, which enhances image security. In image decryption, data extraction and image restoration operations can be performed separately.

Our main contributions are described as follows:

(1) A new room reservation scheme called IRPC is proposed, which can reuse the redundant space that cannot be processed by other compression methods. This scheme greatly improves the utilization rate of the image to elevate the embedding capacity.

(2) A high-capacity RDHEI method based on IRPC is introduced, which can embed a large amount of additional data into images. The proposed method not only protects the content of the image itself but also ensures the covert transmission of additional data.

(3) A six-dimensional chaotic encryption system is used to solve the problem of low image security caused by the single and predictable encryption key. The high security of the encrypted image is proved by a large number of experiments. In the rest of this paper, the related work of this study is introduced in Sect. 2. The detail of the proposed RDHEI method is given in Sect. 3, which mainly includes four parts: (1) image prediction and room reservation; (2) image encryption; (3) data hiding; (4) data extraction and image restoration. In Sect. 4,

experimental results and comparisons with other advanced methods are shown. Finally, a summary of this paper is described in Sect. 5.

2   Related Work

RDHEI mainly includes room reservation technology and image encryption technology. Room reservation is the process of reserving continuous space to embed additional data in the image. According to whether to reserve the room before or after image encryption, RDHEI is classified into two categories: Vacating Room After Encryption (VRAE) [7,14,22,25] and Reserving Room Before Encryption (RRBE) [5,10,13,23]. In 2013, the first RRBE method was proposed by Ma et al. [10]. Specifically, they reserved the room by embedding a portion of the Least Significant Bit (LSB) plane into other parts of the original image. Subsequently, Peneaux et al. [13] proposed a significant RRBE method based on the Most Significant Bit (MSB) plane prediction to generate a label map, which could mark the embeddable position of data in the image. In 2019, a novel RRBE method based on the Block-based MSB Plane Rearrangement (BMPR) scheme [5] is introduced by Chen et al., reserving the room more effectively. To increase the embedding rate (ER) of images, Yin et al. [23] made an improvement on BMPR by using pixel prediction and bit plane rearrangement compression. Although the embedding capacity of the above studies has a relatively good performance, the utilization rate of the redundant room and structural characteristics in the images still has room to grow. In this study, an improved RRBE method that combines the proposed scheme IPRC and existing algorithms is proposed.

3   Proposed Method

The framework of the proposed method is shown in Fig. 1, It can be divided into four parts: (1) The content owner processes the image and reserves the room based on the IRPC mechanism to generate the reserved image. (2) The content owner encrypts the reserved image through the chaotic encryption system to obtain the encrypted image. (3) The data hider embeds additional data into encrypted images to get the marked encrypted image. (4) The receiver extracts additional data or restores images from the marked encrypted image according to the key they have. Section 3.1 shows the process of image preprocessing and room reservation. The image encryption is described in Sect. 3.2. The detail of data embedding in the encrypted image is demonstrated in Sect. 3.3. In the end, Sect. 3.4 presented data extraction and image recovery.

3.1   Room Reservation

Before reserving the room, the original image needs to be preprocessed first, which involves prediction error and bit plane division. The prediction error image generated by the prediction error algorithm can not only restore the original image but also reserve more embedding rooms. The bit plane division is to divide image pixels into 8 bit-planes according to their binary values. The elements of each bit plane are fixed bits of the binary value of the corresponding pixel point, and the bit plane can retain the correlation between adjacent pixels.

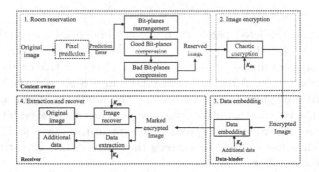

Fig. 1. The framework of the proposed method.

In this study, the Gradient Adjacent Predictor (GAP) [11] prediction algorithm is used for image prediction. As shown in Fig. 2, the predicted pixel $\widehat{x}_i$ of the current point is related to the seven pixels that are close to it. $\widehat{x}_i$ is calculated by Eqs. (1), (2), and (3).

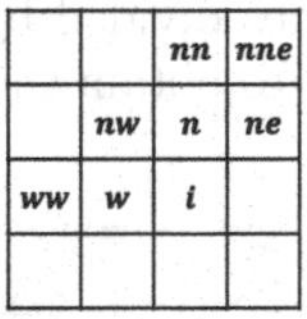

		nn	*nne*
	nw	*n*	*ne*
ww	*w*	*i*	

Fig. 2. Seven contextual pixels of the current pixel in GAP-2.

$$\begin{cases} d = d_v - d_h \\ d_v = |x_w - x_{nw}| + |x_n - x_{nn}| + |x_{ne} - x_{nee}| \\ d_h = |x_w - x_{ww}| + |x_n - x_{nw}| + |x_n - x_{ne}| \end{cases} \qquad (1)$$

$$y = \frac{1}{2}(x_w + x_n) + \frac{1}{4}(x_{ne} - x_{nw}) \qquad (2)$$

$$\widehat{x}_l = \begin{cases} x_w, d > 80 \\ x_n, d < -80 \\ \frac{1}{2}(x_w + y), d \in (32, 80] \\ \frac{1}{2}(x_n + y), d \in [-80, -32) \\ \frac{1}{4}(x_w + 3y), d \in (8, 32] \\ \frac{1}{4}(x_n + 3y), d \in [-32, -8) \\ y, d \in [-8, 8], \end{cases} \qquad (3)$$

where d is used to classify processing situations and y is the parameter to calculate the prediction error. x_a represents the pixel values of the points adjacent to the predicted point and the specific position of a is shown in Fig. 2. If the current pixel is a boundary pixel, special processing is required with Eq. (4).

$$X_{ww} = X_w, X_{ne} = X_{nn} = X_{nn} = X_n \tag{4}$$

Subtract the predicted pixels from the original pixel to get the predicted error ex. Except for the reference pixels in the first row and first column, all pixels of the original image are replaced with their respective ex to generate the prediction error image. Afterward, convert ex into 8-bit binary. MSB plane is used to record the symbol of the ex because the prediction error algorithm will lead to both positive and negative error results. The other seven bits are used to record the absolute value of the prediction error in the range $[0, 127]$, as shown in Eq. (5). When the value of ex exceeds the range $[-127, 127]$, this ex will be recorded as the auxiliary information.

$$ex^k(i, j) = \left\lfloor \frac{ex(i, j)}{2^{k-1}} \right\rfloor \mod 2, k = 1, 2, \ldots, 7, \tag{5}$$

where ex^k represents the k-th bit plane, $k \in [1, 8]$. ex^8 represents the MSB plane and $ex(i, j)$ represents the prediction error for pixel (i, j).

After image preprocessing is completed, a new room reservation scheme based on Image Reprocessing and Polymorphic Compression (IRPC) will be carried out, as shown in Fig. 3. To be specific, rough reservation and refined reservation are used to deal with the two types divided by IRPC, respectively. Finally, all the reserved rooms are connected to obtain the reserved image.

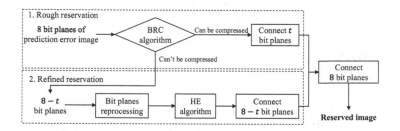

Fig. 3. The framework of IRPC scheme.

Rough Reservation. It introduces a compression algorithm based on the Bit plane Rearrangement and Compression (BRC) scheme [23], which acts on t bit-planes. First, BRC splits t bit-planes of the reserved image into n blocks. For each bit plane, 4 arrangement patterns within or between n blocks are used to obtain 4 different bit streams. Then, these bit streams are compressed, and the shortest one will be selected as the compressed bit plane stream of this bit plane.

Finally, t bit-streams are connected continuously, and the remaining space in t bit-planes is the embeddable room. Except for the MSB plane, the higher bit planes are preferred because they are more correlated than the lower bit planes. For example, if $t=3$, the bit planes that can be performed by rough reservation operation are ex^7, ex^6, and ex^5.

Refined Reservation. There are t bit-planes that have been processed by rough reservation operation, so the remaining $(8\text{-}t)$ bit-planes will be processed by the refined reservation operation next. It consists of three steps as follows:

(1) Bit Plane Reprocessing (BR)

In the remaining $(8\text{-}t)$ bit planes, 3 or 4 bit-planes are selected to form a bit-plane group to generate a secondary image with pixels at $[0, 255]$. Only when $t=6$, the MSB plane will be selected for refined reservation.

Take one bit-plane group as an example, it is divided into two secondary bit-plane groups, as shown in Fig. 4(b). Then, each secondary bit-plane group is split longitudinally. The odd and even columns are combined to form bit plane a_1 and a_2, respectively, as shown in Fig. 4(c). In Fig. 4(d), a_2 is placed in the high position and a_1 in the low position to generate a secondary image, which is one-fourth the size of the original image. There are two secondary images can be obtained at last.

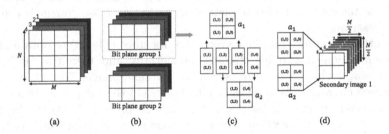

Fig. 4. Reprocessing process: (a) bit plane group; (b) secondary bit plane groups; (c) subdivision of secondary bit plane group; (d) secondary bit plane group details.

(2) Generate Auxiliary Information

A new reservation scheme based on Histogram Exchange (HE) is proposed, which can fully utilize gray-scale histogram features of secondary images. First, the peak point P and the bottom point B in the gray-scale histogram are fetched out to generate the peak and bottom group. For each group, use the order in which P and B appear in the image to embed additional data. Equation (6) is used to determine whether this group can be embedded.

$$C_i = (\text{num}(P(i)) + \text{num}(B(i))) \times (2 \times \text{num}(K)) - \text{length}(Aux(i)), \qquad (6)$$

where $num(x)$ and $length(x)$ represent the number and the length of x.

If $C_i > 0$, the i-th group can vacate more rooms than the amount of auxiliary information needed for recovery, which means this group can be used. The auxiliary information required for this embedding should be recorded to recover the image later. It includes the value of the bottom point and peak point, the number of bottom points, and the position information of each bottom point in the image. After completing the recording, the next group will be determined whether to be embedded. If $C_i \leq 0$, this group cannot be embedded. Then, the whole process is over. The total auxiliary information contains information about each group that can be used.

(3) Histogram Exchange (HE)

In HE, binary auxiliary information and additional data generated by the refined reservation operation will be embedded into the peak and bottom groups. To get a continuously available room, the additional data embedded in HE is a part of the header of the compressed bit stream produced by rough reservation operation. First, pixels in the image with values $P(i)$ and $B(i)$, which represent the i-th peak and bottom group, will be scanned. Then, look up the bits in the binary sequence that need to be embedded. If it is "0", change this pixel to $B(i)$. If it is "1", change this pixel to $P(i)$. Loop embedding until the embeddable room runs out or the embedding is complete. Finally, the processed image is converted into a bit stream.

The other secondary bit plane group repeats steps (1) - (3) to obtain another bit stream. Connect two bit streams sequentially and then turn it into the reserved image, which is the same size as the original image. The reserved image is composed of two or three types of bit planes: (1) the bit planes generated by refined reservation; (2) the bit planes that not be processed; (3) the bit planes generated by rough reservation. The continuous vacated room that can be used is in the low-level planes of the reserved image.

3.2   Image Encryption

In this section, a six-dimensional chaotic encryption system is introduced to encrypt the image. The hyperchaotic sequence generated by Cellular Neural Networks (CNN) system is used as the key [8], and the binary sequence of the key is used to generate the encrypted image. For an image of size $M \times N$, each pixel in the image is treated as a cell. The state equation of the cellular neural network is shown as Eq. (7).

$$C\frac{dx_{ij}(t)}{dt} = -\frac{x_{ij}(t)}{R_x} + \sum_{k,I \in N_{ij}(r)} A_{kI} y_{kI}(t) + \sum_{k,I \in N_{ij}(r)} B_{kI} u_{kI}(t) + I_{ij}, \qquad (7)$$

where x_{ij} is the state variable, u_{kI} is the input variable, y_{kI} is the output variable, C and R_x are the system constant, I_{ij} is the threshold, A is the feedback coefficient matrix, B is the control coefficient matrix.

Set parameters and get a six-dimensional chaotic sequence through calculation. The key matrix H with size $M \times N$ is generated according to the six-dimensional chaotic sequence. The pixel $I_{res}(i,j)$ of the reserved image and corresponding value $H(i,j)$ of H are converted into an 8-bit-binary sequence. Then, the XOR operation is performed to generate pixel points in the encrypted image, as shown in Eq. (8).

$$S_{i,j} = \sum_{i=1}^{8} H_{i,j,k} \oplus I_{i,j,k} \times 2^{k-1}(1 \leq i \leq M, 1 \leq j \leq N, 1 \leq k \leq 8), \quad (8)$$

where $H_{i,j,k}$ and $I_{i,j,k}$ are the values of row i, column j, bit plane k of H and the reserved image, respectively. Finally, the encrypted image is obtained by the above process.

3.3 Data Hiding in Encrypted Image

The additional data is embedded in the vacated room in this section. First, the data to be embedded is encrypted by the key K_d. Then, the last 24 bits of the LSB plane are used to store the number of bits that can be embedded. Finally, the additional data is embedded from the 25th position from the bottom of the LSB plane. After the embeddable room of the current bit plane is used up, embedding continues towards the upper bit plane until the additional data is fully embedded or the embeddable room is exhausted. The structure of bit plane is shown in Fig. 5. After that, the marked encrypted image is obtained.

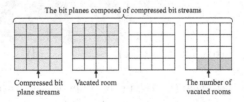

Fig. 5. The structure that makes up the bit plane.

3.4 Data Extraction and Image Recovery

In the proposed method, data extraction and image recovery are independent, and both operations depend entirely on the type of key held by the receiver. There are two cases: (1) If the receiver only owns the data extraction key K_d, the length of additional data should be read from the LSB plane first, then the secret data can be extracted completely. (2) If the receiver only has the image recovery key K_{en}, image decryption should be done first, then the number of bit planes reserved by rough reservation can be extracted from the MSB plane. Subsequently, the original bit plane is recovered from the bit planes according to other auxiliary information. Finally, sort the original bit planes to restore the original image.

4   Experimental Results and Discussion

To evaluate the embedding performance and security of the proposed method, experimental results and analysis are presented in this section. First, the environment and data of this experiment are shown in Sect. 4.1. Then, the embedding performance and reversibility of the method are analyzed in Sect. 4.2. To demonstrate that the proposed method is superior to the most advanced methods [21,23,24], a comparative experiment is designed. The security of the method is analyzed in Sect. 4.3.

4.1   Experimental Environment

As shown in Fig. 6, several classical grayscale images sized 512×512 are used in the experiments: Lena, Jetplane, Baboon, Airplane, and Tiffany. In addition, three data sets commonly used in the RDHEI field, UCID, BOSSbase, and BOWS-2, are also used to demonstrate the embedding performance of the proposed method. The experimental environment is 3.60GHz Inter Core i7-12700k CPU, 16GB Memory, Windows 10, and MATLAB R2021a.

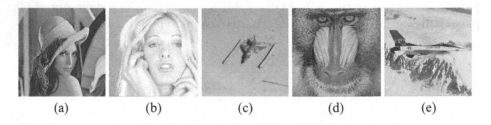

(a)   (b)   (c)   (d)   (e)

Fig. 6. Test images: (a) Lena; (b) Tiffany; (c) Baboon; (d) Airplane; (e) Jetplane

4.2   Embedded Performance Analysis

In this section, the embedding performance of the proposed method will be analyzed and compared with the most advanced RDHEI method in three aspects.

Reversibility Analysis. Peak Signal-to-Noise Ratio (PSNR) and Structural Similarity (SSIM) verify the similarity between the original image and the restored image. When the value of PSNR is $+\infty$ and the value of SSIM is 1, the structure and content of the two images are identical. The experiment shows that the PSNR value of all recovered images is $+\infty$, and the SSIM value is 1, as shown in Table 1.

Table 1. Reversibility analysis of the proposed method.

Test image	UCID	BOSSbase	BOWS-2
Best case	5.4722	7.8401	7.0425
Worst case	0.2716	0.7416	0.6239
Average	2.9594	3.7272	3.6313
PSNR	$+\infty$	$+\infty$	$+\infty$
SSIM	1	1	1

The Embedding Ability of Grayscale Images. The Embedding Rate (ER) is mainly used to show the embedding ability of an image, and the unit of ER is bits per pixel (bpp). Table 2 shows the results of the comparison between [23] and the proposed method in terms of the number of bit planes used and ER. Both [23] and the proposed method introduce BRC for room reservation, but the proposed method utilizes refined processing for further reservation on more bit planes. As the number of bit planes that can reserve room increases, ER gradually increases. Therefore, the proposed method makes better use of the redundant space in the image and has better embedding performance. Obviously, the proposed method outperforms the best existing methods in embedding performance on five grayscale images, as shown in Fig. 7(a).

Table 2. Comparison of the proposed method with the number of bit planes and embedding capacity used in [23]

Test Image	Bit plane's number		Vacated Bits' number		
	[23]	Proposed	[23]	Proposed	Increment
Lena	4	**8**	806114	**848918**	**42804**
Tiffany	4	**8**	825670	**871365**	**45695**
Jetplane	5	**8**	891826	**903554**	**11728**
Airplane	5	**8**	1005781	**1040286**	**34505**
Baboon	3	**7**	362673	**373376**	**10723**
Average	4.25	**7.8**	778412	**807500**	**29088**

The Embedding Ability of Datasets. In order to further analyze the embedding performance of the proposed method, three public image datasets UCID, BOSSBase, and BOWS-2 are tested experimentally. The experimental results are shown in Table 1. The average ER of the proposed method on the three data sets reaches 2.9594, 3.7272, and 2.6313, respectively, which is 0.1bpp higher than [23]. This suggests that the proposed scheme has wider applicability than [23]. Figure 7(b) vividly shows the embedding performance of the proposed method compared with that of [21, 23, 24] in three image datasets.

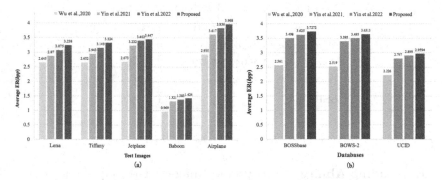

Fig. 7. Comparison of ER (bpp): (a) Gray images; (b) Images data sets.

4.3   Security Analysis

In this section, the security of the encrypted image is verified in two aspects: gray histogram and image correlation.

Histogram Analysis of Gray Value of Image. The grayscale histogram of an image intuitively displays the distribution of grayscale values in [0, 255], which can reflect the characteristics of the image to a certain extent. The encryption algorithm for images should strive to achieve a uniform distribution of grayscale histograms as much as possible. Take Lena image as an example, Fig. 8(a), Fig. 8(b), and 8(c) show the grayscale histograms of the original image, encrypted image, and marked encrypted image, respectively. This shows that the proposed method has strong security, and the original image content is well protected.

Correlation Analysis of Image Pixels. The correlation of adjacent pixels in an image can reflect the degree of correlation between adjacent pixels. A good image encryption algorithm needs to reduce the correlation of neighboring pixels from horizontal, vertical, and diagonal pixels to close to zero. The calculation of the correlation coefficient is shown in Eq. (9).

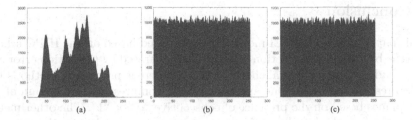

Fig. 8. (a) The histogram of the original image; (b) The histogram of the encrypted image; (c) The histogram of the marked encrypted image.

$$\begin{cases} E(x) = \frac{1}{N} \sum_{i=1}^{N} x_i \\ D(x) = \frac{1}{N} \sum_{i=1}^{N} (x_i - E(x))^2 \\ COV(x, y) = \frac{1}{N} \sum_{i=1}^{N} (x_i - E(x))(y_i - E(y)), \end{cases} \tag{9}$$

where (x, y) is the pixel value and N is the number of pixels selected in the image. The closer the correlation coefficients of the three directions are to 0, the higher the security performance of the image is.

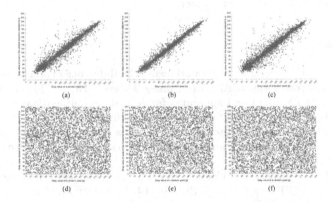

Fig. 9. Correlation point graph of adjacent elements in an image: (a) Original image: horizontal direction; (b) Original image: vertical direction; (c) Original image: diagonal direction; (d) Encrypted image: horizontal direction; (e) Encrypted image: vertical direction; (f) Encrypted image: diagonal direction.

For Lena image, the proposed method is compared with [23] in terms of pixel correlation. As shown in the Fig. 9, the correlation in the horizontal, vertical, and diagonal directions of the proposed method is lower than that of [23], indicating that the proposed method performs better in image security.

5 Conclusion

In this paper, a novel RDHEI method was proposed based on the IRPC scheme. Since the bit planes of prediction error images generated by prediction error algorithms had various structural characteristics, different processing methods were used to reserve the room. One approach was to compress the bit stream of part of bit planes based on the presence of consecutive "0" or "1". The other method was to reprocess the bit planes that cannot be compressed by the above method to generate a secondary image, and then carry out the room reservation method according to the histogram features of the secondary image. Furthermore, this RDHEI method introduced a six-dimensional chaotic system, which enhanced image security greatly. Extensive experiments showed that the proposed method could not only extract data and recover images without loss but also perform better embedding capacity and image security than existing methods. In the future, this work can be considered applied to high-capacity image protection and data hiding in cloud services, medical, and other fields.

Acknowledgements. This paper is funded in part by the National Natural Science Foundation of China (62032019), and the Capacity Development Grant of Southwest University (SWU116007), Chongqing Industrial Control System Information Security Technology Support Center, Chongqing Intelligent Instrument and Control Equipment Engineering Technology Research Center.

References

1. Alattar, A.M.: Reversible watermark using the difference expansion of a generalized integer transform. IEEE Trans. Image Process. **13**(8), 1147–1156 (2004)
2. Barton, J.M.: Method and apparatus for embedding authentication information within digital data. United States Patent, 5 646 997 (1997)
3. Celik, M.U., Sharma, G., Tekalp, A.M.: Lossless watermarking for image authentication: a new framework and an implementation. IEEE Trans. Image Process. **15**(4), 1042–1049 (2006)
4. Celik, M.U., Sharma, G., Tekalp, A.M., Saber, E.: Lossless generalized-LSB data embedding. IEEE Trans. Image Process. **14**(2), 253–266 (2005)
5. Chen, K., Chang, C.C.: High-capacity reversible data hiding in encrypted images based on extended run-length coding and block-based MSB plane rearrangement. J. Vis. Commun. Image Represent. **58**, 334–344 (2019)
6. Fridrich, J., Goljan, M., Du, R.: Invertible authentication. In: Security and Watermarking of Multimedia Contents III, vol. 4314, pp. 197–208. SPIE (2001)
7. Hong, W., Chen, T.S., Wu, H.Y.: An improved reversible data hiding in encrypted images using side match. IEEE Signal Process. Lett. **19**(4), 199–202 (2012)
8. Hu, G., Kou, W., Peng, J., et al.: A novel image encryption algorithm based on cellular neural networks hyper chaotic system. In: 2018 IEEE 4th International Conference on Computer and Communications (ICCC), pp. 1878–1882. IEEE (2018)
9. Kumar, S., Gupta, A., Walia, G.S.: Reversible data hiding: a contemporary survey of state-of-the-art, opportunities and challenges. Appl. Intell. **52**, 7373–7406 (2022)

10. Ma, K., Zhang, W., Zhao, X., Yu, N., Li, F.: Reversible data hiding in encrypted images by reserving room before encryption. IEEE Trans. Inf. Forensics Secur. **8**(3), 553–562 (2013)
11. Ni, B., Bi, W.: New predictor-based schemes for reversible data hiding. Multimedia Tools Appl. **82**(4), 5923–5948 (2023)
12. Ni, Z., Shi, Y.Q., Ansari, N., Su, W.: Reversible data hiding. IEEE Trans. Circuits Syst. Video Technol. **16**(3), 354–362 (2006)
13. Puteaux, P., Puech, W.: An efficient MSB prediction-based method for high-capacity reversible data hiding in encrypted images. IEEE Trans. Inf. Forensics Secur. **13**(7), 1670–1681 (2018)
14. Qian, Z., Zhang, X.: Reversible data hiding in encrypted images with distributed source encoding. IEEE Trans. Circuits Syst. Video Technol. **26**(4), 636–646 (2015)
15. Qin, C., Chang, C.C., Huang, Y.H., Liao, L.T.: An inpainting-assisted reversible steganographic scheme using a histogram shifting mechanism. IEEE Trans. Circuits Syst. Video Technol. **23**(7), 1109–1118 (2012)
16. Shi, Y.Q., Li, X., Zhang, X., Wu, H.T., Ma, B.: Reversible data hiding: advances in the past two decades. IEEE access **4**, 3210–3237 (2016)
17. Shih, F.Y.: Digital Watermarking and Steganography: Fundamentals and Techniques. CRC Press (2017)
18. Tian, J.: Reversible watermarking by difference expansion. In: Proceedings of Workshop on Multimedia and Security, vol. 19. ACM Juan-les-Pins (2002)
19. Tsai, P., Hu, Y.C., Yeh, H.L.: Reversible image hiding scheme using predictive coding and histogram shifting. Sig. Process. **89**(6), 1129–1143 (2009)
20. Wang, J., Chen, X., Ni, J., Mao, N., Shi, Y.: Multiple histograms-based reversible data hiding: framework and realization. IEEE Trans. Circuits Syst. Video Technol. **30**(8), 2313–2328 (2019)
21. Wu, Y., Xiang, Y., Guo, Y., Tang, J., Yin, Z.: An improved reversible data hiding in encrypted images using parametric binary tree labeling. IEEE Trans. Multimedia **22**(8), 1929–1938 (2019)
22. Yin, Z., Niu, X., Zhang, X., Tang, J., Luo, B.: Reversible data hiding in encrypted AMBTC images. Multimedia Tools Appl. **77**, 18067–18083 (2018)
23. Yin, Z., Peng, Y., Xiang, Y.: Reversible data hiding in encrypted images based on pixel prediction and bit-plane compression. IEEE Trans. Dependable Secure Comput. **19**(2), 992–1002 (2020)
24. Yin, Z., Xiang, Y., Zhang, X.: Reversible data hiding in encrypted images based on multi-MSB prediction and Huffman coding. IEEE Trans. Multimedia **22**(4), 874–884 (2019)
25. Zhang, X.: Separable reversible data hiding in encrypted image. IEEE Trans. Inf. Forensics Secur. **7**(2), 826–832 (2011)

All You See Is the Tip of the Iceberg: Distilling Latent Interactions Can Help You Find Treasures

Zhuo Cai[1], Guan Yuan[1]($\boxtimes$), Xiaobao Zhuang[1], Xiao Liu[1], Rui Bing[1],
and Shoujin Wang[2]

[1] China University of Mining and Technology, Xuzhou, China
{czhuo,yuanguan,zhuangxb,liuxiao,bingrui}@cumt.edu.cn
[2] University of Technology Sydney, Sydney, Australia
shoujin.wang@uts.edu.au

Abstract. Recommender systems suffer from data sparsity problem severely, which can be attributed to the combined action of various possible causes like: gradually strengthened privacy protection policies, exposure bias, etc. In these cases, the unobserved items do not always refer to the items that users are not interested in; they could also be imputed to the inaccessibility of interaction data or users' unawareness over items. Thus, blindly fitting all unobserved interactions as negative interactions in the training stage leads to the incomplete modeling of user preferences. In this work, we propose a novel training strategy to distill latent interactions for recommender systems (shorted as DLI). Latent interactions refer to the possible links between users and items that can reflect user preferences but not happened. We first design a False-negative interaction selecting module to dynamically distill latent interactions along the training process. After that, we devise two types of loss paradigms: Truncated Loss and Reversed Loss. The former one can reduce the detrimental effect of False-negative interactions by discarding the False-negative samples in the loss computing stage, while the latter turning them into positive ones to enrich the interaction data. Meanwhile, both loss functions can be further detailed into full mode and partial mode to discriminate different confidence levels of False-negative interactions. Extensive experiments on three benchmark datasets demonstrate the effectiveness of DLI in improving the recommendation performance of backbone models.

Keywords: Recommender systems · Data sparsity · Latent interactions · Truncated Loss · Reversed Loss

1 Introduction

Recommender systems have served as a promising solution to alleviate the information overload problem in many applications [2,3,19]. In real scenarios, the amount of collected interaction data between users and items constitutes a

B. Luo et al. (Eds.): ICONIP 2023, CCIS 1968, pp. 244–257, 2024.
https://doi.org/10.1007/978-981-99-8181-6_19

extremely small portion of the entire interaction space. There are many reasons, such as page space limitation; exposure bias [8] (i.e., recommender systems only study on observed interactions, further exacerbating the unavailability of some items) and so on.

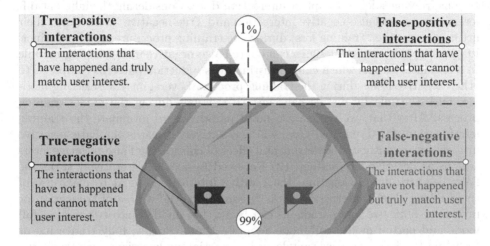

Fig. 1. A schematic diagram of four types of implicit interactions in form of "iceberg".

Many existing methods mainly work on users' history interaction data, while considerable portion of unobserved interactions have not been well explored. This results in the data sparsity issue [17,18]. However, the observed interactions (i.e., positive interactions) are just the tip of the iceberg compared with the whole interaction space (as shown in Fig. 1). Among the unobserved items, there are many that match users' interests, however remain unnoticed or unknown to them. We call this kind of items as latent items (i.e., False-negative items) and the links between these items and users as latent interactions (i.e., False-negative interactions in Fig. 1). These latent interactions may be observational negative but can positively reflect user preferences. Comprehensively considering the aforementioned phenomenon, we argue that it is crucial to distill latent interactions for recommender systems. It helps users find "treasures" (i.e., the latent items that users may be interested in).

Despite the critical importance of mining latent interactions, few research takes latent interactions into account. Ding et al. [6] robustify negative sampling by identity False-negative items, however removing them when performing negative sampling instead of leveraging them to enhance recommendation. Liu et al. [13] attempt to utilize False-negative interactions to enhance the performance of sequential recommendation, however do not discriminate various confidence levels of False-negative interaction. It is evident that not all False-negative interactions possess the same level of confidence. Some interactions are more likely to be False-negatives than others. Therefore, treating all selected False-negative interactions equally is not justified, as it could introduce excessive noise. Some work [4,14] utilizes the explicit rating information to enrich the interaction space. However, the explicit rating information is not always available.

To tackle above problems, we propose a training strategy (DLI) to elegantly distill and leverage latent interactions from the large number of unobserved interactions, relying solely on implicit interaction data. Considering the inherent differences between False-negative interaction and True-negative interactions, their fitting process (i.e., training loss) during the training procedure may be distinct. Inspired by this, we first design a candidate False-negative item selecting module to select latent items which exhibit distinct loss reduction patterns compared to other negative items during the training process. Leveraging the selecting module, DLI dynamically distills False-negative signals during the model training process. After that, we design two novel paradigms to formulate the training loss function: **(1)** Truncated Loss, which iteratively discards the False-negative interactions to mitigate the detrimental effect of considering False-negative interactions as negative interactions; **(2)** Reversed Loss, which iteratively turns the False-negative interactions into the positive interactions so as to make use of them to enrich the interaction data. According to the degree of truncation or reversion, both two loss paradigms can be further divided into two modes: full mode and partial mode. Full mode adopts a thorough way, fully truncating or reversing the False-negative samples, while partial mode assigns each truncated or reversed False-negative sample with a weight so as to discriminate different confidence levels of them. Extensive experiments show the effectiveness of DLI.

The main contributions of this work are as follows:

- We emphasize the significance of latent interactions and give a comprehensive understanding of how to effectively explore latent interactions.
- We design a novel training strategy (DLI) to distill and utilize latent interactions to improve recommendation performance. DLI has four versions, providing fine-grained strategies for handling various recommendation scenarios.
- Experiments on three datasets validate the effectiveness of DLI in improving the recommendation performance of the backbone models.

2   Related Work

Our work focuses on capturing False-negative user-item interactions so as to improve the robustness and efficacy of recommendation models. Gunawardana et al. [7] define the robustness of recommender systems as "the stability of the recommendation in presence of fake information". To enhance the anti-noise ability, some work focuses on reduce the negative impact of False-positive interactions (i.e., noisy interactions). For example, Wu et al. [22] introduce noises into interaction data and try to reconstruct original data with auto-encoder. Prior work on robust learning [9] and curriculum learning [1] reveals that noisy samples are relatively hard to fit so that they have relatively bigger loss values in early training procedure. Inspired by this, Wang et al. [20] design a denoising training strategy to dynamically identify and prune the noisy interactions to improve the robustness of recommender training. In addition to False-positive interactions, we argue that the False-negative interactions are also a kind of "fake information", however not been deeply explored. This motivates us to study on latent interactions (i.e., False-negative interactions) and make the most of them to improve the efficacy of recommender training.

3   Method

3.1   Task Formulation

The task of recommender training is to learn a scoring function s to predict the preference of a user over an item. The task can be formalized as $\hat{y}_{u,i} = s(u, i, \Theta)$, where $\hat{y}_{u,i}$ is the predicted score that represents how a given user u like the item i. Θ is the model parameters learned from the whole training data $\mathcal{D} = \{(u, i, y_{u,i}) | \forall u \in \mathcal{U}, \forall i \in \mathcal{I}\}$, where $\mathcal{U}$ and $\mathcal{I}$ are the user set and item set. $y_{u,i}$ denotes whether user u has any interactions with item i. In this paper, we focus only on implicit feedbacks for simplicity, so the value of $y_{u,i}$ is binary (i.e., $y_{u,i} = \{0, 1\}$). $y_{u,i} = 1$ means user u has interacted with item i and $y_{u,i} = 0$ the opposite. To optimize the model parameters, recommender training aims to minimize a specific loss:

$$\hat{\Theta} = \arg\min_{\Theta} \sum_{(u,i,y_{u,i}) \in \mathcal{D}} \mathcal{L}(y_{u,i}, \hat{y}_{u,i}), \tag{1}$$

where $\mathcal{L}(\cdot, \cdot)$ is the loss function which can be pointwise (e.g., Square Loss [23], Binary Cross-Entropy (BCE) Loss [11,16]) or pairwise (e.g., Bayesian Personalized Ranking (BPR) Loss [15]). In this work, we focus on pointwise loss BCE. $\hat{\Theta}$ is the model parameters that generate the minimum of model loss. In the end of model training, preference scores $\{\hat{y}_{u,i} = s(u, i, \hat{\Theta}) | \forall i \in \mathcal{I} \wedge y_{u,i} = 0\}$ are computed using updated model parameters. And the unobserved items with top-K highest preference scores are recommended to user u.

In real scenarios, there is a considerable proportion of items that are not interacted by certain users but could potentially match their interests (i.e., False-negative items). Under this circumstance, the conventional training strategy given above results in sub-optimal recommendation performance, as the lack of fitting of these latent interactions could lead to incomplete modeling of user preferences. The recommender training with the aim of distilling latent interactions can be formulated as:

$$\Theta^* = \arg\min_{\Theta} \sum_{(u,i,y_{u,i}) \in \mathcal{D}^*} \mathcal{L}_{DLI}(y_{u,i}, \hat{y}_{u,i}), \tag{2}$$

where $\mathcal{L}_{DLI}(\cdot, \cdot)$ is the new loss paradigm aiming to distill and utilize the latent interactions to improve the recommendation performance. Concretely, $\mathcal{L}_{DLI}(\cdot, \cdot)$ could be chosen as Truncated Loss or Reversed Loss. $\mathcal{D}^*$ is the data with False-negative interactions discarded or reversed to positive ones. After obtaining the model parameter Θ^*, the predicted preference scores will be recomputed to yield better recommended item list for users (Fig. 2).

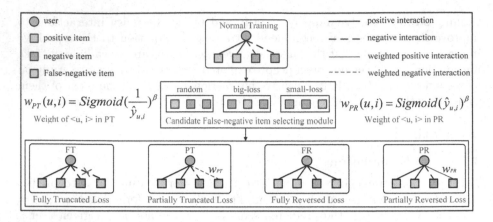

Fig. 2. A toy example of the workflow of DLI. FT, PT, FR and PR are the loss paradigms proposed in this work.

3.2 Distilling Latent Interactions

Candidate False-Negative Item Selecting Module. False-negative interactions (i.e., latent interactions) refer to the items that users are interested in but not interacted with. In contrast, True-negative interactions refer to the items that users have neither expressed interest in nor interacted with. Two types of negative interactions are inherently distinct, inevitably generating different training loss values during the fitting process. According to this property, we first design a candidate False-negative item selecting module to distill latent items. Concretely, our candidate False-negative item selecting module has two versions: (**1**) selecting items with top ranked biggest loss and (**2**) selecting items with top ranked smallest loss.

The selection rate of candidate latent items from the negative samples is an important factor in the training strategy. To distill as many latent items as possible while minimizing the introduction of noisy items, the rate should follows 3 properties: (**1**) the rate should be correlated with the iteration t, indicating that it should adaptively change during the training process; (**2**) the rate should gradually increase along the training process, allowing for the gradual distillation of False-negative interactions from negative interactions; (**3**) the rate should have an upper bound to avoid injecting too many noisy interactions into the training process. Based on these considerations, we define the rate function for determining the proportion of selected False-negative interactions as:

$$r\left(t\right) = \min\left(\frac{r_{max}}{\alpha}t, r_{max}\right), \tag{3}$$

where α is the hyper-parameter that controls the pace at which the selection rate approaches the upper bound r_{max}. A smaller α means a faster pace, while a larger value means a slower pace. The selection rate can increase smoothly along with the interation t until reaching the upper pound r_{max}.

After selecting candidate False-negative items, we design two types of paradigms to formulate the loss function to distill and utilize latent interactions from the negative interactions. Meanwhile, each proposed loss paradigm includes two versions. The four versions of our proposed training strategies can be flexibly adopted to handle various recommendation scenarios. For example, different recommendation datasets encompass diverse tasks with distinct properties. Therefore, a single training strategy may not always be suitable for all recommendation tasks. Our proposed training strategies allow us to select the appropriate version when dealing with different recommendation scenarios.

Our proposed loss paradigms are instantiated on the basis of BCE loss:

$$\mathcal{L}\left(u, i, y_{u,i}\right) = -\left(y_{u,i}\log\left(\hat{y}_{u,i}\right) + \left(1 - y_{u,i}\right)\log\left(1 - \hat{y}_{u,i}\right)\right), \qquad (4)$$

where $y_{u,i}$ and $\hat{y}_{u,i}$ are the label and predicted score of user u over item i. Here, we omit the regularization for brevity.

Truncated Loss. Truncated Loss can be categorized into: Fully Truncated Loss and Partially Truncated Loss, according to whether fully discarding the False-negative samples or not.

Fully Truncated Loss (FT). Not all unobserved items (i.e., negative items) refer to users' negative interest. Hence, FT iteratively discards the samples that are identified as False-negative ones during the training process. This approach dynamically removes the false-negative samples from the loss calculation, mitigating their impact on the training procedure. FT is formulated as:

$$\mathcal{L}_{FT}(u, i, y_{u,i}) = \begin{cases} 0, & i \in Select\left(\mathcal{I}_u^-, r\left(t\right)\right) \\ \mathcal{L}\left(u, i, y_{u,i}\right), \text{otherwise}, \end{cases} \qquad (5)$$

where $\mathcal{I}_u^-$ is the sampled unobserved item set of user u. $Select\left(\cdot, \cdot\right)$ is the candidate False-negative item selecting module, which selects False-negative items from the whole negative samples with a rate of $r(t)$. FT removes the loss of the False-negative items selected by $Select\left(\cdot, \cdot\right)$, while retaining the others.

Partially Truncated Loss (PT). We argue that the selected latent items have different confidence levels to be real False-negative items. Considering this, it is not wise to discard them equally. Hence, instead of discarding them completely, PT assigns False-negative samples smaller weights during the loss calculation. PT is formulated as:

$$\mathcal{L}_{PT}\left(u, i, y_{u,i}\right) = \begin{cases} w_{PT}\left(u, i\right)\mathcal{L}\left(u, i, y_{u,i}\right), i \in Select\left(\mathcal{I}_u^-, r\left(t\right)\right) \\ \mathcal{L}\left(u, i, y_{u,i}\right), \qquad\qquad\text{otherwise}, \end{cases} \qquad (6)$$

where $w_{PT}\left(u, i\right)$ is the weight of False-negative sample $\langle u, i\rangle$. In this case, the False-negative interactions are assigned with smaller weights, instead of being removed. Therefore, the training data is not changed, only the False-negative samples are given smaller weights. Obviously, a higher predicted score $\hat{y}_{u,i}$ of user u over item i means a higher possibility of user u favors item i, which means that this interaction is more prone to be a False-negative interaction. Thus, the negative impact of this interaction is more severe than those with smaller predicted scores. To reduce the negative impact, the False-negative interactions with higher predicted scores should be assigned smaller weights, so that they play smaller roles in loss calculation. To achieve this, $w_{PT}(u, i)$ is defined as:

$$w_{PT}(u, i) = Sigmoid\left(\frac{1}{\hat{y}_{u,i}}\right)^{\beta}, \tag{7}$$

where β is a hyper-parameter to smooth the weight.

Reversed Loss. Since there are substantial latent interactions existing in recommender systems, distilling the False-negative interactions and further reversing them into the positive interactions instead of only removing them plays an essential role on improving recommendation performance. By doing this, users' latent interests can be captured to consummate the modeling of user preferences. This is the inspiration for the design of Reversed Loss. Reversed Loss can be further detailed into Fully Reversed Loss and Partially Reversed Loss.

Fully Reversed Loss (FR). Compared to removing the False-negative interactions in training process, it is more rational to utilize them to enrich the interaction data to enhance the recommendation performance. In FR, the reversed False-negative samples share equal weight with original positive samples (i.e., 1). That is, there is no priority between positive samples, whether they are original positive samples or reversed False-negative samples. Functionally, FR iteratively reverses False-negative samples into positive ones and recomputes the BCE loss:

$$\mathcal{L}_{FR}(u, i, y_{u,i}) = \begin{cases} \mathcal{L}(u, i, 1), & i \in Select(\mathcal{I}_u^-, r(t)) \\ \mathcal{L}(u, i, y_{u,i}), & \text{otherwise,} \end{cases} \tag{8}$$

In FR, the label of False-negative interaction $\langle u, i \rangle$ is set to 1. By doing this, the latent interactions can be distilled and turned into positive ones, thereby densifying the sparse training data. By incorporating these latent interactions, the recommender training can more comprehensively capture users' latent interests.

Partially Reversed Loss (PR). The latent interactions are essentially pseudo positive interactions that we estimate and may have some level of uncertainty. In other words, different latent interactions have various levels of confidence in being real False-negative interactions. Hence, PR assign a corresponding weight to each reversed false-negative sample:

$$\mathcal{L}_{PR}(u, i, y_{u,i}) = \begin{cases} w_{PR}(u, i)\mathcal{L}(u, i, 1), & i \in Select(\mathcal{I}_u^-, r(t)) \\ \mathcal{L}(u, i, y_{u,i}), & \text{otherwise,} \end{cases} \tag{9}$$

where $w_{PR}(u, i)$ is the weight of False-negative sample $\langle u, i \rangle$. Contrary to $w_{PT}(u, i)$, we argue that $w_{PR}(u, i)$ should be an increasing function of predicted score $\hat{y}_{u,i}$ and $w_{PR}(u, i) \in (0, 1)$. It is because a higher $\hat{y}_{u,i}$ means that $\langle u, i \rangle$ has a higher possibility to be a latent interaction, which means it is more like a potential positive interactions. Thus, the $w_{PR}(u, i)$ is defined as:

$$w_{PR}(u, i) = Sigmoid(\hat{y}_{u,i})^{\beta}, \tag{10}$$

where β is the hyper-parameter used to smooth the weights. Different from FR's directly setting False-negative interaction's label to 1, PR additionally matches each reversed sample with a weight..

4   Experiments

4.1   Experimental Settings

Datasets and Evaluation Metrics. To verify our proposed training strategy DLI, we select three public datasets [20]: Yelp[1], Amazon-book[2] and Adressa[3]. We follow prior work [20] to process all datasets. The statistics of datasets are reported in Table 1. Two metrics are chosen to evaluate the performance of DLI: Recall@K (shorted as R@K) and NDCG@K (shorted as N@K), where K is the number of recommended items.

Table 1. Statistics of datasets.

Dataset	# User	# Item	# Interaction	Density
Yelp	45,548	57,396	1,672,520	0.0640%
Amazon-book	80,464	98,663	2,714,021	0.0341%
Adressa	212,231	6,596	419,491	0.0240%

Backbones. We choose two neural network-based backbone models: GMF [11] and NeuMF [11]; and three graph-based backbone models: NGCF [21], LRGCCF [5] and LightGCN [10] to evaluate the effectiveness of DLI.

Specifically, FT, PT, FR and PR are conducted under three modes of candidate False-negative item selecting module (i.e., selecting items randomly, selecting items with large-loss and selecting items with small-loss) respectively, and the best performances are chosen to report.

Other Training Strategy. Apart from comparing our training strategy with backbone models, we also compare it with training strategies that focuses on False-positive interactions: BCE-T [20] and BCE-R [20].

Parameter Settings. The embedding size of each model is set to 32. For NeuMF, the layer number of MLP is set to 3. We select the batch size in {256, 512, 1024, 2048}. Moreover, the learning rate is tuned in range {0.0001, 0.001, 0.01, 0.1}. α is tuned in {1k, 5k, 10k, 20k, 30k, 40k, 50k} to control the speed of approaching the upper bound of False-negative item selecting rate. And the upper bound of False-negative item selecting rate r_{max} is tuned in the range {0.1, 0.2, ..., 0.9}. β is searched in {0.01, 0.05, 0.1, 0.5, 1}. For graph-based models, number of convolution layers is set to 1. The maximum number of epochs is set to 10. Adam [12] is applied to optimize model parameters.

---

[1] https://www.yelp.com/dataset/.
[2] http://jmcauley.ucsd.edu/data/amazon/.
[3] https://www.adressa.no/.

Table 2. Overall performance of two backbones (GMF and NeuMF) trained normally (denoted as "-") and trained by DLI (denoted as FT, PT, FR and PR). *the improvement is significant at p < 0.05.

Datasets	Backbones	Training strategy	Evaluation metrics			
			R@10	R@20	N@10	N@20
Yelp	GMF	-	0.0253	0.0438	0.0184	0.0243
		FT	0.0255	0.0439	0.0184	0.0243
		PT	0.0252	0.0441	0.0183	0.0243
		FR	0.0329	0.0551	0.0239	**0.0310***
		PR	**0.0332***	**0.0552**	**0.0241***	**0.0310***
	Improvement[a]		31.2%	26.0%	31.0%	27.6%
	NeuMF	-	0.0188	0.0352	0.0133	0.0186
		FT	0.0196	0.0356	0.0138	0.0191
		PT	0.0185	0.0339	0.0131	0.0181
		FR	0.0266	0.0455	0.0191	0.0251
		PR	**0.0277***	**0.0482***	**0.0204***	**0.0278***
	Improvement[a]		47.3%	36.9%	53.4%	49.5%
Amazon-book	GMF	-	0.0231	0.0388	0.0159	0.0207
		FT	0.0232	0.0384	0.0160	0.0207
		PT	0.0231	0.0381	0.0159	0.0205
		FR	0.0255	0.0419	**0.0182***	0.0231
		PR	**0.0256***	**0.0420***	**0.0182***	**0.0232***
	Improvement[a]		10.8%	8.2%	14.5%	12.1%
	NeuMF	-	0.0201	0.0344	0.0137	0.0181
		FT	0.0200	0.0344	0.0134	0.0178
		PT	0.0201	0.0347	0.0136	0.0181
		FR	0.0220	0.0369	0.0153	0.0198
		PR	**0.0221***	**0.0375***	**0.0155***	**0.0202***
	Improvement[a]		10.0%	9.0%	13.1%	11.6%
Adressa	GMF	-	0.1473	0.1727	0.0916	0.0991
		FT	0.1512	0.1796	0.0939	0.1024
		PT	0.1348	0.1690	0.0855	0.0960
		FR	**0.1970***	**0.2421***	**0.1232***	**0.1365***
		PR	0.1961	0.2418	0.1226	0.1361
	Improvement[a]		33.7%	40.2%	34.5%	37.7%
	NeuMF	-	0.1970	0.2309	0.1328	0.1428
		FT	0.2027	0.2464	0.1350	0.1476
		PT	0.2051	0.2512	0.1364	0.1498
		FR	**0.2647***	**0.3105***	**0.1634***	**0.1774***
		PR	0.2399	0.3068	0.1530	0.1732
	Improvement[a]		34.4%	34.5%	23.0%	24.2%

[a] The improvement of the best performance of DLI (boldfaced performance) over the performance under normal training (the line of "-").

4.2 Performance Comparison

Performance of GMF and NeuMF Based on DLI Training Strategy.
From the results reported in Table 2, we have the following observations that:

- On all datasets, our training strategies (especially FR and PR) generally outperform vanilla models. In particular, FR improves vanilla GMF by 31.2%, 10.8% and 33.7% in terms of R@10 on three datasets, respectively. The significant improvement verifies the effectiveness of our training strategy DLI, which can be attributed to the distillation of latent interactions. Experimental results also indicate that our proposed training strategy exhibits excellent generalization ability, making it deployable on various models.
- FR and PR consistently outperform FT and PT on both three datasets. It is because Truncated BCE Loss (FT and PT) is insufficient to take advantage of latent interactions to improve recommendation performance. In this case, the latent interactions are only discarded rather than utilized. This also validate that simply removing False-negative items instead of leveraging them when sampling negative samples are sub-optimal. In general, PR outperforms FR slightly, which could be attributed to the rational properties of PR. PR assign the reversed False-negative samples with corresponding weights to discriminate various confidence levels of selected False-negative interactions.

Performance of Graph Based Backbones Trained by DLI. We perform FR training strategy based on three GNN-based models, as shown in Table 3 (with convolution layer set to 1 and learning rate set to 0.01 to save training time). Training graph-based models with the proposed strategy can promote the performance. NGCF model witnesses a larger improvement of DLI. The reason may be that the feature transformation and non-linear activation function make the model hard to train, while our training strategy can facilitate model training.

Table 3. Overall performance of DLI based on graph-based backbones on Yelp dataset. "-" denotes normal training and FR is our proposed training strategy. *the improvement is significant at $p < 0.05$.

Backbone	Training strategy	Evaluation metircs							
		R@10	R@20	R@50	R@100	N@10	N@20	N@50	N@100
NGCF	-	0.0223	0.0388	0.0786	0.1288	0.0155	0.0208	0.0315	0.0428
	FR	**0.0284***	**0.0494***	**0.0947***	**0.1501***	**0.0203***	**0.0270***	**0.0391***	**0.0515***
Improvement		27.4%	27.3%	20.4%	16.5%	30.9%	29.8%	24.1%	20.3%
LRGCCF	-	0.0349	0.0585	0.1085	0.1677	0.0259	0.0334	0.0469	0.0603
	FR	**0.0370***	**0.0601***	**0.1127***	**0.1754***	**0.0270***	**0.0343***	**0.0483***	**0.0625***
Improvement		6.0%	2.7%	3.9%	4.6%	4.2%	2.7%	3.0%	3.6%
LightGCN	-	0.0343	0.0578	0.1087	0.1704	0.0244	0.0320	0.0457	0.0595
	FR	**0.0362***	**0.0592***	**0.1129***	**0.1755***	**0.0257***	**0.0330***	**0.0474***	**0.0614***
Improvement		5.5%	2.4%	3.9%	3.0%	5.3%	3.1%	3.7%	3.2%

Comparison Between Our Training Strategy with False-Positive Interaction Based Training Strategy. The goal of our proposed training strategy is to distill latent interactions (i.e., False-negative interactions) for recommendation. In contrast, there is another training strategy that focuses on alleviating the effect of False-positive interactions (BCE-T and BCE-R) [20]. For BCE-T and BCE-R, we set the maximum drop rate to 0.2 and the number of iterations to reach the maximum drop rate to $3k$. We compare their performances in Table 4. From the results, we can find that: training strategy focus on False-negative interactions outperforms that on False-positive interactions. This is because the number of negative interactions are far higher than that of positive interactions. Hence, the effort focused on False-negative interactions has more potential than that on False-positive interactions.

Table 4. Overall performance of GMF and NeuMF trained by FR and PR compared to that trained by BCE-T and BCE-R. *the improvement is significant at $p < 0.05$.

Datasets	Backbones	Training strategy	Evaluation metrics			
			R@10	R@20	N@10	N@20
Yelp	GMF	BCE-T	0.0259	0.0446	0.0185	0.0245
		BCE-R	0.0257	0.0445	0.0186	0.0246
		FR	0.0329	0.0551	0.0239	**0.0310***
		PR	**0.0332***	**0.0552***	**0.0241***	**0.0310***
	Improvement		28.2%	23.8%	29.6%	26.0%
	NeuMF	BCE-T	0.0216	0.0397	0.0153	0.0212
		BCE-R	0.0188	0.0350	0.0131	0.0182
		FR	0.0266	0.0455	0.0191	0.0251
		PR	**0.0277***	**0.0482***	**0.0204***	**0.0278***
	Improvement		28.2%	21.4%	33.3%	31.1%
Amazon-book	GMF	BCE-T	0.0240	0.0394	0.0170	0.0218
		BCE-R	0.0213	0.0350	0.0147	0.0189
		FR	0.0255	0.0419	**0.0182***	0.0231
		PR	**0.0256***	**0.0420***	**0.0182***	**0.0232***
	Improvement		6.7%	6.6%	7.1%	6.4%
	NeuMF	BCE-T	0.0209	0.0357	0.0145	0.0192
		BCE-R	0.0197	0.0338	0.0132	0.0177
		FR	0.0220	0.0369	0.0153	0.0198
		PR	**0.0221***	**0.0375***	**0.0155***	**0.0202***
	Improvement		5.7%	5.0%	6.9%	5.2%

4.3   Ablation Study

In this section, we fix all hyper-parameters ($r_{max} = 0.2$, $\alpha = 3k$ and $\beta = 0.01$) to investigate the effectiveness of different candidate False-negative item selecting module: (1) selecting items with big-loss and (2) selecting items with small-loss on the basis of FR. From the results shown in Fig. 3, we have the observation that: FR-small-loss consistently has the better performances as compared to FR-big-loss. It indicates that the negative item with small loss is prone to be a False-negative item. Truncating or reversing the interactions with big loss in the early stage of training process may mislead the model, as samples with big loss play a more dominant role than those with small loss in the back-propagation stage. In this case, introducing True-negative interactions into the latent interaction set will degenerate recommender training.

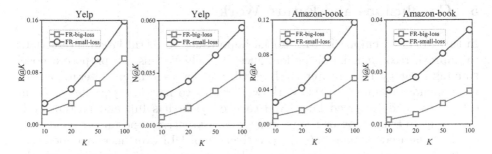

Fig. 3. Recommendation performance of FR on GMF under different candidate False-negative item selecting module. K denotes the number of recommended items.

4.4   Study of Parameter Sensibility

In all four variants of DLI training strategy (FT, PT, FR and PR), r_{max} and α are two hyper-parameters to determine the upper bound and the approaching speed of selected False-negative items. We conduct extensive experiments searching r_{max} in $\{0, 0.2, ..., 0.8\}$ and α in $\{1k, 5k, 10k, 30k, 100k\}$, while keeping other hyper parameters fixed ($\beta = 0.01$). From Fig. 4, we observe that the best performance is achieved when r_{max} and α are in the moderate range. Concretely, FR generally achieves the best performance on Adressa when $r_{max} = 0.6$ and $\alpha = 10k$. R@20 and N@20 firstly rise significantly and then drop as r_{max} and α increase (i.e., convex pattern). This is because low r_{max} and high α make latent interactions under distilled, so that the recommendation performance can't be improved well. And high r_{max} and low α lead to the overabundance of selected False-negative samples, which may introduce lots of noisy interactions and further limit performance. It is worth noting that, R@20 and N@20 under different r_{max} and α settings are much higher than those of $r_{max} = 0$ (i.e., no False-negative interaction is distilled). This verifies the effectiveness of our proposed training strategy DLI.

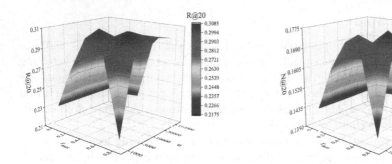

Fig. 4. Performance of FR on basis of NeuMF under different values of hyper-parameter r_{max} and α on Adressa dataset.

5 Conclusions and Future Work

In this work, we emphasize the crucial importance to distill latent interactions to improve recommendation performance. To achieve this, we propose a novel training strategy DLI, to iteratively distill and leverage latent interactions for recommender training. In this work, we design two new loss paradigms: Truncated Loss and Reversed Loss. And each of them has full and partial mode according to truncation or reversion degrees. Extensive experiments conducted on three datasets demonstrate the effectiveness of DLI over backbone models.

In future work, we can extend our work in several potential directions. Firstly, we can explore how to apply DLI to pair-wise loss functions, such as BPR loss. Secondly, we can investigate how to apply DLI to graph-based recommendation models and adapt graph structure accordingly?

Acknowledgment. This work is partially supported by the Natural Science Foundation of China (No. 71774159), China Postdoctoral Science Foundation under Grants 2021T140707, Jiangsu Postdoctoral Science Foundation under Grants 2021K565C, and the Key R&D Program of Xuzhou (No. KC2023178).

References

1. Bengio, Y., Louradour, J., Collobert, R., Weston, J.: Curriculum learning. In: ICML, pp. 41–48 (2009)
2. Cai, Z., Yuan, G., Qiao, S., Qu, S., Zhang, Y., Bing, R.: FG-CF: friends-aware graph collaborative filtering for poi recommendation. Neurocomputing **488**, 107–119 (2022)
3. Cai, Z., Yuan, G., Zhuang, X., Wang, S., Qiao, S., Zhu, M.: Adaptive self-propagation graph convolutional network for recommendation. WWW **26**, 3183–3206 (2023)
4. Chae, D.K., Kang, J.S., Kim, S.W., Choi, J.: Rating augmentation with generative adversarial networks towards accurate collaborative filtering. In: WWW, pp. 2616–2622 (2019)

5. Chen, L., Wu, L., Hong, R., Zhang, K., Wang, M.: Revisiting graph based collaborative filtering: a linear residual graph convolutional network approach. In: AAAI, vol. 34, pp. 27–34 (2020)
6. Ding, J., Quan, Y., Yao, Q., Li, Y., Jin, D.: Simplify and robustify negative sampling for implicit collaborative filtering. Adv. Neural. Inf. Process. Syst. **33**, 1094–1105 (2020)
7. Gunawardana, A., Shani, G., Yogev, S.: Evaluating recommender systems. In: Ricci, F., Rokach, L., Shapira, B. (eds.) Recommender Systems Handbook, pp. 547–601. Springer, New York (2022). https://doi.org/10.1007/978-1-0716-2197-4_15
8. Gupta, S., Wang, H., Lipton, Z., Wang, Y.: Correcting exposure bias for link recommendation. In: ICML, pp. 3953–3963. PMLR (2021)
9. Han, B., et al.: Co-teaching: robust training of deep neural networks with extremely noisy labels. In: Advances in Neural Information Processing Systems, vol. 31 (2018)
10. He, X., Deng, K., Wang, X., Li, Y., Zhang, Y., Wang, M.: LightGCN: simplifying and powering graph convolution network for recommendation. In: SIGIR, pp. 639–648 (2020)
11. He, X., Liao, L., Zhang, H., Nie, L., Hu, X., Chua, T.S.: Neural collaborative filtering. In: WWW, pp. 173–182 (2017)
12. Kingma, D.P., Ba, J.: Adam: a method for stochastic optimization. In: ICLR (Poster) (2015)
13. Liu, X., et al.: UFNRec: utilizing false negative samples for sequential recommendation. In: ICDM, pp. 46–54. SIAM (2023)
14. Marlin, B.M., Zemel, R.S.: Collaborative prediction and ranking with non-random missing data. In: RecSys, pp. 5–12 (2009)
15. Rendle, S., Freudenthaler, C., Gantner, Z., Schmidt-Thieme, L.: BPR: Bayesian personalized ranking from implicit feedback. In: UAI, pp. 452–461 (2009)
16. Song, W., Wang, S., Wang, Y., Wang, S.: Next-item recommendations in short sessions. In: RecSys, pp. 282–291 (2021)
17. Wang, S., Pasi, G., Hu, L., Cao, L.: The era of intelligent recommendation: editorial on intelligent recommendation with advanced AI and learning. IEEE Intell. Syst. **35**(05), 3–6 (2020)
18. Wang, S., Xu, X., Zhang, X., Wang, Y., Song, W.: Veracity-aware and event-driven personalized news recommendation for fake news mitigation. In: WWW, pp. 3673–3684 (2022)
19. Wang, S., Zhang, X., Wang, Y., Liu, H., Ricci, F.: Trustworthy recommender systems. arXiv preprint arXiv:2208.06265 (2022)
20. Wang, W., Feng, F., He, X., Nie, L., Chua, T.S.: Denoising implicit feedback for recommendation. In: WSDM, pp. 373–381 (2021)
21. Wang, X., He, X., Wang, M., Feng, F., Chua, T.S.: Neural graph collaborative filtering. In: SIGIR, pp. 165–174 (2019)
22. Wu, Y., DuBois, C., Zheng, A.X., Ester, M.: Collaborative denoising autoencoders for top-n recommender systems. In: WSDM, pp. 153–162 (2016)
23. Yu, M., et al.: BPMCF: behavior preference mapping collaborative filtering for multi-behavior recommendation. In: Tanveer, M., Agarwal, S., Ozawa, S., Ekbal, A., Jatowt, A. (eds.) Neural Information Processing, ICONIP 2022. Communications in Computer and Information Science, vol. 1792, pp. 407–418. Springer, Singapore (2023). https://doi.org/10.1007/978-981-99-1642-9_35

TRFN: Triple-Receptive-Field Network for Regional-Texture and Holistic-Structure Image Inpainting

Qingguo Xiao, Zhiyuan Han, Zhaodong Liu$^{(\boxtimes)}$, Guangyuan Pan,
and Yanpeng Zheng

School of Automation and Electrical Engineering, Linyi University, Linyi Shandong,
China
`liuzhaodong2017@sina.cn`

Abstract. Image inpainting is a challenging task and has become a hot issue in recent years. Despite the significant progress of modern methods, it is still difficult to fill arbitrary missing regions with both vivid textures and coherent structures. Because of the limited receptive fields, methods centered on convolution neural networks only deal with regular textures but lose holistic structures. To this end, we propose a Triple-Receptive-Field Network (TRFN) that fuses local convolution features, global attention mechanism, and frequency domain learning in this study. TRFN roots in a concurrent structure that enables different receptive fields and retains local features and global representations to the maximum extent. TRFN captures effective representations and generates simultaneously detailed textures and holistic structures by using the concurrent structure. Experiments demonstrate the efficacy of TRFN and the proposed method achieves outstanding performance over the competitors.

Keywords: Image inpainting · Vision transformer · Local convolution · Receptive field

1 Introduction

Image inpainting is studied to deal with completing the missing regions being semantically consistent and visually realistic. It has become a hot topic in the computer vision community and is used in various applications such as photo editing, object removal, and image restoration. Many traditional methods [5,11] search close approximate patches for the plausible reconstruction but perform poorly at obtaining novel realistic textures and semantically coherent structures.

Supported by National Natural Science Foundation of China, grant number: 62103177, Natural Science Foundation of Shandong Province, China, grant number: ZR2023QF097.

In the past few years, Convolutional Neural Networks(CNNs)-based methods have been predominant because of their outstanding feature representation ability. Many generative models using CNNs and Generative Adversarial Networks (GANs) are proposed for image inpainting [13,26]. By learning on large-scale datasets, these networks achieved tremendous success. However, these methods suffer from a common dilemma. They still do not perform well in understanding long-distance feature dependencies and global structures because of the limited receptive fields of CNNs.

To this end, many works begin to explore multiple receptive fields through other operations or neurons except for convolutions. [3] proposes Shift-GCN (Graphic Convolutional Network) structures to obtain flexible receptive fields by shift graph operations and lightweight point-wise convolutions. [19] proposes a graph neural network with adaptive receptive paths to adaptively guide the breadth and depth exploration of the receptive fields. In recent years, transformer has demonstrated power in various vision tasks [7,30] due to the capability of long-distance learning. Some recent works [6,14,27] attempt to adopt transformers to overcome the dilemma of CNNs and have achieved remarkable success.

Therefore, this motivates our work of taking advantage of transformers. We first adopt transformers to enable a non-local role with a global receptive field during the learning. Different from these recent pioneering works [14,27] where CNN and transformer are applied in general sequential form, they are fused to simultaneously capture local and global features in this study. Simultaneously applying convolutional operations and attention mechanisms can solve the shortcomings of the single way. Besides, benefit from [23], which uses Fast Fourier Convolutions (FFCs) that have the image wide receptive field for image inpainting and achieve nice performance, we explore receptive fields from the frequency domain as it allows the model to obtain a completely different perspective of receptive fields.

Formally, this paper proposes a novel Triple-Receptive-Field Network (TRFN) for image inpainting. It is based on an essential component which is Triple-Receptive-Field Module (TRFM). TRFM is a concurrent structure and includes three different receptive fields. Specifically, dense residual blocks are constructed for deep CNN features. Transformer blocks are adopted for the global receptive field and FFCs take charge of another representation from a totally different perspective. The sequential way only captures one kind of feature at one time. The three different receptive fields are embedded in a parallel stream in the proposed network. By the concurrent structure, each channel can fulfill triple feature representations at one time. By a linear projection, the learned representations with different receptive fields are embedded into each other.

The contributions are summarized as follows: (1) A novel triple-receptive-field module is proposed. Three different receptive fields are embedded parallelly to extract better mappings and more representative features. (2) A novel inpainting network is proposed. The network simultaneously generates vivid textures and holistic structures. (3) Extensive results on benchmark datasets demonstrate the proposed method significantly improves model performance and validates the efficacy over the competitors.

2 Related Work

Image Inpainting. Deep learning has brought great performance improvement to image inpainting. Yu et al. [33] introduced the Contextual Attention Network and proposed a two-stage strategy to refine the inpainting process from coarse to fine. The U-Net architecture was applied by [16,18,34] as the variant of auto-encoder architecture to enhance the connection between features in the encoder and decoder. Nazeri et al. [21] proposed an edge-guided image inpainting method to reconstruct reasonable structures. Liu et al. [16] used the equalized features to supplement decoder features for image generation through skip connections and the proposed method is effective to recover structures and textures. Cao et al. [2] proposed a Multi-scale Sketch Tensor inpainting (MST) network with a novel encoder-decoder structure to make reliable predictions of the holistic image structures. Most of the current deep image inpainting networks are based on CNN and suffer from limited receptive fields.

Receptive Field. Receptive fields indicate that a neural unit inside the network corresponds to the size in the original image and describes the relationship between neurons on two functional feature maps. Early receptive fields are extremely limited as networks are restricted to a single branch and only one type of convolution in one layer. Afterward, Szegedy et al. [24] proposed an inception structure to increase network width and receptive fields. The core structure of inception includes multiple branches, which correspond to different receptive fields.

For the limited spatial size of conventional convolutions' receptive field, Yu et al. [32] proposed dilation convolution to enlarge the convolutional receptive fields and has been widely applied in kinds of networks. Residual learning can primarily enlarge the receptive fields for networks as it addresses the problem of training deep networks and larger receptive fields can be achieved by stacking more deep layers.

Transformer [25] shows its great capacity on many tasks in both the NLP and CV communities as it enables global receptive fields and learns long-range interactions. Dosovitskiy et al. [7] first proposed a vision transformer for image recognition and achieved superior performance. Many works are devoted to reducing the time and space complexity for transformers. Esser et al. [8] and Ramesh et al. [22] leverage discrete representation learning for lower computational cost. Wan et al. [28] first used a transformer for low-resolution global reconstruction and then used CNN for further high-quality details. Recently, more methods based on transformers are proposed to deal with kinds of cv tasks. Especially, the combination of CNN and transformer to take both advantages is becoming a trend.

Many works [4,19,23] devote to other operations to exploit more diverse receptive fields. By using Graph Convolutional Networks, Liu et al. [19] proposed a network structure that consists of two complementary functions designed for breadth and depth exploration respectively for learning adaptive receptive fields. Chi et al. [4] use fast Fourier convolutions that have receptive fields covering an entire image to capture complex periodic structures. Suvorov et al. [23] utilized ffc for image inpainting and achieve impressive performance.

In this study, for the first time, we build a triple-receptive-field network that uses local cnn receptive fields for details, transformer receptive fields, and image-wide receptive fields from the frequency domain for global features and context to achieve excellent inpainting performance.

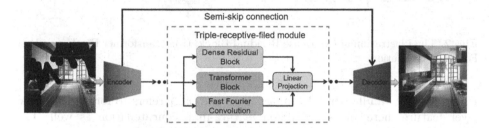

Fig. 1. The overall network architecture.

3   Methods

3.1   Network Architecture

Figure 1 shows the overall architecture of TRFN. TRFN includes three components: encoder, bottleneck, and decoder. The encoder is used to learn features and the decoder translates previous encoded features to desired outputs. [31] indicates that vanilla encoder-decoder architecture is not suitable for preserving subtle details as successive downsamplings inevitably lead to the loss of high-frequency information. We integrate Semi U-Net proposed by [31] and follow a similar network architecture for refine stage. Specifically, the intermediate bottleneck is stacked with triple-receptive-field modules.

3.2   Triple-Receptive-Field Module

As shown in Fig. 2, TRFM enables three different receptive fields and comprises three components: dense residual block, transformer block, and fast Fourier convolution. In the dense residual block, the layer receives the residual of all previous layers in a feed-forward way:

$$x_l = H_l(x_1, ..., x_{l-1}) + x_{l-1} \tag{1}$$

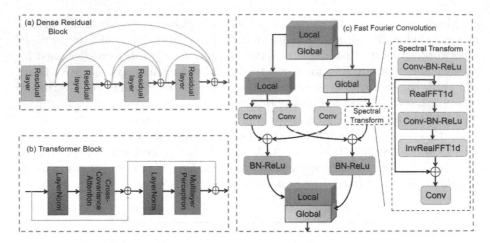

Fig. 2. The illustration of (a): Dense Residual Block, (b): Transformer Block, (c): Fast Fourier Convolution.

Different from vanilla dense learning [10], $x_1, ..., x_{l-1}$ refers to forward residual layer features here and we abandon feature-map concatenation as well. The illustration of the dense residual block is available in Fig. 2 (a).

Due to pairwise interactions between all N token features, self-attention has a computational complexity that scales quadratically in N. As the non-zero part of the eigenspectrum of the Gram matrix and covariance matrix is equivalent, the eigenvectors can be computed in terms of each other. To this end, we leverage cross-variance attention in [1] to minimize the computational cost. The eigendecomposition of either the Gram or covariance matrix can be obtained depending on which of the two matrices is the smallest. The analog Gram matrix $QK^T = XW_qW_k^TX^T$ has the quadratic cost $N \times N$ and N is affected by the input. While the $d_k \times d_q$ cross-covariance matrix $K^TQ = W_k^TX^TXW_q$ can be computed in linear time in the number of elements N. Based on the cross-covariance matrix, the attention is computed as:

$$Attention(K, Q, V) = softmax(K^T\hat{Q}/\tau)V \qquad (2)$$

where $\hat{K}$ and $\hat{Q}$ are normalized matrixes and τ is a learnable temperature parameter that allows for sharper or more uniform distribution of attention weights. We then utilize linear projection, GELU2 activation, and dropout to build Multilayer Perceptron (MLP). MLP is appended in the end to construct the transformer block. The illustration of the transformer block is available in Fig. 2 (b).

We also utilize FFC to enable a receptive field that covers the entire image. FFC divides channels into local and global branches. The local branch uses traditional convolution to update local feature maps. The global branch performs the Fourier transform on the characteristic image and updates it in the frequency domain. To ensure the output is real-valued, Real FFT that has only half of the spectrum is applied and Inverse Real FFT is computed correspondingly to real-valued signals. The outputs of the local and global branches are then fused. The illustration of FFC is available in Fig. 2 (c).

The three components work in a parallel stream. Finally, we use a linear projection to concatenate and fuse the features: $x' = \sum_{i=1}^{3} W_i F_i(x)$ where W_i is the self-learnable weight parameter and $F_i(x)$ is the feature extractor with different receptive fields. TRFM is embedded parallelly with three different receptive fields to generate better mappings and extract more representative features.

3.3 Objective Loss Function

This study adopts a hybrid objective function. It consists of L_1-based pixel loss, adversarial loss, and high receptive field (HRF) perceptual loss [23]. The pixel loss L_{pixel} is based on a normalized L_1 distance:

$$L_{pixel} = E_{I,I_p} \|I_P - I\|_1 \tag{3}$$

where I is the ground truth and Ip is the output. We also adopted feature match loss L_{fm} [29]. It is based on L_1 loss between discriminator features of true and fake samples.

Adversarial loss guides the results to approximate a distribution that is consistent with the real images. It is defined:

$$L_{adv} = E_{I_p,I}[log(D(I_p,I))] + E_{I_p}[log(1 - D[I_p, G(I_p)])] \tag{4}$$

Perceptual loss is used for reconstructing images that are more in line with the visual perception. [23] demonstrates that using a high receptive field perceptual (HRF) loss instead of the conventional one can improve the quality of the inpainting model. Therefore, in this paper, we adopt HRF loss L_{hrf}:

$$L_{hrf} = E([\phi_{hrf}(I) - \phi_{hrf}(G(I_p))]^2) \tag{5}$$

where ϕ_{hrf} is a pretrained semantic segmentation network with a high receptive field.

4   Experiments and Results

The proposed approach is implemented with Pytorch v 1.7.1 and NVIDIA GeForce GTX-3090 GPU on an Ubuntu system. Adam is utilized for training the networks and the optimizer learning rate is set $2e-4$. The weight of L_{pixel}, L_{adv}, L_{fm}, L_{hrf}, and L_{ssim} are set 10, 10, 100, 30, and 0.1 respectively. The evaluation is conducted on two datasets: Places2 [36] and Indoor [12]. For Places2, about 1,800k images from various scenes are used as the training set, and 36,500 images as the validation. For Indoor consisting of buildings and indoor scenes, 20,055 images and 1,000 images are used for training and validation respectively. The weights of L_{pixel}, L_{adv}, L_{fm} and L_{hrf} are set 10, 10, 100 and 30 respectively. We follow the mask setting from [15,23], and [2]. We compare our method with the following algorithms: Partial Convolution(Pconv) [15], Edge Connect(EC) [20], Residual Gather Module(RGM) [17], and Multi-scale Sketch Tensor inpainting(MST) [2].

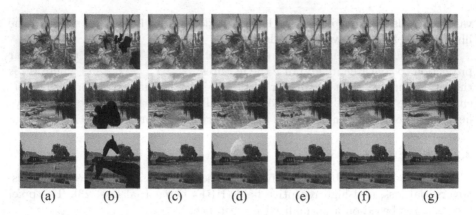

Fig. 3. Visual comparison results on Places2. (a) Ground truth, (b) Input, (c) Pconv, (d) EC, (e) RGM, (f) MST, (g) Ours. Ours achieves the best visual results.

4.1 Results and Analysis

As demonstrated in some previous research, objective metrics that are L_2 based often contradict human judgment. Some meaningless blurring contents can bring small L_2 loss but large perceptual deviations. Therefore, Frechet inception distance (FID) [9] and Learned Perceptual Image Patch Similarity (LPIPS) [35] which pay attention to perceptual metrics are also utilized to make a more thorough evaluation to assess the performance of all compared methods.

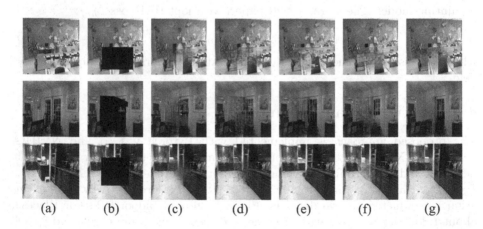

Fig. 4. Visual comparison results on Indoor. (a) Ground truth, (b) Input, (c) Pconv, (d) EC, (e) RGM, (f) MST, (g) Ours. Ours achieves the best visual results.

The quantitative results are shown in Table 1 and the qualitative results are shown in Fig. 3, which is from Places2, and Fig. 4, which is from Indoor containing many structures which are more difficult to complete. It is observed that

Table 1. Quantitative results on Indoor and Places2. ↑ means larger is better and ↓ means lower is better. The best results are bold.

		Pconv	EC	RGM	MST	Ours
Indoor	PSNR↑	23.77	24.07	24.40	24.52	**25.26**
	SSIM↑	0.867	0.884	0.883	0.894	**0.908**
	FID↓	31.102	22.02	25.889	21.65	**18.07**
	LPIPS↓	0.149	0.135	0.129	0.122	**0.104**
Places2	PSNR↑	23.94	23.31	23.91	24.02	**24.78**
	SSIM↑	0.861	0.836	0.869	0.862	**0.891**
	FID↓	7.44	6.21	5.174	**3.53**	3.693
	LPIPS↓	0.146	0.149	0.128	0.137	**0.105**

only Pconv that does not take structure learning into account performs better on Places2 than that on Indoor. For Places2 which includes many natural scenes, Pconv obtains a comparable result but performs poorly on Indoor whose samples consist of complex structures as it is without any extra structure learning strategy. EC constructed an edge generator and RGM utilized structural similarity loss to capture structure information. For MST, both line and edge learning are exploited. Although they improve the results for images with complex structures, the proposed method outperforms all compared methods. TRFN performs superiorly on both Places2 and Indoor as it can generate simultaneously fine details and consistent structures by using the triple-receptive-field module. The advantage is more obvious for large masks and heavily structured images as shown in Fig. 4. As shown in Table 1, for the four metrics including FID and LPIPS that accord with the human perception, the proposed method obtains a superior performance.

4.2 Ablation Study

We study the performance of TRFM and evaluate it with two different networks. One is built with TRFM. The other is built using the previous predominant way which is convolution. The performance of TRFM is shown in Fig. 5. Figure 5 (b) are the results using CNN and Fig. 5 (c) are the results using TRFM. The experiment validates the previous study which is that diverse receptive fields are important for high performance. As demonstrated in Fig. 5, the importance is noticeable. Limited by the receptive fields, the model of CNN fails to learn reasonable holistic structures, and meaningless artifacts are generated especially for large missing regions, while the results using TRFM are much better than those with only convolutions. TRFM enables local receptive fields and global receptive fields simultaneously and can extract effective features. By using TRFM, not only fine textures but also holistic structures are achieved. The same effect is also validated quantitatively. As shown in Table 2, the metrics are significantly improved by TRFM. It is noted that the bottleneck is stacked with only TRFMs.

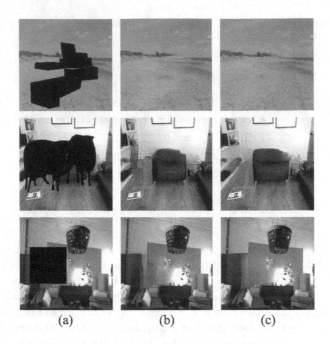

(a)　　　　　　(b)　　　　　　(c)

Fig. 5. Performance of TRFM. (a) Input, (b) Results without TRFM, (c) Results with TRFM.

Pursuing a deeper model is a general trick in many tasks as deeper layers will create more features of various receptive fields. Therefore, deeper models containing more TRFMs can be trialed with more GPU RAMs. Results indicate that the proposed TRFM is an effective structure and helps design more powerful deep models.

Table 2. Ablation studies of TRFM.

	Indoor				Places2			
	PSNR↑	SSIM↑	FID↓	LPIPS↓	PSNR↑	SSIM↑	FID↓	LPIPS↓
CNN	24.53	0.895	30.311	0.128	24.55	0.887	6.455	0.121
TRF	**25.27**	**0.908**	**18.06**	**0.104**	**24.79**	**0.897**	**3.683**	**0.105**

4.3 Object Removal

One application of image inpainting is object removal. The simulation effect of the proposed approach is given in Fig. 6. The to-be-removed object is marked with a light green color. The images after object removal are realistic and look natural. Consistency and continuity can be observed in the results.

Fig. 6. Examples of object removal by our approach.

5   Conclusion

In this study, we propose a Triple-Receptive-Field Network (TRFN) for image inpainting. TRFN roots in TRFM take advantage of local convolutions, global attention mechanism, and frequency domain learning. TRFN enables hierarchical effective feature representations by using different receptive fields simultaneously. Experiments validate the efficacy and demonstrate that the proposed method can obtain both high-quality textures and holistic structures. Moreover, TRFM is a rather effective structure. It can achieve significant improvements and help design more efficient models with multiple and parallel receptive fields.

References

1. Ali, A., et al.: XCiT: cross-covariance image transformers. In: Ranzato, M., Beygelzimer, A., Dauphin, Y., Liang, P., Vaughan, J.W. (eds.) Advances in Neural Information Processing Systems, vol. 34, pp. 20014–20027. Curran Associates, Inc. (2021)
2. Cao, C., Fu, Y.: Learning a sketch tensor space for image inpainting of man-made scenes. In: Proceedings of the IEEE/CVF International Conference on Computer Vision (ICCV), pp. 14509–14518 (2021)
3. Cheng, K., Zhang, Y., He, X., Chen, W., Cheng, J., Lu, H.: Skeleton-based action recognition with shift graph convolutional network. In: Proceedings of the IEEE/CVF Conference on Computer Vision and Pattern Recognition, pp. 183–192 (2020)
4. Chi, L., Jiang, B., Mu, Y.: Fast fourier convolution. Adv. Neural. Inf. Process. Syst. **33**, 4479–4488 (2020)
5. Darabi, S., Shechtman, E., Barnes, C., Goldman, D.B., Sen, P.: Image melding: combining inconsistent images using patch-based synthesis. ACM Trans. Graph. (Proceedings of SIGGRAPH 2012) **31**(4), 82:1–82:10 (2012)
6. Dong, Q., Cao, C., Fu, Y.: Incremental transformer structure enhanced image inpainting with masking positional encoding. In: 2022 IEEE/CVF Conference on Computer Vision and Pattern Recognition (CVPR), pp. 11348–11358 (2022). https://doi.org/10.1109/CVPR52688.2022.01107
7. Dosovitskiy, A., et al.: An image is worth 16 ×16 words: transformers for image recognition at scale. arXiv preprint arXiv:2010.11929 (2020)
8. Esser, P., Rombach, R., Ommer, B.: Taming transformers for high-resolution image synthesis. In: Proceedings of the IEEE/CVF Conference on Computer Vision and Pattern Recognition, pp. 12873–12883 (2021)

9. Heusel, M., Ramsauer, H., Unterthiner, T., Nessler, B., Hochreiter, S.: GANs trained by a two time-scale update rule converge to a local nash equilibrium. In: Advances in Neural Information Processing Systems, vol. 30 (2017)

10. Huang, G., Liu, Z., Van Der Maaten, L., Weinberger, K.Q.: Densely connected convolutional networks. In: Proceedings of the IEEE Conference on Computer Vision and Pattern Recognition, pp. 4700–4708 (2017)

11. Huang, J.B., Kang, S.B., Ahuja, N., Kopf, J.: Image completion using planar structure guidance. ACM Trans. Graph. **33**(4), 129 (2014)

12. Huang, K., Wang, Y., Zhou, Z., Ding, T., Gao, S., Ma, Y.: Learning to parse wireframes in images of man-made environments. In: Proceedings of the IEEE Conference on Computer Vision and Pattern Recognition, pp. 626–635 (2018)

13. Li, J., Wang, N., Zhang, L., Du, B., Tao, D.: Recurrent feature reasoning for image inpainting. In: Proceedings of the IEEE/CVF Conference on Computer Vision and Pattern Recognition, pp. 7760–7768 (2020)

14. Li, W., Lin, Z., Zhou, K., Qi, L., Wang, Y., Jia, J.: MAT: mask-aware transformer for large hole image inpainting. In: Proceedings of the IEEE/CVF Conference on Computer Vision and Pattern Recognition (2022)

15. Liu, G., Reda, F.A., Shih, K.J., Wang, T.-C., Tao, A., Catanzaro, B.: Image inpainting for irregular holes using partial convolutions. In: Ferrari, V., Hebert, M., Sminchisescu, C., Weiss, Y. (eds.) ECCV 2018. LNCS, vol. 11215, pp. 89–105. Springer, Cham (2018). https://doi.org/10.1007/978-3-030-01252-6_6

16. Liu, H., Jiang, B., Song, Y., Huang, W., Yang, C.: Rethinking image inpainting via a mutual encoder-decoder with feature equalizations. arXiv e-prints arXiv:2007.06929 (2020). https://doi.org/10.48550/arXiv.2007.06929

17. Liu, H., Jiang, B., Song, Y., Huang, W., Yang, C.: Rethinking image inpainting via a mutual encoder-decoder with feature equalizations. In: Vedaldi, A., Bischof, H., Brox, T., Frahm, J.M. (eds.) Computer Vision - ECCV 2020, pp. 725–741. Springer International Publishing, Cham (2020)

18. Liu, H., Jiang, B., Xiao, Y., Yang, C.: Coherent semantic attention for image inpainting. In: 2019 IEEE/CVF International Conference on Computer Vision (ICCV), pp. 4169–4178 (2019)

19. Liu, Z., Chen, C., Li, L., Zhou, J., Li, X., Song, L., Qi, Y.: GeniePath: graph neural networks with adaptive receptive paths. In: Proceedings of the AAAI Conference on Artificial Intelligence, vol. 33, pp. 4424–4431 (2019)

20. Nazeri, K., Ng, E., Joseph, T., Qureshi, F., Ebrahimi, M.: EdgeConnect: structure guided image inpainting using edge prediction. In: 2019 IEEE/CVF International Conference on Computer Vision Workshop (ICCVW), pp. 3265–3274 (2019). https://doi.org/10.1109/ICCVW.2019.00408

21. Nazeri, K., Ng, E., Joseph, T., Qureshi, F.Z., Ebrahimi, M.: EdgeConnect: generative image inpainting with adversarial edge learning. arXiv e-prints arXiv:1901.00212 (2019). https://doi.org/10.48550/arXiv.1901.00212

22. Ramesh, A., et al.: Zero-shot text-to-image generation. In: International Conference on Machine Learning, pp. 8821–8831. PMLR (2021)

23. Suvorov, R., et al.: Resolution-robust large mask inpainting with fourier convolutions. In: 2022 IEEE/CVF Winter Conference on Applications of Computer Vision (WACV), pp. 3172–3182 (2022). https://doi.org/10.1109/WACV51458.2022.00323

24. Szegedy, C., Vanhoucke, V., Ioffe, S., Shlens, J., Wojna, Z.: Rethinking the inception architecture for computer vision. In: 2016 IEEE Conference on Computer Vision and Pattern Recognition (CVPR), pp. 2818–2826 (2016). https://doi.org/10.1109/CVPR.2016.308

25. Vaswani, A., et al.: Attention is all you need. In: Guyon, I., Luxburg, U.V., Bengio, S., Wallach, H., Fergus, R., Vishwanathan, S., Garnett, R. (eds.) Advances in Neural Information Processing Systems, vol. 30. Curran Associates, Inc. (2017)

26. Wan, Z., et al.: Bringing old photos back to life. In: CVPR 2020 (2020)

27. Wan, Z., Zhang, J., Chen, D., Liao, J.: High-fidelity pluralistic image completion with transformers. In: 2021 IEEE/CVF International Conference on Computer Vision (ICCV), pp. 4672–4681 (2021). https://doi.org/10.1109/ICCV48922.2021.00465

28. Wan, Z., Zhang, J., Chen, D., Liao, J.: High-fidelity pluralistic image completion with transformers. In: Proceedings of the IEEE/CVF International Conference on Computer Vision, pp. 4692–4701 (2021)

29. Wang, T.C., Liu, M.Y., Zhu, J.Y., Tao, A., Kautz, J., Catanzaro, B.: High-resolution image synthesis and semantic manipulation with conditional GANs. In: Proceedings of the IEEE Conference on Computer Vision and Pattern Recognition, pp. 8798–8807 (2018)

30. Wei, C., Fan, H., Xie, S., Wu, C.Y., Yuille, A., Feichtenhofer, C.: Masked feature prediction for self-supervised visual pre-training. In: Proceedings of the IEEE/CVF Conference on Computer Vision and Pattern Recognition, pp. 14668–14678 (2022)

31. Xiao, Q., Li, G., Chen, Q.: Image inpainting network for filling large missing regions using residual gather. Expert Syst. Appl. **183**, 115381 (2021)

32. Yu, F., Koltun, V., Funkhouser, T.: Dilated residual networks. In: Proceedings of the IEEE Conference on Computer Vision and Pattern Recognition, pp. 472–480 (2017)

33. Yu, J., Lin, Z., Yang, J., Shen, X., Lu, X., Huang, T.S.: Generative image inpainting with contextual attention. In: Proceedings of the IEEE Conference on Computer Vision and Pattern Recognition, pp. 5505–5514 (2018)

34. Zeng, Y., Fu, J., Chao, H., Guo, B.: Learning pyramid-context encoder network for high-quality image inpainting. arXiv e-prints arXiv:1904.07475 (2019). https://doi.org/10.48550/arXiv.1904.07475

35. Zhang, R., Isola, P., Efros, A.A., Shechtman, E., Wang, O.: The unreasonable effectiveness of deep features as a perceptual metric. In: Proceedings of the IEEE Conference on Computer Vision and Pattern Recognition, pp. 586–595 (2018)

36. Zhou, B., Lapedriza, A., Khosla, A., Oliva, A., Torralba, A.: Places: a 10 million image database for scene recognition. IEEE Trans. Pattern Anal. Mach. Intell. **40**(6), 1452–1464 (2018)

PMFNet: A Progressive Multichannel Fusion Network for Multimodal Sentiment Analysis

Jiaming Li[1], Chuanqi Tao[1,2(✉)], and Donghai Guan[1,2]

[1] College of Computer Science and Technology, Nanjing University of Aeronautics and Astronautics, Nanjing, China
{ljiaming,taochuanqi,dhguan}@nuaa.edu.cn
[2] Key Laboratory of Ministry of Industry and Information Technology for Safety-Critical Software, Nanjing University of Aeronautics and Astronautics, Nanjing, China

Abstract. The core of multimodal sentiment analysis is to find effective encoding and fusion methods to make accurate predictions. However, previous works ignore the problems caused by the sampling heterogeneity of modalities, and visual-audio fusion does not filter out noise and redundancy in a progressive manner. On the other hand, current deep learning approaches for multimodal fusion rely on single-channel fusion (horizontal position/vertical space channel), and models of the human brain highlight the importance of multichannel fusion. In this paper, inspired by the perceptual mechanisms of the human brain in neuroscience, to overcome the above problems, we propose a novel framework named Progressive Multichannel Fusion Network (PMFNet) to meet the different processing needs of each modality and provide interaction and integration between modalities at different encoded representation densities, enabling them to be better encoded in a progressive manner and fused over multiple channels. Extensive experiments conducted on public datasets demonstrate that our method gains superior or comparable results to the state-of-the-art models.

Keywords: Multimodal sentiment analysis · MLP · Multichannel fusion

1 Introduction

With the development of social media and the improvement of video quality, multimodal sentiment analysis (MSA) has attracted more and more attention in recent years [1]. MSA is an extension of traditional text-based sentiment analysis, including other modalities that provide supplementary semantic information, such as audio and visual.

To obtain robust modal representations, MISA [2] decomposes modal features in the joint space to present modally invariant and specific representations. Self-MM [3] uses a multimodal task and three unimodal subtasks to obtain dense unimodal representations. MMCL [4] devises unimodal contrastive coding to acquire more robust unimodal representations. However, these works almost treat inputs from all modalities identically, ignoring that they may need different levels of processing due to either their inherent properties or the initial level of abstraction of their input representations. The sampling heterogeneity of the three modalities makes the audio and visual modalities naturally noisy and redundant, while the text modality is relatively pure, moreover, the textual modality is usually extracted by powerful feature extractors, like Bert [5], compared to the fuzzy feature extractors COVAREP [6] and Facet [7] which extract audio and visual features that may be inherently noisy. And we will validate this point in the subsequent experiments. Thus, audio and visual features that do not progressively remove noise and redundancy may disturb the inference of sentiment.

In neuroscience, neural computation is related to the integration of synaptic inputs across a neuron's dendritic arbor [8, 9]. Traditionally, neural network fusion mechanisms have been modeled as single-channel (horizontal position channel or vertical space channel) processing. e.g. It is common to use Transformer [10] to fuse multimodal representations on the horizontal position channel [11, 12]. However, recent studies have highlighted that this model is too simple and that dendrites can integrate synaptic inputs across multiple channel dimensions, such as local, global, different locations, temporal scales, and spatial scales channels [13–15]. It provides a strong impetus to explore the extension of multimodal fusion to multiple channels.

For the aforementioned issue, in this paper, we propose a novel framework named Progressive Multichannel Fusion Network (PMFNet). Specifically, we devise asymmetric progressive fusion coding (APFC) with an efficient bimodal feature absorption module to filter inherent noise contained in the audio and visual modalities and acquire more robust fused representations in a progressive manner. Besides, symmetric multichannel fusion (SMCF) with a pseudo siamese attention network which consists of a module called sym-MLP, is presented to fuse modalities, successfully capturing multichannel cross-modal dynamics globally. The novel contributions of our work can be summarized as follows:

- We propose a novel progressive multichannel fusion network that adapts to the different processing needs of each modality and provides interaction and integration between modalities of different representation densities in a progressive manner.
- We design SMCF, which uses a multichannel mechanism to simulate neuronal dendrites fusing information over multiple channels to learn common and interactive features of high semantic density at the global level.
- We conduct extensive experiments on two publicly available datasets and gain superior or comparable results to the state-of-the-art models.

2 Related Work

2.1 Multimodal Fusion in MSA

Multimodal sentiment analysis (MSA) aims at integrating and processing multimodal data, such as acoustic, visual, and textual information, to comprehend varied human emotions [1].

The core is the fusion strategy. Zadeh et al. [16] used LSTM to model each modality in the temporal dimension. Zadeh et al. [17] proposed a Tensor Fusion Network to explicitly capture unimodal, bimodal, and trimodal interactions through a 3-fold Cartesian product from modality embedding. Tsai et al. [18] used cross-modal attention to capture the bimodal interactions. Hazarika et al. [2] used mode-invariant and mode-specific representations by projecting each modality into two different subspaces to reduce the gap between modalities and capture modality. Han et al. [12] designed an innovative end-to-end bimodal fusion network that conducts fusion and separation on pairs of modality representations. Paraskevopoulos et al. [19] proposed a feedback module named MMLatch that allows modeling top-down cross-modal interactions between higher and lower level architectures.

2.2 Transformer

Transformer is a sequence-to-sequence model. The attention mechanism of the transformer makes it possible to correlate and process multimodal data. Multimodality can improve the performance of previous unimodal tasks (e.g. Machine Translation [20], Speech Recognition [21] etc.) or open up new frontiers (e.g. Image Captioning [22], Visual Question Answering [23] etc.) Multimodal transformers have been widely used for a variety of tasks, including cross-modal sequence generation [24], image segmentation [25] and video captioning [26]. Nagrani et al. [11] proposed MBT which is a new transformer architecture for audio-visual fusion that restricts the cross-modal attention flow to the latter layers of the network. As for the MSA task, Tsai et al. [18] designed a transformer-based structure to capture the connection between any two modalities. Yuan et al. [27] proposed a transformer-based framework to process and improve the robustness of models in the case of random missing in nonaligned modality sequences.

2.3 Multi-layer Perceptron

MLP-based architectures, which consist of linear layers followed by activation, normalization, and dropout, have recently been found to reach comparable results to convolutional and transformer-based methods. Tolstikhin et al. [28] proposed the groundbreaking MLP-Mixer to utilize token-mixing and channel-mixing MLPs to effectively merge information in both the spatial and channel domains. Touvron et al. [29] followed a ResNet-like structure and simplified the token-mixing MLP to a single fully connected layer. Liu et al. [30] designed gMLP to process spatial features by adding spatial gating units. Chen et al. [31] proposed CycleMLP, which introduces cycle MLP layers as a replacement for spatial MLPs. Guo et al. [32] utilized Hire-MLP blocks to facilitate the learning of hierarchical representations. The MLP framework for vision-and-language fusion was first investigated in [33], demonstrating that MLP is effective in multimodal fusion.

3 Methodology

3.1 Problem Definition

The input to the model is unimodal raw sequences $X_m \in R^{l_m \times d_m}$ from the same video clip, where l_m is the sequence length and d_m is the dimension of the representation vector of modality $m \in \{t, v, a\}$, respectively. The goal of the MSA task is to learn a mapping $f(X_t, X_v, X_a)$ to infer the sentiment core $\hat{y} \in R$.

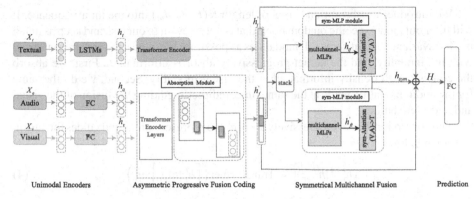

Fig. 1. The overall architecture of PMFNet.

3.2 Unimodal Encoders

We first encode the input X_m of the multimodal sequence into the length representations as h_m. Specifically, we use BERT to encode input sentences to obtain the hidden representations of textual modality as the original features of the text. The original text features are fed into LSTMs for the purpose of consolidating the textual contextual temporal information into the length representation of h_t. For visual and audio features, we use a fully connected network to obtain the representations. The context-aware feature encoding for each modality can be formulated as follows:

$$h_t = \text{LSTM}\left(X_t; \theta_t^{LSTM}\right) \in R^{l_t \times d_t} \tag{1}$$

$$h_m = W_m X_m + b_m \in R^{l_m \times d_m} \quad m \in \{v, a\} \tag{2}$$

3.3 Asymmetric Progressive Fusion Coding

Due to their inherent complexity or the initial level of abstraction of their input representations, the three modalities may need different levels of processing, so we propose asymmetric progressive fusion coding with an efficient bimodal feature absorption module, which focuses on the visual and audio modalities. As shown in Fig. 1. Specifically, following [11], we adopt a progressive fusion strategy that allows the visual and audio modalities to remove noise and redundancy in the fusion process in a progressive mode, thus obtaining a pure and effective fusion representation.

Firstly, we feed the visual and audio modalities into the Transformer encoder separately to encode them at the same level as the textual modality and initially remove noise and redundancy.

$$h_m^{l+1} = \text{Transformer}\left(h_m^l, h_m^l, h_m^l\right) \quad m \in \{v, a\} \tag{3}$$

Secondly, it is easier for predictive networks to make conditional predictions when high-dimensional embeddings are compressed into a more compact representation space [34]. Inspired by [11], we process the visual and audio modalities step by step.

We introduce an absorption kernel of length $K(K \ll d_m)$ into the input sequence. It will filter out mode-specific random noise that is not relevant to our task and keep as much mode-invariant content across all modalities as possible. We concat the absorption kernel with the modality, and the kernel progressively absorbs information. First, we absorb the concise, high-density information in the visual modality. Second, we do the same for the audio modality, while at the same time, indirectly guiding the fusion of the visual and audio modalities.

The absorption kernel progressively absorbs the representations and guides the fusion as shown below:

$$\left[h_v^{l+1} | h{'}_{ker}^{l+1} \right] = \text{Transformer}\left(\left[h_v^l | h_{ker}^l \right]; \theta_v \right) \tag{4}$$

$$\left[h_a^{l+1} | h_{ker}^{l+1} \right] = \text{Transformer}\left(\left[h_a^l | h{'}_{ker}^l \right]; \theta_a \right) \tag{5}$$

Finally, the textual modality is fed into the Transformer encoder to obtain a high semantic density representation, where the encoding of the textual modality is done in an asymmetric structure with respect to the visual and audio modalities.

$$h_t^{l+1} = \text{Transformer}\left(h_t^l, h_t^l, h_t^l \right) \tag{6}$$

3.4 Symmetric Multichannel Fusion

The symmetric multichannel fusion module utilizes a multichannel mechanism to model synaptic integration signals in neuronal cells to learn commonalities and interactive features across different high semantic density modalities.

Based on MLP-Mixer [28], we design two sym-MLP as the core in a pseudo siamese structure [35], where the sym-MLP module includes horizontal position channel MLP, vertical space channel MLP, and symmetric attention mechanism, as shown in Fig. 2.

Pseudo siamese sym-MLP means that two sym-MLP have the same architecture but their parameters are not shared. This allows them to process the different inputs in the same way while with unique neuron weights according to their own computation.

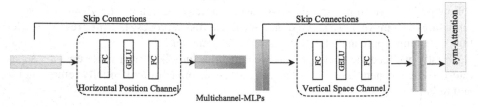

Fig. 2. Illustration of sym-MLP.

Specifically, we stack the dense high-dimensional fusion representation $h{'}_f$ and the high-dimensional dense pure representation $h{'}_t$ into a multichannel representation, and feed them into two Multichannel-MLPs modules in the symmetric form of $stack(h'_f, h'_t)$, $stack(h'_t, h'_f)$, making them fuse in the horizontal position channel and vertical space channel in a multichannel manner.

$$h'_{tf} = \text{Multichannel} - \text{MLPs}\left(stack(h'_t; h'_f)\right) \qquad (7)$$

$$h'_{ft} = \text{Multichannel} - \text{MLPs}\left(stack(h'_f; h'_t)\right) \qquad (8)$$

Then we obtain multichannel fusion features with cross-modal reinforcement of each other, and the features are fed into the sym-Attention module in a weighted manner to further integrate local contextual cross-modal information to obtain the final high-density representation, where $k, s \in \{tf, ft\}, k \neq s$, $W_{\{k,s\}}^{\{Q,K,V\}}$ are learnable projection matrices used to construct keys $h'_s W_s^K$, queries $h'_k W_k^Q$, values $h'_s W_s^V$, α and β are weighted hyper-parameters that adjust the impact of different attentions.

$$a_{k,s} = \text{softmax}\left(\frac{\left(h'_k W_k^Q\right)\left(h'_s W_s^K\right)^T}{\sqrt{d}}\right)\left(h'_s W_s^V\right) + h'_k \qquad (9)$$

$$h_{sym} = \alpha \cdot a_{k,s} + \beta \cdot a_{s,k} \qquad (10)$$

3.5 Prediction

Finally, we obtain the final representations $H = h'_f + h'_t + h_{sym}$ and feed H into a layer of a fully connected network to obtain the final sentiment prediction $\hat{y}$.

$$\hat{y} = W_H H + b_H \in R^{d_{out}} \qquad (11)$$

where d_{out} is the output dimensions, W_H is the weight vectors, b_H is the bias.

4 Experiments

4.1 Datasets and Metrics

CMU-MOSI [36] is one of the most popular benchmark datasets for MSA. The dataset contains 2199 short monologue video clips taken from 93 YouTube movie review videos. The utterances are manually annotated with a sentiment score from −3 (strongly negative) to 3 (strongly positive).

CMU-MOSEI [37] contains 22852 annotated video segments (utterances) from 1000 distinct speakers and 250 topics gathered from online video-sharing websites. The dataset is gender balanced. The videos are transcribed and properly punctuated. Each utterance is annotated with a sentiment intensity from [−3, 3]. The train, validation and test set contain 16322, 1871, and 4659 samples respectively.

We evaluate the absolute error (MAE) and Pearson correlation (Corr) for regression, binary accuracy (Acc-2), F1 score, and 7 class accuracy (Acc-7) from −3 to 3. For binary classification, we consider [−3, 0) labels as negative and (0, 3] as positive.

4.2 Implementation Details

We acquire COVAREP, Facet features and raw text from the public CMU-MultimodalSDK. Specifically, for the visual modality, we use Facet to obtain 35-dimensional features. For the audio modality, we use COVAREP to obtain 74-dimensional features. For the text modality, we use BERT to obtain 768-dimensional features. The model is trained with a single Nvidia GeForce RTX 2080 Ti GPU.

4.3 Baselines

TFN [17] introduces a multi-dimensional tensor by calculating the outer product among different modalities to capture unimodal, bimodal and trimodal interactions. LMF [38] leverages low-rank weight tensors to reduce the complexity of tensor fusion. MFN [16] utilizes a delta-memory attention network with a multi-view gated memory to explicitly accounts for the cross-view interaction. MFM [39] is a cycle-style generative-discriminative model which presents jointly optimize multimodal discriminative factors and modality-specific generative factors. RAVEN [40] employs nonverbal information to shift word representations with the attention-based model. MulT [18] extends three sets of Transformers with directional pairwise cross-modal attention which lately adapts streams from one modality to another. MCTN [41] learns joint multimodal representations by translating between modalities. ICCN [42] uses deep canonical correlation analysis to learn correlations between text, audio, and video modalities. PMR [43] introduces a message hub that can encourage a more efficient multimodal fusion to exchange information with each modality. SPT [44] samples from a longer multimodal input sequence by using a sequence of hidden states. TCSP [45] treats the textual modality as the core and aims to use the shared and private modules to fuse the textual features with two non-textual features.

4.4 Results

Table 1. Unimodal experiments on the CMU-MOSEI dataset.

Model	Acc-2	Acc-7	F1
LSTM(V)	64.25	40.48	61.58
LSTM(A)	64.42	40.07	61.10
LSTM(T)	82.14	47.86	81.53
APFC	66.41	42.15	64.22

First, we perform unimodal experiments to validate the question we posed earlier that the information density of the three modality representations differs significantly. Specifically, we feed the input features into LSTMs to obtain representations with temporal properties. The results in Table 1 illustrate that the text modality features are relatively pure and semantically rich, while the visual and audio modalities differ significantly from the text modality in scores, demonstrating that they are noisy or redundant and therefore require different levels of asymmetric progressive processing for the three modalities.

Then, after being processed by our proposed progressive fusion module, the fused visual-audio representations improve the model's performance, demonstrating that the fused representations are now relatively pure and information-rich. And then after being processed by the multichannel fusion module, the results in Table 2 show that our PMFNet achieves the best performance and outperforms all the baselines on the CMU-MOSEI dataset and achieves competitive results in main metrics on the CMU-MOSI dataset. Besides, finding that the model does not work the best on the CMU-MOSI dataset, we attribute it to the small data size of the dataset that is insufficient for training the model. The above results demonstrate that the framework is effective.

Table 2. Results on CMU-MOSEI. (B) means textual features are based on Bert; [1] means the results provided by [2] and [2] is from the original paper.

Model	CMU-MOSEI					CMU-MOSI				
	MAE	Corr	Acc-2	F1	Acc-7	MAE	Corr	Acc-2	F1	Acc-7
TFN[1] (B)	0.593	0.700	82.5	82.1	50.2	0.901	0.698	80.8	80.7	34.9
LMF[1](B)	0.623	0.677	82.0	82.1	48.0	0.917	0.695	82.5	82.4	33.2
MFN[1]	-	-	76.0	76.0	-	0.965	0.632	77.4	77.3	34.1
RAVEN[1]	0.614	0.662	79.1	79.5	50.0	0.915	0.691	78.0	76.6	33.2
MFM[1](B)	0.568	0.717	84.4	84.3	51.3	0.877	0.706	81.7	81.6	35.4
MulT[1](B)	0.580	0.703	82.5	82.3	51.8	0.871	0.698	83.0	82.8	40.0
ICCN[1](B)	0.565	0.713	84.2	84.2	51.6	**0.860**	**0.710**	83.0	83.0	39.0
MCTN[1]	0.609	0.670	79.8	80.6	49.6	0.909	0.676	79.3	79.1	35.6
PMR[2]	-	-	83.3	82.6	52.5	-	-	**83.6**	**83.4**	**40.6**
SPT[2]	-	-	82.6	82.8	-	-	-	82.8	82.9	-
TCSP[2]	0.576	0.715	82.8	82.7	-	0.908	**0.710**	80.9	81.0	-
PMFNet(ours)	**0.547**	**0.748**	**84.8**	**84.8**	**53.2**	0.994	0.642	80.6	80.5	31.56

4.5 Ablation Study

We perform comprehensive ablation experiments on CMU-MOSEI to analyze the contributions of the APFC and SMCF modules in PMFNet. Table 3 shows that for APFC, when removing the core absorption kernel the results show a small amount of decline, demonstrating its effectiveness in capturing useful information. Removing the entire APFC module leads to a significant decline of 2.3% in Acc-7, 1.3% in F1, and 1.1% in Acc-2, implying the importance of progressive fusion with the absorption kernel. For

SMCF, replacing the sym-MLP with a Transformer layer hurts the performance, demonstrating the effectiveness of the multichannel fusion strategy. Removing the sym-MLP module results in a sharp decline of 3.0% in Acc-7, 1.7% in Acc-2, and 1.6% in F1, implying the effectiveness of the multichannel fusion module for cross-modal fusion.

Table 3. Ablation study of PMFNet on CMU-MOSEI dataset. Note that "rp sym-MLP " denotes replacing sym-MLP with Transformer Encoder.

Model	MAE	Corr	Acc-2	F1	Acc-7
w/o Absorption kernel	0.570	0.735	84.3	84.3	52.1
w/o APFC	0.577	0.724	83.7	83.5	50.9
rp sym-MLP	0.575	0.733	84.1	84.2	51.4
w/o sym-MLP	0.592	0.714	83.1	83.2	50.2
PMFNet(ours)	**0.547**	**0.748**	**84.8**	**84.8**	**53.2**

4.6 Effect of Kernel Numbers

We conduct experiments on the MOSEI dataset to investigate the impact of absorption kernel sizes in APFC. We evaluate sizes ranging from 1 to 8 and evaluated performance using Acc-2 and F1 metrics. Figure 3 illustrates the results. Our model achieves the best performance when the kernel size is 4. A possible reason for this is that excessively small kernels may overlook valuable information, while overly large kernels risk absorbing noisy and redundant data, compromising effectiveness. Therefore, selecting a moderate kernel size within a narrow range is advisable.

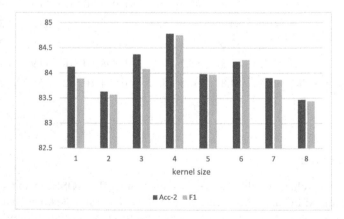

Fig. 3. Experimental results with different absorption kernel sizes on the MOSEI dataset.

5 Conclusion

In this paper, we propose a novel framework named Progressive Multichannel Fusion Network (PMFNet) for multimodal representations to capture intra- and inter-modality dynamics simultaneously, enabling them to encode purely and fuse efficiently over multiple channels. PMFNet is motivated by recent advances in brain models of perceptual mechanisms subject to neuroscience, where human brain neurons internally integrate signals from multiple channels in a progressive manner to form multichannel fusion signals.

Specifically, we devise asymmetric progressive fusion coding (APFC) to filter inherent noise contained in the audio and visual modalities and acquire more robust fused representations in a progressive manner, and symmetric multichannel fusion (SMCF) is presented to fuse multichannel cross-modal dynamics globally.

For future work, we will explore the way to communicate and fuse multimodal data over more channels and extend the network applicability to missing modalities. We also consider the introduction of external knowledge in the fusion process.

References

1. Morency, L.P., Mihalcea, R., Doshi, P.: Towards multimodal sentiment analysis: harvesting opinions from the web. In: Proceedings of the 13th International Conference on Multimodal Interfaces, pp. 169–176 (2011)
2. Hazarika, D., Zimmermann, R., Poria, S.: MISA: modality-invariant and –specific representations for multimodal sentiment analysis. In: Proceedings of the 28th ACM International Conference on Multimedia, pp. 1122–1131 (2020)
3. Yu, W., Xu, H., Yuan, Z., Wu, J.: Learning modality-specific representations with self-supervised multi-task learning for multimodal sentiment analysis. In: Proceedings of the AAAI Conference on Artificial Intelligence, vol. 35, pp. 10790–10797 (2021)
4. Lin, R., Hu, H.: Multimodal contrastive learning via uni-Modal coding and cross-Modal prediction for multimodal sentiment analysis. In: Findings of the Association for Computational Linguistics, EMNLP 2022, pp. 511–523 (2022)
5. Devlin, J., Chang, M., Lee, K., Toutanova, K.: BERT: pre-training of deep bidirectional transformers for language understanding. In: Proceedings of the 2019 Conference of the North American Chapter of the Association for Computational Linguistics: Human Language Technologies, vol. 1, pp. 4171–4186 (2019)
6. Degottex, G., Kane, J., Drugman, T., Raitio, T., Scherer, S.: COVAREP – a collaborative voice analysis repository for speech technologies. In: 2014 IEEE International Conference on Acoustics, Speech and Signal Processing, pp. 960–964 (2014)
7. Facial expression analysis. https://imotions.com/
8. Magee, J.: Dendritic integration of excitatory synaptic input. Nat. Rev. Neurosci. **1**, 181–190 (2000)
9. Branco, T., Häusser, M.: The single dendritic branch as a fundamental functional unit in the nervous system. Curr. Opin. Neurobiol. **20**(4), 494–502 (2010)
10. Vaswani, A., et al.: Attention is all you need. In: Proceedings of the 31st International Conference on Neural Information Processing Systems, vol. 30, pp. 6000–6010 (2017)
11. Nagrani, A., Yang, S., Arnab, A., Jansen, A., Schmid, C., Sun, C.: Attention bottlenecks for multimodal fusion. In: Advances in Neural Information Processing Systems, vol. 34, pp. 14200–14213 (2021)

12. Han, W., Chen, H., Gelbukh, A., Zadeh, A., Morency, L.P., Poria, S.: Bi-bimodal modality fusion for correlation-controlled multimodal sentiment analysis. In: Proceedings of the 2021 International Conference on Multimodal Interaction, pp. 6–15 (2021)
13. Williams, S.: Spatial compartmentalization and functional impact of conductance in pyramidal neurons. Nat. Neurosci. **7**, 961–967 (2004)
14. Ran, Y., Huang, Z., Baden, T., et al.: Type-specific dendritic integration in mouse retinal ganglion cells. Nat. Commun. **11**, 2101 (2020)
15. Li, S., Liu, N., Zhang, X., Zhou, D., Cai, D.: Bilinearity in spatiotemporal integration of synaptic inputs. PLoS Comput. Biol. **10**(12), e1004014 (2014)
16. Zadeh, A., Liang, P.P., Mazumder, N., Poria, S., Cambria, E., Morency, L.P.: Memory fusion network for multi-view sequential learning. In: Proceedings of the AAAI Conference on Artificial Intelligence, pp. 5634–5641 (2018)
17. Zadeh, A., Chen, M., Poria, S., Cambria, E., Morency, L.P.: Tensor fusion network for multi-modal sentiment analysis. In: Proceedings of the 2017 Conference on Empirical Methods in Natural Language Processing, pp. 1103–1114 (2017)
18. Tsai, Y.H., Bai, S., Liang, P.P., Kolter, J.Z., Morency, L.P., Salakhutdinov, R.: Multimodal transformer for unaligned multimodal language sequences. In: Conference of the Association for Computational Linguistics, vol. 1, pp. 6558–6569 (2019)
19. Paraskevopoulos, G., Georgiou, E., Potamianos, A.: MMLatch: bottom-up top-down fusion for multimodal sentiment analysis. In: IEEE International Conference on Acoustics, Speech and Signal Processing, pp. 4573–4577 (2022)
20. Caglayan, O., Madhyastha, P.S., Specia, L., Barrault, L.: Probing the need for visual context in multimodal machine translation. In: Proceedings of the 2019 Conference of the North American Chapter of the Association for Computational Linguistics: Human Language Technologies, vol. 1, pp.4159–4170 (2019)
21. Paraskevopoulos, G., Parthasarathy, S., Khare, A., Sundaram, S.: Multimodal and multiresolution speech recognition with transformers. In: Proceedings of the 58th Annual Meeting of the Association for Computational Linguistics, pp. 2381–2387 (2020)
22. You, Q., Jin, H., Wang, Z., Fang, C., Luo, J.: Image captioning with semantic attention. In: IEEE Conference on Computer Vision and Pattern Recognition, pp. 4651–4659 (2016)
23. Agrawal, A., Lu, J., Antol, S., Mitchell, M., et al.: VQA: visual question answering. In: IEEE International Conference on Computer Vision, pp. 2425–2433 (2015)
24. Seo, P.H., Nagrani, A., Schmid, C.: Look before you speak: visually contextualized utterances. In: 2021 IEEE/CVF Conference on Computer Vision and Pattern Recognition, pp. 16872–16882 (2021)
25. Ye, L., Rochan, M., Liu, Z., Wang, Y.: Cross-modal self-attention network for referring image segmentation. In: IEEE/CVF Conference on Computer Vision and Pattern Recognition, pp. 10494–10503 (2019)
26. Sun, C., Myers, A., Vondrick, C., Murphy, K., Schmid, C.: VideoBERT: a joint model for video and language representation learning. In: IEEE/CVF International Conference on Computer Vision, pp. 7463–7472 (2019)
27. Yuan, Z., Li, W., Xu, H., Yu, W.: Transformer-based feature reconstruction network for robust multimodal sentiment analysis. In: Proceedings of the 29th ACM International Conference on Multimedia, pp. 4400–4407 (2021)
28. Tolstikhin, I.O., et al.: MLP-mixer: an all-MLP architecture for vision. In: Advances in Neural Information Processing Systems, vol. 34, pp. 24261–24272 (2021)
29. Touvron, H., Bojanowski, P., Caron, M., Cord, M., El-Nouby, A., Grave, E., et al.: ResMLP: feedforward networks for image classification with data-efficient training. IEEE Trans. Pattern Anal. Mach. Intell. **45**(4), 5314–5321 (2021)
30. Liu, H., Dai, Z., So, D.R., Le, Q.V.: Pay attention to MLPs. In: Advances in Neural Information Processing Systems, vol. 34, pp. 9204–9215 (2021)

31. Chen, S., Xie, E., Ge, C., Liang, D., Luo, P.: CycleMLP: a MLP-like architecture for dense prediction. In: International Conference on Learning Representations (2022)
32. Guo, J., et al.: Hire-MLP: vision MLP via hierarchical rearrangement. In: IEEE/CVF Conference on Computer Vision and Pattern Recognition, pp. 816–826 (2022)
33. Nie, Y., et al.: MLP architectures for vision-and-language modeling: an empirical study. arXiv preprint arXiv:2112.04453 (2021)
34. Oord, A.V., Li, Y., Vinyals, O.: Representation learning with contrastive predictive coding. arXiv preprint arXiv:1807.03748 (2018)
35. Bromley, J., et al.: Signature verification using a "siamese" time delay neural network. In: Proceedings of the 6th International Conference on Neural Information Processing Systems, pp. 737–744 (1993)
36. Zadeh, A., Zellers, R., Pincus, E., Morency, L.P.: Multimodal sentiment intensity analysis in videos: facial gestures and verbal messages. IEEE Intell. Syst. **31**(6), 82–88 (2016)
37. Zadeh, A., Liang, P.P., Poria, S., Cambria, E., Morency, L.P.: Multimodal language analysis in the wild: CMU-MOSEI dataset and interpretable dynamic fusion graph. In: Proceedings of the 56th Annual Meeting of the Association for Computational Linguistics, vol. 1, pp. 2236–2246 (2018)
38. Liu, Z., Shen, Y., Lakshminarasimhan, V. B., Liang, P.P., Zadeh, A., et al.: Efficient low-rank multimodal fusion with modality-specific factors. In: Proceedings of the 56th Annual Meeting of the Association for Computational Linguistics, pp. 2247–2256 (2018)
39. Tsai, Y.H., Liang, P.P., Zadeh, A., Morency, L.P., Salakhutdinov, R.: Learning factorized multimodal representations. In: International Conference on Learning Representations (2019)
40. Wang, Y., Shen, Y., Liu, Z., Liang, P.P., Zadeh, A., Morency, L.P.: Words can shift: dynamically adjusting word representations using nonverbal behaviors. In: Proceedings of the AAAI Conference on Artificial Intelligence, pp. 7216–7223 (2019)
41. Pham, H., Liang, P.P., Manzini, T., Morency, L.P., et al.: Found in translation: learning robust joint representations by cyclic translations between modalities. In: Proceedings of the AAAI Conference on Artificial Intelligence, vol. 33, pp. 6892–6899 (2019)
42. Sun, Z., Sarma, P.K., Sethares, W.A., et al.: Learning relationships between text, audio, and video via deep canonical correlation for multimodal language analysis. In: Proceedings of the AAAI Conference on Artificial Intelligence, vol. 34, pp. 8992–8999 (2020)
43. Lv, F., Chen, X., Huang, Y., Duan, L., Lin, G.: Progressive modality reinforcement for human multimodal emotion recognition from unaligned multi-modal sequences. In: IEEE Conference on Computer Vision and Pattern Recognition, pp. 2554–2562 (2021)
44. Cheng, J., Fostiropoulos, I., Boehm, B.W., Soleymani, M.: Multimodal phased transformer for sentiment analysis. In: Proceedings of the 2021 Conference on Empirical Methods in Natural Language Processing, pp. 2447–2458 (2021)
45. Wu, Y., Lin, Z., Zhao, Y., Qin, B., Zhu, L.: A text-centered shared-private framework via cross-modal prediction for multimodal sentiment analysis. In: Findings of the Association for Computational Linguistics, ACL-IJCNLP 2021, pp. 4730–4738 (2021)

Category-Wise Meal Recommendation

Ming Li[1], Lin Li[1(✉)], Xiaohui Tao[2], Qing Xie[1], and Jingling Yuan[1]

[1] Wuhan University of Technology, Wuhan, China
{liming7677,cathylilin,felixxq,yil}@whut.edu.cn
[2] The University of Southern Queensland, Darling Heights, Australia
Xiaohui.Tao@unisq.edu.au

Abstract. Meal recommender system, as an application of bundle recommendation, aims to provide courses from specific categories (e.g., appetizer, main dish) that are enjoyed as a meal for a user. Existing bundle recommendation methods work on learning user preferences from user-bundle interactions to satisfy users' information need. However, users in food scenarios may have different preferences for different course categories. It is a challenge to effectively consider course category constraints when predicting meals for users. To this end, we propose a model **CMRec: C**ategory-wise Meal **Rec**ommendation model. Specifically, our model first decomposes interactions and affiliations between users, meals, and courses according to category. Secondly, graph neural networks are utilized to learn category-wise user/meal representations. Then, the likelihood of user-meal interactions is estimated category by category. Finally, our model is trained by a category-wise enhanced Bayesian Personalized Ranking loss. CMRec outperforms state-of-the-art methods in terms of Recall@K and NDCG@K on two public datasets.

Keywords: Meal recommendation · Category-wise representation · Graph neural networks

1 Introduction

Bundle recommendation systems predict users' preferences for a bundle of items [22], which has achieved great results such as product bundle recommendation [14,15,18,19,22], music list recommendation [3,4,8,20], and travel package recommendation [9,12,13]. Meal recommendation [1,6,11] is an application of bundle recommendation in food scenarios. It has great practical value, such as recommending meals for gourmets on recipe sharing sites. Usually a meal consists of multiple courses and each course in the meal comes from a specific category. For example, the basic three-course meals[1] shown in Fig. 1 comes from three categories: appetizer, main dish, and dessert. The task of meal recommendation is to provide a meal to a user with the category-constrained characteristic.

Advanced general bundle recommendation models can be applied to the meal recommendation. Their focus is learning user preferences from their historical

---

[1] https://en.wikipedia.org/wiki/Full-course_dinner.

B. Luo et al. (Eds.): ICONIP 2023, CCIS 1968, pp. 282–294, 2024.
https://doi.org/10.1007/978-981-99-8181-6_22

user-bundle interaction information. DAM [5] uses the factorized attention mechanism and multi-task framework to capture bundle-wise association and collaborative signals. AttList [7] learns bundle-wise user preferences by aggregating interacted bundles of users through hierarchical attention networks. BGCN [4] and MIDGN [21] use graph convolution networks to learn multi-hop collaborative information. Although it is intuitive to use these to solve meal recommendation problem, they pay more attention to meal-wise user preferences and do not take full advantage of category information.

Users may have different preferences for different course categories, which means that considering the category level can more accurately estimate the user's preferences. In the example shown in Fig. 1, we can learn from the user's review history on the left that he gives high ratings to meals with seafood as appetizers and chicken as main dishes. In the candidate meal on the right, $meal_1$ and meal $meal_2$ with the same meal-wise feature "Chicken, Seafood", have different category-wise features about appetizers and main dishes. When only considering meal-wise user preference, the $meal_1$ will be recommended to him regardless of whether its appetizer and main dish are his enjoyment. Compared with the meal-wise user preference, the category-wise user preference can help us provide the user with the more desirable $meal_2$.

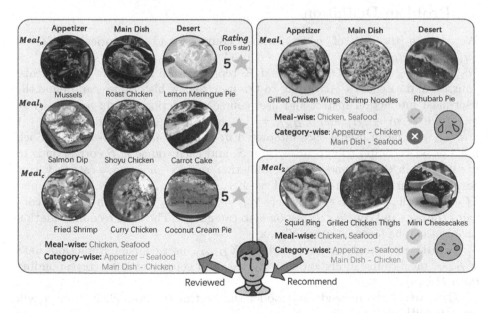

Fig. 1. An example shows that a user has different preferences for different course categories, and category-wise user preferences contribute to an accurate recommendation.

In this paper, we propose a novel model **CMRec: C**ategory-wise Meal **Rec**ommendation. Specifically, our CMRec first utilizes a graph to model interactions and affiliations between users, meals, and courses, and decompose it according to categories. Graph neural networks are used to learn category-wise user and meal representations at both course and meal levels. Next, for each category, user and meal representations at these two levels are considered together to predict the likelihood of user-meal interactions. Then, the final prediction will combine interaction estimations on all categories. Finally, meal-wise and category-wise Bayesian Personalized Ranking (BPR) losses are designed to train our model at meal and category granularities, respectively. We conduct comparative experiments with several state-of-the-art meal and general bundle recommendation methods on two datasets. The results show that our CMRec outperforms all baselines in Recall@K and NDCG@K.

The contributions delivered by this work are summarized below: (1) We emphasize the category-constrained characteristic of the meal and utilize category information in meal recommendation, which is the first ever attempt at meal recommendation according to literature. (2) We propose the model CMRec. It captures category-wise user preferences at both meal and course levels for category-wise interaction prediction, and is trained by a category-wise enhanced BPR loss. (3) Extensive experimental results show that our proposed CMRec outperforms existing state-of-the-art baselines by 12.3%–14.9% and 4.2%–10.9% in Recall@K and NDCG@K on two datasets.

2   Problem Definition

We denote the set of users, meals, courses, and categories as $\mathcal{U} = \{u_1, u_2, \cdots, u_D\}$, $\mathcal{M} = \{m_1, m_2, \cdots, m_O\}$, $\mathcal{S} = \{s_1, s_2, \cdots, s_N\}$, and $\mathcal{C} = \{c_1, c_2, \cdots, c_K\}$ where D, O, N, and K represent the number of users, meals, courses, and categories. According to the user review history, we can define the user-meal interaction matrix and user-course interaction matrix as $\mathbf{X}_{D \times O} = \{x_{um} | u \in \mathcal{U}, m \in \mathcal{M}\}$ and $\mathbf{Y}_{D \times N} = \{y_{us} | u \in \mathcal{U}, s \in \mathcal{S}\}$, in which $x_{um} = 1$ and $y_{us} = 1$ mean user u once interacted m and user u once observed s. Each meal in $\mathcal{M}$ is composed of courses from $\mathcal{S}$, and each course in $\mathcal{S}$ corresponds to a category in $\mathcal{C}$. The meal-course affiliation matrix and course-category corresponding matrix are represented as $\mathbf{Z}_{O \times N} = \{z_{ms} | m \in \mathcal{M}, s \in \mathcal{S}\}$ and $\mathbf{H}_{N \times K} = \{h_{sc} | s \in \mathcal{S}, c \in \mathcal{C}\}$, where $z_{ms} = 1$ and $h_{sc} = 1$ mean meal m contains course s and course s corresponds to category c. The meal recommendation problem can be formulated as follows:

Input: user-meal interaction data $\mathbf{X}_{D \times O}$, user-course interaction data $\mathbf{Y}_{D \times N}$, meal-course affiliation data $\mathbf{Z}_{O \times N}$, and course-category corresponding data $\mathbf{H}_{N \times K}$.

Output: a recommendation model that estimates how likely user u will interact with meal m.

3   Our Proposed Model

3.1   Category-Wise Heterogeneous Graphs Construction

The interaction and affiliation between users, meals, and courses can be explicitly modeled by a heterogeneous graph To make it easier to learn user preferences for

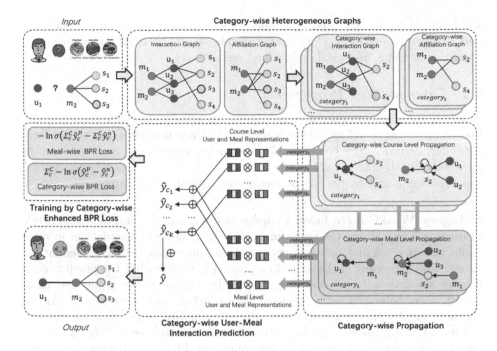

Fig. 2. The framework of the proposed CMRec model. Starting from the upper left corner and going clockwise, CMRec estimates the probability $\hat{y}$ of user u_1 interacting with meal m_2. For a clear display, we divide the heterogeneous graph into two kinds of relationships and draw them separately (Best view in color). (Color figure online)

each category, this heterogeneous graph is decomposed into multiple category-wise heterogeneous graphs with more focused category information. We first build an undirected graph $\mathcal{G} = \{\mathcal{V}, \mathcal{E}\}$ where the nodes set $\mathcal{V}$ consists of user nodes $u \in \mathcal{U}$, meal node $m \in \mathcal{M}$, and course nodes $s \in \mathcal{S}$. In addition, the set of its edges $\mathcal{E}$ consists of user-meal interaction edges (u, b) with $x_{um} = 1$, user-course interaction edges (u, i) with $y_{us} = 1$, and meal-course affiliation edges (m, s) with $z_{ms} = 1$. The top middle of Fig. 2 shows the constructed heterogeneous graph. Then, we decompose the heterogeneous graph $\mathcal{G}$ into category-wise heterogeneous graphs $\mathcal{G}^C = \{\mathcal{G}_1^C, \mathcal{G}_2^C, \cdots, \mathcal{G}_K^C\}$ according to categories. For example, for category c_1, we remove all item nodes that do not belong to category c_1 and their affiliation edges with meals in $\mathcal{G}$. Meanwhile other nodes and edges are reserved to form the category-wise heterogeneous graph $\mathcal{G}_1^C$. For each category-wise graph, meal information and user interaction information about this category are separated from the overall heterogeneous graph for subsequent learning of category-wise user and meal representations.

One-hot encoding is used to encode all graph nodes and them are compressed to dense real-value vectors as follows:

$$p_u = P^T v_u^U, q_u = Q^T v_s^S, r_m = R^T v_m^M \tag{1}$$

where $v_u^U, v_s^S, v_m^M \in \mathbb{R}^N$ denote the one-hot feature vectors for user u, course s, and meal m, respectively. P, Q, and R denote the matrices of user embedding, course embedding, and meal embedding, respectively.

3.2 Category-Wise Propagation

After constructing category-wise heterogeneous graphs, category-wise propagation at course level and meal level is adopted to learn user preferences. Precisely speaking, category-wise course level propagation focuses on user's preferences for courses with different categories, and category-wise meal level propagation captures user's preferences for the meals that have different categories.

Category-Wise Course Level Propagation. A meal is composed of multiple courses, and each affects the user's final choice. Bundle recommendations in some scenarios, are given high user tolerance, such as a music playlist recommendation. A user who likes most of the songs in a playlist is likely to like the playlist, even if he or she does not like some songs in this list. However, we argue that users have a lower tolerance for a meal recommendation, and users usually do not choose a meal with some food course they do not like. This situation requires our model to capture user preferences for course category level. What's more, a user may have different preferences for courses of different categories, as shown in Fig. 1. It is necessary for meal recommendation to learn user's preferences for courses with different categories

It is general to learn user preferences by aggregating the information of the user's neighbors on the graph. We use an embedding propagation layer between users and courses on each category-wise heterogeneous graph to capture user preferences for courses and the representations of themselves. By aggregating the representation of each category course in the meal, we can also obtain the category-wise representations of the meal at course level. Moreover, multi-hop collaborative information can be learned by stacking propagation layers. In a category-wise graph, we formulate propagation-based and pooling-based embedding updating rules can be formulated as follows:

$$
\begin{aligned}
&For \quad \mathcal{G}_c^C \in \mathcal{G}^C \\
&p_{u,1,c}^{(0)} = p_u, \quad q_{s,1,c}^{(0)} = q_s, \\
&p_{u,1,c}^{(l+1)} = \sigma(p_{u,1,c}^{(l)} + \sum_s^{\mathcal{N}_u^c} q_{s,1,c}^{(l)})W_1^{(l+1)} + b_1^{l+1}), \\
&q_{s,1,c}^{(l+1)} = \sigma(q_{s,1,c}^{(l)} + \sum_u^{\mathcal{N}_s^c} p_{u,1,c}^{(l)})W_1^{(l+1)} + b_1^{l+1}), \\
&r_{m,1,c}^{(l+1)} = \sum_s^{\mathcal{N}_m^c} q_{s,1,c}^{(l+1)},
\end{aligned}
\tag{2}
$$

where $\mathcal{G}_c^C$ is the category-wise heterogeneous graph for category c, $p_{u,1,c}^{(x)}$ means the user u representation at the course level x-th propagation layer in $\mathcal{G}_c^C$, and $q_{s,1,c}^{(x)}$ and $r_{m,1,c}^{(x)}$ are analogous. W_1^x and b_1^x are the course level shared learned weight and bias at x-th layer, respectively. σ is non-linear activation function *LeakyReLU*.

For each category, such a propagation layer is stacked L times and outputs L category-wise user and meal representations. We concatenate all layers' output to combine the information received from neighbors of different depths. The representation of course level users and meals for all categories is as follows:

$$p^*_{u,1,c_1} = p^{(0)}_{u,1,c_1}||\cdots||p^{(L)}_{u,1,c_1}, \quad r^*_{m,1,c_1} = r^{(0)}_{m,1,c_1}||\cdots||r^{(L)}_{m,1,c_1},$$
$$\cdots \tag{3}$$
$$p^*_{u,1,c_K} = p^{(0)}_{u,1,c_K}||\cdots||p^{(L)}_{u,1,c_K}, \quad r^*_{m,1,c_K} = r^{(0)}_{m,1,c_K}||\cdots||r^{(L)}_{m,1,c_K},$$

Category-Wise Meal Level Propagation. A meal is made up of courses of specific categories, and different categories of courses contribute differently to the meal. For example, the appetizer is mainly to open the user's taste buds, and the main dish, as the most important course, brings enjoyment and satiety to the user. At the same time, users may have different attitudes towards different meal categories. For example, in Fig. 1, the user enjoys an appetizer of seafood and a main course of chicken in meals. It is necessary to learn category-wise user and meal representations from the meal level for meal recommendation.

For each category-wise heterogeneous graph, we obtain the course level user and meal representations about this category through course level propagation. These representations are used as the initial values of user and meal nodes in the graph. In this way, user-meal propagation is forced to focus on information of a certain category. In addition, we follow BGCN [4] to propagate information between meals based on the overlap intensity of courses in meals. The embedding updating rules for user u and meal m at meal level can be formulated as follows:

$$For \quad \mathcal{G}^C_c \in \mathcal{G}^C$$
$$p^{(0)}_{u,2,c} = FC_u(p^*_{u,1,c}), \quad r^{(0)}_{m,1,c} = FC_m(r^*_{m,1,c}),$$
$$p^{(l+1)}_{u,2,c} = \sigma(p^{(l)}_{u,2,c} + \textstyle\sum_m^{\mathcal{N}^c_u} r^{(l)}_{m,2,c})W^{(l+1)}_2 + b^{l+1}_2),$$
$$r^{(l+1)}_{m,2,c} = \sigma(r^{(l)}_{m,2,c} + \textstyle\sum_u^{\mathcal{N}^c_m} p^{(l)}_{u,2,c} + \textstyle\sum_{m'}^{\mathcal{O}^c_m} \beta_{mm',c} \cdot r^{(l)}_{m',2,c})W^{(l+1)}_2 + b^{l+1}_2), \tag{4}$$

where $FC_u(\cdot)$ and $FC_m(\cdot)$ are fully connected layers for dimensional transformation, and $\mathcal{G}^C_c$ is a category-wise heterogeneous graph. $p^{(x)}_{u,2,c}$ means the user u representation at the meal level x-th propagation layer in $\mathcal{G}^C_c$, and $r^{(x)}_{m,2,c}$ is analogous. W^x_2 and b^x_2 are the meal level shared learned weights and bias at x-th layer. $\mathcal{O}^c_m$ is the overlap intensity matrix for category c, and $\beta_{mm',c}$ represents the overlap intensity in category c between meals m and m' after normalization.

Similarly, the outputs of all layers are concatenated as the meal level user and meal representations. The formula is as follows:

$$p^*_{u,2,c_1} = p^{(0)}_{u,2,c_1}||\cdots||p^{(L)}_{u,2,c_1}, \quad r^*_{m,2,c_1} = r^{(0)}_{m,2,c_1}||\cdots||r^{(L)}_{m,2,c_1},$$
$$\cdots \tag{5}$$
$$p^*_{u,2,c_K} = p^{(0)}_{u,2,c_K}||\cdots||p^{(L)}_{u,2,c_K}, \quad r^*_{m,2,c_K} = r^{(0)}_{m,2,c_K}||\cdots||r^{(L)}_{m,2,c_K},$$

3.3 Category-Wise Interaction Prediction

When a user selects a meal, he usually carefully considers the courses at each category for good taste. Our model combines category-wise user and meal representations from both levels to estimate the likelihood of user-meal interactions. Finally, all categories are considered together to obtain the final estimate. The formula for calculating the probability of interaction between user u and meal m is shown in Eq. 6, where $\otimes$ is the inner product.

$$
\begin{aligned}
\hat{y}_{c_1} &= (p^*_{u,1,c_1} \otimes r^*_{m,1,c_1}) + (p^*_{u,2,c_1} \otimes r^*_{m,2,c_1}), \\
\hat{y}_{c_2} &= (p^*_{u,1,c_2} \otimes r^*_{m,1,c_2}) + (p^*_{u,2,c_2} \otimes r^*_{m,2,c_2}), \\
&\qquad \cdots \\
\hat{y}_{c_K} &= (p^*_{u,1,c_K} \otimes r^*_{m,1,c_K}) + (p^*_{u,2,c_K} \otimes r^*_{m,2,c_K}), \\
\hat{y} &= \hat{y}_{c_1} + \hat{y}_{c_1} + \cdots + \hat{y}_{c_K},
\end{aligned}
\tag{6}
$$

3.4 Category-Wise Enhanced Training

As described above, our model conducts preference learning and prediction at category-wise. To give our model an additional optimization for all operations on each category, we combine the category-wise BPR loss with regular meal-wise BPR loss to optimize our model at both meal and category granularities. The category-wise enhanced BPR loss we define is shown in Eq. 7.

$$
\begin{aligned}
Loss_m &= -ln\sigma(\textstyle\sum_c^{\mathcal{C}} \hat{y}_c^p - \sum_c^{\mathcal{C}} \hat{y}_c^n), \\
Loss_c &= \textstyle\sum_c^{\mathcal{C}} - ln\sigma(\hat{y}_c^p - \hat{y}_c^n), \\
Loss &= Loss_m + Loss_c + \lambda \cdot ||\Theta||^2,
\end{aligned}
\tag{7}
$$

where $Loss_m$ and $Loss_c$ are meal-wise and category-wise BPR Loss, respectively. $\hat{y}_c^p$ and $\hat{y}_c^n$ are the likelihood of the user interacting with the positive and negative sampling meals in category c. Positive and negative sampling meals come from observed and unobserved user-meal interaction, respectively [4,7,21]. $\lambda \cdot ||\Theta||^2$ is the L_2 regularization to prevent over-fitting.

4 Experiments

4.1 Datasets and Metrics

Since our work concerns on category information in a bundle, we conduct our experiments on the following two publicly available datasets. **MealRec** [11]: it is the largest public dataset in the field of meal recommendation, which comes from a recipe sharing platform[2]. A meal here comprises course recipes from three categories (i.e., starter, main course, and dessert). **Clothing** [17]: it is a public clothing bundle dataset constructed from the Amazon e-commerce dataset.

---

[2] https://www.allrecipes.com/.

Each bundle is a combination of several clothes. 1,206 item categories in the original data are summarized into 5 categories, namely clothing, shoes, jewelry, accessories, and others. Their statistics is shown in Table 1. MealRec has a more extensive data scale and denser data density than Clothing. The user-bundle interaction density of MealRec is 0.77%, which is 7.7 times that of Clothing. For both datasets, we divide the training, validation, and test set in a ratio of 8:1:1 by following baselines.

Table 1. Statistics for two experimental datasets.

Dataset	U	B	I	C	U-B	U-I	B-I	U-I density	U-B density
MealRec	1,575	3,817	7,280	3	46,767	151,148	11,451	1.30%	0.77%
Clothing	965	1,910	4,487	5	1,912	6,326	6,326	0.14%	0.10%

Two widely used metrics [4, 7, 21] Recall@K and NDCG@K are employed to evaluate the top-K recommendation performance. Recall measures the ratio of test bundles within the top-K ranking list, and NDCG accounts for the position of the hits by assigning higher scores to those at top ranks.

4.2 Baselines and Hyper-parameters Setting

In addition to traditional meal recommendation models, we also consider several advanced general bundle recommendation models because the meal recommendation is a subset of bundle recommendation. The baselines consist of a traditional CF model (CFR), two attention models (DAM and AttList), and two graph-based models (BGCN and MIDGN). Our model will be extensively compared with these various types of advanced models. All baselines are as follows: **CFR** [16], a general food recommendation framework that uses collaborative filtering to learn both user latent vectors and food latent vectors. **DAM** [5], a multi-task (user-item and user-bundle) attention model for bundle recommendation. It uses the attention mechanism to aggregate items to obtain the representation of bundle. **AttList** [7], a hierarchical self-attentive recommendation model leverages the hierarchical structure of items, bundles, and users to learn bundle-wise user preferences. **BGCN** [4], a graph-based model that utilizes graph convolutional networks to aggregate user historical interaction bundle information to learn bundle-wise user preferences. **MIDGN** [21], a novel graph-based model which disentangles user interactions and bundles according to the user's intents, and presents the user and bundle at a more granular level.

We have reproduced CFR based on their reported framework design and experimental settings. The other four baselines are implemented using their open-source codes, and the parameters follow the original settings in their works. Following all baselines, we implement our model through PyTorch and train it with Adam optimizer. Our hardware environment is Tesla P100 with 16G memory. The mini-batch size is fixed at 128, commonly used in baselines. The learning rate is set to $3e-4$ and L_2 regularization term is tuned as $1e-7$. The embedding size of all models is set to 64.

Table 2. Performance comparison of different models on MealRec and Clothing.

Dataset	Metrics	CFR	DAM	AttList	BGCN	MIDGN	CMRec	% Improv
MealRec	Recall@20	0.1330	0.2038	0.3047	0.2267	0.2055	**0.3424**	12.3%
	NDCG@20	0.0758	0.1147	0.1904	0.1415	0.1178	**0.2171**	14.0%
	Recall@40	0.1833	0.2851	0.3972	0.3131	0.2866	**0.4561**	14.8%
	NDCG@40	0.0888	0.1364	0.2150	0.1629	0.1385	**0.2472**	14.9%
	Recall@80	0.2571	0.3932	0.5071	0.4095	0.3767	**0.5718**	12.7%
	NDCG@80	0.1051	0.1620	0.2402	0.1854	0.1594	**0.2753**	14.6%
Clothing	Recall@20	0.1042	0.2443	0.2637	0.4517	0.4721	**0.5028**	6.5%
	NDCG@20	0.0944	0.1796	0.2048	0.3079	0.3214	**0.3566**	10.9%
	Recall@40	0.1369	0.3068	0.2802	0.4773	0.4931	**0.5142**	4.2%
	NDCG@40	0.1091	0.1923	0.2082	0.3133	0.3237	**0.3589**	10.8%
	Recall@80	0.1568	0.3550	0.3076	0.4943	0.5135	**0.5540**	7.8%
	NDCG@80	0.1525	0.2038	0.2128	0.3165	0.3341	**0.3656**	9.4%

4.3 Overall Performance

The performance comparison of our model CMRec and other state-of-the-art models on two datasets are reported in Table 2. Based on those results, we have the following observations.

Our Model Outperforms All Baselines. Our model significantly outperforms the recent model MIDGN on MealRec by 12.3%–14.9% on all metrics. On the Clothing dataset, our model achieves 10.9%, 10.8%, and 9.4% improvement on NDCG@20, 40, and 80, respectively. A higher NDCG@K value indicates that the test bundles have higher positions in the ranking list, which means higher recommendation quality. Due to learning category-wise user preferences, our model achieves a significant improvement over others.

The Data Density Affects the Performance of Graph-Based Recommendation Models. Through the propagation mechanism to learn the information of neighbors, graph-based models (MIDGN and BGCN) alleviate the impact of data sparseness and perform better than the attention model (DAM and AttList) on Clothing. However, when the dataset is denser, more neighbor information will be aggregated, and more noise may also be introduced simultaneously. AttList uses the attention mechanism in aggregation to consider the principle that different items and bundles have different importance to users, achieving better results than BGCN and MIDGN on the dense dataset MealRec. Our model decomposes user interactions and learns user preferences in each category, reducing the impact of item noise on capturing users' preferences.

It is Important to Capture Category-Wise User Preference from User Interaction History. Different from DAM largely relying on the user ID to represent a user, AttList, BGCN, and MIDGN estimate a user bundle-wise preference by aggregating the interacted bundles of the user through hierarchical attention networks and graph convolution networks, respectively. The latter three models achieve better results than CRF and DAM on the two datasets. However, these models do not effectively consider that users may have different preferences for different item categories. Our CMRec captures category-wise user preferences to predict on each category, achieving the best performance.

4.4 Ablation Study

Next, we investigate the effects of several key designs through the ablation study on MealRec. The design and results of the ablation study are shown in Table 3.

Table 3. Ablation study on MealRec.

Models		MW	S-CW	M-CW	SM-CW	**CMRec**
Designs	meal-wise	✓				
	category-wise		✓	✓	✓	✓
	course level	✓	✓		✓	✓
	meal level	✓		✓	✓	✓
	cate-loss					✓
Metrics	Recall@20	0.1857	0.2547	0.253	0.3138	**0.3424**
	NDCG@20	0.1071	0.1602	0.1493	0.1952	**0.2171**
	Recall@40	0.2655	0.3470	0.3501	0.4139	**0.4561**
	NDCG@40	0.1278	0.1840	0.1745	0.2209	**0.2472**
	Recall@80	0.3743	0.4631	0.4592	0.5345	**0.5718**
	NDCG@80	0.1525	0.2103	0.1994	0.2490	**0.2753**

Effectiveness of Category-Wise User Preferences. Both propagating at course level and meal level, SM-CW that learns category-wise user preferences is 68.9% higher than MW that learns meal-wise user preferences in terms of Recall@20. This result demonstrates the importance of learning the different preferences of users in different categories. **Effectiveness of User Preferences at Course Level and Meal Level.** Both S-CW and M-CW have a considerable improvement over MW. SM-CW that learns user preferences both at course and meal levels, has been further improved. It is verified that it is necessary to learn category-wise preferences at both course and meal levels, and the information learned at these two levels is complementary. **Effectiveness of the category-wise enhanced BPR loss.** CMRec outperforms SM-CW in terms of all metrics. This improvement is largely contributed to the category-wise enhanced BPR loss, which gives our model an additional optimization for learning preferences and predicting interactions on each category.

5   Related Work

More and more research is being conducted on bundle recommendations, with significant progress being made on many real-world applications, such as product bundle recommendations [2,14,18,19,22], music playlist recommendations [3,4,8,20], and travel package recommendations [9,12,13], etc. Among bundle recommendation systems, product bundle recommendations are the most widely studied. Association analysis techniques [2] and traditional recommendation techniques like CF [14] are applied to this task. Many researchers [4,10,19,21] have also made significant progress on applying GCN in bundle recommendation. BGCN [4], for instance, utilizes the graph neural network's power to learn the representation of users and bundles from complex structures. However, when the above bundle methods are applied to meal recommendation, they pay more attention to meal-wise user preferences than category-wise ones, and do not consider that users may have different preferences for different categories.

Meal recommendation as an application of bundle recommendation in food scenarios. Elsweiler [6] attempted to incorporate health and nutrition into the meal recommendation problem DIETOS [1] has provided personalized dietary recommendations by analyzing the consumption data of healthy people and diet-related chronic disease patients. The aforementioned methods mainly make recommendations to users for health or specific nutritional needs, but pay less attention to users' personalized preferences. CFR [16] is a general collaborative learning framework of personalized recommendation, which introduces rich contextual information. CCMR [11] as an attention model utilizes the user's interactions and meal composition to learn user and meal representations. The above methods are important explorations in meal recommendation, but compared with graph-based methods, these methods do not efficiently model user-meal-course interactions and affiliations. Mining useful information from these complex relationships is the key to improving recommendation quality.

6   Conclusion and Future Work

In order to effectively consider course category constraints to meal recommendation, we propose a novel model CMRec: category-wise meal recommendation. Our model decomposes user interactions on category-wise and predicts user-meal interactions in each category after learning category-wise preferences. Moreover, category-wise enhanced BPR loss is used to optimize the model. Comprehensive experiments are conducted with several state-of-the-art meal and bundle recommendation methods on two datasets. The results demonstrate the effectiveness of CMRec. In the future, we will try to consider the rich descriptive information of meals and courses, such as pictures, text, nutritional information, etc.

Acknowledgements. This work is partially supported by NSFC, China (No. 62276 196).

References

1. Agapito, G., Simeoni, M., Calabrese, B., Guzzi, P.H., Fuiano, G., Cannataro, M.: DIETOS: a recommender system for health profiling and diet management in chronic diseases. In: Proceedings of the 11th ACM International Conference on Recommender Systems, RecSys 2017, vol. 1953, pp. 32–35 (2017)
2. Agrawal, R., Srikant, R., et al.: Fast algorithms for mining association rules. In: Proceedings of the 20th Conference on Very Large Data Bases, VLDB 1994, vol. 1215, pp. 487–499 (1994)
3. Cao, D., Nie, L., He, X., Wei, X., Zhu, S., Chua, T.: Embedding factorization models for jointly recommending items and user generated lists. In: the 40th Conference on Research and Development in Information Retrieval, SIGIR 2017, pp. 585–594 (2017)
4. Chang, J., Gao, C., He, X., Jin, D., Li, Y.: Bundle recommendation with graph convolutional networks. In: Proceedings of the 43rd International Conference on Research and Development in Information Retrieval, SIGIR 2019, pp. 1673–1676 (2019)
5. Chen, L., Liu, Y., He, X., Gao, L., Zheng, Z.: Matching user with item set: collaborative bundle recommendation with deep attention network. In: Proceedings of the 28th International Joint Conference on Artificial Intelligence, IJCAI 2019, pp. 2095–2101 (2019)
6. Elsweiler, D., Harvey, M.: Towards automatic meal plan recommendations for balanced nutrition. In: Proceedings of the 9th ACM Conference on Recommender Systems, RecSys 2015, pp. 313–316 (2015)
7. He, Y., Wang, J., Niu, W., Caverlee, J.: A hierarchical self-attentive model for recommending user-generated item lists. In: Proceedings of the 28th ACM International Conference on Information and Knowledge Management, CIKM 2019, pp. 1481–1490 (2019)
8. He, Y., Zhang, Y., Liu, W., Caverlee, J.: Consistency-aware recommendation for user-generated item list continuation. In: Proceedings of the 13th ACM International Conference on Web Search and Data Mining, WSDM 2020, pp. 250–258 (2020)
9. Herzog, D., Sikander, S., Wörndl, W.: Integrating route attractiveness attributes into tourist trip recommendations. In: Proceedings of the 23th ACM International Conference on World Wide Web, WWW 2019, pp. 96–101 (2019)
10. Li, C., et al.: Package recommendation with intra- and inter-package attention networks. In: Proceedings of the 44th International ACM SIGIR Conference on Research and Development in Information Retrieval, SIGIR 2021, pp. 595–604 (2021)
11. Li, M., Li, L., Xie, Q., Yuan, J., Tao, X.: MealRec: a meal recommendation dataset. CoRR abs/2205.12133 (2022)
12. Lim, K.H., Chan, J., Leckie, C., Karunasekera, S.: Personalized tour recommendation based on user interests and points of interest visit durations. In: Proceedings of the 24th International Joint Conference on Artificial Intelligence, IJCAI 2015, pp. 1778–1784 (2015)
13. Lim, K.H., Chan, J., Leckie, C., Karunasekera, S.: Personalized trip recommendation for tourists based on user interests, points of interest visit durations and visit recency. Knowl. Inf. Syst. 54(2), 375–406 (2018)
14. Liu, G., Fu, Y., Chen, G., Xiong, H., Chen, C.: Modeling buying motives for personalized product bundle recommendation. ACM Trans. Knowl. Discov. Data (TKDD) 11(3), 1–26 (2017)

15. Liu, Y., Xie, M., Lakshmanan, L.V.S.: Recommending user generated item lists. In: Proceedings of the 8th Conference on Recommender Systems, RecSys 2014, pp. 185–192 (2014)
16. Min, W., Jiang, S., Jain, R.C.: Food recommendation: framework, existing solutions, and challenges. IEEE Trans. Multim. 22(10), 2659–2671 (2020)
17. Sun, Z., Yang, J., Feng, K., Fang, H., Qu, X., Ong, Y.S.: Revisiting bundle recommendation: datasets, tasks, challenges and opportunities for intent-aware product bundling. In: Proceedings of the 45th International ACM SIGIR Conference on Research and Development in Information Retrieval, SIGIR 2022, pp. 2900–2911 (2022)
18. Wan, M., Wang, D., Liu, J., Bennett, P., McAuley, J.J.: Representing and recommending shopping baskets with complementarity, compatibility and loyalty. In: Proceedings of the 27th ACM International Conference on Information and Knowledge Management, CIKM 2018, pp. 1133–1142 (2018)
19. Wang, X., Liu, X., Liu, J., Wu, H.: Relational graph neural network with neighbor interactions for bundle recommendation service. In: Proceedings of 2021 IEEE International Conference on Web Services, ICWS 2021, pp. 167–172 (2021)
20. Yang, H., Zhao, Y., Xia, J., Yao, B., Zhang, M., Zheng, K.: Music playlist recommendation with long short-term memory. In: Proceedings of Database Systems for Advanced Applications 24th International Conference, DASFAA 2019, vol. 11447, pp. 416–432 (2019)
21. Zhao, S., Wei, W., Zou, D., Mao, X.: Multi-view intent disentangle graph networks for bundle recommendation. arXiv preprint arXiv:2202.11425 (2022)
22. Zhu, T., Harrington, P., Li, J., Tang, L.: Bundle recommendation in ecommerce. In: Proceedings of the 37th International ACM SIGIR Conference on Research and Development in Information Retrieval, SIGIR 2014, pp. 657–666 (2014)

A Data-Free Substitute Model Training Method for Textual Adversarial Attacks

Fenghong Chen and Zhidong Shen[✉]

Key Laboratory of Aerospace Information Security and Trusted Computing, Ministry of Education, School of Cyber Science and Engineering, Wuhan University, Wuhan, China
shenzd@whu.edu.cn

Abstract. BERT and other pre-trained language models are vulnerable to textual adversarial attacks. While current transfer-based textual adversarial attacks in black-box environments rely on real datasets to train substitute models, obtaining the datasets can be challenging for attackers. To address this issue, we propose a data-free substitute training method (DaST-T) for textual adversarial attacks, which can train substitute models without the assistance of real data. DaST-T consists of two major steps. Firstly, DaST-T creates a special Generative Adversarial Network (GAN) to train substitute models without any real data. The training procedure utilizes samples synthesized at random by the generative model, where labels are generated by the attacked model. In particular, DaST-T designs a data augmenter for the generative model to facilitate rapid exploration of the entire sample space, thereby accelerating the performance of substitute model training. Secondly, DaST-T applies existing white-box textual adversarial attack methods to the substitute model to generate adversarial text, which is then migrated to the attacked model. DaST-T can effectively address the issue of limited access to real datasets in black-box textual adversarial attacks. Experimental results on text classification tasks in NLP show that DaST-T can achieve superior attack performance compared to other baselines of black-box textual adversarial attacks while requiring fewer sample queries.

Keywords: Textual adversarial attack · Substitute training · BERT

1 Introduction

Deep neural networks have proven to be successful in the field of Natural Language Processing (NLP). Specifically, the emergence of large pre-trained models such as BERT has significantly improved the performance of many NLP tasks and achieved new state-of-the-art [3]. However, recent studies have shown that BERT and its fine-tuned model are susceptible to interference of textual adversarial attacks, which can lead to incorrect predictions.

Black-box textual adversarial attacks are more practical in real-world scenarios compared to white-box attacks. While white-box attacks rely on gradient information to modify specific words or sentences [2,15], black-box attacks can only access the output classification probabilities or labels of the attacked model. Current research on black-box attacks focuses on query-based and substitution-based attacks. Query-based attacks use numerous queries to randomly replace words in the text [10,17], while substitution-based attacks generate adversarial text using a substitution model with decision boundaries similar to the attacked model. The substitute model is trained with the training datasets of the attacked model or real data with labels from attacked model [16]. However, both training datasets and real data are strictly protected, making it difficult for attackers to directly obtain. Therefore, it is important to develop methods that can train substitute models without any real data, which can help assess the risks faced by pre-trained models in NLP more comprehensively.

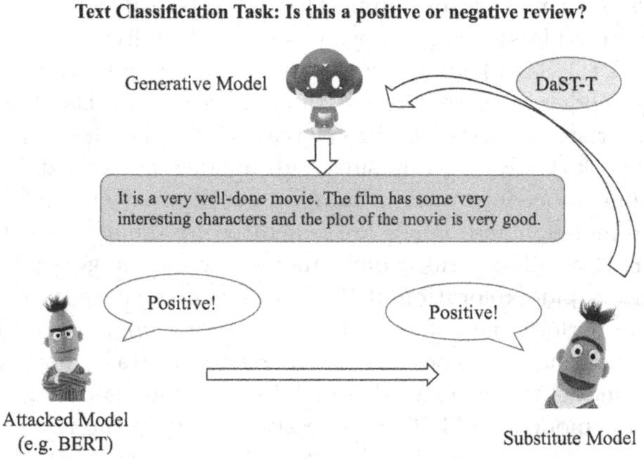

Fig. 1. DaST-T trains substitute model with samples synthesized by generative model, where labels of synthetic samples are generated by attacked model.

In this paper, we propose a data-free substitute training method for textual adversarial attacks, DaST-T. Inspired by transfer-based adversarial attacks in the field of image domain [18,21], we consider using synthetic data to train substitute models. However, due to the discreteness, semantic nature, and logical structure of text samples, these methods cannot be directly applied. To cope with this problem, DaST-T designs a special Generative Adversarial Network (GAN). As is shown in Fig. 1, the generative model synthesizes samples randomly, which are then predicted by the attacked model and substitute model. The output of the attacked model is used as labels of synthetic samples to guide the training of substitute models. The generative model then combines the output of the substitute model to back-propagate and update weights. The process is cycled until the substitute model can replicate the decision boundaries of the attacked model. In particular, a data augmenter for the generative model is designed to cope with the limitation of the distribution of the synthetic samples. Since training datasets used for DaST-T are randomly generated by the generative model, it's more difficult for defenders to track and target.

DaST-T is evaluated on two medium-length document-level text classification datasets, i.e., IMDB and Yelp, and one sentence-level text classification dataset, AG News. Experimental results show that DaST-T can successfully fool downstream fine-tuned BERT models with a higher attack rate and lower query numbers. Additionally, the semantic similarity of adversarial text is higher compared to existing baselines for black-box textual adversarial attacks.

The contributions of this paper are summarized as follows:

- We introduce DaST-T, a substitute training method for discrete text data, leveraging a specially designed GAN to address the challenge of limited access to real datasets encountered in current alternative attacks.
- DaST-T incorporates a data augmenter for the generative model, aiding the generator in discerning the complete distribution of synthetic samples, thereby enhancing the training efficacy and effectiveness of substitute models.
- Experimental results reveal the effectiveness of DaST-T and show its superior performance when compared to existing baselines in the field.

2   Related Works

Inspired by adversarial attack methods in the field of computer vision, early work on white-box textual adversarial attacks concentrated on gradient-based methods. Goodfellow et al. [14] apply Fast Gradient Sign Method (FGSM) to text classification tasks in NLP by computing minimum perturbation and adding to a continuous word embedding vector to generate adversarial text. Subsequent researchers [1,4] proposed rule-based heuristics, including disordering words, misspelling, adding or deleting words, etc., which are designed manually by humans. While other researchers [19] have focused their work on improving the semanticity of the generated adversarial text while reducing the number of perturbated words to improve the success rate of attack.

In the black-box settings, some previous works [6,13] are based on greedy algorithms that rank words in a sentence by computing importance scores and later greedily scrambling words. However, these algorithms have limitations in terms of attack performance and efficiency. To overcome this problem, recent research [9,12] has proposed query-efficient black-box attack methods and efficient local search algorithms. Meanwhile, some researchers have attempted to apply transfer-based adversarial attack methods to the text domain. Li et al. [10] first demonstrated experimentally that adversarial text is migratory, i.e., an adversarial text generated on one classifier can also mislead another classifier. Wang et al. [16] used the teacher-student approach to train substitute models. However, this method is based on real datasets which are hard to obtain. Based on this, DaST-T is proposed to overcome the difficulty in acquiring real datasets.

3 Method

3.1 Attack Scenario

Given a text classification model $T : X \rightarrow Y$, the model maps text x in dataset X to a corresponding classification label y and then forms a set Y. In this paper, the attack scenario assumes that attackers do not know any details of attacked model T, such as model architecture, loss function, gradient information, etc., but can only access the output probability of attacked model. Attackers add a small perturbation ϵ to legitimate document $(x, y) \in (X, Y)$ to get adversarial text $\overline{x}$, which will make attacked model change its classification label from y to y', that is $T(\overline{x}) = y', y' \neq y$. Since the problem studied in this paper is untargeted attacks, the objective can be formulated as:

$$\min_{\epsilon} \|\epsilon\| \text{ subject to } \underset{y_i}{\mathrm{argmax}} T\left(y_i \mid \bar{X}\right) \neq y \text{ and } \|\epsilon\| \leq \mathrm{r} \tag{1}$$

where ϵ and r are samples of perturbation and upper limit of perturbation, respectively. For attacking machine learning system which is hard to detect, r is set to a small value in attack methods. $\overline{X}$ is the generated adversarial text, which can cause the attacked model T to output wrong labels.

3.2 Adversarial Substitute Attacks

Adversarial substitute attacks refer to the training of a substitute model in a black-box environment to generate adversarial text, which is then migrated to the target model. The theoretical basis of such attacks is the transferability property of adversarial samples, i.e., adversarial text generated on one model can work as well for another model. Adversarial substitute attacks do not require numerous queries, as in query-based black-box attacks, thereby reducing the time required for an attack and preventing being tracked and targeted. To achieve this, the core of adversarial substitute attacks is to train a substitute model with decision boundaries as similar as possible to the target model. While current methods mainly use the same training dataset as the target model or numerous real data marked by the target model to train substitute models, in some real-world tasks, obtaining the target model's training dataset or a large amount of labeled real data is difficult. Therefore, we propose a data-free adversarial substitute attacks method, which is introduced in the last subsection and shown in Fig. 2.

3.3 DaST-T

Data-Free Substitute Model Training Phase. As shown in Fig. 2, a special Generative Adversarial Network (GAN) is designed to help train substitute models without real data. The generative model is a pre-trained text generation model that is trained on a massive corpus of text from books, articles, and websites. The generative model consists of 24 decoder layers, each of which contains a multi-headed self-attentive mechanism capable of simultaneously attending

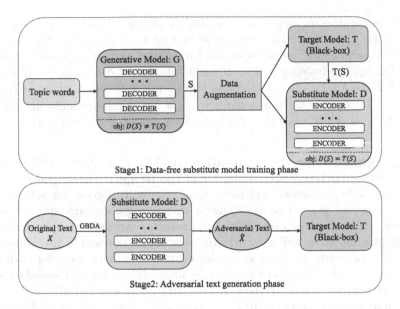

Fig. 2. Process of Data-free Substitute Training Method for Textual Adversarial Attacks (DaST-T)

to different locations in the input and capturing word dependencies. And each encoder layer contains a feedforward neural network consisting of two fully connected layers with ReLU activation functions.

In contrast to this, the substitute model as the discriminator in GAN consists of 12 stacked encoder layers. Each encoder layer employs a masked multi-head self-attention mechanism, allowing the model to attend to different positions in the input sequence and capture word relationships. Additionally, each encoder layer includes a feed-forward neural network with non-linear activation functions. In this two-player game between generator and discriminator, target model plays the role of a referee. This special GAN can effectively speed up the training of substitute models, and help substitute models learn the decision boundaries of the target model well.

Using binary classification as an example for analysis, the value function of both sides of the equation can be described as follows:

$$\max_G \min_D V_{G,D} = d(D(S), T(S)) \tag{2}$$

where $d(D, T)$ is a measure of the output distance between the substitute model and the target model. The attack scenario in this study is set up as black-box attack based on the output probability, and this measure can be represented as

$$d(T, D) = \|D(S), T(S)\|_F \tag{3}$$

where $D(S)$ and $T(S)$ represent output probabilities of the substitute model and target model respectively.

The generative model creates synthetic samples, preprocesses and augments them, and then trains the generator and discriminator through separate back-propagation steps based on target and substitute model classifications. During the training period, the loss function of discriminator D is set to $\mathcal{L}_D = CE(D(\hat{X}), T(\hat{X}))$, and the loss function of generative model G is set to $\mathcal{L}_G = e^{-CE(D(\hat{X}), T(\hat{X}))}$. Discriminator reaches the global optimal solution when and only when $\forall x \in \hat{X}, D(x) = T(x)$ at which time $L_D = 0$, $L_G = 1$.

Data Augmenter for the Generative Model. While data augmentation is commonly used in image processing, it can also be applied to language tasks. In this paper, data augmenter performs synonym substitution on synthetic samples generated by the generative model using an existing knowledge base, WordNet [5]. Specifically, data augmenter first preprocesses synthetic samples to remove redundant symbols, as well as to perform word separation and lexical annotation. Next, 10% of words are randomly selected from synthetic samples each time. For selected words, WordNet identify a set of candidate words. Candidate words replace the original words in the text after checking their lexicality.

Data augmenter increases the diversity of synthetic samples, thereby expanding the synthetic sample set in the entire sample space. This expansion helps the substitute model learn the decision boundaries of the target model faster in the whole sample space.

Adversarial Text Generation Phase. Once DaST-T obtains the substitute model, existing white-box attack methods can be used against the substitute model to generate the adversarial text that will be migrated to attack the target model. Here, the GBDA method [7] is used to generate adversarial texts.

GBDA enables the sampling optimization problem of finding adversarial samples to be converted into a continuously derivable parameter optimization problem by introducing Gumbel-softmax. Then, by adding fluency constraints and BERT-Score similarity constraints, the semantics of adversarial text is motivated to be as consistent as possible with the original text. The final optimization objective function is obtained as follows:

$$L(\Theta) = E_{\tilde{\pi} \sim \tilde{P}_{\theta}} l(e(\tilde{\pi}), y; h) + \lambda_{1m} NLL_g(\pi) + \lambda_{sim} \rho_g(x, \pi) \qquad (4)$$

where $\lambda_{1m}, \lambda_{\text{sim}} > 0$ are hyperparameters. $NLL_g(\pi)$ denotes the fluency constraint, as shown in Eq. 5. $\rho_g(x, \pi) = 1 - R_{BERT}(x, \pi)$ denotes the BERT Score similarity constraint, as shown in Eq. 6.

$$NLL_g(\pi) := -\sum_{i=1}^{n} \log p_g(\pi_i \mid \pi_1 \cdots \pi_{i-1}) \qquad (5)$$

$$R_{BERT}(x, x') = \sum_{i=1}^{m} w_i \max_{j=1,\ldots,m} v_i^T v_j' \qquad (6)$$

4   Experiments

4.1   Datasets

DaST-T is applied to text classification tasks of NLP and evaluated on three different datasets.

IMDB is a document-level sentiment classification dataset of movie reviews, including both positive and negative categories.

Yelp is a document-level text classification dataset for binary classification. Following Zhang et al. [20], comments on Yelp are rated 1 and 2 as negative and 4 and 5 as positive.

AG News is a sentence-level text classification dataset, which includes four categories: world, sports, business, and science.

4.2   Baselines

We choose three black-box textual adversarial attack methods as baselines. All of the compared baseline methods are used the same underlying language model for subsequent evaluations.

TextFooler [8] is a black-box textual adversarial attack method, which finds synonyms of each word by computing the similarity between word embeddings to construct adversarial text with high semantic similarity.

BERT-Attack [11] is a black-box adversarial text attack method that uses BERT to attack fine-tuned BERT model by using a pre-trained masked language model for masking prediction to obtain a set of candidate words with high semantic similarity.

BBA [9], which incorporates a categorical kernel with automatic relevance determination to dynamically learn the importance score for each categorical variable in an input sequence. It achieves lower attacked accuracy while minimizing perturbation size and query requirements.

4.3   Automatic Evaluation Metrics

The initial metric is the original accuracy (Original ACC) of the target model on the original datasets. Attacked accuracy (Attacked Acc) is the core metric used to measure the successfulness of textual adversarial attacks, which represents the accuracy of the target model on adversarial text. The difference between original accuracy and attacked accuracy reflects the effect of the attack intuitively. Query number, another key metric, refers to the average number of queries used per sample during black-box adversarial attacks. A lower query number indicates higher performance of the attack method and less likelihood of being tracked and targeted. Semantic similarity, using the Universal Sentence Encoder, is an important metric to evaluate the quality of generated adversarial text. A higher semantic similarity indicates stronger semantic consistency and less likelihood of the adversarial text being detected by humans.

Table 1. Performance comparison of DaST-T and other baselines when attacking fine-tuned BERT models.

Dataset	Method	Original Acc (%)	Attacked Acc (%)	Query Number	Cosine Sim
IMDB	TextFooler [8]	93.0	13.6	1134	0.86
	BERT-Attack [11]		11.4	454	0.86
	BBA [9]		6.8	339	0.91
	DaST-T (ours)		5.1	117	0.92
Yelp	TextFooler [8]	97.3	6.6	743	0.74
	BERT-Attack [11]		5.1	273	0.77
	BBA [9]		4.6	319	0.8
	DaST-T (ours)		4.0	44	0.83
AG News	TextFooler [8]	95.1	12.5	357	0.57
	BERT-Attack [11]		10.6	213	0.69
	BBA [9]		6.8	154	0.66
	DaST-T (ours)		6.8	108	0.69

4.4 Attacking Results

As shown in Table 1, DaST-T successfully attacks fine-tuned BERT models and achieves better performance on the IMDB, Yelp, and AG News datasets. In terms of attacked accuracy, DaST-T is able to successfully fool fine-tuned BERT model. Compared with baselines, DaST-T is able to further reduce attacked accuracy, which demonstrates that it has better attack effectiveness. In addition, DaST-T shows an improvement in semantic similarity compared to baselines, which indicates that the adversarial text generated by the DaST-T method better preserves the semanticity of the original text. In particular, the average query number of per sample for DaST-T is significantly lower than that of baselines. The appearance of such a big improvement is due to the fact that DaST-T trains substitute models without the help of any real data.

4.5 Ablations and Discussions

Effects of the Hyperparameter Top-K. Top-k of the generative model is one of the most important hyperparameters of DaST-T. The generative model synthesizes samples through multiple rounds, while top-k refers to the number of words generated by the generative model in each round. As is shown in Fig. 3, attacked accuracy of the target model tends to decrease when the value of top-k is between 1 and 3, while verbs increase since the value of top-k is between 3 and 10. This is because when the value of top-k is either too large or too small, the logicality and quality of synthetic samples become poor, which in turn leads to worse attack effects and an increase in attacked accuracy. As a result, it can be demonstrated that the top-k of the generative model directly influences the performance of DaST-T.

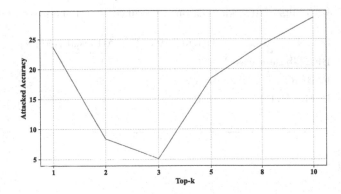

Fig. 3. Attacked accuracy with different top-k of the generative model on IMDB

Effects of Data Augmentation. To improve the performance of DaST-T, data augmentation is introduced to synthetic samples of the generative model. This subsection will explore the effects of data augmentation. The comparison of attacked accuracy of the target model with and without data augmentation on IMDB and Yelp datasets is shown in Table 2. Data augmentation can decrease attacked accuracy of the target model by nearly 2.4% on IMDB, and 1.3% on the Yelp dataset from Table 2. This suggests that introducing data augmentation for the generative model can help improve the quality of synthetic samples used for training, which in turn enhances the performance of DaST-T.

Table 2. Comparison of attacked accuracy with and without data augmentation

Dataset	Method	Original Acc (%)	Attacked Acc (%)
IMDB	Base	93.0	7.5
	Data-augmentation		5.1
Yelp	Base	95.1	5.3
	Data-augmentation		4.0

Evaluation on Various Fine-Tuned BERT Models. To thoroughly assess the transformation of this method, DaST-T is applied to a variety of fine-tuned BERT models. Table 3 shows the results of attacking various fine-tuned BERT models on IMDB. As seen in Table 3, DaST-T decreases the original accuracy of various fine-tuned BERT models from over 90% to less than 6%, which means fine-tuned BERT models are commonly threatened by data-free substitute training attacks.

Table 3. Results of attacking various fine-tuned BERT models on IMDB

Model	Original Acc (%)	Attacked Acc (%)
BERT-base-uncased	93.0	5.1
RoBERTa	91.4	5.8
DistilBERT	92.8	5.2

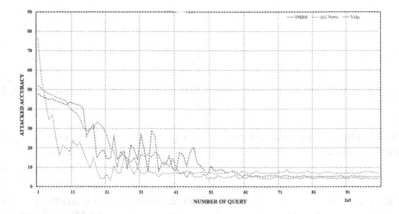

Fig. 4. Attack Accuracy Convergence of DaST-T on IMDB, AG News and Yelp.

Training Convergence. Figure 4 shows the attack accuracy curves of attacked model generated by DaST-T on IMDB, AG News and Yelp. Their attack accuracy converges after 60,000, 35,000 and 50,000 queries, respectively. The convergence speed of DaST-T on different datasets varies because of the average length of samples are not the same. This suggests that the average length of the dataset affects the speed of training convergence.

Examples of Synthetic Samples. This subsection shows some examples of samples synthesized by generative model and gives the output labels and substitute models of the attacked models. As shown in Table 4, the generative model can synthesize samples based on topic words. The topic words are displayed at the beginning of the synthesized text. In addition, for the same synthetic samples, the substitute model can give the same labels as attacked model.

Table 4. Examples of synthetic samples with output labels of the attacked model and substitute model

Synthetic samples	Attacked model	Substitute model
Movie review: The film was a bit of an oddity, but it's still a fun movie to watch. I've seen the trailer and I think it was a great movie. I'm not a huge fan, so it's a little bit like watching a film with a bunch of other people. I think the movie is a great one. I think the story was great, and I'm really glad that the movie is a little more of a story than it was before because it was a lot more of a mystery.	Positive	Positive
Hotel comments: The hotel exceeded all expectations with its impeccable service, luxurious amenities, and views. The staff was attentive and friendly, ensuring a comfortable stay. The rooms were spacious and elegantly furnished, offering a relaxing ambiance. From the moment I stepped in, I knew it would be a memorable experience.	Positive	Positive
Sports news: In a thrilling sports showdown that had spectators on the edge of their seats, Team Lightning made a sensational comeback to secure a remarkable victory against their formidable rivals, Team Thunder, in a nail-biting match. The game, held at the renowned Thunderdome Stadium, was filled with intense action and incredible displays of skill from both sides.	Sports	Sports

5 Conclusion

In this paper, we proposed DaST-T, a substitute training method specifically designed for black-box textual adversarial attacks on BERT. By leveraging a specially designed GAN, DaST-T enables the training of substitute models using synthetic samples in the absence of real data. The incorporation of data augmentation further enhances the training process and improves the accuracy of the substitute model. Experimental results show that DaST-T surpasses the performance of the baselines in terms of attack effectiveness. Notably, further ablation studies reveal that even when the training datasets are well protected, downstream fine-tuned BERT models remain vulnerable to textual adversarial attacks. The contributions of this work include effectively addressing the challenge of acquiring real datasets with labels from the attacked model, which is a major limitation in existing substitute-based attacks. To the best of our knowledge, we are the first to apply the concept of data-free substitute training to textual adversarial attacks. In future work, we will explore the applicability of DaST-T to other NLP tasks and investigate defense strategies against data-free substitute training.

Acknowledgements. The work of this article has been supported by the Key R&D projects in Hubei Province under Grant No.2022BAA041.

References

1. Alzantot, M., Sharma, Y., Elgohary, A., Ho, B.J., Srivastava, M., Chang, K.W.: Generating natural language adversarial examples. arXiv preprint arXiv:1804.07998 (2018)
2. Cheng, M., Yi, J., Chen, P.Y., Zhang, H., Hsieh, C.J.: Seq2Sick: evaluating the robustness of sequence-to-sequence models with adversarial examples. In: Proceedings of the AAAI Conference on Artificial Intelligence, vol. 34, pp. 3601–3608 (2020)
3. Devlin, J., Chang, M.W., Lee, K., Toutanova, K.: BERT: pre-training of deep bidirectional transformers for language understanding. arXiv preprint arXiv:1810.04805 (2018)
4. Ebrahimi, J., Rao, A., Lowd, D., Dou, D.: HotFlip: white-box adversarial examples for text classification. arXiv preprint arXiv:1712.06751 (2017)
5. Fellbaum, C.: WordNet: An Electronic Lexical Database. MIT Press, Cambridge (1998)
6. Garg, S., Ramakrishnan, G.: BAE: BERT-based adversarial examples for text classification. arXiv preprint arXiv:2004.01970 (2020)
7. Guo, C., Sablayrolles, A., Jégou, H., Kiela, D.: Gradient-based adversarial attacks against text transformers. arXiv preprint arXiv:2104.13733 (2021)
8. Jin, D., Jin, Z., Zhou, J.T., Szolovits, P.: Is BERT really robust? A strong baseline for natural language attack on text classification and entailment. In: Proceedings of the AAAI Conference on Artificial Intelligence, vol. 34, pp. 8018–8025 (2020)
9. Lee, D., Moon, S., Lee, J., Song, H.O.: Query-efficient and scalable black-box adversarial attacks on discrete sequential data via Bayesian optimization. In: International Conference on Machine Learning, pp. 12478–12497. PMLR (2022)
10. Li, J., Ji, S., Du, T., Li, B., Wang, T.: TEXTBUGGER: generating adversarial text against real-world applications. arXiv preprint arXiv:1812.05271 (2018)
11. Li, L., Ma, R., Guo, Q., Xue, X., Qiu, X.: BERT-attack: adversarial attack against BERT using BERT. arXiv preprint arXiv:2004.09984 (2020)
12. Liu, S., Lu, N., Chen, C., Tang, K.: Efficient combinatorial optimization for word-level adversarial textual attack. IEEE/ACM Trans. Audio Speech Lang. Process. **30**, 98–111 (2021)
13. Maheshwary, R., Maheshwary, S., Pudi, V.: A strong baseline for query efficient attacks in a black box setting. arXiv preprint arXiv:2109.04775 (2021)
14. Miyato, T., Dai, A.M., Goodfellow, I.: Adversarial training methods for semi-supervised text classification. arXiv preprint arXiv:1605.07725 (2016)
15. Tan, S., Joty, S., Kan, M.Y., Socher, R.: It's morphin'time! Combating linguistic discrimination with inflectional perturbations. arXiv preprint arXiv:2005.04364 (2020)
16. Wang, B., Xu, C., Liu, X., Cheng, Y., Li, B.: SemAttack: natural textual attacks via different semantic spaces. arXiv preprint arXiv:2205.01287 (2022)

17. Wang, X., Jin, H., He, K.: Natural language adversarial attacks and defenses in word level (2019). arXiv preprint arXiv:1909.06723
18. Yu, M., Sun, S.: FE-DaST: fast and effective data-free substitute training for black-box adversarial attacks. Comput. Secur. **113**, 102555 (2022)
19. Zang, Y., et al.: Word-level textual adversarial attacking as combinatorial optimization. arXiv preprint arXiv:1910.12196 (2019)
20. Zhang, X., Zhao, J., LeCun, Y.: Character-level convolutional networks for text classification. Adv. Neural Inf. Process. Syst. **28** (2015)
21. Zhou, M., Wu, J., Liu, Y., Liu, S., Zhu, C.: DAST: data-free substitute training for adversarial attacks. In: Proceedings of the IEEE/CVF Conference on Computer Vision and Pattern Recognition, pp. 234–243 (2020)

Detect Overlapping Community via Graph Neural Network and Topological Potential

Xiaohong Li[✉], Qixuan Peng, Ruihong Li, and Xingjun Guo

College of Computer Science and Engineering, Northwest Normal University, Lanzhou 730070, China
{xiaohongli,2021222181}@nwnu.edu.cn

Abstract. Overlapping community structure is an important characteristic of real complex networks, the goal of the overlapping community detection is to resolve the modular with the information contained in the networks. However, most existing methods based on deep learning techniques directly utilize the original network topology or node attributes, ignoring the importance of various edge information. Inspired by the effective representation learning capability of graph neural network and the ability of topological potential to measure the intimacy between nodes, we propose a novel model, named DOCGT, for overlapping community detection. This model deconstructs the original graph into a first-order graph and a second-order graph, and builds a set of graph neural network modules based on the Bernoulli-Poisson (BP) model, and then uses its advantages to independently learn the node embedding representation of different orders. To this end, we introduce the concept of topological potential matrix. It can not only effectively merge the above embeddings, but also integrate abundant edge information into the entire model. This fused embedding matrix can help us get the final community structure. Experimental results on real datasets show that our method can effectively detect overlapping community structures.

Keywords: Overlapping community detection · Graph neural network · Topological potential matrix

1 Introduction

Community detection, as a classic graph partition problem, reveals the features and connections between different community members. A community is intuitively viewed as a group of nodes with more connections amongst its members than between its members and the remainder of the network [1]. In networks with different functions (such as social networks [2] and biological networks [3]), these groups of nodes (i.e. communities) are often interpreted as different meanings. Through community detection, we can analyze the importance, relevance, and evolution of communities in detail, and identify research trends. It is well

B. Luo et al. (Eds.): ICONIP 2023, CCIS 1968, pp. 308–320, 2024.
https://doi.org/10.1007/978-981-99-8181-6_24

known that real world networks overlap. Although some methods for identifying overlapping and hierarchically nested communities in networks have existed in the past [4–6], most of them can only find the shallow associations.

In recent years, we notice a significant development of graph deep learning techniques for community detection. For example, some approaches attempted to focus on graph representation. To generate smoothing features, AGE [7] introduces a Laplacian smoothing filter, adaptively selects positive and negative samples according to the similarity of node pairs, and uses the final node representation for simple community division. CommDGI [8] applies k-means to node clustering to achieve joint optimization of the linear combination of DGI objectives, mutual information and modularity. All of the above methods are applied to non-overlapping communities, which aim to assign each node to a community. Although graph deep learning has achieved great success in non-overlapping community detection, applying it directly to overlapping communities does not achieve great results. Compared to non-overlapping community detection, overlapping community detection seeks to soft-assign nodes to potentially multiple communities. BIGCLAM [4], as a classic traditional model of overlapping community detection, describes community detection as a variant of non-negative matrix factorization (NMF) to maximize the likelihood of the model. NOCD [9] inherits the community concept of BIGCLAM and is a rare GNN that explicitly optimizes overlapping community detection. It achieves effective community division by maximizing the possibility of Bernoulli-Poisson model. DMoN [10], as an advanced non-overlapping community detection model, has also tried to apply it to overlapping communities due to its nature of maximizing modularity. At the time of writing, UCoDe [11] proposes a state-of-the-art method that can both detect overlapping and non-overlapping communities, and can compete with NOCD, especially in overlapping community detection.

Recently, some GNNs methods that explicitly optimize for community detection have attracted our interests. In particular, graph convolutional network (GCN) act as the first-order local approximation of spectral graph convolution. Each convolutional layer only processes first-order neighborhood information [12]. By stacking several convolutional layers, multi-order information is passed. Based on the sparsity of the real-world network, the advantage of GCN is that it can naturally integrate the node attribute information into the graph for learning, but it ignores the impact of various edge information. Especially in the field of community detection, multi-layer stacking can make GCN results overly smooth and lose the community details [9–11]. To solve these problems, we develop an overlapping community detection model via graph neural network and topological potential (DOCGT). Specifically, we contribute:

1. DOCGT deconstructs the original graph into a first-order graph and a second-order graph, and attempts to independently learn the embedding representation of nodes through a set of graph neural network modules.
2. DOCGT introduces the concept of topological potential matrix, which can not only fuse the node embedding information with different orders, but also integrate the rich edge information into the model.

3. DOCGT shows that it can not only recovers community affiliations accurately, but also reconstructs the overlapping scale of the original community as much as possible.

2   Preliminaries

2.1   Topological Potential

Field theory as a classical mathematical model for describing the non-contact interactions between objects, it can be used to describe the interaction and the association among network nodes. Each node is regarded as a field source, and these nodes interact with each other [13]. Given a graph $G = (V, E)$, where $V = \{v_1, v_2, \cdots, v_N\}$ represent its node set and $E = \{(v_i, v_j), v_i, v_j \in V\}$ represent its edge set, the topological potential of node v_i can be defined as follows:

$$\varphi(v_i) = \sum_{l=1}^{s} [m(v_l) \times \exp(-(\frac{d_{il}}{\sigma})^2)] \tag{1}$$

where v_l indicates the node within the influence scope of node v_i, s represents the total number of nodes within the influence scope, $1 \le s \le n - 1$, $1 \le l \le s$, $m(v_l)$ refers to the mass of node v_l, d_{il} denotes the hops between nodes v_i and v_l, and impact factor σ used to control the influence scope of nodes where the maximum scope is $[3\sigma/\sqrt{2}]$ hops.

2.2   Bernoulli-Poisson Model

The Bernoulli-Poisson (BP) model [4,9] as a traditional graph generative model that allows three types of communities: Non-overlapping, Nested and Overlapping. The model suggests that the emergence of communities is due to the close connections between nodes. Given the affiliations $\boldsymbol{Z} \in \mathbb{R}_{\ge 0}^{N \times C}$, adjacency matrix entries A_{ij} are sampled i.i.d as

$$A_{ij} \sim Bernoulli(1 - \exp(-(\boldsymbol{Z}_i \cdot \boldsymbol{Z}_j^T))) \tag{2}$$

where $\boldsymbol{Z}_i$ is the i's row of the matrix $\boldsymbol{Z}$, which represents the affiliations between node i and each community. Intuitively, the more communities nodes i and j have in common (i.e. the higher the dot product $(\boldsymbol{Z}_i \cdot \boldsymbol{Z}_j^T)$ is), the more likely they are to be connected by an edge.

3   Proposed Model

For the processing of BP models, unlike traditional methods, which regard affiliation matrix $\boldsymbol{Z}$ as a free variable for optimization, the introduction of graph neural networks further improves its processing speed and performance. Inspired by [9], it uses GCN to generate $\boldsymbol{Z}$, which achieves the best performance by stacking two layers. This proves the importance of second-order information from another perspective, but ignores the influence of differences among different types of neighbors on experimental results. To further explore the effect of edge type differences on community detection, our model integrates edge information into the entire node embedding by deconstructing the original graph, and finally realizes the division of overlapping communities.

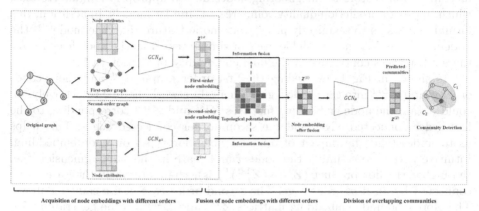

Fig. 1. Overall Framework of the Overlapping Community Detection Model.

3.1   Acquisition of Node Embeddings with Different Orders

Definition 1 (First-order Graph). *Given a original graph $G = (V, E)$, with its node set V and edge set E. Based on this, we generate a first-order graph $G^{1st} = (V, E^{1st})$ on which the edges are represented only as the connections of nodes in the original graph to their first-order neighbors.*

Definition 2 (Second-order Graph). *Given a original graph $G = (V, E)$, with its node set V and edge set E. Based on this, we generate a second-order graph $G^{2nd} = (V, E^{2nd})$ on which the edges are represented only as the connections of nodes in the original graph to their second-order neighbors.*

The traditional BP model links the generation of edges with the affiliation matrix $Z \in \mathbb{R}_{\geq 0}^{N \times C}$. The probability of node connection is proportional to the strength of shared members (the number of public communities of two nodes). In order to support a potentially infinite number of communities and more accurately mine the hidden connections in the original graph, we further extend the dimension of the affiliation matrix $Z \in \mathbb{R}_{\geq 0}^{N \times D}$, which generalizes the probability density in the model. This change further generalizes the notion of affiliations, and the core idea of the entire BP model shifts from looking for more public communities between pairs of nodes to looking for similarities in more dimensions between pairs of nodes. It allows us to take advantage of the BP model in the field of community detection and apply it to node embedding representations.

To implement our idea, we build a set of shallow graph neural network module based on BP model, which can effectively learn node embedding under different order neighbor information. The process of generating Z using GCN is as follows

$$\boldsymbol{Z} := GCN_\theta(\boldsymbol{A}, \boldsymbol{X}) \tag{3}$$

Among them, the input of the module includes an adjacency matrix $\boldsymbol{A}$ and a node attribute matrix $\boldsymbol{X}$. As verified in [9], the GNN outputs similar community affiliation vectors for neighboring nodes due to appropriate inductive bias, which improves prediction quality compared to simpler models. In addition, this formula allows us to seamlessly incorporate node features into the model. If the node attribute $\boldsymbol{X}$ is not available, we can simply use $\boldsymbol{A}$ as the node feature. All these advantages contribute to the final community prediction task.

As shown in Fig 1, by feeding the first-order graph G^{1st} and the second-order graph G^{2nd} into the shallow graph neural network module respectively, we could obtain the node embedding matrices $\boldsymbol{Z}^{1st}$ and $\boldsymbol{Z}^{2nd}$ corresponding to the first-order connected edges and the second-order connected edges. This helps us to understand the impact of different neighbor types on node embedding. Intuitively, the more similar the nodes i and j are in the same dimension (i.e. the higher the dot product $(\boldsymbol{Z}_i^{1st} \cdot (\boldsymbol{Z}_j^{1st})^T)$ is), the more likely they are to be connected by an first-order edge. The second-order case is the same as above. Therefore, the node embedding matrices $\boldsymbol{Z}^{1st}$ and $\boldsymbol{Z}^{2nd}$ can be generated with the shallow graph neural network module, which are expressed as

$$\boldsymbol{Z}^{1st} := GCN_{\theta 1}(\boldsymbol{A}^{1st}, \boldsymbol{X}) \tag{4}$$

$$\boldsymbol{Z}^{2nd} := GCN_{\theta 2}(\boldsymbol{A}^{2nd}, \boldsymbol{X}) \tag{5}$$

Unlike the traditional approaches of directly optimizing the node embedding matrices $\boldsymbol{Z}^{1st}$ and $\boldsymbol{Z}^{2nd}$, we search for neural network parameters θ^1 and θ^2 that minimize the negative log-likelihood

$$L_1(\boldsymbol{Z}^{1st}) = - \log P(\boldsymbol{A}^{1st}|\boldsymbol{Z}^{1st})$$
$$= - \sum_{(i,j)\in E^{1st}} \log(1 - \exp(-\boldsymbol{Z}_i^{1st} \cdot (\boldsymbol{Z}_j^{1st})^T))$$
$$+ \sum_{(i,j)\notin E^{1st}} (\boldsymbol{Z}_i^{1st} \cdot (\boldsymbol{Z}_j^{1st})^T) \tag{6}$$

$$L_1(\boldsymbol{Z}^{2nd}) = - \log P(\boldsymbol{A}^{2nd}|\boldsymbol{Z}^{2nd})$$
$$= - \sum_{(i,j)\in E^{2nd}} \log(1 - \exp(-\boldsymbol{Z}_i^{2nd} \cdot (\boldsymbol{Z}_j^{2nd})^T))$$
$$+ \sum_{(i,j)\notin E^{2nd}} (\boldsymbol{Z}_i^{2nd} \cdot (\boldsymbol{Z}_j^{2nd})^T) \tag{7}$$

For the convenience of description, we simplify it as follows

$$L_1(\widetilde{Z}) = - \log P(\widetilde{A}|\widetilde{Z}) \tag{8}$$

$$\widetilde{\theta} = \arg\min L_1(GCN_{\widetilde{\theta}}(\widetilde{A}, \boldsymbol{X})) \tag{9}$$

where $\widetilde{Z} = \{\boldsymbol{Z}^{1st}, \boldsymbol{Z}^{2nd}\}$ denotes the set of multi-order node embedding matrix, $\widetilde{A} = \{\boldsymbol{A}^{1st}, \boldsymbol{A}^{2nd}\}$ represents the set of multi-order adjacency matrix, $\widetilde{\theta} = \{\theta^1, \theta^2\}$.

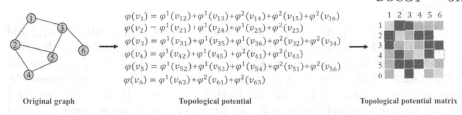

Original graph   Topological potential   Topological potential matrix

Fig. 2. Generation of Topological Potential Matrix without Node Mass.

3.2 Fusion of Node Embeddings with Different Orders

Inspired by the topological potential in the graph, which acts as a short-range field with a limited range of influence over each node, controlled by the parameter σ. Although the best value of σ varies on different datasets, its influence range is always maintained at about two hops [13]. Coincidentally, many GNN-based community detection methods [9,10] also keep their optimal stacking layers around two layers. This further confirms the experimental results of the BP model in graph neural networks. For better integrate $\boldsymbol{Z}^{1st}$ and $\boldsymbol{Z}^{2nd}$, we need to distinguish their importance of each other in the whole network. Topological potential is a traditional method commonly used in community detection, where $\varphi(v_i)$ denotes the sum of the potential of nodes within the influence scope of node v_i, so we try to take it apart and get the topological potential component between each node pair. The details are as follows

$$\varphi(v_{ij}) = [m(v_j) \times \exp(-(\frac{d_{ij}}{\sigma})^2)] \tag{10}$$

Compared to Eq. (1), If node $v_j \in v_l$, its means nodes v_j within the influence scope of node v_i, where $\varphi(v_{ij})$ denotes the topological potential component of node v_i acting on node v_j. Otherwise, its means nodes v_j without the influence scope of node v_i, where $\varphi(v_{ij}) = 0$. Thus, we propose a new concept of the topological potential matrix $\boldsymbol{T} \in \mathbb{R}^{N \times N}$, which consists of topological potential components of interaction between nodes. Obviously, when we ignore the node mass, $\boldsymbol{T}$ is a symmetric matrix. To make the concept of topological potential matrix easier to understand, we re-describe its concept and give a small example to illustrate its generation process.

Definition 3 (Topological Potential Matrix). *Given a original graph $G = (V, E)$, we can obtain the topological potential matrix $\boldsymbol{T} \in \mathbb{R}^{N \times N}$ ($T_{ij} = \varphi(v_{ij})$), which consists of topological potential components of interaction between nodes. Each component in the matrix represents the topological potential value of the two nodes under the interaction force. It is a symmetric matrix.*

As shown in Fig. 2, we show the generation process from the original graph to the topological potential matrix. For example, the topological potential value of node v_1 is equal to the sum of the topological potential components in its field (we control its influence range within two hops). It includes the topological potential components $\varphi(v_{12})$ and $\varphi(v_{13})$ in the first-order range and the topological potential components $\varphi(v_{14})$, $\varphi(v_{15})$ and $\varphi(v_{16})$ in the second-order range. Obviously, according to the definition of topological potential, $\varphi(v_{12})$ and $\varphi(v_{21})$ in the red box represent the topological potential components with different nodes as the center. When considering node mass, $\varphi(v_{12}) \neq \varphi(v_{21})$. In order to generalize the application of the topological potential matrix, we choose the average value of the two. In addition, in this paper we will no longer consider the topological potential components outside the two-hop range (such as $\varphi(v_{46})$ in the orange box).

Then, $\boldsymbol{Z}^{1st}$ and $\boldsymbol{Z}^{2nd}$ are connected by the topological potential matrix $\boldsymbol{T}$ as follows

$$Z^{(1)} = T \times (Z^{1st} + Z^{2nd}) \tag{11}$$

where $\boldsymbol{Z}^{1st} \in \mathbb{R}^{N \times D}$, $\boldsymbol{Z}^{2nd} \in \mathbb{R}^{N \times D}$. $\boldsymbol{Z}^{(1)}$ represents the fused node embedding matrix. The emergence of topological potential matrix can not only realize the fusion of node embeddings with different orders, but also integrate more fine-grained information into the whole model.

3.3   Division of Overlapping Communities

After completing the above two steps, we get the node embedding matrix $\boldsymbol{Z}^{(1)}$ that can more comprehensively and truly reflect the features in original graph. In order to achieve more accurate overlapping community division, we continue to use the advantages of BP model in this regard, and feed the obtained node embedding matrix $\boldsymbol{Z}^{(1)}$ into a novel graph neural network module to get the final community affiliation $\boldsymbol{Z}^{(2)}$

$$Z^{(2)} := GCN_\theta(A, Z^{(1)}) \tag{12}$$

Similar to the first step, we search for neural network parameters θ that minimize the negative log-likelihood

$$\begin{aligned} L_2(Z^{(2)}) &= -\log P(A|Z^{(2)}) \\ &= -\sum\nolimits_{(i,j)\in E} \log(\mathcal{C}(1 - \exp(-Z_i^{(2)} \cdot (Z_j^{(2)})^T))) \\ &\quad + \sum\nolimits_{(i,j)\notin E} (Z_i^{(2)} \cdot (Z_j^{(2)})^T) \end{aligned} \tag{13}$$

$$\theta = \arg\min L_2(GCN_\theta(A, Z^{(2)})) \tag{14}$$

where $\mathcal{C}(\mathcal{C} = \exp(-T_{ij}))$. This is done to alleviate the impact caused by abundant information. We add the penalty term to the original loss function for more accurate community detection. The loss function searches for neural network parameters θ by minimize the negative log-likelihood function.

Therefore, the loss function of our model can be summarized as follows

$$L = L_1 + L_2 \tag{15}$$

4   Experiments

4.1   Experimental Setup

Datasets. For a more comprehensive comparison, we select two widely-used datasets. The detailed statistics are summarized in Table 1. **Facebook** [14] is a collection of small (50–800 nodes) self-networks in the Facebook graph. We selected six networks commonly used in Facebook to evaluate the performance of the proposed framework. **CS** [15] is a co-authorship network, constructed from the Microsoft Academic Graph. Communities correspond to their research areas and node attributes are based on keywords of the papers by each author.

Baselines. We compare the proposed model with four sets of baselines. **BIG-CLAM** [4] as a classic model, is based on the Bernoulli-Poisson model and uses coordinate ascent to learn Z. **NOCD** [9] combines a BP probabilistic model and a two-layer GCN to obtain the final community structure by minimizing the negative log-likelihood of BP and then setting a threshold to continuously identify and remove weak affiliations. **DMoN** [10] is originally proposed for non-overlapping communities, but it can be used for overlapping community detection, because its essence is to maximize modularity through GCN, and it is not associated with disjoint clusters, which can generate soft cluster assignments. **UCoDe** [11] is a GNN for community detection with a single contrastive loss by maximize modularity while explicitly allowing overlaps among communities.

Evaluation Protocol. As mentioned in [9], commonly used metrics to quantify the agreement between true and detected communities, such as Jaccard and F1 scores, can give arbitrarily high scores for community assignments that are completely uninformative. As an alternative, we report overlapping normalized information (**NMI**) [16]. On the basis of retaining **NMI**, in order to be able to better verify the recovery of most models in the overlapping part of the community, we introduce the community overlap rate (**COR**) to evaluate the ability of the model to identify the correct cluster members. It inherits ideas from [17]. As shown in formula Eq.(16)

$$COR = \frac{\sum\limits_{k=1}^{C} |C_k^p|}{\sum\limits_{k=1}^{C} |C_k^{gt}|} \tag{16}$$

where $|C_k^p|$ and $|C_k^{gt}|$ represents the number of nodes in the predicted community k and the ground-truth community k in the graph respectively.

Implementation. The proposed model contains two sets of GCNs. The first set of GCNs is responsible for extracting node embedding representation with different orders, and only one layer of graph convolution structure is set. We

Table 1. Dataset statistics. K stands for 1000. **N** represents the number of nodes, **E** denotes the number of edges, **D** indicates the dimension of the attribute matrix, and **C** expresses the number of communities.

DATASET	Fb-348	Fb-414	Fb-686	Fb-698	Fb-1684	Fb-1912	CS
N	224	150	168	61	786	747	22.0K
E	3.2K	1.7K	1.6K	210	14K	30K	96.8K
D	21	16	9	6	15	29	7.8K
C	14	7	14	13	17	46	18

fix the size of the first layer to be 128 dimensions. The second set of GCNs is responsible for detecting community structure, and we keep the size of its output layer as the number of communities k. We independently sample and process the first-order graph and second-order graph in the first set of GCNs. The learning rate is set to 10^{-3}. Threshold α is kept at 0.5 [18]. In the experiments in this paper, we temporarily choose to ignore the node mass difference in the topological potential, and only discuss the effects of first-order and second-order information. We repeat each experiment 20 times, train the model for 500 epochs, and report the average results for each run.

Table 2. Recovery of ground-truth communities, measured by NMI (in %). Results for **DOCGT** are averaged over 20 initializations. Best performer in bold; second best underlined.

	Fb-348	Fb-414	Fb-686	Fb-698	Fb-1684	Fb-1912	CS
BIGCLAM	21.5	41.6	13.2	38.3	30.1	24.3	0.0
NOCD-X	31.6	50.3	<u>19.5</u>	33.7	29.0	37.6	39.2
DMoN-X	18.2	38.5	10.5	18.3	5.4	23.0	21.5
UCoDe-X	27.1	47.2	**21.9**	23.6	19.2	27.8	23.7
DOCGT-X	30.4	49.2	19.2	42.6	<u>35.5</u>	37.5	41.3
NOCD-G	**33.8**	<u>54.5</u>	16.1	<u>43.4</u>	34.2	<u>39.2</u>	<u>48.7</u>
DMoN-G	24.6	41.3	11.4	38.1	20.7	25.6	30.4
UCoDe-G	22.7	42.6	12.9	40.2	29.5	25.3	32.9
DOCGT-G	<u>32.7</u>	**56.0**	18.6	**45.8**	**39.7**	**40.6**	**51.2**

4.2 Performance of Community Detection

Recovery of Ground-Truth Communities. We provide results for both forms of each algorithm in NOCD, DMoN, UCoDe and DOCGT in Tables 2. BIG-CLAM serves as a reference. The input to GCN in model-X is node attributes;

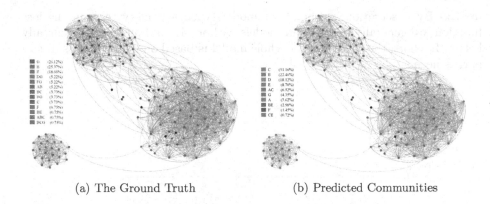

(a) The Ground Truth   (b) Predicted Communities

Fig. 3. Visualization on the Fb-414 Dataset.

the input to model-G is the concatenation of adjacency attributes and node attributes. Table 2 shows how well different methods recover the ground-truth communities. We find that DOCGT-G performs a little better, followed by NOCD-G in most datasets, and our method generally outperforms the second-best method by one to three percentage points in overlapping NMI values. Figure 3 shows the community detection results of our model on the Fb-414 network.

Importance of Edge Information of Different Orders and Penalty Terms. To better verify the importance of edge information of different orders, we conduct an ablation study on the dataset, and the results are listed in Table 3. As an effective overlapping community method based on BP model and graph neural network, NOCD is used as a control group to verify the effectiveness of our model. Our model attempts to incorporate more rich edge information in the graph to improve the detection accuracy of overlapping communities. Overall, we find that this move will lead to fluctuations in the accuracy of the results, especially on some small datasets. This is why we introduce the penalty term $\mathcal{C}$. The entire penalty term is constructed on the basis of the topological potential

Table 3. Ablation study on Datasets. Subscript p means penalty term.

	Fb-348	Fb-414	Fb-686	Fb-698	Fb-1684	Fb-1912	CS
NOCD-X	31.6	50.3	**19.5**	33.7	29.0	37.6	39.2
DOCGT-X	29.6	47.3	18.5	37.9	34.2	35.9	39.8
DOCGT-X$_p$	30.4	49.2	<u>19.2</u>	42.6	<u>35.5</u>	37.5	41.3
NOCD-G	**33.8**	<u>54.5</u>	16.1	43.4	34.2	<u>39.2</u>	<u>48.7</u>
DOCGT-G	31.6	50.5	17.1	<u>44.3</u>	34.9	38.4	48.6
DOCGT-G$_p$	<u>32.7</u>	**56.0**	18.6	**45.8**	**39.7**	**40.6**	**51.2**

matrix. By observation, we find that models with a penalty term in the loss function are generally better than models without it, and model-G is generally better. In short, the success of the whole model is based on many aspects. Every part is necessary.

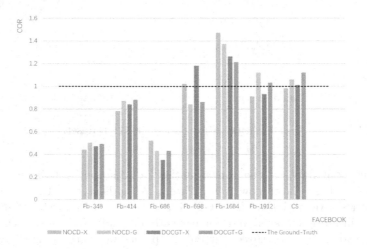

Fig. 4. COR Index of NOCD and DOCGT on Datasets. (The closer the COR is to the ground truth, the better the scale of overlapping communities can be restored.)

4.3   Significance of Recovering Overlapping Scale

As we mentioned earlier, a good overlapping community detection algorithm must be able to restore the original graph structure to the greatest extent possible, whether it is the affiliation of the community or the scale of the overlap. In the experimental test, how to measure a good community affiliation often corresponds to a higher overlapping NMI value, and a good overlap scale needs to be able to truly show the gap between the predicted and the ground truth scale, which is why we introduce COR as an evaluation indicator. As shown in Fig. 4, we plot the COR curves of the best models (DOCGT and NOCD) based on the conclusions in Table 3. We find that when real-world networks are small, sparse information leads to gaps in the recovery performance (especially the three datasets Fb-686, Fb-689, and Fb-1684). When the network scale is gradually expanded, various models based on graph neural network can achieve good results.

5   Conclusion

In this paper, we propose an overlapping community detection model via graph neural network and topological potential (DOCGT). It independently learns node embeddings of different orders by deconstructing the original graph, and introduces a topological potential matrix to integrate edge information into the entire embedding information, and finally realizes the division of overlapping communities through graph neural networks. Experimental results show that the fusion of edge information helps the whole method to perform well on overlapping NMI metrics and outperform the state-of-the-art methods.

References

1. Su, X., et al.: A comprehensive survey on community detection with deep learning. IEEE Trans. Neural Netw. Learning Syst., 1–21 (2022)
2. Wu, L., Zhang, Q., Chen, C.H., Guo, K., Wang, D.: Deep learning techniques for community detection in social networks. IEEE Access **8**, 96016–96026 (2020)
3. Rahiminejad, S., Maurya, M.R., Subramaniam, S.: Topological and functional comparison of community detection algorithms in biological networks. BMC Bioinform. **20**(1), 1–25 (2019)
4. Yang, J., Leskovec, J.: Overlapping community detection at scale: a nonnegative matrix factorization approach. In: Proceedings of the sixth ACM International Conference on Web Search and Data Mining, pp. 587–596 (2013)
5. Xie, J., Kelley, S., Szymanski, B.K.: Overlapping community detection in networks: the state-of-the-art and comparative study. ACM Comput. Surv. (CSUR) **45**(4), 1–35 (2013)
6. Ni, L., Luo, W., Zhu, W., Hua, B.: Local overlapping community detection. ACM Trans. Knowl. Discov. Data (TKDD) **14**(1), 1–25 (2019)
7. Cui, G., Zhou, J., Yang, C., Liu, Z.: Adaptive graph encoder for attributed graph embedding. In: Proceedings of the 26th ACM SIGKDD International Conference on Knowledge Discovery & Data Mining, pp. 976–985 (2020)
8. Zhang, T., Xiong, Y., Zhang, J., Zhang, Y., Jiao, Y., Zhu, Y.: CommDGI: community detection oriented deep graph infomax. In: Proceedings of the 29th ACM International Conference on Information & Knowledge Management, pp. 1843–1852 (2020)
9. Shchur, O., Günnemann, S.: Overlapping community detection with graph neural networks. Comput. Sci. **50**(2.0), 49–2 (2019)
10. Tsitsulin, A., Palowitch, J., Perozzi, B., Müller, E.: Graph clustering with graph neural networks. Methods **20**, 31 (2023)
11. Moradan, A., Draganov, A., Mottin, D., Assent, I.: UCoDe: unified community detection with graph convolutional networks. arXiv preprint arXiv:2112.14822 (2021)
12. Welling, M., Kipf, T.N.: Semi-supervised classification with graph convolutional networks. In: J. International Conference on Learning Representations (ICLR 2017) (2016)
13. Zhi-Xiao, W., Ze-chao, L., Xiao-fang, D., Jin-hui, T.: Overlapping community detection based on node location analysis. Knowl.-Based Syst. **105**, 225–235 (2016)
14. Mcauley, J., Leskovec, J.: Discovering social circles in ego networks. ACM Trans. Knowl. Discov. Data (TKDD) **8**(1), 1–28 (2014)

15. Wang, K., Shen, Z., Huang, C., Wu, C.H., Dong, Y., Kanakia, A.: Microsoft academic graph: when experts are not enough. Quant. Sci. Stud. **1**(1), 396–413 (2020)
16. Jing, B., Park, C., Tong, H.: HDMI: high-order deep multiplex infomax. In: Proceedings of the Web Conference 2021, pp. 2414–2424 (2021)
17. Chen, Z., Li, L., Bruna, J.: Supervised community detection with line graph neural networks. In: International Conference on Learning Representations (2020)
18. He, H., Garcia, E.A.: Learning from imbalanced data. IEEE Trans. Knowl. Data Eng. **21**(9), 1263–1284 (2009)

DeFusion: Aerial Image Matching Based on Fusion of Handcrafted and Deep Features

Xianfeng Song[1], Yi Zou[1(✉)], Zheng Shi[1], Yanfeng Yang[1], and Dacheng Li[2]

[1] South China University of Technology, Guangzhou, China
zouyi@scut.edu.cn
[2] Gosuncn Technology Group CO., LTD, Guangzhou, China

Abstract. Machine vision has become a crucial method for drones to perceive their surroundings, and image matching, as a fundamental task in machine vision, has also gained widespread attention. However, due to the complexity of aerial images, traditional matching methods based on handcrafted features lack the ability to extract high-level semantics and unavoidably suffer from low robustness. Although deep learning has potential to improve matching accuracy, it comes with the high cost of requiring specific samples and computing resources, making it infeasible for many scenarios. To fully leverage the strengths of both approaches, we introduce DeFusion, a novel image matching scheme with a fine-grained decision-level fusion algorithm that effectively combines handcrafted and deep features. We train generic features on public datasets, enabling us to handle unseen scenarios. We use RootSIFT as prior knowledge to guide the extraction of deep features, significantly reducing computational overhead. We also carefully design preprocessing steps by incorporating drone attitude information. Eventually, as evidenced by our experimental results, the proposed scheme achieves an overall 2.5–6x more correct matches with improved robustness when compared to existing methods.

Keywords: Feature Fusion · Image Matching · Neural Network

1 Introduction

In recent years, the advancements in *Unmanned Aerial Vehicle* (UAV) technology have led to its gradual integration into various national economic industries around the world, such as security, agriculture and logistics. In addition, advances in both vision sensors and image processing technology have established machine vision as the fundamental method for UAVs to perceive their surroundings. Particularly, image matching is considered essential in the field of machine vision for object detection [1,2] and image stitching [3,4].

Handcrafted feature based image matching algorithms are mainly based on expert knowledge and provide strong interpretability. Nevertheless, they lack the ability to extract high-level features that are especially important in tasks such

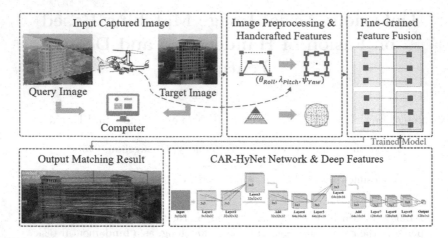

Fig. 1. Overall system architecture. By inputting the *Target Image* and the *Query Image* captured by the drone into the computer, the process begins with *Image Preprocessing and Handcrafted feature extraction* (Section §3). Subsequently, CAR-HyNet is employed to extract deep features (Section §4). Finally, a fine-grained feature fusion algorithm merges and matches the two types of features, yielding the final result (Section §5).

as aerial images, which are often affected by illumination, attitude, rotation and other factors. On the other hand, the recent developments of image matching algorithms based on deep learning have dramatically improved performance and matching accuracy due to their strong feature extraction capability for complex features such as morphology and texture [5]. However, it requires a large amount of specific samples and computing resources for training and inference, which greatly limits its application. It is therefore popular in many fields to combine traditional machine vision techniques with deep learning. This is especially useful where fast implementation is required to provide more reliable feature point matching pairs in image matching [6].

We summarize our major contributions of this paper as below.

(1) We *design* a preprocessing method using drone attitude information. For 2D objects, we take advantage of the drone attitude information to perform an inverse perspective transformation. This improves feature detection while avoiding high latency of simulated perspective transformation.

(2) We *propose* a novel deep learning architecture, named the *Coordinate Attention Residual HyNet* (CAR-HyNet), based on the HyNet [7] architecture. By incorporating coordinate attention, sandglass structure, and residual structure, we effectively enhance the performance of the model.

(3) We *introduce* a novel approach for fusing handcrafted and deep features at decision level. We use RootSIFT [8] to generate handcrafted features and use them as prior knowledge of CAR-HyNet to extract image patches and generate deep features. Finally, we use a fine-grained fusion method to efficiently fuse these two features.

The architecture of the system is illustrated in Fig. 1. Typically, the drone hovers in the air, takes a picture of the ground as *Query Image* and stores current attitude information in *Exchangeable Image File Format* (Exif). The computer on the ground takes the *Target Image* and *Query Image* as inputs to perform the image matching task using the method proposed in this paper.

More precisely, this paper is organized as follows. We provide a detailed analysis of the related work in Sect. 2. In Sect. 3, we describe the image preprocessing. Next, in Sect. 4, we go through details of the proposed CAR-HyNet network, followed by the feature fusion method in Sect. 5. We describe experimental setup and comparative studies in Sect. 6. We further share our thoughts regarding the limitations of current work and areas for future exploration in Sect. 7. Finally, we conclude this paper in Sect. 8.

2 Related Work

2.1 Image Matching

Image matching is the process of comparing two images with or without rotation and scaling by a specific algorithm to find the regions with the greatest similarity, in order to determine the geometric relationship between two images [9]. Region-based image matching algorithms perform image matching by comparing differences directly at pixel level or converting images to other information domains for similarity matching. Feature-based image matching algorithms have been widely studied due to their ability to reduce the impact of noise or deformation by selecting invariant features or significant regions for matching.

2.2 Using Handcrafted Features

Handcrafted feature-based image matching is generally divided into three steps: feature point extraction, feature descriptor generation, and feature matching and filtering. Lowe et al. [10] propose the *Scale Invariant Feature Transform* (SIFT) algorithm in 1999. The algorithm uses the Difference-of-Gaussian (DoG) method to approximate LoG, which speeds up feature extraction. The algorithm has invariance to scaling, rotation and translation, as well as a certain degree of illumination and affine invariance.

Many advances have been made based on the SIFT. For instance, the SURF [11] incorporates box filtering and image integration to accelerate gradients. The FAST detector is suitable for real-time video processing, while the BRIEF [12] employs binary descriptors. The ORB [13] is invariant to rotation and scale, and the KAZE [14] preserves edge information. Although these algorithms have improved detection speed, SIFT is still widely used in practice due to its advantages in invariance and robustness to illumination and affine transformations [15].

2.3 Using Deep Features

Deep learning can extract higher level semantic features from images compared to handcrafted features and has been applied in image matching. Verdie et al. [16]

introduce the *Temporally Invariant Learned Detector* (TILDE), which demonstrates robustness against changes in lighting and weather conditions. Yi et al. [17] propose an end-to-end algorithm called the *Learned Invariant Feature Transform* (LIFT). Tian et al. [18] show that L2-Net can be combined directly with SIFT by matching features at L2 distance. HardNet [19] is proposed based on L2-Net to maximize the distance between the nearest positive and negative samples. Subsequently, Luo et al. [20] introduce the geometric similarity measure and propose the *Geometry Descriptor* (GeoDesc). Tian et al. [21] propose the SOSNet by introducing second order constraints into feature descriptors and the HyNet [7] which further enhances feature representation.

2.4   Combining Handcrafted Features with Deep Features

In recent years, researchers have focused on the relationship between handcrafted and deep features [22]. Combining multiple features can often achieve superior performance over a single feature. Barroso et al. [23] propose *Key.Net*, which combines handcrafted features with CNN. Zhou et al. [24] combine CNN with color feature HSV, shape feature HOG, and local feature SIFT for image classification. Rodriguez et al. [25] propose SIFT-AID by combining SIFT and CNN to produce affine invariant descriptors. However, the proposed algorithm is time consuming due to simulated perspective transformations. Song et al. [26] propose a multi-data source deep learning object detection network (MS-YOLO) based on millimeter-wave radar and vision fusion. Nevertheless, most existing methods focus on the direct combination of handcrafted and deep features, which inevitably leads to inferior results after feature fusion.

3   Attitude-Oriented Image Preprocessing

In aerial scenes, the UAV may be at a tilt angle, causing the shape of the target in the oblique image to undergo geometric changes compared to the rectified image, resulting in perspective transformation. Most feature matching algorithms are not robust to perspective transformation. A more classical and widely used approach is to simulate perspective transformation by generating multiple images for matching [27], as shown in Fig. 2. Although the number of feature matches under perspective changes can be greatly improved by matching multi-view images separately, it is inefficient and imposes high latency.

Note that two images in space can be transformed using a transformation matrix and UAV attitude information is available. Therefore, for image matching of 2D targets, we propose to correct the oblique image to a bird's eye view using the attitude based inverse perspective transformation to improve the performance of feature point extraction and matching. More importantly, this approach does not incur high latency from simulating viewpoints as it only performs transformation and matching once. However, for 3D object image matching, inverse perspective transformation is less effective and therefore not recommended.

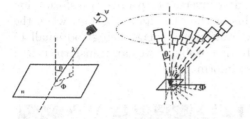 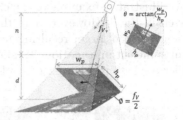

Fig. 2. Simulate perspective transformation [28].

Fig. 3. Perspective transformation model.

We provide an analytical model of perspective transformation. Suppose that $F(w_p, h_p, \theta, \phi, f_V)$ represents the perspective transformation matrix, where w_p and h_p are the pixel width and pixel height of the image, θ is the rotation of the camera, ϕ is the rotation of the image plane, and f_V is the vertical perspective. Figure 3 shows the aforementioned variables and their relationships.

With the center of the image as the center of the circle, upward as the positive direction of the y-axis, rightward as the positive direction of the x-axis, and inward as the positive direction of the z-axis, the coordinates of the four endpoints of the image are defined as: $(-\frac{w_p}{2}, \frac{h_p}{2}, 0)$, $(\frac{w_p}{2}, \frac{h_p}{2}, 0)$, $(\frac{w_p}{2}, -\frac{h_p}{2}, 0)$, $(-\frac{w_p}{2}, -\frac{h_p}{2}, 0)$. Define the perspective transformation matrix F as below,

$$F = PTR_\phi R_\theta, \tag{1}$$

where R_θ is the rotation matrix around the z-axis, R_ϕ is the rotation matrix around the x-axis, T is the translation matrix that moves the coordinate system along the z-axis, and P is the projection matrix of the vertical field of view f_V.

To calculate T, we define $d = \sqrt{w_p^2 + h_p^2}$ as the side length of the square containing any rotated portion of the image. As shown in Fig. 3, using f_V from camera parameters, we calculate $h = \frac{d}{2\sin(\frac{f_V}{2})}$, which describes the degree of translation of the object along the negative z-axis. Thus, the matrix T is

$$T = \begin{bmatrix} 1 & 0 & 0 & 0 \\ 0 & 1 & 0 & 0 \\ 0 & 0 & 1 & -h \\ 0 & 0 & 0 & 1 \end{bmatrix} = \begin{bmatrix} 1 & 0 & 0 & 0 \\ 0 & 1 & 0 & 0 \\ 0 & 0 & 1 & -\frac{d}{2\sin(\frac{f_V}{2})} \\ 0 & 0 & 0 & 1 \end{bmatrix}. \tag{2}$$

Correspondingly, the projection matrix P is given by

$$P = \begin{bmatrix} \cot\left(\frac{f_V}{2}\right) & 0 & 0 & 0 \\ 0 & \cot\left(\frac{f_V}{2}\right) & 0 & 0 \\ 0 & 0 & -\frac{(f+n)}{f-n} & -\frac{2fn}{f-n} \\ 0 & 0 & -1 & 0 \end{bmatrix}, \tag{3}$$

where $n = h - \frac{d}{2}$ and $f = h + \frac{d}{2}$. The perspective transformation matrix can be obtained by substituting Equation (3) into Equation (1).

An illustrative example of the effect after inverse perspective transformation is given in Fig. 4. However, note that this method has its limitations when the tilt angle is too large, resulting in the transformed image being too small to retain sufficient information. To alleviate this problem, we can transform it in a smaller angle to prevent excessive loss of information.

(a)   (b)   (c)   (d)

Fig. 4. Inverse perspective transformation at (a) 0°, (b) 30°, (c) 45°, and (d) 60°.

4   Design and Improvement of the CAR-HyNet Network

By combining handcrafted features with deep features, more accurate and adaptable features can be extracted and described. The convolutional network HyNet [7] evolves from L2-Net [18] and introduces a new triplet loss function from the perspective of optimizing feature descriptors, which makes the image match up to the state-of-the-art.

4.1   CAR-HyNet Network Structure

To address the challenges in aerial image processing, we propose a new improved multi-channel *Coordinate Attention Residual HyNet* (CAR-HyNet) network based on HyNet, as shown in Fig. 5. More precisely, we introduce *Coordinate Attention* (CoordAtt) [29] and design a *Coordinate Attention Sandglass Network* (CA-SandGlass), and modify HyNet to apply CA-SandGlass for aerial image processing. In addition, we take full advantage of RGB three-channel as inputs to further improve overall image matching performance.

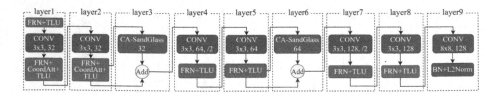

Fig. 5. Overall network structure of CAR-HyNet.

Coordinate Attention Sandglass Network. We notice that conventional convolutional operations can only capture local positional relationships, a significant drawback for processing aerial images from UAVs. To address this limitation, we propose using *Coordinate Attention* (CoordAtt) [29] by embedding position information into channel attention to capture remote dependencies for accurate feature descriptors.

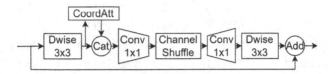

Fig. 6. Structure of CA-SandGlass.

Furthermore, since SandGlass [30] is a lightweight module. Considering that CoordAtt focuses on long-range dependencies and SandGlass focuses on feature information at different scales, combining these two techniques allows the network to generate more comprehensive and discriminative feature representations. Additionally, as the residual connection needs to be built on high-dimensional features, we further combine them to form the CA-SandGlass block. The structure is shown in Fig. 6.

Increasing Nonlinearity of the Model. The original HyNet structure is the same as L2-Net, which consists of 6 feature extraction layers and 1 output layer. In contrast, we propose adding 2 layers of CA-SandGlass to increase non-linearity. Our experiments indicate that adding more than 2 layers of CA-SandGlass does not result in any improvement in matching accuracy. To avoid potential adverse effects such as gradient dispersion, we connect the 2 CA-SandGlass layers to the backbone using a residual connection. Overall, this design offers the best balance between performance and complexity.

RGB Three-Channel Image Input. Another important improvement we introduced in CAR-HyNet is leveraging the full RGB three-channel as inputs. Compared to grayscale images, color images contain much richer information at a negligible computational cost. The absence of color information in processing can result in incorrect matching, particularly in regions with the same grayscale and shape but different colors.

4.2 CAR-HyNet Performance Evaluation

To evaluate the performance of the proposed CAR-HyNet, we compare it with several widely used models. For fairness, we use the unified Brown [31] dataset for evaluation, which includes Liberty (LIB), Notre Dame (ND), and Yosemite

(YOS), and experimental results for existing models are taken from their papers. We perform standard *False Positive Rate at 95%* (FPR95) measurements across 6 training and test sets, as shown in Table 1. As we can see, CAR-HyNet outperforms the other models with a notable improvement in detection performance.

Table 1. Patch verification performance on the Brown dataset (FPR@95) [7].

Train	ND	YOS	LIB	YOS	LIB	ND	Mean
Test	LIB		ND		YOS		
SIFT [32]	29.84		22.53		27.29		26.55
TFeat [33]	7.39	10.13	3.06	3.80	8.06	7.24	6.64
L2-Net [18]	2.36	4.70	0.72	1.29	2.57	1.71	2.23
HardNet [19]	1.49	2.51	0.53	0.78	1.96	1.84	1.51
DOAP [34]	1.54	2.62	0.43	0.87	2.00	1.21	1.45
SOSNet [21]	1.08	2.12	0.35	0.67	1.03	0.95	1.03
HyNet [7]	0.89	**1.37**	0.34	0.61	0.88	0.96	0.84
CAR-HyNet	**0.77**	1.53	**0.30**	**0.57**	**0.69**	**0.64**	**0.75**

5   Decision Level Fusion for Image Matching

One challenge we face in this work is how to fuse handcrafted features with deep features appropriately. The vast majority of the existing literature focuses on feature level fusion using weighted fusion of multiple features. For example, color features, corner features, histogram features, and convolution features of the image are fused directly using different weights. However, directly weighting different features for superposition ignores the fact that different features have different degrees of sensitivity in different scenarios. This inescapably leads to poor performance as some features inadvertently suppress others, introducing a large number of incorrect matches.

To this end, we propose a new fine-grained decision level fusion method that combines handcrafted features with deep features. The method fully considers the correlation of different feature extraction methods on feature points, effectively improving the number of correct matching pairs.

5.1   Extracting Handcrafted Features

To prepare for decision level fusion, we first extract handcrafted features using the RootSIFT algorithm [8]. RootSIFT is an extended mapping algorithm to the *Scale-Invariant Feature Transform* (SIFT) algorithm, but achieves a higher number of correct matches of feature descriptors. In our tests with DEGEN-SAC and NNDR=0.85, the number of correct matches of feature descriptors is improved by approximately 19.4% after RootSIFT mapping.

5.2   Extracting Deep Features

To achieve fine-grained control over the correspondence of feature points during the fusion process and reduce the computational workload of deep learning, we use handcrafted features as prior knowledge for deep feature extraction. We first reconstruct the scale pyramid of the color image based on the image processed in the previous section and feature points extracted by RootSIFT. We then generate patches by intercepting the image at a size of 64 × 64 in the corresponding scale space and then scaling them down to 32 × 32.

Since data augmentation techniques can introduce noisy data and CNNs are not invariant to rotation [35], we further rotate patches to primary orientation. Finally, we feed patches into CAR-HyNet to eventually generate 128-dimensional deep features. This method fully leverages the feature points extracted by Root-SIFT, mitigates the lack of rotational invariance in CNNs, and successfully generates deep features for fusion.

5.3   Fine-Grained Decision Level Feature Fusion

Next, we present the design of the proposed algorithm for decision level fusion. As a high level fusion, decision level fusion offers global optimal decision with high accuracy and flexibility [36].

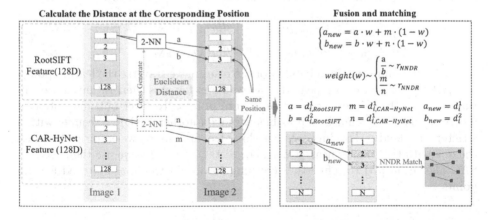

Fig. 7. Fine-grained Decision level fusion using Euclidean distance. *NNDR* represents *Nearest Neighbor Distance Ratio*, *2-NN* represents the nearest neighbor and second-nearest neighbor based on Euclidean distance.

An illustrative example of the algorithm flow is shown in Fig. 7. For a pair of input images to be matched, denoted as *Image1* and *Image2*, we first extract their RootSIFT feature descriptors $D^1_{RootSIFT}$, $D^2_{RootSIFT}$, and then extract the CAR-HyNet feature descriptors $D^1_{CAR-HyNet}$, $D^2_{CAR-HyNet}$ with rotation and scale invariance, respectively. We calculate the Euclidean distance between the feature descriptors of each feature point in the two images under RootSIFT and CAR-HyNet respectively, and calculate the two nearest points, $d_{i,RootSIFT}^m$ and $d_{i,CAR-HyNet}^m$, for each feature point, where $m = 1, 2$ indicates the first and second nearest neighbors. For the two nearest neighboring points of each feature point, we find the distances of these two points at the corresponding positions of the CAR-HyNet feature points by traversing the RootSIFT feature points in turn. We then use the NNDR method to determine the success of the matching from the two feature extraction algorithms.

To retain more implicit matching point pairs, we use the NNDR threshold α with a lenient strategy. The Euclidean distance d of the feature descriptor is calculated in Equation (4), where dim represents the dimension of the feature descriptor, $D_{type,k}^i$ represents the k-th dimensional value of the feature descriptor of the i-th feature point of $type = RootSIFT, CAR - HyNet$, m represents the m-th closest distance. When $i = j$, it is further simplified as $d_{i,type}^m$.

$$d_{ij,type}^m = \sqrt{\sum_{k=1}^{dim}(D_{type,k}^i - D_{type,k}^j)} \tag{4}$$

Furthermore, using the weight $w \in [0, 1]$ to fuse the Euclidean distances of two points and generate new distances as the nearest neighbor $d_{i\_new}^1$ and second nearest neighbor distance $d_{i\_new}^2$ of the feature point. In our experiments, we set w to 0.75. The fusion equation is calculated as Equation (5).

$$\begin{cases} d_{i\_new}^1 = d_{i,RootSIFT}^1 \cdot w + d_{i,CAR-HyNet}^1 \cdot (1-w) \\ d_{i\_new}^2 = d_{i,RootSIFT}^2 \cdot w + d_{i,CAR-HyNet}^2 \cdot (1-w) \end{cases} \tag{5}$$

To further improve the reliability of the results, we traverse the feature points of CAR-HyNet again to repeat the above steps for cross-generation and eventually merge them into a new set of feature points. To filter out any potential incorrect matches as much as possible, we take another NNDR screening with a stricter threshold α. For duplicate matching feature points, we empirically retain the one with the smaller nearest neighbor distance. Finally, we use the DEGENSAC algorithm [37] to refine the screening to obtain the final matched feature points.

5.4   Fine-Grained Decision Level Feature Fusion Evaluation

We present our evaluation of the proposed decision level fusion method. We test the performance of different algorithms under direct weighted fusion and our method. For a fair comparison, we select RootSIFT+HyNet, RootSIFT+HardNet and RootSIFT+CAR-HyNet for our experiments at different heights and perspectives. The results are shown in Table 2.

Table 2. Correct matching numbers of different feature fusion methods with different heights and perspectives (NNDR threshold=0.85, fusion weight=0.75), where DWF refers to traditional *Direct Weighted Fusion* and DLF refers to our proposed *Decision Level Fusion*. Note that due to limited space, we only show results of 9 typical images of different heights and perspectives respectively.

Algorithm	Method	Heights						Perspectives					
		d1	d3	d5	d6	d7	d9	a1	a3	a5	a7	a9	a15
HardNet	DWF	1013	196	87	48	35	18	295	165	189	160	109	20
	DLF	**1086**	**230**	**94**	**56**	**44**	**20**	**351**	**216**	**236**	**206**	**175**	**85**
HyNet	DWF	1003	192	83	47	35	18	296	170	188	159	108	23
	DLF	**1073**	**227**	**93**	**57**	**41**	**22**	**358**	**221**	**238**	**208**	**181**	**84**
CAR-HyNet	DWF	1027	216	91	57	35	18	400	275	267	223	208	32
	DLF	**1099**	**244**	**109**	**64**	**46**	**22**	**483**	**341**	**308**	**269**	**292**	**89**

As can be seen from Table 2, the improved feature fusion method shows excellent performance with all three algorithms. Moreover, our proposed RootSIFT+CAR-HyNet method outperforms other methods and yields a higher number of correct matches.

6   Experiments

In this section, we first describe our experimental setup, datasets, and evaluation criteria. We conduct matching experiments on real world aerial images, and compare the proposed method with existing methods in detail.[1]

Since CNNs are not rotation invariant, which leads to detect more matching feature points only when two images are similar. To make the results more comparable, patches in the following experiments are generated after the operations described in Sect. 5.2. The utilization of different combinations of algorithms is presented in Table 3.

Table 3. Combination of different algorithms

Algorithm	Feature points	Feature descriptor	Image pyramid
SIFT	SIFT	SIFT	Gray
KAZE	KAZE	KAZE	Gray
KeyNet	KeyNet	KeyNet	Gray
HardNet	RootSIFT	HardNet	Gray
HyNet	RootSIFT	HyNet	Gray
CAR-HyNet	RootSIFT	CAR-HyNet	Color

---

[1] Our code is publicly available on Github: https://github.com/songxf1024/DeFusion.

6.1 Lab Setup and Datasets

For the experimental setup, we use a Supermicro server in our lab with an Intel(R) Xeon(R) Gold 6230 CPU, NVIDIA RTX 3090 GPU, and 128GB of memory. The software environment consists of Ubuntu 20.04, Python 3.7 and Pytorch framework. We use a DJI Mini2 to capture aerial images over an open area and the target object in the target detection task is a cloth of size 100cm×177cm. The original images captured by the DJI Mini2 are 4000×3000 in size and compressed to 800×600. Due to space limitations, we only display a selection of typical images, as shown in Fig. 8.

We choose the widely used Brown [31] and HPatches [38] datasets for the joint training of CAR-HyNet. The Brown dataset includes three sub-datasets: Liberty (LIB), Notre Dame (ND), and Yosemite (YOS). We employ standard *False Positive Rate at 95%* (FPR95) as the evaluation metric. In addition, we employ *Nearest Neighbor Distance Ratio* (NNDR) as the feature matching method. For practical drone applications, we set the NNDR threshold at 0.85. Meanwhile, to prevent too many feature points, we employ a feature intensity filter to retain the first 4000 feature points.

(a) *Target detection* at different perspectives in image a1, a7 and a17

(b) *Target detection* at different heights in image d1, d5 and d9

(c) *Target detection* at different rotations in image r2, r5 and r7

(d) *Building* at different perspectives in image h2, h5 and h9

(e) *Aerial* at different perspectives in image ha1, ha5 and ha9

(f) *Aerial* at different rotations in image hr1, hr2 and hr6

Fig. 8. Samples from aerial image datasets, captured by the DJI Mini2.

6.2 Impact of NNDR Thresholds and Fusion Weights

In our experiments, we observe that different NNDR thresholds and feature fusion weights have a significant impact on matching performance. To understand the potential correlations, we conduct a series of experiments using a typical set of images from the dataset. Figure 9 represents the trend of the number of matches and accuracy of CAR-HyNet with different weights ($\frac{RootSIFT}{CAR-HyNet} = \frac{w}{1-w}$), where *fine* represents the number of correct matches, and *coarse* represents the maximum number of matches including incorrect matches. In addition, Fig. 10 represents the influence of different algorithms by NNDR thresholds in perspective and height scenarios. To ensure practical applicability for UAVs, we empirically set the NNDR threshold at 0.85 and the feature fusion weight at 0.75 after extensive experimental comparisons, as it provides an appropriate balance between fusing handcrafted and deep features, enabling us to maximize the number of correct matches while achieving high matching accuracy as much as possible.

Another observation we find is that, as shown in Fig. 10(b), the matching accuracy of the proposed method is highest when the NNDR threshold is set to a small value, and then decreases as the NNDR threshold increases. This can be explained by the fact that with an increased threshold, the proposed method is able to include more potential matching points. Therefore, the matching accuracy begins to decrease even though the number of correct matches remains maximal.

6.3   Time Consumption Evaluation

In addition, we evaluate the time consumption of different algorithms with a maximum limit of 4000 feature points, as shown in Table 4. We use *a0* and *a4* from our dataset for the evaluation. Experiments show that the proposed method offers the best performance in time and number of correct matches. As shown in Table 4, it takes 5500ms for ASIFT to detect 260 correct matches, while the proposed method takes 1096ms to detect 300 correct matches. Also note that other methods in comparison take about 1100ms, but with less than 120 correct matches. By analyzing our method, we also note an interesting observation, where the stages of *generate patches* and *feature matching* accounted for most of the time, as shown in Table 5.

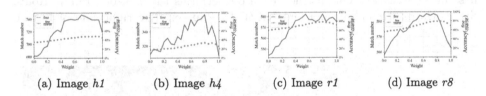

(a) Image *h1*	(b) Image *h4*	(c) Image *r1*	(d) Image *r8*

Fig. 9. Number of matches and matching accuracy at different feature fusion weights ($\frac{RootSIFT}{CAR-HyNet} = \frac{w}{1-w}$), where *fine* is the number of correct matches and *coarse* is the maximum number of matches including incorrect ones.

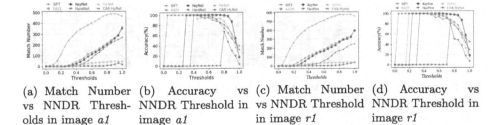

(a) Match Number vs NNDR Thresholds in image *a1* (b) Accuracy vs NNDR Threshold in image *a1* (c) Match Number vs NNDR Threshold in image *r1* (d) Accuracy vs NNDR Threshold in image *r1*

Fig. 10. Performance on sample image *a1* and *r1* at different NNDR thresholds.

Table 4. Time-consumption and number of correct matches of different algorithms with a maximum limit of 4000 feature points.

Algorithm	Elapsed Time(ms)	Correct Matches
SIFT	1166.7	70
ASIFT	5006.2	274
KAZE	898.0	79
KeyNet	493.6	5
HardNet	1575.8	117
HyNet	1605.0	122
CAR-HyNet	1096.1	**300**

Table 5. Time elapsed at each stage of CAR-HyNet with a maximum limit of 4000 feature points.

Stage	NMS	Preprocess	IPM	Filter	RootSIFT	Pyramid	Patches	CAR-HyNet	Match
Elapsed Time(ms)	0.001	77.638	7.318	0.005	147.221	83.511	340.984	59.418	380.013

6.4 Overall Performance Evaluation

Figure 11 provides an overall comparative perspective of the proposed method along with other algorithms and the actual matching effect of our method. As shown on the left side of Fig. 11, the proposed method can provide significantly better overall matching performance than other algorithms with robustness in all scenarios. Our method even outperforms the well-performing Root-SIFT+HardNet combination and KAZE by 2.5x-6x. The right side of Fig. 11

(a) Performance in *rota-*(b) Actual effect in image *r2* (c) Actual effect in image *r6*
tion

(d) Performance in *build-*(e) Actual effect in image *h2* (f) Actual effect in image *h8*
ing match

Fig. 11. Overall matching performance comparison and actual matching results in the real world.

provides a real world perspective of the actual matching results for the testing images. Due to space limitations, we only give a part of the results here, but experimental results on other data show the same trend.

7 Discussion and Future Work

The proposed method shows excellent performance in image matching. However, we also notice that there are several areas for potential improvement. We share our thoughts below for discussion.

Rotation Invariance. We rotate patches according to the primary orientation of feature points computed by SIFT, which gives our method rotation invariant that can be compared with SIFT. However, the accuracy of the primary orientation in SIFT is inherently inaccurate, implying errors in the matching of certain feature points. Since rotation is very common in the real world, we plan to investigate methods to improve rotation invariance and reduce errors caused by primary orientation.

Real-Time. Currently, the proposed method takes an average of 1 s to complete an image matching, which is acceptable for offline applications but too slow to meet real-time operational demands. In future work, we plan to investigate the use of faster feature descriptors, dimension reduction techniques, employing lightweight models, and other strategies to minimize matching latency.

Operating Platform . Our experiments are conducted on a computer in the lab. In future work, we plan to explore transferring the system to a drone platform, thereby achieving an end-to-end and real-time image matching system.

8 Conclusion

In this paper, we propose a novel image matching scheme. The proposed image preprocessing improves detection performance by using drone attitude information. We design the CAR-HyNet network that is more suitable for feature representation and generate deep features using SIFT as prior knowledge. Finally, we propose a fine-grained decision level fusion algorithm to effectively combine handcrafted features and deep features. Experimental results show that our proposed RootSIFT+CAR-HyNet combination provides the best overall matching performance. The effectiveness of our method is further demonstrated through experiments, where it takes an average of only 1 s for 4000 feature points and achieves 2.5–6x more matches than existing methods. In addition, it is trivial to generalize the proposed method to other datasets. In summary, we believe that the proposed image matching scheme shows great potential. As we shared in the previous section, there are still several aspects for further exploration, and we hope this paper will pave the way for more active exploration in the field of image matching.

Acknowledgment. This research was supported in part by the South China University of Technology Research Start-up Fund No. X2WD/K3200890, as well as partly by the Guangzhou Huangpu District International Research Collaboration Fund No. 2022GH13. Any opinions, findings, and conclusions or recommendations expressed in this publication are those of the authors and do not necessarily reflect the views of the sponsoring agencies.

References

1. Sharma, M., Singh, H., Singh, S., Gupta, A., Goyal, S., Kakkar, R.: A novel approach of object detection using point feature matching technique for colored images. In: Singh, P.K., Kar, A.K., Singh, Y., Kolekar, M.H., Tanwar, S. (eds.) Proceedings of ICRIC 2019. LNEE, vol. 597, pp. 561–576. Springer, Cham (2020). https://doi.org/10.1007/978-3-030-29407-6_40
2. Rashid, M., Khan, M.A., Sharif, M., Raza, M., Sarfraz, M.M., Afza, F.: Object detection and classification: a joint selection and fusion strategy of deep convolutional neural network and sift point features. Multimedia Tools Appl. **78**(12), 15751–15777 (2019)
3. Jiayi, M., Huabing, Z., Ji, Z., Yuan, G., Junjun, J., Jinwen, T.: Robust feature matching for remote sensing image registration via locally linear transforming. IEEE Trans. Geosci. Remote Sens. **53**(12), 6469–6481 (2015)
4. Ravi, C., Gowda, R.M.: Development of image stitching using feature detection and feature matching techniques. In: 2020 IEEE International Conference for Innovation in Technology (INOCON), pp. 1–7. IEEE (2020)
5. Bochkovskiy, A., Wang, C.-Y., Liao, H.-Y.M.: YOLOv4: Optimal speed and accuracy of object detection. CoRR, abs/2004.10934 (2020)
6. O'Mahony, N., et al.: Deep learning vs. traditional computer vision. In: Science and information conference, pp. 128–144. Springer (2019)
7. Tian, Y., Laguna, A.B., Ng, T., Balntas, V., Mikolajczyk, K.: HyNet: learning local descriptor with hybrid similarity measure and triplet loss. Adv. Neural Inf. Process. Syst. **33**, 7401–7412 (2020)
8. Arandjelović, R., Zisserman, A.: Three things everyone should know to improve object retrieval. In: 2012 IEEE Conference on Computer Vision and Pattern Recognition, pp. 2911–2918. IEEE (2012)
9. Pérez-Lorenzo, J., Vázquez-Martín, R., Marfil, R., Bandera, A., Sandoval, F.: Image Matching Based on Curvilinear Regions. na (2007)
10. Lowe, D.G.: Object recognition from local scale-invariant features. In: Proceedings of the Seventh IEEE International Conference on Computer Vision, vol. 2, pp. 1150–1157. IEEE (1999)
11. Bay, H., Tuytelaars, T., Van Gool, L.: SURF: speeded up robust features. In: Leonardis, A., Bischof, H., Pinz, A. (eds.) ECCV 2006. LNCS, vol. 3951, pp. 404–417. Springer, Heidelberg (2006). https://doi.org/10.1007/11744023_32
12. Calonder, M., Lepetit, V., Strecha, C., Brief, F.P.: Binary robust independent elementary features. In: Proceedings of the European Conference on Computer Vision, pp. 778–792
13. Rublee, E., Rabaud, V., Konolige, K., Orb, G.B.: An efficient alternative to sift or surf. In: Proceedings of International Conference on Computer Vision, pp. 2564–2571

14. Alcantarilla, P.F., Bartoli, A., Davison, A.J.: KAZE features. In: Fitzgibbon, A., Lazebnik, S., Perona, P., Sato, Y., Schmid, C. (eds.) ECCV 2012. LNCS, vol. 7577, pp. 214–227. Springer, Heidelberg (2012). https://doi.org/10.1007/978-3-642-33783-3_16

15. Efe, U., Ince, K.G., Alatan, A.A.: Effect of parameter optimization on classical and learning-based image matching methods. In: Proceedings of the IEEE/CVF International Conference on Computer Vision, pp. 2506–2513 (2021)

16. Verdie, Y., Yi, K., Fua, P., Lepetit, V.: TILDE: a temporally invariant learned detector. In Proceedings of the IEEE Conference on Computer Vision and Pattern Recognition, pp. 5279–5288 (2015)

17. Yi, K.M., Trulls, E., Lepetit, V., Fua, P.: LIFT: learned invariant feature transform. In: Leibe, B., Matas, J., Sebe, N., Welling, M. (eds.) ECCV 2016. LNCS, vol. 9910, pp. 467–483. Springer, Cham (2016). https://doi.org/10.1007/978-3-319-46466-4_28

18. Tian, Y., Fan, B., Wu, F.: L2-Net: deep learning of discriminative patch descriptor in Euclidean space. In Proceedings of the IEEE Conference on Computer Vision and Pattern Recognition, pp. 661–669 (2017)

19. Mishchuk, A., Mishkin, D., Radenovic, F., Matas, J.: Working hard to know your neighbor's margins: local descriptor learning loss. In: Advances in Neural Information Processing Systems, vol. 30 (2017)

20. Luo, Z., et al.: GeoDesc: learning local descriptors by integrating geometry constraints. In: Proceedings of the European Conference on Computer Vision (ECCV), pp. 168–183 (2018)

21. Tian, Y., Yu, X., Fan, B., Wu, F., Heijnen, H., Balntas, V.: SOSNet: second order similarity regularization for local descriptor learning. In: Proceedings of the IEEE/CVF Conference on Computer Vision and Pattern Recognition, pp. 11016–11025 (2019)

22. Liang, Z., Yi, Y., Qi, T.: SIFT Meets CNN: a decade survey of instance retrieval. IEEE Trans. Pattern Anal. Mach. Intell. **40**(5), 1224–1244 (2017)

23. Barroso-Laguna, A., Riba, E., Ponsa, D., Mikolajczyk, K.: Key. net: keypoint detection by handcrafted and learned CNN filters. In: Proceedings of the IEEE/CVF International Conference on Computer Vision, pp. 5836–5844 (2019)

24. Tianyu, Z., Zhenjiang, M., Jianhu, Z.: Combining CNN with hand-crafted features for image classification. In: 2018 14th IEEE International Conference on Signal Processing (ICSP), pp. 554–557. IEEE (2018)

25. Rodríguez, M., Facciolo, G., von Gioi, R.G., Musé, P., Morel, J.-M., Delon, J.: SIFT-AID: boosting sift with an affine invariant descriptor based on convolutional neural networks. In 2019 IEEE International Conference on Image Processing (ICIP), pp. 4225–4229. IEEE (2019)

26. Song, Y., Zhengyu, X., Xinwei, W., Yingquan, Z.: MS-YOLO: object detection based on yolov5 optimized fusion millimeter-wave radar and machine vision. IEEE Sens. J. **22**(15), 15435–15447 (2022)

27. Yu, G., Jean-Michel, M.: ASIFT: an algorithm for fully affine invariant comparison. Image Process. Line **1**, 11–38 (2011)

28. Morel, J.-M., Guoshen, Yu.: ASIFT: a new framework for fully affine invariant image comparison. SIAM J. Img. Sci. **2**(2), 438–469 (2009)

29. Hou, Q., Zhou, D., Feng, J.: Coordinate attention for efficient mobile network design. In: Proceedings of the IEEE/CVF Conference on Computer Vision and Pattern Recognition, pp. 13713–13722 (2021)

30. Zhou, D., Hou, Q., Chen, Y., Feng, J., Yan, S.: Rethinking bottleneck structure for efficient mobile network design. In: Vedaldi, A., Bischof, H., Brox, T., Frahm, J.-M. (eds.) ECCV 2020. LNCS, vol. 12348, pp. 680–697. Springer, Cham (2020). https://doi.org/10.1007/978-3-030-58580-8_40
31. Winder, S.A.J., Brown, M.: Learning local image descriptors. In: 2007 IEEE Conference on Computer Vision and Pattern Recognition, pp. 1–8. IEEE (2007)
32. Lowe, D.G.: Distinctive image features from scale-invariant keypoints. Int. J. Comput. Vis. **60**(2), 91–110 (2004)
33. Balntas, V., Riba, E., Ponsa, D., Mikolajczyk, K.: Learning local feature descriptors with triplets and shallow convolutional neural networks. In: British Machine Vision Conference (BMVC), vol. 1, pp. 3 (2016)
34. He, K., Lu, Y., Sclaroff, S.: Local descriptors optimized for average precision. In: Proceedings of the IEEE Conference on Computer Vision and Pattern Recognition, pp. 596–605 (2018)
35. Kim, J., Jung, W., Kim, H., Lee, J.: CyCNN: a rotation invariant CNN using polar mapping and cylindrical convolution layers. arXiv preprint arXiv:2007.10588 (2020)
36. Gunatilaka, A.H., Baertlein, B.A.: Feature-level and decision-level fusion of non-coincidently sampled sensors for land mine detection. IEEE Trans. Pattern Anal. Mach. Intell. **23**(6), 577–589 (2001)
37. Chum, O., Werner, T., Matas, J.: Two-view geometry estimation unaffected by a dominant plane. In 2005 IEEE Computer Society Conference on Computer Vision and Pattern Recognition (CVPR 2005), vol. 1, pp. 772–779. IEEE (2005)
38. Balntas, V., Lenc, K., Vedaldi, A., Mikolajczyk, K.: HPatches: a benchmark and evaluation of handcrafted and learned local descriptors. In: Conference on Computer Vision and Pattern Recognition (CVPR) (2017)

Unsupervised Fabric Defect Detection Framework Based on Knowledge Distillation

Haotian Liu⬛, Siqi Wang⬛, Chang Meng⬛, Hengyu Zhang⬛,
Xianjing Xiao⬛, and Xiu Li$^{(\boxtimes)}$⬛

Tsinghua Shenzhen International Graduate School,
Tsinghua University, Beijing, China
{liu-ht21,wangsq21,mengc21,zhang-hy21,xxj21}@mails.tsinghua.edu.cn,
li.xiu@sz.tsinghua.edu.cn

Abstract. Fabric defect detection is a critical task in the textile industry. Efficient and accurate automated detection schemes, such as computer vision fabric quality inspection, are urgently needed. However, traditional feature-based methods are often limited and difficult to implement universal solutions in industrial scenarios due to their specificity towards certain defect types or textures. Meanwhile, machine learning methods may face difficulties in harsh industrial production environments due to insufficient data and labels. To address these issues, we propose an unsupervised defect detection framework based on knowledge distillation, which includes a visual localization module to assist with the detection task. Our approach significantly improves classification and segmentation accuracy compared to previous unsupervised methods. Besides, we perform a comprehensive set of ablation experiments to determine the optimal values of different parameters. Furthermore, our method demonstrates promising performance in both open databases and real industrial scenarios, highlighting its high practical value.

Keywords: Unsupervised Learning · Industrial Defect Detection · Knowledge Distillation

1 Introduction

Automated fabric defect detection has been studied for many years [2]. Researchers have used various methods to analyze fabric texture characteristics, such as structure, statistics, spectrum, and model, to identify differences between defective and standard textures and locate defects accurately [8,14].

H. Liu and S. Wang—Both authors contributed equally to this research.

B. Luo et al. (Eds.): ICONIP 2023, CCIS 1968, pp. 339–351, 2024.
https://doi.org/10.1007/978-981-99-8181-6_26

Currently, fabric defect detection algorithms can be classified into two categories: traditional visual methods and deep learning methods. Traditional algorithms use filtering [7,8] and morphological operators [17,20] to process low-level image features, achieving high-speed defect detection with few samples. However, these methods have limited receptive fields and rely on manual parameter tuning and thresholding, or require high computational costs, making them ineffective for segmenting diverse and complex types of fabric defects.

Deep learning methods use complex models like VGGNet [18] and ResNet [9] to extract rich multi-scale features from complex plane figures. Fabric defect detection algorithms based on deep learning can be categorized as supervised or unsupervised.

Supervised deep learning algorithms for fabric defect detection have improved significantly in terms of accuracy and speed compared to traditional texture contrast methods [11,19]. However, they require a large number of defective samples as training sets, which may not be feasible in industrial production scenarios where most samples are flawless and defects are difficult to cover.

Unsupervised learning models do not need labeled ground truth or human involvement [1,10,12,13,16], making them appropriate for various industrial situations. Advanced unsupervised techniques like AE-L2 [4] and VAE-grad [6] classify samples by comparing the similarity of the representation learned from defective and non-defective samples. However, these methods may not explicitly consider both global and local information, which could lead to decreased robustness. Additionally, these methods may not perform well in certain defect detection tasks that involve high global consistency defects, such as those found in fabrics.

In this paper, we present an efficient unsupervised learning framework for fabric defect detection in industrial settings. Our approach is inspired by distillation methods, where a teacher network with a large number of parameters provides prior knowledge to a student network with a smaller number of parameters to distinguish defective samples from non-defective ones. To ensure uniform feature representation across different scales, we include an auxiliary loss function in the distillation network. We also propose a defect localization model based on Gaussian statistical analysis, which calculates the Mahalanobis distance between each pixel's embedding and the mean value of positive samples, to assist manual detection in real-world scenarios.

Our main contributions are as follows:

1. We propose an unsupervised learning-based fabric detection framework that fully utilizes flawless fabric samples. Our approach uses a single-class unsupervised method to identify flawless fabric samples and localize suspected flaws, achieving outstanding performance in separating over 90% of flawless and flawed samples without manual re-inspection.
2. We introduce a defect-free sample classification model based on knowledge distillation, which uses a teacher-student network structure without complex parameter settings or expensive regional training. Our approach relies on the similarity of activation values of layer feature vectors to classify samples

with consistent texture backgrounds, without involving modelling defects or texture analysis.

3. We propose a defect localization model based on Gaussian statistical analysis of feature space. This model achieves excellent visualization results in locating pixel-level fabric defects.

4. Our method performs well in both open datasets and real industrial scenarios, indicating its high application value.

2 Related Work

Up to now, the existing fabric defect detection algorithms can be roughly divided into three categories: traditional feature-based detection methods, supervised deep learning methods, and semi-supervised or unsupervised deep learning methods.

2.1 Traditional Feature-Based Methods

Feature-based methods are the mainstream technical route for early fabric defect detection solutions. These methods' most dominant low-level features include texture features, grayscale features, etc. Techniques including rule-based inference, histogram statistics, wavelet analysis, and filtering [7,14,20] are widely used for defect detection based on fabric texture. While the processing of grayscale features mainly uses techniques such as morphological operations and grey-level co-occurrence matrix [8].

These methods are highly interpretable and independent of a large number of samples. However, these algorithms are often accompanied by a large number of hyperparameters requiring constantly adjusting according to the actual application, which makes the model sensitive and inconvenient to apply in the real industry world. Moreover, traditional algorithms tend to focus on regional information which is easily affected by factors such as environmental changes, image rotation, and texture patterns. The poor robustness leads to inaccurate localization of defects.

2.2 Supervised-Based Methods

With the development of deep learning techniques, many neural network models are used for fabric defect detection tasks, such as GoogleNet [19], Mobile-Unet [11], DCNN [22], and S-YOLOv3 [21].

For fabric defect detection, the advantages of using supervised deep learning techniques are their significant improvement of detection accuracy and the resistance to environmental noise during interference. However, supervised methods often require many samples severed as training datasets, which is highly cost-consuming. Furthermore, most of the samples in the natural production process are flawless, and it is challenging to cover all types of defects. The reliability of the trained model is not guaranteed when the sample size is unbalanced.

2.3 Semi-supervised-Based or Unsupervised-Based Methods

Current work on fabric defect detection rarely uses semi-supervised or unsupervised learning. For anomaly detection, some semi-supervised or unsupervised methods use autoencoder or adversarial training for one-class sample learning [1,10,13,16]. Recent works also use transfer learning methods to extract sub-networks from pre-trained models for anomaly detection [15]. AE-L2 [4] and VAE-grad [6], the autoencoder-based methods, use only normal samples for reconstruction training and exploit the difference in the model's ability to reconstruct normal and abnormal samples to achieve the defect detection task.

3 Framework

3.1 Fabric Classification Based on Knowledge Distillation

To address the scarcity of professionally annotated data in industrial defect detection tasks, we propose an unsupervised learning approach based on knowledge distillation. This method allows us to meet production requirements by reducing storage demands and training periods, which are typically high for deep learning networks with complex parameters. By leveraging the knowledge-rich teacher network, which extracts inherent laws from gathered samples, the lightweight student network can preserve the majority of knowledge.

We utilize large databases, such as ImageNet, to pre-train the knowledge-rich teacher network and establish standard reference patterns for normal samples. The student network is trained to produce consistent outputs with the teacher network for normal samples, using a step-by-step adjustment process that monitors the output of both networks at various scales, covering low to high-level features across different layers. During testing, the outputs of the teacher and student networks will be similar for normal samples, but different for abnormal samples, due to the teacher network's greater familiarity with defects. Therefore, it is crucial to set an appropriate threshold to differentiate between different types of samples.

Inspired by the article [15], our fabric detection framework is shown in Fig. 1. The VGG-16 architecture is used as the backbone for both the teacher and student networks. However, the student network has fewer channels than the teacher network in each layer, except for the layer before the pooling layer.

In order to fully transfer knowledge, we define the network loss function $Loss_{total}$, which consists of two components: distance loss $Loss_{dis}$ and direction loss $Loss_{dir}$. The purpose of $Loss_{dis}$ is to minimize the Euclidean distance between the output of the teacher network and the student network in each intermediate layer, which can be formulated as

$$Loss_{dis} = \sum_{i=1}^{N_{layer}} \frac{1}{N_i} \sum_{j=1}^{N_i} (layer_t^{(i)}(j) - layer_s^{(i)}(j))^2, \tag{1}$$

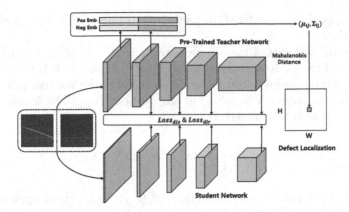

Fig. 1. We propose a framework for fabric defect detection that utilizes knowledge distillation methods for classification and Mahalanobis distance for defect localization. Our approach employs a mixed target loss function for feature maps of different scales to ensure similarity between the student and teacher network outputs for positive samples. Feature embeddings from the teacher network are used for defect localization by calculating the Mahalanobis distance.

where N_{layer} is the number of intermediate layers and N_i is the total number of neurons at layer i. Besides, $layer_t^{(i)}(j)$ and $layer_s^{(i)}(j)$ respectively represent the value of the $i - layer$ of the teacher network and the $i - layer$ of the student network. As for $Loss_{dir}$, since the feature vectors have high dimensions, the two vectors with the same Euclidean distance may obtain drastically different outcomes following the activation function. Therefore, cosine similarity between vectors is taken into account, and the formula is defined as

$$Loss_{dir} = 1 - \sum_i \frac{layer_t^{(i)^T} \cdot layer_s^{(i)}}{\|layer_t^{(i)^T}\| \cdot \|layer_s^{(i)}\|}. \tag{2}$$

Hence, the total loss function of the network $Loss_{total}$ is formulated as

$$Loss_{total} = \lambda Loss_{dis} + \gamma Loss_{dir}, \tag{3}$$

where λ and γ are used to balance the ratio of the two losses. This loss function enables the student network to comprehend the normal samples' knowledge and filter the redundant information.

3.2 Defect Localization Based on Gaussian Statistical Analysis

The knowledge distillation-based model utilizes feature vectors obtained during the training process to classify input images. However, this method alone is inadequate for detecting defective areas in industrial production. To address this limitation, we propose a localization module that employs an outlier detection algorithm to accurately locate defects. Unlike traditional approaches, our algorithm does not require labels or parameter training, making it both efficient and effective.

Considering textures of the samples are uniform and of high similarity, feature vectors of the positive sample $X_{ij} = x_{ij}^k, k \in \{1, 2, \cdots, K\}$, where K is the number of normal samples, can be assumed to fulfill the multivariate Gaussian distribution assumption $N(\mu_{ij}, \Sigma_{ij})$ in all positions (i, j). The expression of its covariance matrix is formulated as

$$\Sigma_{ij} = \frac{1}{N-1} \sum_{k=1}^{N} (x_{ij}^k - \mu_{ij})(x_{ij}^k - \mu_{ij})^T, \tag{4}$$

where μ_{ij} is the mean of positive samples at (i, j). In practical application, the regular term αI is usually added after Eq. (4), where I is the identity matrix, to make the covariance matrix full rank and invertible because the sample size of the dataset may be smaller than the feature dimension. In our experiments, α is set to 0.01. In this way, we obtain the Gaussian distribution of each position of the positive samples under the deep semantic representation.

Inspired by [5], we use Mahalanobis distance as the criterion for detecting outliers. The feature embeddings are created when the test images are fed into the distillation network. The Mahalanobis distances between the embedding of negative samples and the embedding of their pixel-level matching points in positive samples are calculated to determine the category. Mahalanobis distance is computed as

$$M_{ij} = \sqrt{(x_{ij} - \mu_{ij})^T \Sigma_{ij}^{-1} (x_{ij} - \mu_{ij})}. \tag{5}$$

The matrix can be directly visualized as a heat map. The location with significant heat is where the defect is, as the distribution of anomalous samples is typically away from the center. As opposed to Euclidean distance, Mahalanobis distance takes the relationship between representations into account. By selecting a suitable threshold, the distance matrix can be binarized to create a segmentation mask, which is then compared with the ground truth to verify the validity of our method. We use AUROC as the evaluation metric for our algorithm.

4   Experiments

4.1   Datasets

Real-scene Datasets of Fabric Defect contain high-definition images of both defective and defect-free fabrics. Each dataset has distinct types of defects and is manually labeled with pixel-precise defect segmentation annotation masks. Specifically, our collection includes four sets of fabric data, denoted as No. 1 to No. 4, representing red-denim, dark-denim, blue-rough-denim, and blue-fine-denim, respectively, as illustrated in Fig. 2. The corresponding details with subsets are shown in Table 1.

Table 1. The details of datasets.

Fig. 2. Examples of the datasets.

Datasets	Positive of Train	Positive of Test	Negative of Test
1:Red Denim	41	10	81
2:Dark Denim	41	10	48
3:Blue Rough Denim	41	10	71
4:Blue Fine Denim	41	10	81
5:Carpet	280	28	89
6:Leather	245	32	92
7:Wood	247	19	60

MVTecAD [3] is a widely used benchmark for common industrial anomaly detection, comprising over 5000 high-resolution images across fifteen object and texture categories. Each category includes a set of defect-free training images and a test set consisting of both defective and non-defective images. For evaluating our model, we select the Carpet, Leather, and Wood categories from the dataset, as they share similar textures with fabrics and can provide more robust and convincing test results. These categories are labeled as datasets No. 5, No. 6, and No. 7 in Fig. 2.

4.2 Comparison Baselines

To demonstrate the effectiveness of our proposed defect detection framework, we compare it with two baseline methods listed below:

- **AE-L2** [4]: The l^2-Autoencoder trains an AE using L2 loss to extract features for defect detection. This method utilizes the idea that by learning the potential features of normal inputs, the reconstruction accuracy of abnormal inputs will be inferior to that of normal inputs.
- **VAE-grad** [6]: The VAE-grad method employs VAE instead of AE for feature extraction and utilizes gradient-based reconstruction technology for defect localization.

4.3 Implementation Details

We adopt VGG-16 pre-trained on ImageNet-1k as the backbone for feature learning on defect-free samples. Multi-scale features generated by concatenating feature vectors from four pooling layers are employed for defect localization. Our model employs a mixture of distance loss and direction loss with $\lambda = 0.5$ and $\gamma = 1$ as the loss function, and we use the Adam optimizer with a learning rate of 0.001 and a batch size of 64 for optimization. The input image resolution is scaled to 128×128 to constrain FLOPs. We train our model on GPU for 200 epochs, and report classification and segmentation AUROC, as well as test set time efficiency in subsequent sections. Figure 3 presents a detailed summary of the results for localizing anomalous and normal features.

Table 2. Overall performance comparison of classification and segmentation on seven datasets. Boldface denotes the highest AUROC score.

Datasets	AE-L2		VAE-grad		Ours	
	CLS AUROC	SEG AUROC	CLS AUROC	SEG AUROC	CLS AUROC	SEG AUROC
1: Red Denim	0.5566	0.5405	0.8528	0.5282	**0.9926**	**0.9640**
2: Dark Denim	0.5146	0.5240	0.5499	0.5195	**0.9896**	**0.9343**
3: Blue Rough Denim	0.7859	0.5167	0.7500	0.5179	**0.9972**	**0.9907**
4: Blue Fine Denim	0.7191	0.5056	0.8568	0.5054	**0.9877**	**0.9863**
5: Carpet	0.5002	0.5878	0.6717	0.7350	**0.8644**	**0.9595**
6: Leather	0.5611	0.7497	0.7122	0.9252	**0.9874**	**0.9926**
7: Wood	0.7423	0.7311	0.8977	0.8382	**0.9605**	**0.9122**
Mean	0.6257	0.5936	0.7559	0.6528	**0.9685**	**0.9628**

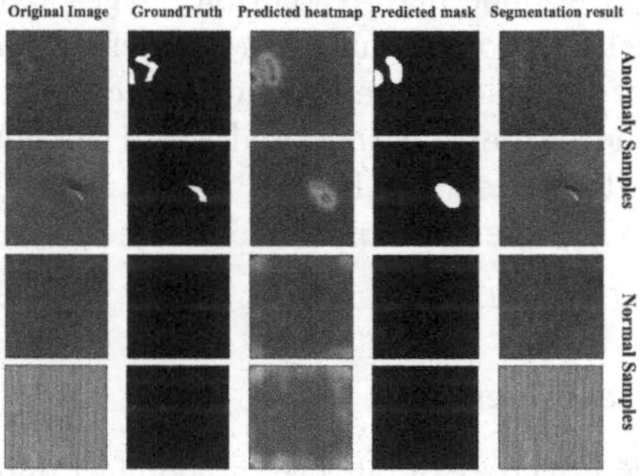

Fig. 3. Our detailed results in localizing anomalous features and normal features in MVTecAD (The second and fourth rows) and our real-scene datasets (The first and third rows).

4.4 Overall Comparison

Results on Fabric Classification. Table 2 summarizes the performance of our proposed method on seven subsets, achieving an average classification AUROC of 0.9685. In comparison, AE-L2 lacks sensitivity to fabric defects while VAE-grad lacks robustness. The performance of VAE-grad is acceptable for datasets No. 1, No. 4, and No. 7 but rapidly degrades on other datasets. Our model performs well on large area dotted defects but slightly degrades for minor dotted defects, possibly due to their similar color with the background texture. It is recommended to adjust the confidence threshold according to specific purposes in practical applications.

Results on Defect Segmentation. Table 2 demonstrates the defect segmentation performance of our proposed method, with an average segmentation AUROC of 0.9628. Our model shows remarkable results in production scenarios, whereas AE-L2 and VAE-grad perform poorly. Although VAE-grad shows promising results on dataset No. 6 and No. 7, it fails to generalize on the datasets from real scenarios. Our model generalizes well on a wide range of defect types and is more robust and effective in defect localization due to its ability to focus on global image information and resist rotation-shift.

4.5 Indepth Study

Impact of Different Backbones. To increase the credibility of our results, we compare the performance of different backbones for distillation, including VGG-16 and ResNet-18. For the classification task, both networks are pre-trained on ImageNet-1k, and the student network has the same number of channels as the teacher network in the last block before each layer, with a decreased number of channels to 16 in the remaining blocks. As shown in Fig. 4(a), VGG-16 performs better than ResNet-18 for dotted defects on small-scale datasets. However, ResNet-18 outperforms VGG-16 by an average of 3 percent in complex

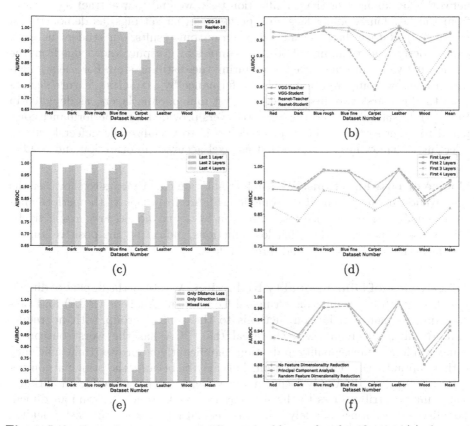

Fig. 4. Indepth study on impact of different backbones for classification (a), for segmentation (b), the number of feature extraction layers for classification (c), for segmentation (d), different loss functions (e), dimensionality reduction (f).

situations where multiple defects appear simultaneously. For the segmentation task, we compare the feature extraction effects of VGG and Resnet as teacher and student networks, respectively. As depicted in Fig. 4(b), the VGG-Teacher model achieves the highest average AUROC of 0.96127 for simpler background textures, while the ResNet-Teacher model outperforms others with an average AUROC of 0.94358 for more complex and challenging datasets. VGG is well-suited for simple textured backgrounds and efficiently handles samples with solid colors, while ResNet provides a more comprehensive representation of knowledge. In the Teacher-Student model, the teacher model generally outperforms the student model, possibly because the student model cannot perfectly distill the knowledge from complex samples, resulting in bias due to imperfect learning. However, for simpler textured samples, the student model occasionally performs surprisingly well. Therefore, the choice of model should depend on the specific requirements.

Choice of the Number of Layers to Extract Features. In this study, we investigate the most effective feature extraction technique using VGG-16 as the network's backbone. For the classification task, we find that extracting features from the last four pooling layers outperforms other settings, as demonstrated in Fig. 4(c). Rather than focusing solely on output results, our student network gains a better understanding of positive samples by grasping knowledge from the teacher network's intermediate layers during the distilling step. For the segmentation task, we compare the model's performance by extracting feature vectors from the first, first two, first three, and first four pooling layers, as depicted in Fig. 4(d). Results show that the first two layers provide better performance compared to other settings. Since our objective is to visualize the defect location, selecting a shallow layer can offer better pixel-accuracy information and results.

Choice of Loss Functions. We conduct ablation experiments to showcase the effectiveness of our proposed loss function. Figure 4(e) displays the performance of the model using distance loss, direction loss, and mixed loss, respectively. Mixed loss outperforms the other loss functions, indicating the positive impact of considering both distance and direction loss.

Effectiveness of Dimensionality Reduction. We study the impact of dimensionality reduction, comparing Principal Component Analysis (PCA) and random dimensionality reduction methods to the control group with no reduction. Results, shown in Fig. 4(f), reveal that PCA has the worst outcomes, while random dimensionality reduction significantly improves time efficiency with acceptable accuracy degradation. This may be because the principal components obtained by PCA only focus on vectors with the most significant variance which may contribute less to the segmentation task. However, random dimensionality reduction occasionally selects beneficial features for the task, yielding better performance. Nevertheless, this approach reduces the algorithm's interpretability. In industrial defect detection, an appropriate reduction step should balance accuracy and efficiency, or original feature vectors should be used.

Table 3. Ablation Study on different ways to calculate the Mahalanobis distance matrix. AUROC and inferring time of pixel-level and mean-value strategies on the seven test sets for segmentation are shown in the table. Boldface denotes the highest AUROC score.

Datasets	Pixel-Level		Mean-Value	
	AUROC	Infer Time/s	AUROC	Infer time/s
1:Red Denim	**0.92426**	2.7648	0.92338	0.0244
2:Dark Denim	0.89493	3.1989	**0.91024**	0.0271
3:Blue Rough Denim	**0.96969**	2.9859	0.96793	0.0248
4:Blue Fine Denim	**0.96074**	2.4634	0.96061	0.0251
5:Carpet	**0.92698**	2.0049	0.87357	0.0209
6:Leather	**0.97194**	2.0555	0.97023	0.0235
7:Wood	**0.84264**	3.0018	0.82500	0.0250
Mean	**0.92731**	2.6393	0.91870	0.0244

Choice of Ways to Calculate the Mahalanobis Distance Matrix. Since the fabric texture in our datasets is uniform, positive samples have a similar pixel distribution. This premise leads us to replace each pixel's representation with the mean value's representation for normal samples, as it reduces computational burden. We present and compare both strategies in Table 3, with the input images scaled to 64 × 64 for increased processing speed. While the pixel-level approach performs better with a slightly higher average AUROC, the mean-value strategy is highly time efficient, reducing the average test time per image from seconds to milliseconds. Additionally, the pixel-level method requires perfect alignment and has a significant bias in cases of rotation or offset. In contrast, calculating the mean value's representation will not be affected by image rotation or offset, which is of great practical value in industrial production.

5 Conclusion

In this paper, we propose an unsupervised framework for fabric detection, which distinguishes negative samples from positive samples with the knowledge distillation method and further localizes defects by calculating the distance matrix based on multivariate Gaussian distribution. Results on fabric classification and defect segmentation, along with ablation experiments, are presented to study parameter influence. The proposed method achieves high accuracy and efficiency on both industrial and real-scene fabric datasets, with user-friendly visual results provided to aid detection procedures.

Acknowledgements. This work was partly supported by the Science and Technology Innovation 2030-Key Project (Grant No. 2021ZD0201404), Key Technology Projects in Shenzhen (Grant No. JSGG20220831110203007) and Aminer. Shen-Zhen.ScientificSuperBrain.

References

1. Akcay, S., Atapour-Abarghouei, A., Breckon, T.P.: GANomaly: semi-supervised anomaly detection via adversarial training. In: Jawahar, C.V., Li, H., Mori, G., Schindler, K. (eds.) ACCV 2018. LNCS, vol. 11363, pp. 622–637. Springer, Cham (2019). https://doi.org/10.1007/978-3-030-20893-6_39
2. Alruwais, N., Alabdulkreem, E., Mahmood, K., Marzouk, R.: Hybrid mutation moth flame optimization with deep learning-based smart fabric defect detection. Comput. Electr. Eng. **108**, 108706 (2023). https://doi.org/10.1016/j.compeleceng.2023.108706
3. Bergmann, P., Fauser, M., Sattlegger, D., Steger, C.: Mvtec ad-a comprehensive real-world dataset for unsupervised anomaly detection. In: Proceedings of the IEEE/CVF Conference on Computer Vision and Pattern Recognition, pp. 9592–9600 (2019)
4. Bergmann, P., Löwe, S., Fauser, M., Sattlegger, D., Steger, C.: Improving unsupervised defect segmentation by applying structural similarity to autoencoders. arXiv preprint arXiv:1807.02011 (2018)
5. Defard, T., Setkov, A., Loesch, A., Audigier, R.: PaDiM: a patch distribution modeling framework for anomaly detection and localization. In: Del Bimbo, A., et al. (eds.) ICPR 2021. LNCS, vol. 12664, pp. 475–489. Springer, Cham (2021). https://doi.org/10.1007/978-3-030-68799-1_35
6. Dehaene, D., Frigo, O., Combrexelle, S., Eline, P.: Iterative energy-based projection on a normal data manifold for anomaly localization. In: International Conference on Learning Representations (2019)
7. H, Z.H.: Fabric defect detection based on improved weighted median filtering and k-means clustering. J. Textiles **40**(12), 50–56 (2019)
8. Hanbay, K., Talu, M.F., Özgüven, Ö.F., Öztürk, D.: Fabric defect detection methods for circular knitting machines. In: 2015 23nd Signal Processing and Communications Applications Conference (SIU), pp. 735–738. IEEE (2015)
9. He, K., Zhang, X., Ren, S., Sun, J.: Deep residual learning for image recognition. In: Proceedings of the IEEE Conference on Computer Vision and Pattern Recognition, pp. 770–778 (2016)
10. Hu, G., Huang, J., Wang, Q., Li, J., Xu, Z., Huang, X.: Unsupervised fabric defect detection based on a deep convolutional generative adversarial network. Text. Res. J. **90**(3–4), 247–270 (2020)
11. Jing, J., Wang, Z., Rtsch, M., Zhang, H.: Mobile-unet: an efficient convolutional neural network for fabric defect detection. Textile Res. J. 004051752092860 (2020)
12. Mei, S., Wang, Y., Wen, G.: Automatic fabric defect detection with a multi-scale convolutional denoising autoencoder network model. Sensors **18**(4), 1064 (2018)
13. Perera, P., Nallapati, R., Bing, X.: OcGAN: one-class novelty detection using GANs with constrained latent representations. In: 2019 IEEE/CVF Conference on Computer Vision and Pattern Recognition (CVPR) (2019)
14. Pourkaramdel, Z., Fekri-Ershad, S., Nanni, L.: Fabric defect detection based on completed local quartet patterns and majority decision algorithm. Expert Syst. Appl. **198**, 116827 (2022)
15. Salehi, M., Sadjadi, N., Baselizadeh, S., Rohban, M.H., Rabiee, H.R.: Multiresolution knowledge distillation for anomaly detection. In: Proceedings of the IEEE/CVF Conference on Computer Vision and Pattern Recognition, pp. 14902–14912 (2021)

16. Schlegl, T., Seeböck, P., Waldstein, S.M., Langs, G., Schmidt-Erfurth, U.: f-AnoGAN: fast unsupervised anomaly detection with generative adversarial networks. Med. Image Anal. **54**, 30–44 (2019)
17. Shumin, D., Zhoufeng, L., Chunlei, L.: Adaboost learning for fabric defect detection based on hog and SVM. In: 2011 International Conference on Multimedia Technology, pp. 2903–2906. IEEE (2011)
18. Simonyan, K., Zisserman, A.: Very deep convolutional networks for large-scale image recognition. arXiv preprint arXiv:1409.1556 (2014)
19. W, L.: Deep learning-based algorithm for online detection of fabric defects. Comput. Appl. **39**(7), 2125 (2019)
20. Yin, G.L.: Fabric defect detection method based on edge detection and wavelet analysis. J. Jiangnan Univ. Nat. Sci. Ed. **10**(6), 5 (2011)
21. Z, J.: Real-time defect detection algorithm for fabric based on s-yolov3 model. Adv. Lasers Optoelectron. **57**(16), 161001 (2020)
22. Z, Z.: Deep learning in fabric defect detection. Foreign Electron. Meas. Technol. **38**(8), 110–116 (2019)

Data Protection and Privacy: Risks and Solutions in the Contentious Era of AI-Driven Ad Tech

Amanda Horzyk[✉]

University of Edinburgh, Edinburgh EH8 9YL, Scotland, UK
amandahorzyk@outlook.com
https://www.linkedin.com/in/amanda-horzyk-73819016a

Abstract. Internet Service Providers (ISPs) exponentially incorporate Artificial Intelligence (AI) algorithms and Machine Learning techniques to achieve commercial objectives of extensive data harvesting and manipulation to drive customer traffic, decrease costs, and exploit the virtual public sphere through innovative Advertisement Technology (Ad-Tech). Increasing incorporation of Generative AI to aggregate and classify collected data to generate persuasive content tailored to the behavioral patterns of users questions their information security and informational self-determinism. Significant risks arise with inappropriate information processing and analysis of big data, including personal and special data protected by influential data protection and privacy regulation worldwide. This paper bridges the inter-disciplinary gap between developers of AI applications and their socio-legal impact on democratic societies. Accordingly, the work asks how AI Behavioural Marketing poses inadequately addressed data and privacy protection risks, followed by an approach to mitigate them. The approach is developed through doctrinal research of regulatory frameworks and court decisions to establish the current legislative landscape for AI-driven Ad-Tech. The work exposes the pertinent risks triggered by algorithms by analyzing the discourse of academics, developers, and social groups. It argues that understanding the risks associated with Information Processing and Invasion is seminal to developing appropriate industry solutions through a cumulative layered approach. This work proves timely to address these contentious issues using conventional and non-conventional approaches and aspires to promote a pragmatic collaboration between developers and policymakers to address the risks throughout the AI Value Chain to safeguard individuals' data protection and privacy rights.

Keywords: Multimedia information processing · Information security · Web search and mining · AI-driven advertisement technologies · Data protection and privacy · Conventional and non-conventional approaches

B. Luo et al. (Eds.): ICONIP 2023, CCIS 1968, pp. 352–363, 2024.
https://doi.org/10.1007/978-981-99-8181-6_27

1   Introduction

AI-driven behavioural advertising brings convenience tech to its verge as privacy and data concerns swell. Whether known or unknown pivotal industry players, Internet Service Providers (ISPs) facilitate and mediate our Internet use [18]. A presumably free communication service, where in fact, our personal data is the price. These ISPs are under the watch of Data Protection Supervisors and Internet Communications Offices in the EU and UK as they continue to shape the contemporary public sphere [49]. Essentially, our virtual society that forms public opinion through nearly universal access where individual and public needs are expressed [49]. However, this space is not an independent one; it is rather tainted by the corporate objectives of its hosts. ISPs' commercial agenda motivates extensive data mining and manipulation used in autonomous Ad Tech, which even the developers find disconcerting [19]. The drive to cost-reduction through Machine Learning ('ML') in Generative AI already permeates the online market [34]. This trend exacerbated the nuances of privacy and data protection, with billions of pounds invested in AI-Generative models behind, for instance, Meta's Advantage+ (an algorithm that automatically generates and adjusts ads to increase sales and customer traffic) [33]. AI algorithms, with a frequently contested definition, are characterised as models with the ability to perform tasks that ordinarily require human intelligence, knowledge, or skill. They are trained on mass quantities of data, in this context—users' personal data. AI-driven advertisements target persuasive content at individuals whose data was collected, aggregated, and analysed to create psychological profiles from their digital footprints [5]. These are often based on Machine Learning ('ML') models, which target online advertising through a series of most effective persuasive language techniques tailored to those psychometric user profiles using Recommendation Systems [4].

This paper critically discusses how AI Behavioural Marketing poses inadequately addressed data and privacy protection risks in Information Processing, followed by an approach to mitigate them in the UK and beyond. It derives frameworks from the legal doctrine of the EU, UK, and US. The EU established the widely adopted standard floor of the GDPR; whereas the UK is exemplified as a model country that responded to those measures but meanwhile is independent of the political union. It recognises the emerging risks and shortcomings of regulatory frameworks and industry solutions that attempt to re-engineer the internet accordingly. The discourse is separated into three following sections. First, the work situates the discussion among current regulatory frameworks and philosophical perspectives in Sect. 2. Section 3 dissects the risks inferred from legal doctrine and those highlighted by academics, developers, and social groups. Section 4 proposes a cumulative approach to address these risks by assessing both conventional and non-conventional solutions.

2   Context: Regulatory Frameworks and Philosophical Perspectives

To appropriately pursue the discussion on what undermines individuals' rights in immersive AI Ad Tech, it is crucial to explain the value of and relationship between privacy and data protection and the legal frameworks governing them. The work focuses on the jurisdictions of the UK and EU; however, a few illustrative US cases help reveal common misconceptions in the field. It undertakes a paternalistic approach to Internet control, as corporations continue to breach our protected rights at a massive scale daily [6].

Invasive techniques incorporated by modern politics and Big Tech marketers were revealed by the Cambridge Analytica scandal. In response, the UK's Parliament, in an interim report, presented the public as unwitting victims of personal data abuse that was "in the public domain, unprotected, and available for use by different players" [5]. If human rights constitute a regime of principles, problems, players, and processes that mold how individual rights are protected [15], then the power balance between these actors must be observed [48]. As Solove reiterates, "a society without privacy protection would be suffocating," as dismissing safeguards by organisations inevitably leads to harm [48]. The harm of power imbalance; the violation of trust. Customers are powerless when companies abuse their data without restraints [9,48]. However, rights to protection can also be trivialised by courts, as in the US court case of Smith v Chase [45], where plaintiffs could not establish injury when the bank sold their customer information (breaching the privacy policy). The court purported that offering services that class action members could freely decline did not constitute actual harm [45]. Often the stale box of actual bodily or psychological harm and idleness caused by the 'nothing to hide' argument may, in the first instance, hinder many from recognising the social value of privacy [48]. On the contrary, the development of the Human Rights Convention and secondary regulation seems to convey an interest in protecting its citizenry—at least, in part.

The distinction between privacy and data protection is not only symbolic [22]. Several international and domestic frameworks tackle individual vulnerability of the right to a private and family life, home, and communications, such as the European Convention on Human Rights (hereafter 'Convention' or 'ECHR') article 8 [7]. The Strasbourg court interprets the right to privacy as designed to also guard personal information against being collected, stored, or disclosed without a legitimate justification [1,28]. ECHR guidance goes as far as interpreting art. 8 as a form of informational self-determinism [10]. To clarify, the breach could involve aggregating and disseminating albeit neutral data already available in the public domain and the individual's right to determine how it is collected and processed [50]. For instance, extensive data mining of user data to train digital targeting technologies or language learning models without users' prior consent.

European precedent established that the data protection concept should also apply to extensive personal data processing by employers, service providers, and advertisers [22]. Notably, even if the ECHR is primarily designed to protect rights from state interference, secondary legislation targeted at private parties is similar. According to further case law, the larger the scope of the recording system (higher quantity and sensitivity of personal data) prone to disclosure, the stricter the approach should be with each stage of information processing [22]. Accordingly, the General Data Protection Regulation ('GDPR') followed in 2016 with a ripple effect on the global Internet economy and model architecture [46]. Even if the initial territorial scope was intended to protect EU citizens and residents, it currently affects data governance wherever it is processed, stored, or communicated. Consequently, the Court of Justice of the European Union ('CJEU') did not eliminate the possibility for national Data Protection Authorities to require worldwide solutions, including the possibility of global delisting by search engines to protect the right of erasure [11]. The relatively recent GDPR imposed a high-risk system of accountability upon Big Tech companies—a long-awaited 'chilling effect' of fines up to 4% of annual worldwide turnover or €20 million, whichever is greater [20]. Arguably, it restores the perceived value of personal information that costs whenever mishandled by powerful players. It pertains to setting potentially global data protection standard floors [42].

Privacy scholarship frequently addresses what privacy is, as equating it to data protection would be incorrect. This paper favours Solove's pluralistic approach to privacy over traditional narrow or too vague definitions [47]. For the purpose of this paper in the context of Autonomous Ad Tech, the subsequent discussion is limited to two categories of Solove's taxonomy; namely (A) Information Processing and (B) Invasion. They involve ISPs at the application layer capable of aggregating, manipulating, and selling collected personal data and directly interfering with individuals' lives and decision-making. The former focuses on the grounds to legitimate data processing by ISPs, including Social Platforms (e.g., Meta, Instagram), Search (e.g., Google), and Cloud Services (e.g., Amazon, Walmart) [11]. The latter incorporates a socio-political angle of liberal democracy to regulate automated decision-making. The following section dissects the distinct risks of both categories.

3   Risks to Data Protection and Privacy Triggered by AI Ad Tech

Interactive customer experiences enabled by AI-targeted sales pose important questions. The paper analyses two contentious issues of this behavioural advertising: Information Processing (algorithms processing high quantities of data enabling user-specific profiling) and Invasion (embodiment of Information Processing displayed in Ads that adjust to user's current or predicted behaviour patterns). Both issues of input and output trigger distinct vulnerabilities of users subject to them. An ankle monitor is a useful analogy of the baseline mechanisms of control; surveillance that instead, customers are voluntarily opting-in [21]. Social media platforms, search engines, 'habit-tracking' apps, and parole ankle monitors have this in common—tracking technology, claiming to *know* their users, capable of monitoring and adjusting their behaviour even if the list begins with 'luxury' convenience and ends with tactic surveillance [21].

Privacy protection under GDPR and the EU Charter constitutes a fundamental right [11]. Any interference with this right by collecting, storing, or disclosing information relating to private life requires justification [40]. As such, the Convention requires one or more legitimate aims and is "necessary in a democratic society" to achieve them [28]. However, these frameworks and industry solutions often fail to address the plurality of privacy harm that requires different sensitivity case-by-case. This paper argues that this sensitivity is developed by an in-depth understanding of the risks involved. This includes existing legal measures that can affect and require changes to the design and architectures of various generative models and recommendation systems that developers should be aware of.

3.1 Information Processing

Particularly troubling questions arise from User Profiling and Real-Time Bidding ('RTB') practices. Under GDPR art 4(4), profiling refers to the automated data processing to evaluate an individual's attributes associated with, e.g., work performance, personal preferences, interests, behaviour, movement, and location. Whereas RTB information broadcasting is about individuals' behaviour using cookies or similar mechanisms for recipients to analyse or predict data subjects' future behaviour from collated data points [27]. ICCL reports that 26–32 billion RTB broadcasts are processed daily from the UK alone to platform recipients worldwide [27].

The Federal Trade Commission Chair illustrates this disturbing scenario: digital technologies collect hyper-granular level user data for firms - tracking the purchased, unpurchased, and viewed items, keystrokes, and even "how long their mouse hovered" over any specific product [23]. These practices, as in the EU Case Vectuary v CNIL (French Data Protection Authorities), are "business as usual" [29]; involving the monetization of neutral or special data collected from 32000 partner apps and users' screen time through RTB [54]. This is not an unordinary case post-GDPR, which questions the effectiveness of 'privacy by design and default' measures [54]. Particularly if special category data is abused, including race, beliefs, health, and biometric data [51]. The Convention requires domestic regulators to guarantee appropriate safeguards under Art 8, as without them, internet users remain powerless [50]. In the UK, current protective mechanisms include the UK GDPR, the Data Protection Act ('DPA') 2018, and the Privacy and Electronic Communications Regulations ('PECR'). However, neither is perfect, and risks associated with Information Processing arise from these standards; others are highlighted by academics, regulators, and social actors.

The regulatory focus of all three instruments is protecting personal data information primarily from disclosure [10,22]. Yet, AI Ad Tech sheds light on other risks to privacy, arguably more important to end users. Informed users can fear that courts will not render their personal data sufficiently private; as precedent focuses on data that recedes to the past [41] or that is systematically collected [32]. Moreover, there is a risk of meaninglessness of consent. Despite a high standard of consent to be intelligible, clear, and plain [11], it is collected using cookies which is a very weak mechanism [28], or is ultimately not required if it is 'strictly necessary' for the operation of the software or web protocol [35]. UK's Information Commissioner's Office ('ICO') highlights that systematic profiling through "intrusive tracking technologies such as cookies" breaches UK privacy laws; especially as most ecosystem participants (users, model developers, and hosts) do not fully understand the privacy issues at stake [30].

Academics and regulatory bodies identify risks beyond power imbalance. They highlight that customers' data is prone to abuse through (a) millions of data points sold by brokers or disclosed to third parties without users' explicit consent; (b) processed for unknown secondary use; (c) aggregated through cross-platform internal or external tracking; (d) exclusion, by lack of transparency on data use, and (e) data leaks allowing employers, governments and stalkers to leverage victim's information [26,47]. Finally, current public debates on AI-driven technology alarm that Big Tech companies rapidly develop accounting for 'technical and ethical debt'—the cost of future fixes instead of responsible solutions [16]. Both debts stem from releasing software not ready for release in the technological race, including Microsoft and Google rapidly developing product alternatives to OpenAI's Chat GPT. Faulty development and limited testing can result in models that exacerbate issues of harmful biases, misinformation, extensive data mining, and deception [16].

3.2   Invasion

AI Ad tech discourse also considers the practical impact on users through Invasion. Fidler identifies that the extent of privacy and regulatory debates correlates with the involvement of civil and political rights [15]. Invasion relates to the risk of individuals' power imbalance (to participate freely in democratic societies) versus digital marketing giants (following commercial agendas) [46]. With the quickest-ever adopted technology in history (Generative AI like ChatGPT) [12], reports on subconscious political and commercial persuasion transpire [3,46]. To illustrate, the Cambridge Analytica scandal exposed the power of this technology when the former CEO reflected that psychological analysis of Facebook users allowed them to "play on the fears" of voters using AI-targeted campaigns to persuade them into a different political leaning [5]. The state-of-art of automated influence through digital nudges and recommender systems has already enticed socio-political discourse [43]. However, the existing debate does not consider the interplay of the risks revealed by the regulatory landscape itself with those highlighted by developers, academics, and end users.

Data Protection and Digital Information (No 2) Bill, currently in Parliament, reveals another risk, that is, the disconcerting approach to data protection [24]. Namely, to notify data subjects *after* a decision was made by AI Ads incorporating automated decision-making, allowing users to contest the decision that could significantly affect them [24]. A likely outcome of this proposal is not labeling content as AI-manipulated but informing users of significant, often subconscious influence post factum. Despite a positive shift forbidding controllers from cross-platform data aggregation without users' knowledge and objection mechanisms, instead of preventing harm, regulators prefer to inform users once an algorithm was deployed and their data harvested [24].

Another fundamental invasive risk, ICO exposes, is discriminatory targeting based on protected characteristics or proxy data [26]. The Working Party ('WP29') on Protection and Privacy warns this risk materialises in profiling and automated decisions that "perpetuate existing stereotypes and social segregation" without appropriate measures [53]. This is due to training datasets that recede to the past without counter-fit examples to challenge the significance of, for instance, gender, age, or race in predictive analysis of user behavior. Accordingly, AI systems can perpetuate "racism, sexism, ableism, and other forms of discrimination" that activist groups wish to address [2].

Industry leaders also reassert the issues of subconscious manipulation, predictive analysis, and effective persuasion leveraged by AI tools. Active persuasion can encompass intentional propaganda of misinformation, biased newscasts, and political content, manipulating unaware ISP users [13]. AI-generated ads, customized for a particular audience, can optimize corporate performance [39], meanwhile, draining customers' consent to personalized content. Specifically, what raises those privacy concerns is the use of data "beyond normally foreseeable" [10].

AI algorithms often incorporate optimization and fitness functions designed to maximize their performance continuously to meet an objective(s) within predetermined parameters. Developers warn that if the goal entails driving user traffic or interaction, it can generate custom ads designed to 'push your buttons', as they embody the most persuasive promotional tactics [39]. An associated risk was revealed by a games company that chose to opt out of Meta's Advantage+ Generative Ad Tech. The reason was the loss of control over the ML Black Box model, driving customer traffic but accidentally encouraging hateful and inflammatory comments [33].

Ultimately, the risks of Invasion lead to an exacerbated corruption of the public sphere through unprecedented marketing. If unaddressed, they can inevitably infringe on the rights of individuals to make informed, independent political choices or be free from the undue influence of subconscious commercial manipulation.

4   A Critical Approach to Mitigating the Risks

The previous section exposing risks of AI Ad Tech associated with (A) Information Processing and (B) Invasion is seminal to forming an approach to mitigate them. This paper argues that the solutions proposed by regulators, academics, and developers can create transferable, cumulative layers of protection based on the UK, even if none is perfect. Proposed solutions are divided into conventional and non-conventional approaches.

4.1 Conventional

There are existing provisions in the UK GDPR that were overlooked and must be reiterated in relation to AI Ad Tech and the autonomous processing and decision-making it involves. A new emphasis can be placed on the rights of data subjects enshrined in the instrument and a need to change the regulatory approach. Firstly, ISP users must be made aware of their Right to Object under Articles 21(2) and (3) to data processing (including profiling) for the purposes of direct marketing, especially if they involve automated means [53]. This right to object binds controllers to stop processing the personal data concerned. Similarly, Art 22 (1) creates a prohibition on fully automated individual decision-making, and WP29 also clarified that recital 71 in the GDPR suggests that behavioural advertising can have a legal or similarly significant effect on data subjects [11]. This means that controllers have the duty to inform users (a) that they are engaging in this activity, (b) provide meaningful insight about the logic involved (c) explain the significance and predicted consequences [36]. These requirements are pivotal for AI developers as it requires communication transparency (when and where the algorithm is used and data is harvested), an adequate level of explainability of the algorithm deployed (which can be difficult with black box models), and a risk assessment of the output on the user and the public domain. Ultimately, these provisions also require ISPs to ascertain a higher standard of consent to AI Ad Tech and clearly communicate its use by, for instance, labelling AI-generated content. As Porter also suggests, the accountability system could further entail a mechanism of "duty of care" on Big Tech companies to adhere to stricter Ad Tech rules proactively, similar to the approach proposed in the coming Online Safety Bill [31].

The recommended regulatory approach should tackle the "underlying drivers of harm upstream throughout the AI Value Chain"; from data harvesting to information processing and, finally, to the output of generative and recommended advertisements [31]. It should undertake a harm-based approach, looking at possible recipients, especially vulnerable groups and minors [31]. Recent research suggests that users increasingly expect autonomy along the lines of informational self-determinism [28]. As such, technical solutions are suggested to re-engineer the architecture of consent, for instance, by increasing user control via browser settings [26], opting out of RTB and Third Party Cookies by default [28], and increasing regulatory pressure for firms to shift to 'contextual advertising' to deviate from intrusive RTB practices [37]. Contextual advertising still involves effective techniques of advertising, for instance, "adverts for dresses being shown on a bridal website," rather than targeting ads based on extensive data processing and specific-user behaviour prediction [37]. MEPs also stress that regulation should involve a global conversation that is timely in effect rather than introducing a framework with a 'Brussels effect' influencing other jurisdictions after years [31]. Conventional approaches are at the forefront of public debate; these proposals are currently challenged and devised in Parliaments and Councils. Their principles and requirements are likely to be implemented first. The coming law will likely begin to constrain the free and extensive datasets used currently to train various models in many industries, followed by requirements to adjust algorithm architectures and applications, anonymize or withdraw personal or special data from the training dataset, and in many instances, will instruct re-training the models altogether. However, non-conventional solutions should also be considered.

4.2 Non-conventional

These solutions propose non-mainstream ideas that can be effective. Even if an absolute ban on algorithmic processing or leaving the internet market to self-regulate are quite controversial, developing Independent Standards of certification may be nonetheless feasible [14,52]. IAB Europe's self-styled Transparency and Consent Framework (TCF) is an independent set of technical specifications and policies that developers, marketers, publishers, and relevant ISPs may wish to adhere to [25]. Its goal is to standardize the requirements to facilitate lawful data processing, consent mechanisms, and automated-decision making practices for firms and algorithm developers [25]. It would encourage uniformity and labelling of websites that adhere to good data processing practices. However, it is not faultless, as in 2022, it was found in breach of the GDPR by Belgian's data protection authorities - placing all followers in jeopardy [54].

Three further ideas were proposed to re-engineer the approach to AI Ad Tech regulation. First, to treat personal data as property [38]. Ritter and Mayer proposed a new construct to redefine how we perceive our data by establishing individuals' rights to "license, transfer, use, modify, and destroy digital information assets" [38]. However, despite the merits of this approach of the clear concepts of ownership, due to the untraceable quantity of data points held by unspecified ISPs and Data brokers, it seems practically unfeasible. Second, as proposed by Zuckerberg and other Social Platforms, is to directly counteract the consequences of Ad Tech rather than to restrict it in the first place (rather than adjusting the training dataset or the model architecture) [17]. This solution would involve another sophisticated artificial algorithm and significant human review to protect against, for instance, election interference. Such a system remains counter-intuitive, as it would rely on processing subject data further to infer who was subject to what content. It would focus on fixing the technical and ethical deficits after potential harm is caused at scale with no guarantee of uniform results and sustainable deployment of the former algorithm designed to meet its skewed objectives. The third involves ISPs phasing out third-party cookies completely, like Google and Apple, using software changes or data tracking apps, tracking the trackers [44]. Arguably, this can equip individuals with a sense of data self-determinism as they would know where, when, and by whom their data is kept.

5   Concluding Remarks

Significant privacy and data protection risks arise with the exponential use of AI-driven Ad Tech capable of unethical Information Processing and Invasive practices. While legislative bodies attempt to regulate ISPs who collect, aggregate and manipulate data, they fall short of preventing the corruption of the public sphere and protecting individuals' informational self-determinism. ISPs continue to find loopholes, including RTB and confusing consent mechanisms, to pursue their commercial agenda. Hence, this paper critically discussed how AI Behavioural Marketing poses inadequately addressed privacy and data protection risks. The work argued that an in-depth understanding of those risks is pivotal in developing pragmatic approaches to protect individual rights at stake. It proposed a cumulative layered approach of implementing conventional and non-conventional solutions to re-engineer the development and deployment of algorithms. Feilser asks [16], "how can someone know what societal problems might emerge before the technology is fully developed?"—It remains a challenge; however, as we approach new dimensions of technology, we cannot only ethically speculate the risks but look back at the bottom-line of harm: the commercial agenda of ISPs and the power imbalance of various actors in our society.

References

1. Amann v Switzerland. European Court of Human Rights Case: 27798/95 (2000). ECHR 2000-II
2. Algorithmic Justice League. http://www.springer.com/lncs. Accessed 10 May 2023
3. Bailenson, J., Yee, N.: Digital chameleons: automatic assimilation of nonverbal gestures in immersive virtual environments. Psychol. Sci. **16**(10), 814–819 (2005)
4. Braca, A., Dondio, P.: Persuasive communication systems: a machine learning approach to predict the effect of linguistic styles and persuasion techniques. J. Syst. Inf. Technol. **25**(2), 160–191 (2023). https://doi.org/10.1108/JSIT-07-2022-0166
5. Commons Select Committees: Disinformation and 'fake news' Interim Report. UK Parliament, London, United Kingdom (2018)
6. CMS: GDPR Enforcement Tracker. https://www.enforcementtracker.com/. Accessed 26 Apr 2023
7. Convention for the Protection of Human Rights and Fundamental Freedoms (1950)
8. Data Protection Act (2018)
9. Dyer v Northwest Airlines Corp. US Federal Court Case: 24029/07 (2004). 334 F Supp (2nd Series) 1196
10. ECHR: Guide on Article 8 of the Convention - Right to respect for private and family life 40/172 Council of Europe, Strasbourg, France (2022)
11. Regulation (EU) 2016/679 General Data Protection Regulation of the European Parliament and of the Council of 27 April 2016 on the protection of natural persons with regard to the processing of personal data and on the free movement of such data (2016)
12. Edwards, B.: ChatGPT sets record for fastest-growing user base in history (2023). https://arstechnica.com/information-technology/2023/02/chatgpt-sets-record-for-fastest-growing-user-base-in-history-report-says/. Accessed 5 May 2023
13. Ellul, J.: Propaganda: The formation of Men's Attitudes. Knopf Doubleday Publishing Group, New York (2021)
14. Fagan, F.: Systemic social media regulation. Duke Law Technol. Rev. **16**, 393–439 (2017)

15. Fidler, D.: Cyberspace and human rights. In: Tsagourias N., Buchan R. (eds.) Research Handbook on International Law and Cyberspace 2021. Edward Elgar Publishing, Cheltenham (2021)
16. Fiesler, C.: AI has social consequences, but who pays the price? Tech companies' problem with 'ethical debt' (2023). https://medium.com/the-conversation/ai-has-social-consequences-but-who-pays-the-price-tech-companies-problem-with-ethical-debt-d685bab5859f. Accessed 3 May 2023
17. Frenkel, S., Isaac, M.: Facebook 'Better Prepared' to fight election interference, Mark Zuckerberg Says. The New York Times, San Francisco (2018)
18. Guadamuz, A.: Law, policy and the internet. In: Edwards, L. (ed) Internet Regulation. Bloomsbury Publishing, London (2018)
19. Goldman, S.: Walmart VP confirms retailer is building on GPT-4, says generative AI is 'as big a shift as mobile' (2023). https://venturebeat.com/ai/walmart-vp-confirms-retailer-is-building-on-gpt-4-says-generative-ai-is-as-big-a-shift-as-mobile/. Accessed 20 Apr 2023
20. Google LLC v Commission nationale de l'informatique et des libertés (CNIL). European Court of Justice Case: C-73/07 (2016). ECR I-09831
21. Gilliard, C.: The Rise of 'Luxury Surveillance'. The Atlantic, Brooklyn (2022)
22. Hannover v Germany. European Court of Human Rights Case: 59320/00 (2004). Reports of Judgments and Decisions 2004-VI
23. Hendrix, J.: FTC Chair Lina Khan Addresses. Tech Policy Press, New York (2022)
24. House of Commons: Explanatory Notes on the Data Protection and Digital Information (2). UK Parliament, London, United Kingdom (2023)
25. IAB Europe: TCF - Transparency & Consent Framework. https://iabeurope.eu/transparency-consent-framework. Accessed 2 May 2023
26. ICO: Response of the Information Commissioner's Office to the consultation on the Online Advertising Programme. Upholding Information Rights. Cheshire, United Kingdom (2022)
27. Irish Council for Civil Liberties: The Biggest Data Breach' Report on scale of Real-Time Bidding data broadcasts. ICCL, Dublin, Ireland (2022)
28. Keller, P.: After Third Party Tracking: Regulating the Harms of Behavioural Advertising Through Consumer Data Protection. Available at SSRN 4115750 (2022). https://doi.org/10.2139/ssrn.4115750
29. Lomas, N.: How a small French privacy ruling could remake adtech for good (2018). https://techcrunch.com/2018/11/20/how-a-small-french-privacy-ruling-could-remake-adtech-for-good/. Accessed 28 Apr 2023
30. Lomas, N.: Behavioural advertising is out of control, warns UK watchdog (2019). https://techcrunch.com/2019/06/20/behavioural-advertising-is-out-of-control-warns-uk-watchdog/. Accessed 29 Apr 2023
31. Lomas, N.: EU lawmakers eye tiered approach to regulating generative AI (2023). https://techcrunch.com/2023/04/21/eu-ai-act-generative-ai/?guccounter=1. Accessed 10 May 2023
32. MM v UK. European Court of Human Rights Case: 24029/07 (2012). ECHR 2012
33. Murphy, H., Criddle, C.: Meta's AI-driven advertising system splits marketers. Financial Times, London, San Francisco (2023)
34. Okudaira, K.: Meta to debut ad-creating generative AI this year, CTO says (2023). https://asia.nikkei.com/Business/Technology/Meta-to-debut-ad-creating-generative-AI-this-year-CTO-says. Accessed 21 Apr 2023
35. The Privacy and Electronic Communications (EC Directive) Regulations (2003)

36. Porter, J.: The UK's tortured attempt to remake the internet, explained (2023). https://www.theverge.com/23708180/united-kingdom-online-safety-bill-explainer-legal-pornography-age-checks. Accessed 10 May 2023
37. Rendle, J.: Digital advertising - state of play in the UK (2023). https://www.taylorwessing.com/en/global-data-hub/2023/april-digital-advertising-and-data-privacy/digital-advertising-state-of-play-in-the-uk. Accessed 10 May 2023
38. Ritter, J., Mayer, A.: Regulating data as property: a new construct for moving forward. Duke Law Technol. Rev. **16**, 220–277 (2018)
39. Rosenberg, L.: The hidden dangers of generative advertising (2023). https://venturebeat.com/ai/the-hidden-dangers-of-generative-advertising/. Accessed 4 May 2023
40. Rotaru v Romania. European Court of Human Rights Case: 28341/95 (2000). ECHR 2000-V
41. Rotaru v The Republic of Moldova. European Court of Human Rights Case: 56386/10 (2010). ECHR 2010
42. Samonte, M.: Google v CNIL Case C-507/17: The Territorial Scope of the Right to be Forgotten Under EU Law. https://europeanlawblog.eu/2019/10/29/google-v-cnil-case-c-507-17-the-territorial-scope-of-the-right-to-be-forgotten-under-eu-law/
43. Susser, D., Grimaldi, V.: Measuring automated influence: between empirical evidence and ethical value. In Proceedings of the 2021 AAAI/ACM Conference on AI, Ethics, and Society (AIES '21), New York, Association for Computing Machinery (2021). https://doi.org/10.1145/3461702.3462532
44. Scott, G., Tene, O., Loose, A.: ICO Issues Opinion on Data Protection and Privacy Expectations for Online Advertising Proposals. https://www.goodwinprivacyblog.com/2021/12/08/ico-issues-opinion-on-data-protection-and-privacy-expectations-for-online-advertising-proposals/
45. Smith v Chase Manhattan. US Federal Court Case: 24029/07 (2002). 741 NYS (2nd Series) 100
46. Solon, O.: How Europe's 'breakthrough' privacy law takes on Facebook and Google. The Guardian, San Francisco (2018)
47. Solove, D.: A taxonomy of privacy. Univ. Pa. Law Rev. **1**, 477–564 (2006)
48. Solove, D.: "I've got nothing to hide"' and other misunderstandings of privacy. San Diego Law Rev. **44**, 745–772 (2007)
49. Soules, M.: Jürgen Habermas and the public sphere. Glob. Media Commun. **3**(2), 201–214 (2007)
50. Tietosuojavaltuutettu v Satakunnan Markkinapörssi Oy and Satamedia Oy. European Court of Justice Case: C-507/17 (2008). ECR I-09831
51. Regulation (EU) 2016/679 Of The European Parliament And Of The Council of 27 April 2016. United Kingdom General Data Protection Regulation (2016). ECLI:EU:C:2019:15
52. Walz, A., Firth-Butterfield, K.: Implementing Ethics Into Artificial Intelligence: A Contribution, from a Legal Perspective, to the Development of an AI Governance Regime **18**, 176–231 (2019)
53. Working Party 29: Guidelines on Automated individual decision-making and Profiling for the purposes of Regulation 2016/679. In Data Protection and Privacy. Belgium (2017)
54. Vectuary v CNIL. European Litigation Chamber of the Data Protection Authority: Complaint relating to Transparency & Consent Framework (2022). DOS-2019-01377

Topic Modeling for Short Texts via Adaptive Pólya Urn Dirichlet Multinomial Mixture

Mark Junjie Li[1], Rui Wang[1], Jun Li[2], Xianyu Bao[2], Jueying He[1],
Jiayao Chen[1(✉)], and Lijuan He[2]

[1] College of Computer Science and Software Engineering, Shenzhen University,
Shenzhen, China
jj.li@szu.edu.cn,
{2210273056,hejueying2020,chenjiayao2021}@email.szu.edu.cn
[2] Shenzhen Academy of Inspection and Quarantine, Shenzhen, China

Abstract. Inferring coherent and diverse latent topics from short texts is crucial in topic modeling. Existing approaches leverage the Generalized Pólya Urn (GPU) model to incorporate external knowledge and improve topic modeling performance. While the GPU scheme successfully promotes similarity among words within the same topic, it has two major limitations. Firstly, it assumes that similar words contribute equally to the same topic, disregarding the distinctiveness of different words. Secondly, it assumes that a specific word should have the same promotion across all topics, overlooking the variations in word importance across different topics. To address these limitations, we propose a novel Adaptive Pólya Urn (APU) scheme, which builds topic-word correlation according to the external and local knowledge, and the Adaptive Pólya Urn Dirichlet Multinomial Mixture (APU-DMM) model that uses the topic-word correlation as an adaptive weight to promote topic inference process. Our extensive experimental study on three benchmark datasets shows the superiority of our model in terms of topic coherence and topic diversity over the eight baseline methods (The code is available at https://github.com/ddwangr/APUDMM).

Keywords: Topic Modeling · Short Texts · Dirichlet Multinomial Mixture · Pólya Urn model

1 Introduction

Short texts have become a prevalent form of information due to the proliferation of Internet media. Understanding the core themes within this information is crucial for various research, including user interest profiling and event detection. Topic Modeling is a method used to condense extensive collections of documents into concise summaries known as latent topics, which consist of related words. Traditional topic models, such as Latent Dirichlet Allocation (LDA) [2] and

© The Author(s), under exclusive license to Springer Nature Singapore Pte Ltd. 2024
B. Luo et al. (Eds.): ICONIP 2023, CCIS 1968, pp. 364–376, 2024.
https://doi.org/10.1007/978-981-99-8181-6_28

Non-negative Matrix Factorization (NMF) [22], have shown great success in discovering latent semantics in normal-sized documents. However, modeling short texts presents unique challenges in identifying co-occurring topic patterns due to the limited number of words in each document, leading to noisy and incoherent topics.

A simple strategy is to assume that each document is assigned to only one topic. The word co-occurrence information corresponding to each topic is improved. The Dirichlet Multinomial Mixture (DMM) model [17] is the first model to apply this assumption, and [24] implements the inference process of DMM based on Gibbs sampling. Many approaches have focused on using word embeddings to retrieve semantic information to reduce data sparsity further. LF-DMM [16] introduces a Bernoulli distribution, which determines whether words are generated from a multinomial distribution or word embedding representations. Based on word embedding, [8–11] use Generalized Pólya Urn (GPU) to incorporate word correlation into the sampling process of DMM. However, there remain two significant limitations in GPU-based topic models, which reckon that: (1) similar words should have the same contribution to an identical topic; (2) a specific word should have the same promotion to all different topics. For instance, it would be inaccurate to claim that similar words of *god*, such as *morality*, *church*, *sin*, and *atheist* have equal contributions to a *religion* topic. For the latter, the relevance of *church* to a *religion* topic and a *architecture* topic is different. In other words, different words should contribute distinctively to different topics during the GPU-enhanced process.

We propose a novel Adaptive Pólya Urn scheme (APU) to solve the above-mentioned limitations. The APU model introduces a comprehensive global semantic coherence measure incorporating external and local knowledge to capture the topic-word correlation. This measure can construct an adaptive promotion matrix that encourages adaptive weights for different words within the Pólya Urn scheme. Based on APU, we propose Adaptive Pólya Urn Dirichlet Multinomial Mixture (APU-DMM) for topic modeling on short texts. On three real-world datasets (Tweet, SearchSnippets, GoogleNews), APU-DMM enables the discovery of more coherent and diverse topics compared to state-of-the-art models. The main contributions of this paper can be concluded as follows:

- We propose the Adaptive Pólya Urn (APU), a novel scheme that integrates external and local knowledge to capture the topic-word correlation.
- We introduce the Adaptive Pólya Urn scheme Dirichlet Multinomial Mixture (APU-DMM), which uses the topic-word correlation to promote adaptive weights for different words during the topic inference process.
- Experimental results demonstrate that our model outperforms many state-of-the-art models, exhibiting its ability to discover more coherent and diverse topics.

The rest of this paper is organized as follows. Section 2 introduces the related research work. Section 3 describes our model in detail. Experimental settings and results are presented in Sect. 4. Finally, we conclude Sect. 5.

2   Related Work

The fast growth of short texts has raised the interest in developing appropriate methods to deal with the lack of enough word co-occurrence information. We briefly review the related work following [20].

2.1   Global Word Co-occurrences Based Methods

Due to the adequacy of global word co-occurrences, [6] introduces the Biterm Topic Model (BTM), which helps group the correlated words. BTM assumes that the two words comprising a biterm have the same topic derived from various topics on the entire dataset. WNTM [27] utilizes global word co-occurrence to build up a word co-occurrence network and learns the distribution over topics using LDA. However, WNTM fails to express the deep meaning among words due to a lack of semantic distance measures. CoFE [1] reconstructs the new text by the global co-occurrence frequency of words and applies the LDA model to the new text. Such methods can only indirectly construct the probability distributions of topics and short texts. Although the global word co-occurrence information is more abundant than the short texts, these models are not complete solutions to data sparsity.

2.2   Self-aggregation Based Methods

The self-aggregation based models merge short texts into long pseudo documents to extract the hidden topics. SATM, first proposed by [21], considers each short text as a sample from a hidden long pseudo-document and merges them to use Gibbs sampling for topic extraction without relying on metadata or auxiliary information. However, SATM is prone to overfitting and is computationally expensive. To improve the performance of SATM, PTM [26] presents the concept of the pseudo document to combine short texts to tackle data sparsity implicitly. PYSTM [18] parameterizes the number of pseudo-long texts by sampling from the Dirichlet process. However, non-semantic related short texts will likely be aggregated in one long document, bringing plenty of non-semantic word co-occurrence.

2.3   Dirichlet Multinomial Mixture Based Methods

Based on the assumption that each document is sampled from a single topic, [24] proposes a collapsed Gibbs Sampling algorithm for Dirichlet Multinomial Mixture (DMM) [17]. Many approaches use word embedding to retrieve semantic information with less sparsity. LF-DMM [16] uses word embeddings to estimate the probability of words. GPU-DMM [11], GPU-PDMM [10], and MultiKE-DMM [9] use the Generalized Pólya Urn (GPU) model to generate topics according to word embeddings. [12] proposes TSSE-DMM over short texts mitigating data sparsity problems by the semantic improvement mechanism. [5] proposes a Multi-GPU model, which allows a ball to be transferred from one urn to another, enabling multi-urn interactions. A Weighted Pólya Urn scheme is proposed in [13], incorporated into the LDA framework. However, these GPU schemes are complicated and inappropriate for the simple assumption in DMM models. Moreover, the topic-word correlation is ignored in the GPU-guided short text modeling.

3   APU-DMM

The proposed Adaptive Pólya Urn Dirichlet Multinomial Mixture (APU-DMM) will be described in this section. We first introduce a semantic coherence measurement to capture the global topic-word correlation. Then, we present how APU-DMM incorporates the topic-word correlation with the APU scheme to form the adaptive weights of those similar-color balls back to the urn.

3.1   Topic-Word Correlation

We first review the two limitations in the GPU scheme discussed in Sect. 1: (1) assuming that similar words hold equal importance to the current topic; (2) a specific word should have the same promotion for all different topics. However, each word is supposed to have distinct semantic coherence to different topics. Different from measuring the topic-word correlation by word embeddings in [23], we propose a global semantic coherence measure to capture the topic-word correlation by external and local knowledge. After the Burn-In phase of Gibbs sampling, the global semantic coherence is calculated at the end of each iteration based on the intermediate topic sampling results. For each word w, its correlation value with the representative words of a topic z is based on the following equation,

$$\mathbb{C}^i_{w,z} = \sum_{v \in TW^i_z} \Phi^i_{z,v} \cdot (\tau(w,v) + LC(w,v)) \tag{1}$$

where $TW^i_z = \{v^i_{z,1}, v^i_{z,2}, ..., v^i_{z,M}\}$ is the Top M representative words (sorted by the topic-word probability in the descending order) in the topic z and $\Phi^i_{z,v}$ is the probability of word v in topic z at ith iteration. Here, we adopt cosine word similarity $\tau(w,v)$ to capture the semantic similarity between w and v in external knowledge, while $LC(w,v)$ to measure the word correlation of w and v in the original dataset. Here, We use a method similar to Pointwise Mutual Information (PMI) [15] for $LC(w,v)$. LC of two words is defined as follows:

$$LC(w,v) = \left[\log \frac{p(w,v)}{p(w) \cdot (v)} \right]_+ \tag{2}$$

where $p(w)$ and $p(w,v)$ denote the probability of the word w and both words occurring in a random document of the current dataset, respectively.

In our case, it measures the extent to which two words tend to co-occur. A positive LC value implies a semantic correlation of words in the local corpus, which differs from word similarity but is important for evaluating the local semantic information. In other words, under the premise of short texts, if both words can appear within the same document, it indicates their semantic relevance. In conclusion, $\mathbb{C}^i_{w,z}$ can capture the global semantic coherence of w with the topic z in the ith iteration.

3.2 Adaptive Pólya Urn

We propose the Adaptive Generalized Pólya Urn (APU) scheme, which allows similar balls to be put back into the urn with different promotional contributions related instead of returning the similar balls with the same weight to the urn in the GPU scheme. Following [11], for all word pairs in vocabulary, if the semantic similarity score is higher than a predefined threshold ϵ, the word pair is saved into a collection $\mathbb{M}$, i.e., $\mathbb{M}_v = \{w | \tau(w, v) > \epsilon\}$. Then, the adaptive promotion matrix $\mathbb{A}$ in APU concerning each word pair is defined below,

$$\mathbb{A}^i_{w,z} = \begin{cases} \mu^i_{w,z} & \text{if } w \in \mathbb{M}_v \\ 0 & \text{otherwise} \end{cases} \tag{3}$$

$$\mu^i_{w,z} = \exp\{p^i(z|w) \cdot \mathbb{C}^i_{w,z}\} \tag{4}$$

When the current word v is assigned to topic z in the ith iteration, the similar words in $\mathbb{M}_v$ will be assigned to topic z with a promotional weight of $\mu^i_{w,z}$. This exp function yields a larger promotional weight if word w has a larger topic-word correlation value with the topic z in terms of $\mathbb{C}^i_{w,z}$ and a higher topic probability z given the word w in term of $p(z|w)$. In this way, the two considerations discussed before are solved appropriately in our APU scheme with examples shown in Sect. 4.4.

3.3 Model Inference

As APU-DMM replaces the GPU scheme in GPU-DMM with the proposed APU scheme, the topic inference process is similar to GPU-DMM except for the calculation of the global Topic-Word Correlation and the adaptive promotion matrix at the end of each iteration after the Burn-In process. Based on collapsed Gibbs sampling, APU-DMM uses the following conditional probability distribution to infer its topic (Find more details in [11]):

$$p(z_d = k | \mathbf{Z}_{\neg d}, \mathbf{D}) \propto$$

$$\frac{m_{k,\neg d} + \alpha}{D - 1 + K\alpha} \times \frac{\prod_{w \in W} \prod_{j=1}^{n_d^w}(\tilde{n}^w_{k,\neg d} + \beta + j - 1)}{\prod_{i=1}^{n_d}(\tilde{n}_{k,\neg d} + V\beta + i - 1)} \tag{5}$$

where $\tilde{n}^w_{k,\neg d}$ is the number of word w in the topic k, $\tilde{n}_{k,\neg d}$ is the number of words assigned to topic k and $m_{k,\neg d}$ is the number of documents belong to topic k. $\neg d$ is that document d is excluded from the counting. We can obtain the posterior distribution Φ in Eq. 1 as follows: $\Phi_{z,v} = \frac{\tilde{n}^w_k + \beta}{\sum_w^V \tilde{n}^w_k + V\beta}$.

Algorithm 1: Gibbs sampling of APU-DMM

Input: Topic number K, α, β, Documents $\mathcal{D}$, and iteration I
Output: Posterior topic-word distribution ϕ
1 Compute LC according to documents $\mathcal{D}$;
2 Initialize counting variables and topic-Word Correlation $\mathbb{C}$;
3 **for** *each iteration* $i \in 1, 2, \ldots, I$ **do**
4   updateTopicTopWords();
5   updateTopicWordCorrelation(); /* See Eq. (1) */
6   **for** *each short document* $d \in \mathcal{D}$ **do**
7   ratioCounter(d,-1);
8   **for** *each word* $w_{d,n} \in d$ **do**
9 | Sampling topic assignment z_d; /* See Eq. (5) */
10   **end**
11   ratioCounter(d,1);
12   **end**
13 **end**

The detail of the Gibbs sampling process of APU-DMM is described in Algorithm 1. At first, APU-DMM calculates the local word correlation (LC) according to the short texts and initializes the counting variable, similar to the initialization process in DMM (Lines 1–2). In each iteration of Gibbs sampling, we first get the top words in every topic and calculate the topic-word correlation $\mathbb{C}$ for each topic and top word, based on Eq. (1). Then, the topic of each document d is sampled based on the conditional distribution in Eq. (5) (Lines 6–10). During this process, APU-DMM subtracts the corresponding counts for each word w in the previous iteration by calling the function *ratioCounter()* (Line 7). This function uses $\mathbb{C}$ as an adaptive weight to update the counting rather than a fixed weight in GPU-DMM. With the new sampled topic, the corresponding counts for each word w are added through function *ratioCounter()* (Line 11).

We analyze the time complexity of APU-DMM inspired by GPU-DMM. We denote the time complexity of GPU-DMM in an iteration as $O(B)$. Then APU-DMM has a time complexity of $O(B+KVM)$, where K is the number of topics, V is the size of the vocabulary, and M is the number of top words in Eq. (1). KVM is the computation required for calculating the global Topic-Word Correlation Eq. (1) and the adaptive promotion matrix Eq. (3). Thus, APU-DMM does not add too much computational cost to GPU-DMM.

4   Experiment

4.1   Experiment Setup

Datasets. The performances are reported over three text datasets[1]. Tweet [25] dataset consists of 2,472 tweets related to 89 queries. The relevance

---

[1] https://github.com/qiang2100/STTM.

Table 1. Statistics on the four datasets. Label: the number of true labels; Doc: the total number of documents; Len: the average length of each document

Dataset	Label	Doc	Len	Vocab
Tweet	89	2,472	8.56	5,098
SearchSnippets	8	12,340	10.72	5,581
GoogleNews	152	11,108	6.23	8,110

between tweets and queries is manually labeled in the 2011 and 2012 microblog tracks at TREC. SearchSnippets [19] belongs to 8 domains, Business, Computers, Culture-Arts, Education-Science, Engineering, Politics-Society, Sports, and Health, respectively. GoogleNews dataset [17] is downloaded from the Google news site and crawled the titles and snippets of 11,108 news articles belonging to 152 clusters. We present the key information of the datasets that are summarized in Table 1.

Baselines and Parameters Settings. We compare our APU-DMM against the following eight state-of-the-art topic models specific to short texts, namely LDA (see footnote 1), BTM(see footnote 1), DMM (see footnote 1), GPU-DMM (see footnote 1), TSSE-DMM[2], MultiKE-DMM[3], SeaNMF[4], and PYSTM[5]. The hyperparameters of each model are set to the recommended values in their papers. We set $\alpha = 50/K$ and $\beta = 0.01$ for our APU-DMM. For the threshold ϵ, we set ϵ as 0.7 on SearchSnippets and 0.8 for Tweet and GoogleNews. We run the Gibbs sampling process of each model in the experiment for 1000 times and report the average results over 5 times. For our APU-DMM, we have an additional 500 Burn-In iterations. We use the pre-trained 300-dimensional word2vec[6] embeddings.

4.2 Evaluation

To evaluate topic quality, we focus on two commonly used measures, topic coherence, and topic diversity. Topic coherence is a quantitative measure of the interpretability of a topic. In computing topic coherence, an external dataset is needed to score word pairs using term co-occurrence. Here, we adopt Pointwise Mutual Information[7] (PMI) to evaluate the topic coherence over English Wikipedia articles[8]. Instead of using a sliding window, we consider a whole document to identify co-occurrence. Given a topic k and its top T words with highest probabilities $(w_1, w_2, \ldots, w_T)$, the PMI of topic k is:

$$
PMI(k) = \frac{2}{T(T-1)} \sum_{1 \leq i \leq j \leq T} log \frac{p(w_i, w_j)}{p(w_i)p(w_j)}
\tag{6}
$$

---

[2] https://github.com/PasaLab/TSSE.
[3] https://github.com/hjyyyyy/MultiKEDMM.
[4] https://github.com/tshi04/SeaNMF.
[5] https://github.com/overlook2021/PYSTM.
[6] https://code.google.com/p/word2vec.
[7] https://github.com/jhlau/topic_interpretability.
[8] http://deepdive.stanford.edu/opendata/.

Table 2. Topic quality comparison when $K = 50$. The best values are marked in boldface. For each baseline, $\uparrow$ or $\downarrow$ indicates that the baseline performs better or worse than our APUDMM significantly according to t-tests with 5% significance level.

Model	Tweet			SearchSnippets			GoogleNews		
	PMI	TU	TQ	PMI	TU	TQ	PMI	TU	TQ
LDA	1.180↓	0.859↑	1.014↓	1.377↓	0.829↓	1.142↓	1.166↓	0.846↑	0.986↓
BTM	1.204↓	0.756↑	0.910↓	1.445↓	0.694↓	1.003↓	1.263↓	0.848↑	1.071↓
DMM	1.233↓	0.731	0.901↓	1.438↓	0.652↓	0.938↓	1.245↓	0.829	1.032↓
GPU-DMM	1.587↓	0.728	1.155↓	1.350↓	0.754↓	1.018↓	1.489↓	0.841↑	1.252↓
TSSE-DMM	1.287↓	0.714↓	0.919↓	1.410↓	0.514↓	0.725↓	1.229↓	0.457↓	0.562↓
MultiKE-DMM	1.451↓	0.726	1.053↓	1.573↓	0.725↓	1.140↓	1.460↓	0.837	1.222↓
SeaNMF	1.311↓	**0.966**↑	1.266	1.651↓	0.847	1.398↓	1.396↓	**0.962**↑	1.343↓
PYSTM	1.207↓	0.867↑	1.046↓	1.446↓	**0.879**↑	1.271↓	1.235↓	0.859↑	1.061↓
APU-DMM	**1.745**	0.732	**1.277**	**2.318**	0.862	**1.998**	**1.742**	0.822	**1.432**

where $p(w_i)$ is the probability that word w_i appears in a document, and $p(w_i, w_j)$ is the probability that words w_i and w_j appear in the same document. The final result is computed as the average value on the top 10 words for all topics.

The topic diversity corresponds to the average probability of each word occurring in one of the other topics of the same model [3]. We use the topic unique metric (TU) [14] for topic diversity evaluation. For the top T words of topic K, it is defined as

$$TU(k) = \frac{1}{T} \sum_{i=1}^{T} \frac{1}{cnt(w_i)} \tag{7}$$

where $cnt(w_i)$ is the total number of times that word w_i appears in the top T words of all topics. TU close to 0 indicates redundant topics; TU close to 1 indicates more varied topics. T is set to 15 in our experiments.

We use the product of topic diversity and topic coherence as the **overall quality** of a model's topics [4,7], Topic quality (TQ) = PMI × TU. For all metrics, t-tests are performed to show whether the performance difference between APU-DMM and each baseline is statistically significant in terms of mean values over each metric. We consider the pairwise performance difference significant if the P-value obtained from the t-test is lower than 0.05.

4.3 Result Analysis

Topic Quality. The overall quality of topic coherence and topic uniqueness is presented in Table 2 and Table 3 when $K = 50$ and $K = 100$, respectively. First, we can observe that our APU-DMM achieves the best TQ score among all datasets, which suggests that our APU-DMM reaches a more comprehensive topic quality than other baselines. Although LDA outperforms several short

Table 3. Topic quality comparison when $K = 100$.

Model	Tweet			SearchSnippets			GoogleNews		
	PMI	TU	TQ	PMI	TU	TQ	PMI	TU	TQ
LDA	1.197↓	0.821↑	0.983↓	1.399↓	0.773↑	1.081↓	1.210↓	0.784↑	0.949↓
BTM	1.185↓	0.653↑	0.774↓	1.511↓	0.604↓	0.913↓	1.298↓	0.733↑	0.951↓
DMM	1.225↓	0.608↓	0.745↓	1.441↓	0.545↓	0.785↓	1.280↓	0.708↑	0.906↓
GPU-DMM	1.491↓	0.572↓	0.853↓	1.333↓	0.637↓	0.849↓	1.529↓	0.697	1.066↓
TSSE-DMM	1.262↓	0.632	0.798↓	1.425↓	0.450↓	0.641↓	1.246↓	0.813↑	1.013
MultiKE-DMM	1.431↓	0.617	0.883↓	1.619↓	0.617↓	0.999↓	1.454↓	0.713	1.037↓
SeaNMF	1.247↓	**0.930↑**	**1.160**	1.751↓	**0.819↑**	1.434↓	1.387↓	**0.928↑**	**1.287↑**
PYSTM	1.176↓	0.760↑	0.894↓	1.490↓	0.787↑	1.173↓	1.276↓	0.791↑	1.009↓
APU-DMM	**1.750**	0.621	1.087	**2.209**	0.744	**1.643**	**1.753**	0.685	1.201

Table 4. Results of ablation experiments when $K = 50$.

Model	Tweet			SearchSnippets			GoogleNews		
	PMI	TU	TQ	PMI	TU	TQ	PMI	TU	TQ
APU w/o TWC	1.587↓	0.728	1.155↓	1.350↓	0.754↓	1.018↓	1.489↓	**0.841↑**	1.252↓
APU w/o LC	1.716↓	0.730	1.253	2.214↓	0.850↓	1.882↓	1.657↓	0.827	1.370↓
APU-DMM	**1.745**	**0.732**	**1.277**	**2.318**	**0.862**	**1.998**	**1.742**	0.822	**1.432**

text topic modeling in the TQ score, it is prone to generate noisy and incoherent topics due to the poor PMI score. If the PMI score is low, the higher value in TU is meaningless since a high TU value can be expected to be achieved by randomly assigning words to a topic. To our knowledge, PYSTM is a state-of-the-art topic model in self-aggregation based models. However, as shown in Table 2 and Table 3, PYSTM is unstable when modeling different data sets. Non-semantic related short texts will likely be aggregated in one long document during inference. Knowledge-enhanced topic models, such as GPU-DMM and MultiKE-DMM, generally outperform other baselines on all datasets, whereas MultiKE-DMM performs better due to the multi-knowledge introduction. This is because external knowledge incorporated with the GPU scheme helps alleviate the data sparsity in short texts and improve topic interpretability. Although SeaNMF performs better than our model in the TU score when $K = 100$, the topic interpretability is worse than ours. In a nutshell, the proposed APU-DMM leverages global topic-word correlation to encourage adaptive weights of different words, resulting in a great balance of topic coherence and topic diversity.

Ablation Study. To test the effectiveness of incorporating topic-word correlation in our APU-DMM, we present the results of model variants on the three datasets, as shown in Table 4, in which APU w/o TWC represents APU-DMM without topic-word correlation and APU w/o LC represents APU-DMM without local word correlation, respectively. APU w/o TWC means that APU-DMM

Table 5. Examples of the Correlation of Similar Words to the Same Topic

Word	Top 5 words in topic z	Topic-Word Correlation
clinic	football sport basketball hockey soccer	1.011
ballroom	football sport basketball hockey soccer	1.044
gym	football sport basketball hockey soccer	1.138
virus	disease infection tuberculosis illness infections	1.224
influenza	disease infection tuberculosis illness infections	1.315
hepatitis	disease infection tuberculosis illness infections	1.315

Table 6. Examples of the Correlation of a Word to Different Topics

Word	Top 5 words in topic z	Correlation
singer	movie pictures movies film photos	1.025
	music jazz guitar songs band	1.228
gym	medical nutritional nutrition diet healthcare	1.061
	football sport basketball hockey soccer	1.138
therapists	industry business corporate businesses legislative	1.004
	medical nutritional nutrition diet healthcare	1.187
virus	nanotechnology science scientific study researchers	1.0092
	disease infection tuberculosis illness infections	1.2237

degenerates to a normal GPU-DMM model. APU w/o LC captures the topic-word correlation in Eq. (1), ignoring the LC calculation. Generally speaking, it can be observed that removing each part leads to performance degradation. APU w/o LC achieves better topic quality in all datasets than APU w/o TWC, demonstrating that our model can address GPU-DMM limitations. Furthermore, APU-DMM attains the best topic quality, indicating the effectiveness of integrating external and local knowledge.

4.4 Examples of Topic-Word Correlation

In this part, we validate the ability of our APU-DMM to differentiate topic-word correlation, characterized by the correlation of similar words to the same topic and the correlation of a word to different topics.

Table 5 shows some examples of the correlation of similar words to the same topic on SearchSnippets. For instance, the word *gym* is correlated with *clinic* and with *ballroom* simultaneously. Under a topic with Top 5 words including *football, sport, basketball, hockey, soccer*, the word *gym* is more correlated to this topic, whereas *clinic* and *ballroom* are both less correlated to this topic.

Table 6 shows examples of a word's correlation to different topics on Search-Snippets. We take word *virus* as the example. A topic with Top 5 words *nanotechnology, science, scientific, study, researchers* may denote a 'science' topic, and another topic with Top 5 words *disease, infection, tuberculosis, illness, infections* may denote a specific 'infectious health' topic. The word *virus* correlates more closely to the 'infectious health' topic than the 'science' topic.

From the last column at this Table 5 and Table 6, we obverse that the correlation of similar words to the same topic or the correlation of a word to different topics are distinct, which again indicates the effectiveness of our APU-DMM in capturing the subtle correlation between words and topic.

5   Conclusion

This paper proposes a new topic model for short texts that differentiates the subtle topic-word correlation, Adaptive Pólya Urn Dirichlet Multinomial Mixture (APU-DMM) model. APU-DMM incorporates a new Adaptive Pólya Urn (APU) scheme that takes full advantage of subtly correlating words to different topics when boosting the inference processing of DMM. We demonstrate the effectiveness of APU-DMM through extensive experiments using three real-world short-text datasets. We use quantitative and qualitative analysis to show that APU-DMM is a novel and effective topic model for short texts capable of producing more coherent and prosperous topics than existing state-of-the-art alternatives. In future work, we will explore methods that can more accurately express the word correlation in the current dataset. We hope this study can shed some light on short text topic modeling and related applications.

Acknowledgement. This work was supported by the National Key R&D Program of China: Research on the applicability of port food risk traceability, early warning and emergency assessment models (No.: 2019YFC1605504).

References

1. Bicalho, P.V., Pita, M., Pedrosa, G., Lacerda, A., Pappa, G.L.: A general framework to expand short text for topic modeling. Inf. Sci. **393**, 66–81 (2017)
2. Blei, D.M., Ng, A.Y., Jordan, M.I.: Latent Dirichlet allocation. J. Mach. Learn. Res. **3**(Jan), 993–1022 (2003)
3. Burkhardt, S., Kramer, S.: Decoupling sparsity and smoothness in the Dirichlet variational autoencoder topic model. JMLR **20**, 131:1–131:27 (2019)
4. Chen, J., Wang, R., He, J., Li, M.J.: Encouraging sparsity in neural topic modeling with non-mean-field inference. In: Koutra, D., Plant, C., Gomez Rodriguez, M., Baralis, E., Bonchi, F. (eds.) ECML PKDD 2023. LNCS, vol. 14172, pp. 142–158. Springer, Cham (2023). https://doi.org/10.1007/978-3-031-43421-1_9
5. Chen, Z., Liu, B.: Mining topics in documents: standing on the shoulders of big data. In: The 20th ACM SIGKDD International Conference on Knowledge Discovery and Data Mining, KDD, pp. 1116–1125. ACM (2014)
6. Cheng, X., Yan, X., Lan, Y., Guo, J.: BTM: topic modeling over short texts. IEEE Trans. Knowl. Data Eng. **26**(12), 2928–2941 (2014)
7. Dieng, A.B., Ruiz, F.J.R., Blei, D.M.: Topic modeling in embedding spaces. Trans. Assoc. Comput. Linguist. **8**, 439–453 (2020)

8. Guo, Y., Huang, Y., Ding, Y., Qi, S., Wang, X., Liao, Q.: GPU-BTM: a topic model for short text using auxiliary information. In: 5th IEEE International Conference on Data Science in Cyberspace, DSC, pp. 198–205. IEEE (2020)

9. He, J., Chen, J., Li, M.J.: Multi-knowledge embeddings enhanced topic modeling for short texts. In: Tanveer, M., Agarwal, S., Ozawa, S., Ekbal, A., Jatowt, A. (eds.) ICONIP 2022. LNCS, vol. 13625, pp. 521–532. Springer, Cham (2022). https://doi.org/10.1007/978-3-031-30111-7_44

10. Li, C., Duan, Y., Wang, H., Zhang, Z., Sun, A., Ma, Z.: Enhancing topic modeling for short texts with auxiliary word embeddings. ACM Trans. Inf. Syst. **36**(2), 11:1–11:30 (2017)

11. Li, C., Wang, H., Zhang, Z., Sun, A., Ma, Z.: Topic modeling for short texts with auxiliary word embeddings. In: Proceedings of the 39th International ACM SIGIR, pp. 165–174 (2016)

12. Mai, C., Qiu, X., Luo, K., Chen, M., Zhao, B., Huang, Y.: TSSE-DMM: topic modeling for short texts based on topic subdivision and semantic enhancement. In: Advances in Knowledge Discovery and Data Mining - 25th Pacific-Asia Conference, PAKDD, vol. 12713, pp. 640–651 (2021)

13. Mimno, D.M., Wallach, H.M., Talley, E.M., Leenders, M., McCallum, A.: Optimizing semantic coherence in topic models. In: Proceedings of the 2011 Conference on Empirical Methods in Natural Language Processing, EMNLP. pp. 262–272. ACL (2011)

14. Nan, F., Ding, R., Nallapati, R., Xiang, B.: Topic modeling with Wasserstein autoencoders. In: Proceedings of the 57th Conference of the Association for Computational Linguistics, pp. 6345–6381. ACL (2019)

15. Newman, D., Lau, J.H., Grieser, K., Baldwin, T.: Automatic evaluation of topic coherence. In: Human Language Technologies: Conference of the North American Chapter of the Association of Computational Linguistics, pp. 100–108 (2010)

16. Nguyen, D.Q., Billingsley, R., Du, L., Johnson, M.: Improving topic models with latent feature word representations. Trans. Assoc. Comput. Linguist. **3**, 299–313 (2015)

17. Nigam, K., McCallum, A., Thrun, S., Mitchell, T.M.: Text classification from labeled and unlabeled documents using EM. Mach. Learn. **39**(2/3), 103–134 (2000)

18. Niu, Y., Zhang, H., Li, J.: A Pitman-Yor process self-aggregated topic model for short texts of social media. IEEE Access **9**, 129011–129021 (2021)

19. Phan, X.H., Nguyen, M.L., Horiguchi, S.: Learning to classify short and sparse text & web with hidden topics from large-scale data collections. In: Proceedings of the 17th International Conference on World Wide Web, pp. 91–100. ACM (2008)

20. Qiang, J., Qian, Z., Li, Y., Yuan, Y., Wu, X.: Short text topic modeling techniques, applications, and performance: a survey. IEEE Trans. Knowl. Data Eng. **34**(3), 1427–1445 (2022)

21. Quan, X., Kit, C., Ge, Y., Pan, S.J.: Short and sparse text topic modeling via self-aggregation. In: Proceedings of the Twenty-Fourth International Joint Conference on Artificial Intelligence, IJCAI, pp. 2270–2276. AAAI Press (2015)

22. Shahnaz, F., Berry, M.W., Pauca, V.P., Plemmons, R.J.: Document clustering using nonnegative matrix factorization. Inf. Process. Manag. **42**(2), 373–386 (2006)

23. Wang, R., Zhou, D., He, Y.: Optimising topic coherence with weighted Po'lya Urn scheme. Neurocomputing **385**, 329–339 (2020)

24. Yin, J., Wang, J.: A Dirichlet multinomial mixture model-based approach for short text clustering. In: The 20th ACM SIGKDD International Conference on Knowledge Discovery and Data Mining, KDD, pp. 233–242. ACM (2014)

25. Zubiaga, A., Ji, H.: Harnessing web page directories for large-scale classification of tweets. In: Proceedings of the 22nd International Conference on World Wide Web, pp. 225–226 (2013)
26. Zuo, Y., et al. Topic modeling of short texts: a pseudo-document view. In: ACM SIGKDD International Conference on Knowledge Discovery and Data Mining, pp. 2105–2114. ACM (2016)
27. Zuo, Y., Zhao, J., Xu, K.: Word network topic model: a simple but general solution for short and imbalanced texts. Knowl. Inf. Syst. 48(2), 379–398 (2016)

Informative Prompt Learning for Low-Shot Commonsense Question Answering via Fine-Grained Redundancy Reduction

Zhikai Lei, Jie Zhou, Qin Chen$^{(\boxtimes)}$, Qi Zhang, and Liang He

School of Computer Science and Technology, East China Normal University,
Shanghai, China
{kausal,qzhang}@stu.ecnu.edu.cn, {jzhou,qchen,lhe}@cs.ecnu.edu.cn

Abstract. Low-shot commonsense question answering (CQA) poses a big challenge due to the absence of sufficient labeled data and commonsense knowledge. Recent work focuses on utilizing the potential of commonsense reasoning of pre-trained language models (PLMs) for low-shot CQA. In addition, various prompt learning methods have been studied to elicit implicit knowledge from PLMs for performance promotion. Whereas, it has been shown that PLMs suffer from the redundancy problem that many neurons encode similar information, especially under a small sample regime, leading prompt learning to be less informative in low-shot scenarios. In this paper, we propose an informative prompt learning approach, which aims to elicit more diverse and useful knowledge from PLMs for low-shot CQA via fine-grained redundancy reduction. Specifically, our redundancy-reduction method imposes restrictions upon the fine-grained neuron-level to encourage each dimension to model different knowledge or clues. Experiments on three benchmark datasets show the great advantages of our proposed approach in low-shot settings. Moreover, we conduct both quantitative and qualitative analyses, which shed light on why our approach can lead to great improvements.

Keywords: Commonsense question answering · Low-shot learning · Redundancy in Pre-trained language models

1 Introduction

Commonsense question answering (CQA) is a type of QA that relies on various commonsense knowledge, such as spatial relations, scientific facts, and social situations [18]. Taking Fig. 1 as an example, for the question *"Where did Puppigerus live?"*, one can choose the correct answer *"ocean"* from several options (e.g., *bog, ocean, land,* and *marsh*) with his commonsense.

Though some researches exhibit effectiveness in integrating the QA model with external commonsense knowledge bases such as ConceptNet [17] and ATOMIC [13] for CQA, the knowledge bases usually suffer from noise and sparsity, where the truly relevant knowledge needed to fill the gap between the

B. Luo et al. (Eds.): ICONIP 2023, CCIS 1968, pp. 377–390, 2024.
https://doi.org/10.1007/978-981-99-8181-6_29

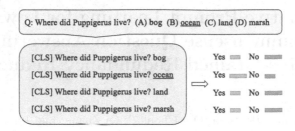

Fig. 1. General framework for CQA using PLMs.

question and the answer could be missing [21]. Therefore, researchers start to investigate the potential of pre-trained language models (PLMs) [4,8] to serve as implicit knowledge bases for CQA [6], and the general framework is shown in Fig. 1. However, these approaches still require thousands or tens of thousands of examples for fine-tuning, which are not available in the low-shot scenarios where only several examples are given, and the performance will severely decline with insufficient training data [15].

To resolve the data deficiency problem, prompt learning has been proposed to enhance the commonsense reasoning ability of PLMs for low-shot CQA. Some researchers converted questions to cloze-format for better extraction of implicit knowledge [5]. Whereas, it has been shown that PLMs suffer from the problem of redundancy in representations [3]. Specifically, the correlations between two neurons are very high, indicating that they encode similar information and therefore are redundant. Considering that the feature dimensions are usually small in low-shot scenarios to avoid over-fitting [23], the redundancy problem makes prompt learning less informative for low-shot CQA, which severely hurts the performance.

Motivated by the *redundancy-reduction* principle in neuroscience that the goal of sensory processing is to recode highly redundant sensory inputs into a factorial code with statistically independent components [1], we propose an informative prompt learning approach for low-shot CQA via fine-grained redundancy reduction. First, we introduce an effective prompt-based framework to better elicit implicit commonsense knowledge from PLMs for CQA. For the prompt template, we rewrite the question sentence as a declarative sentence and inject the option with a suffix "right? [MASK]" as shown in Fig. 2, and the answer depends on whether the [MASK] is predicted as *yes* or *no* according to the prompt-based representation. In order to make the prompt learning more informative to include various relevant knowledge for CQA, we present a fine-grained redundancy reduction method, which treats the [MASK] representations for a question with different options as twins, and imposes restrictions that the correlation matrix computed from the twins should be as close to the identity matrix as possible. In other words, different dimensions should represent different views or clues for CQA, which makes the representations more informative and less redundant. We also provide a theoretical analysis for our fine-grained redundancy reduction method in an information theory view, indicating that our method can retain as much useful information as possible and reduce the task-irrelevant noise.

Extensive experiments are performed on three well-known benchmark datasets, namely CommonsenseQA, OpenBookQA, and SocialIQA. The results demonstrate the effectiveness of our proposed approach. In particular, we outperform the state-of-the-art approaches in both few- and zero-shot settings in most cases. Also, we analyze how our approach reduces redundancy and visualize the representations for a better understanding.

The main contributions of our work can be summarized as follows:

- To the best of our knowledge, it is the first attempt to explore how to make prompt learning more informative by fine-grained redundancy reduction, which aims to make full use of each hidden dimension to elicit as much useful implicit knowledge as possible from PLMs for downstream tasks in low-shot scenarios.
- We investigate the potential of our proposed informative prompt learning approach for low-shot CQA, and also verify the effectiveness in an information theory view by bridging the gap between the redundancy reduction loss and information bottleneck objective.
- We conduct elaborate analyses of the experimental results on three benchmark datasets, which show the great advantages of our approach.

2   Related Work

2.1   Commonsense Question Answering

Commonsense question answering (CQA) relies on various commonsense knowledge for answer prediction [18]. Some previous works focus on leveraging extra knowledge bases (e.g., ConceptNet [17] and ATOMIC [13]) for CQA. Other researchers adopted Graph Neural Networks (GNN) to model the knowledge in knowledge bases [22]. However, these methods suffer from the knowledge sparsity problem that the truly relevant knowledge for CQA could be missing in existing knowledge bases.

Recently, many researchers turn to utilizing the implicit commonsense knowledge stored in PLMs for CQA [6]. For low resources settings, prompt learning has been proposed to better elicit knowledge from PLMs for downstream tasks. Some works designed prompt templates to induce knowledge from PLMs [5].

Although PLM-based prompt learning achieves good performance in CQA, recent studies reveal that PLMs suffer from the redundancy problem, leading prompt learning to be less informative by encoding similar and irrelevant information [3,7]. Therefore, we focus on more informative prompt learning to capture more task relevant clues while forgetting the irrelevant information.

2.2 Redundancy in PLMs

The related works about model compression demonstrate that large-scale neural models can be compressed into smaller ones with a little performance decline [12], indicating that many parameters can be removed for downstream tasks. Later works explicitly investigate the redundancy problem from different perspectives. Some researchers investigated the redundancy problem in neuron-level and layer-level, demonstrating that up to 85% of the neurons and 11 layers are highly correlated with similar information, which can be removed for specific downstream tasks [3].

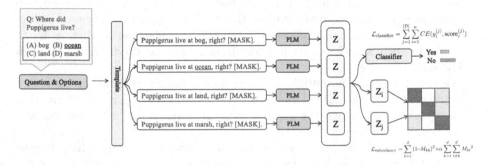

Fig. 2. Overview of our approach. We use a pre-trained language model for prompt template encoding to obtain the [MASK] representation Z, which is then used to compute the cross-entropy classifier loss $\mathcal{L}_{\text{classifier}}$ and the fine-grained redundancy reduction loss $\mathcal{L}_{\text{redundancy}}$ respectively.

Different from previous studies, we mainly focus on urging each dimension to be less correlated and encode diverse information instead of removing the layers or neurons, aiming to elicit different knowledge or clues from PLMs in low-shot scenarios. In particular, we present a fine-grained redundancy reduction method and apply it to low-shot CQA within the prompt-based framework.

3 Approach

For ease of understanding, we first provide some notations used in this paper. Given a set of question and option pairs $\{q^{(j)}, O^{(j)}\}_{j=1}^{|\mathcal{D}|}$, where $|\mathcal{D}|$ is the number of samples, $O^{(j)} = \{o_1^{(j)}, o_2^{(j)}, \ldots, o_n^{(j)}\}$ is the option set for question $q^{(j)}$, and n is the number of options. The CQA task aims to select the exact correct answer from the options.

We propose an informative prompt-based approach for low-shot CQA via fine-grained redundancy-reduction, and the overview is shown in Fig. 2. Firstly, to better utilize the implicit knowledge in the PLMs, we design a prompt-based framework by transforming the questions into a natural sentence with filled options. Then, we present a fine-grained redundancy reduction method, which treats the representations of a question with different options ($q^{(j)}$, $O^{(j)}$) as twins, and encourages each dimension to elicit different knowledge from PLMs while mitigating the redundancy to make prompt learning more informative. Finally, we give an analysis of the redundancy reduction method in an information theory view, which serves as a warranty for the effectiveness of our approach.

3.1 Prompt-Based Framework

Prompt learning has obtained great success in various low-shot NLP tasks. Motivated by its effectiveness, we present a prompt-based framework for low-shot CQA. As the question-option pair does not resemble the masked word prediction in pre-training, we transform the pair into natural sentences with the cloze question and options to better extract commonsense knowledge from the PLMs. Specifically, we first use the syntactic-based rewriting translation method [5] to transform the question $q^{(j)}$ to a cloze question $c^{(j)}$. For example, the natural question "*Where did Puppigerus live?*" will be translated into "*Puppigerus live at [OPTION]*".

For each option $o_i^{(j)} \in O^{(j)}$, the [OPTION] token in cloze question $c^{(j)}$ is replaced with a specific option $o_i^{(j)}$, denoted as $c_i^{(j)}$. Then, we obtain the prompt template sentence $s_i^{(j)}$ by using pattern $\mathcal{P}(.)$ described in Eq. 1,

$$s_i^{(j)} = \mathcal{P}(c_i^{(j)}) = c_i^{(j)}, \text{right? [MASK]}. \tag{1}$$

where [MASK] is the mask token in the PLM. For the above cloze question "*Puppigerus live at [OPTION]*", the final prompt template sentence $s_i^{(j)}$ is "*Puppigerus live at ocean, right? [MASK]*" when option $o_i^{(j)} = $ "*ocean*".

Next, we feed the prompt template sentence $s_i^{(j)} \in \{s_1^{(j)}, \ldots, s_n^{(j)}\}$ to a masked language model MLM, which predicts the probability of each token t appearing at [MASK]:

$$P(t|s_i^{(j)}) = P_{MLM}\left([MASK] = t|s_i^{(j)}\right) \tag{2}$$

Each option can be regarded as a binary classification issue, so we associate the score with the probability of "yes" and "no" tokens appearing at [MASK]. The score for option $o_i^{(j)}$ is calculated as:

$$\text{score}_i^{(j)} = \frac{\exp\left(P(yes|s_i^{(j)})\right)}{\exp\left(P(yes|s_i^{(j)})\right) + \exp\left(P(no|s_i^{(j)})\right)} \tag{3}$$

Cross-entropy loss is applied to each option, and the training objective of classification becomes:

$$\mathcal{L}_{\text{classifier}} = \sum_{j=1}^{|\mathcal{D}|} \sum_{i=1}^{n} CE(y_i^{(j)}, \text{score}_i^{(j)})$$

where $y_i^{(j)}$ takes the value of 1 if $o_i^{(j)}$ is the correct option. Otherwise, $y_i^{(j)}$ is 0.

3.2 Fine-Grained Redundancy Reduction

Existing studies have proved that a significant portion of neurons in PLMs encode similar information, which is redundant and irrelevant for downstream tasks, especially under a small sample regime [3,7]. Motivated by the *redundancy-reduction* principle in neuroscience that the goal of sensory processing is to recode highly redundant sensory inputs into a factorial code with statistically independent components [1], we present a fine-grained redundancy-reduction method to make prompt learning more informative. We first calculate a correlation matrix M in the dimensionality level for hidden representations of the twins (a question with different options) [24]. Then, we restrict the correlation matrix M to be as close to the identity matrix as possible. In this way, different dimensions attempt to encode different information and the redundant information will be mitigated, encouraging prompt learning to elicit more diverse and relevant knowledge from PLMs for CQA.

Specifically, given a batch of samples $B = \{q^{(j)}, O^{(j)}\}_{j=1}^{|B|}$, we get the prompt-based sentences $S = \{s_1^{(j)}, s_2^{(j)}, \ldots, s_n^{(j)}\}_{j=1}^{|B|}$, where $|B|$ is the number of samples in the batch. Then, S is fed into MLM to extract the hidden representations of [MASK], denoted as $\mathbf{Z} \in \mathbb{R}^{|B| \times n \times d}$, where d denotes the representation dimension. For each question, we randomly choose two options and get the corresponding [MASK] representations, denoted as $\mathbf{Z}^1$ and $\mathbf{Z}^2 \in \mathbb{R}^{|B| \times d}$ respectively. We calculate the correlation matrix $M \in \mathbb{R}^{d \times d}$ along the batch dimension as:

$$M_{kt} = \frac{\sum_{b=1}^{|B|} \mathbf{Z}_{b,k}^1 \mathbf{Z}_{b,t}^2}{\sqrt{\sum_{b=1}^{|B|} \left(\mathbf{Z}_{b,k}^1\right)^2} \sqrt{\sum_{b=1}^{|B|} \left(\mathbf{Z}_{b,t}^2\right)^2}} \tag{4}$$

where b is the batch index, and k, t are the indexes of dimension d.

Finally, the redundancy reduction objective is formulated as:

$$\mathcal{L}_{\text{redundancy}}(\mathbf{Z}^1, \mathbf{Z}^2) = \sum_{k=1}^{d} (1 - M_{kk})^2 + \alpha \sum_{k=1}^{d} \sum_{t \neq k}^{d} M_{kt}^2 \tag{5}$$

where α is a hyperparameter defined to balance the two terms.

The total loss is formulated by combining the classifier loss with the redundancy reduction loss:

$$\mathcal{L} = \mathcal{L}_{\text{classifier}} + \beta \mathcal{L}_{\text{redundancy}} \tag{6}$$

where β is a constant to trade off the two losses.

3.3 Rethinking Fine-Grained Redundancy Reduction in an Information Theory View

We provide an analysis of our fine-grained redundancy reduction objective function in an information theory view. Specifically, we find this objective function can be regarded as the information bottleneck objective [19,20], which aims to learn a representation Z by maximizing the mutual information between Z and output Y and minimizing the mutual information between Z and input X. Thus, the information bottleneck objective helps the model forget the task-irrelevant information while retaining the task-relevant information. In our method, the information bottleneck objective aims to learn a representation that retains as much information about sample X as possible while keeping the least information about the specific option X_i applied to that sample X (Fig. 3). In this way, we can learn an invariant informative representation with non-redundant information.

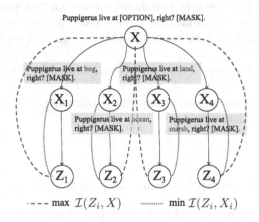

Fig. 3. Rethinking fine-grained redundancy reduction in an information theory view.

The objective above can be formulated as:

$$
\begin{aligned}
\mathcal{L}_{IB} &= \mathcal{I}(Z_i, X_i) - \lambda \mathcal{I}(Z_i, X) \\
&= (H(Z_i) - H(Z_i|X_i)) - \lambda (H(Z_i) - H(Z_i|X)) \\
&= (H(Z_i) - 0) - \lambda (H(Z_i) - H(Z_i|X)) \\
&= H(Z_i|X) + \frac{1 - \lambda}{\lambda} H(Z_i)
\end{aligned}
\tag{7}
$$

where $\mathcal{I}(.,.)$ refers to mutual information, and $H(.)$ represents entropy. Mutual information is transferred to entropy according to the definition. X_i represents the sample X with option o_i. $H(Z_i|X_i)$ is equal to zero because the encoder is determined, so Z_i is fixed for the same X_i. Since the overall scaling is not important, we rearrange the equation in the last line.

It is hard to calculate the entropy of high-dimensional signals. To simplify this problem, we assume that Z_i belongs to a Gaussian distribution, the entropy of which is determined by the log determinant of the covariance matrix [2]. So the objective becomes:

$$
\mathcal{L}_{IB} = \mathbb{E}_X log|M_{Z_i|X}| + \frac{1 - \lambda}{\lambda} log|M_{Z_i}|
\tag{8}
$$

It has been determined that after several simplifications and approximations, Eq. 8 can set up a bridge with Eq. 5 [24].

4　Experimental Setup

4.1　Datasets

To evaluate the effectiveness of our approach, we conduct experiments on three widely used CQA datasets.

CommonsenseQA [18] requires commonsense for answer prediction, which is about the concepts and relations from ConceptNet. As test labels are not public, we divide the training set into a new training set and test set following [9].

OpenBookQA [11] requires scientific facts and knowledge for question answering.

SocialIQA [14] focuses on evaluating the emotional and social intelligence of models in everyday situations.

4.2　Baselines

The following recent advanced baselines are used for comparison:

- **ALBERT** [8]: We use the ALBERT-xxlarge-v2 version as the base model.
- **Self-talk** [16]: It generates the background knowledge from PLMs (GPT2-Large and ALBERT-xxlarge-v2) for unsupervised CQA.
- **Cloze translation based prompts** [5]: The natural questions are converted into cloze-style sentences as prompts to better solicit commonsense knowledge from PLMs with four translation methods, namely syntactic-based rewriting (**Syntactic-based**), unsupervised seq2seq (**Unsup. Seq2Seq**), supervised seq2seq (**Sup. Seq2Seq**) and supervised sequence tagging (**Sup. Tag**).
- **Methods with Ensemble**: **Ensemble** simply sums the probabilities for each answer from different methods, and takes the answer with the highest probability [5]. **Consistency** uses the ensemble answer as the pseudo label and encourages the consistency between different predictions with self-training [5].
- **Methods with Knowledge Base**: [10] uses knowledge-driven data construction (**KDC**) for zero-shot CQA, which generates question and answer pairs based on external knowledge bases, and then uses PLMs for fine-tuning on the generated data.

4.3   Implementation Details

We use ALBERT-xxlarge-v2 as our basic model, which has 235M parameters. Experiments are performed with a single RTX 3090, and it takes about 4–32 s to train an epoch. The learning rate is set to $5e - 6$ and the batch size is 4. The hidden dimension is 1024. For the fine-grained redundancy reduction loss, we search the parameters α and β from $\{1e - 3, 2e - 3\}$. In few-shot settings, we use the AdamW algorithm for optimization and early stopping to avoid overfitting. To reduce the serendipity, we employ three different seeds to sample 16/32/64/128 instances for few-shot settings and report the best performance. In zero-shot settings, we use our model to obtain the pseudo labels for the data, which is then used for fine-tuning iteratively until the model converges as demonstrated in the previous study [15].

5   Experimental Analysis

5.1   Main Results

Performance in Few-Shot Settings. The ALBERT model is used as a basic baseline as it is the base of our approach. In addition, given that our approach does not rely on extra knowledge base or ensemble methods, we also include one of the state-of-the-art single-model based baseline (as shown in Table 2), i.e., Syntactic-based [5] that utilized syntactic rules to transform natural questions to cloze format for prompt learning. The results are shown in Table 1.

We observe that our approach consistently outperforms the basic baseline and the recent advanced Syntactic-based method in all few-shot settings with substantial improvements. It is also interesting to find that the improvement tends to be larger with fewer training samples, which exhibits the great advantage of our informative prompt learning in scenarios with very few samples. This finding is in line with previous studies that the large deep neural networks are redundant with high correlations in each layer, especially when training with small samples [23]. Thus, our approach is more effective by performing fine-grained redundancy reduction to decrease the correlations between each feature dimension, which makes prompt learning to be more informative in few-shot scenarios.

Performance in Zero-Shot Settings. We further investigate the effectiveness of our approach by comparing with the recent competitive baselines in zero-shot settings, and the results are shown in Table 2. All the baseline results are reported in previous studies [5], which can be divided into three groups: methods with single model, methods with ensemble, and methods with knowledge base.

From Table 2, we have the following observations. **First**, our proposed approach achieves overwhelming advantages over the single-model-based baselines. In particular, we achieve an average improvement of up to 14.95% compared

Table 1. The accuracy (%) performance in few-shot settings. We implement all the baselines following the same setting as ours. The best result on each dataset is marked in bold.

Method	CommonsenseQA		SocialIQA		OpenBookQA		Average	
	dev	test	dev	test	dev	test	dev	test
16 Exapmles								
Base (ALBERT)	30.14	26.91	35.26	35.66	32.00	27.20	32.47	29.92
Syntactic-based	59.13	56.49	49.85	47.84	44.80	41.60	51.26	48.64
Ours	**63.23**	**59.23**	**54.71**	**52.07**	**50.80**	**50.20**	**56.25**	**53.83**
w/o $\mathcal{L}_{\text{redundancy}}$	49.80	44.64	50.26	50.81	46.40	44.20	48.82	46.55
32 Exapmles								
Base (ALBERT)	31.12	27.32	36.49	34.40	32.60	27.00	33.40	29.57
Syntactic-based	58.07	56.57	51.48	50.31	46.20	42.40	51.92	49.76
Ours	**63.39**	**60.03**	**57.01**	**54.27**	**51.20**	**52.00**	**57.20**	**55.43**
w/o $\mathcal{L}_{\text{redundancy}}$	51.92	49.80	52.71	50.85	46.20	47.20	50.28	49.28
64 Exapmles								
Base (ALBERT)	33.99	32.63	39.36	37.46	35.60	28.00	36.32	32.70
Syntactic-based	61.10	60.03	54.40	49.78	48.60	47.00	54.70	52.27
Ours	**65.36**	**60.92**	**57.83**	**55.89**	**54.20**	**54.00**	**59.13**	**56.94**
w/o $\mathcal{L}_{\text{redundancy}}$	56.18	51.25	56.86	55.35	48.60	50.20	53.88	52.27
128 Exapmles								
Base (ALBERT)	41.61	39.89	44.11	40.29	36.80	28.00	40.84	36.06
Syntactic-based	65.19	61.80	59.88	55.71	55.20	46.00	60.09	54.50
Ours	**67.08**	**63.74**	**60.39**	**57.55**	**58.60**	**57.00**	**62.02**	**59.43**
w/o $\mathcal{L}_{\text{redundancy}}$	59.38	53.91	59.42	55.76	51.40	52.40	56.73	54.02

with the best single-model-based baseline (i.e., Syntactic-based). **Second**, we also outperform the methods relying on ensemble or knowledge bases in most cases, which further verifies the effectiveness of our approach that encourages different feature dimensions to elicit different knowledge from PLMs by redundancy reduction during prompt learning.

Ablation Studies. We conduct ablation studies to examine the effectiveness of each component of our approach. The results are shown in Table 1. We observe that the performance drops significantly after removing the fine-grained redundancy reduction component from our approach in all cases. In addition, when further removing our prompt-based framework, our approach degenerates into the basic baseline Base (ALBERT). It is notable that there are large margins between the performance of model without redundancy reduction (denoted

Table 2. The accuracy (%) performance in zero-shot settings. For the baseline Consistency, we report the best accuracy score with different random seeds [5]. The best result on each dataset is marked in bold.

Method	CommonsenseQA		SocialIQA		OpenBookQA		Average	
	dev	test	dev	test	dev	test	dev	test
Base (ALBERT)	31.14	28.52	41.71	40.47	31.80	33.00	34.88	34.00
Self-talk (GPT2)	31.53	29.74	45.34	44.47	28.40	30.80	35.09	35.00
Self-talk (ALBERT)	15.89	17.49	26.25	26.48	22.20	19.40	21.45	21.12
Unsup. Seq2Seq	43.49	42.86	40.94	38.80	40.00	39.20	41.48	40.29
Sup. Tag	50.86	48.51	41.53	40.78	39.00	38.60	43.80	42.63
Sup. Seq2Seq	51.60	49.00	44.73	41.41	39.00	39.80	45.11	43.40
Syntactic-based	50.94	48.67	44.11	42.00	41.60	39.80	45.55	43.49
Methods with Ensemble								
Ensemble	54.62	51.57	44.11	42.04	41.00	39.20	46.58	44.27
Consistency	64.21	61.43	55.12	55.58	50.80	50.60	56.71	55.87
Methods with Knowledge Base								
KDC (RoBERTa)	68.63	**66.88**	56.04	51.93	34.80	38.00	53.16	52.27
KDC (ALBERT)	66.50	64.87	51.02	52.28	45.40	48.00	54.31	55.05
Ours	**69.04**	60.19	**57.37**	**56.34**	**58.40**	**58.80**	**61.60**	**58.44**

as w/o $\mathcal{L}_{\text{redundancy}}$) and Base (ALBERT), indicating the effectiveness of our prompt-based framework for CQA.

5.2 Further Analysis

Quantitative Analysis for Redundancy Reduction. To investigate whether our approach helps reduce redundancy in the fine-grained neuron-level, we calculate the Pearson product-moment correlation based on the representation of each neuron for the [MASK] tokens [3]. Afterwards, we obtain the correlation matrix $corr$, $corr \in \mathbb{R}^{d \times d}$. $corr[i][j]$ denotes the correlation between the neuron i and j. For the limited space, we average the correlation matrix over all the three benchmark datasets in 16-shot scenarios, and show the results in Fig. 4(a). We observe that our approach obtains lower correlations in the neuron-level representations compared with the baseline without redundancy reduction.

Qualitative Analysis for Redundancy Reduction. We visualize the first 20 feature representations with and without our redundancy reduction method in Fig. 4(b) and 4(c). Each color means a unique feature dimension, and the three dots in the same color represents the question with three options respectively. We observe that the representation without redundancy reduction seems relatively messy, and there are no clear boundaries between different dimensions, indicating that they represent similar information. In contrast, our redundancy reduction method narrows the gap within the same dimension and expands the distance between different dimensions, which can include more diverse and useful clues for informative prompt learning.

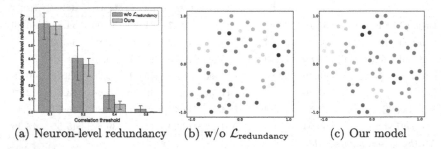

(a) Neuron-level redundancy   (b) w/o $\mathcal{L}_{\text{redundancy}}$ (c) Our model

Fig. 4. (a) illustrates the percentage of neuron-level redundancy with different correlation thresholds. (b) and (c) show the first 20 feature representations (total number is 1024) visualized based on a batch of samples from the test set of SocialIQA. Each color means a unique feature dimension, and the three dots in the same color represent a question with three options. (Color figure online)

6 Conclusions and Future Work

In this paper, we propose an informative prompt learning approach for low-shot CQA, which guides different feature dimensions to elicit different knowledge or clues from PLMs via our fine-grained redundancy reduction method. We also bridge the gap between the redundancy reduction loss and information bottleneck objective, which further verifies the effectiveness of our approach in the information theory view. The experimental results on three benchmark datasets show the overwhelming superiority of our approach over the general prompt learning without any redundancy reduction. In particular, we outperform the recent advanced baselines that rely on ensemble or extra knowledge bases in most cases. It is also interesting to find that the improvement tends to be larger with fewer training samples, indicating the great advantage of our approach in scenarios with a sample scarcity problem. In the future, we would like to explore the effectiveness of our approach for other low-shot tasks, and investigate more strategies for redundancy reduction.

Acknowledgements. This research is funded by the National Key Research and Development Program of China (No. 2021ZD0114002), the National Nature Science Foundation of China (No. 61906045), and the Science and Technology Commission of Shanghai Municipality Grant (No. 22511105901, No. 21511100402).

References

1. Barlow, H.: Possible principles underlying the transformations of sensory messages. Sens. Commun. **1**, 217–233 (1961)
2. Cai, T.T., Liang, T., Zhou, H.H.: Law of log determinant of sample covariance matrix and optimal estimation of differential entropy for high-dimensional gaussian distributions. J. Multivar. Anal. **137**, 161–172 (2015)
3. Dalvi, F., Sajjad, H., Durrani, N., Belinkov, Y.: Analyzing redundancy in pre-trained transformer models. In: EMNLP, pp. 4908–4926. ACL (2020)
4. Devlin, J., Chang, M., Lee, K., Toutanova, K.: BERT: pre-training of deep bidirectional transformers for language understanding. In: NAACL-HLT, pp. 4171–4186. ACL (2019)
5. Dou, Z., Peng, N.: Zero-shot commonsense question answering with cloze translation and consistency optimization. ArXiv (2022)
6. Huang, Z., Wu, A., Zhou, J., Gu, Y., Zhao, Y., Cheng, G.: Clues before answers: generation-enhanced multiple-choice QA. ArXiv (2022)
7. Kovaleva, O., Romanov, A., Rogers, A., Rumshisky, A.: Revealing the dark secrets of BERT. In: EMNLP/IJCNLP, pp. 4364–4373. ACL (2019)
8. Lan, Z., Chen, M., Goodman, S., Gimpel, K., Sharma, P., Soricut, R.: ALBERT: a lite BERT for self-supervised learning of language representations. In: ICLR 2020, Addis Ababa, Ethiopia, 26–30 April 2020. OpenReview.net (2020)
9. Lin, B.Y., Chen, X., Chen, J., Ren, X.: KagNet: knowledge-aware graph networks for commonsense reasoning. In: EMNLP/IJCNLP, pp. 2829–2839. ACL (2019)
10. Ma, K., Ilievski, F., Francis, J., Bisk, Y., Nyberg, E., Oltramari, A.: Knowledge-driven data construction for zero-shot evaluation in commonsense question answering. In: AAAI (2021)
11. Mihaylov, T., Clark, P., Khot, T., Sabharwal, A.: Can a suit of armor conduct electricity? A new dataset for open book question answering. In: EMNLP, pp. 2381–2391. ACL (2018)
12. Sanh, V., Debut, L., Chaumond, J., Wolf, T.: DistilBERT, a distilled version of BERT: smaller, faster, cheaper and lighter. ArXiv (2019)
13. Sap, M., et al.: ATOMIC: an atlas of machine commonsense for if-then reasoning. In: AAAI, pp. 3027–3035. AAAI Press (2019)
14. Sap, M., Rashkin, H., Chen, D., Bras, R.L., Choi, Y.: Social IQA: commonsense reasoning about social interactions. In: EMNLP, pp. 4462–4472. ACL (2019)
15. Schick, T., Schütze, H.: Exploiting cloze-questions for few-shot text classification and natural language inference. In: EACL, pp. 255–269. ACL (2021)
16. Shwartz, V., West, P., Bras, R.L., Bhagavatula, C., Choi, Y.: Unsupervised commonsense question answering with self-talk. In: EMNLP, pp. 4615–4629. ACL(2020)
17. Speer, R., Chin, J., Havasi, C.: ConceptNet 5.5: an open multilingual graph of general knowledge. In: Singh, S., Markovitch, S. (eds.) AAAI, pp. 4444–4451. AAAI Press (2017)
18. Talmor, A., Herzig, J., Lourie, N., Berant, J.: CommonsenseQA: a question answering challenge targeting commonsense knowledge. In: NAACL-HLT, pp. 4149–4158. ACL (2019)
19. TISHBY, N.: The information bottleneck method. In: Proceedings of the 37th Annual Allerton Conference on Communications, Control and Computing, 1999, pp. 368–377 (1999)

20. Tishby, N., Zaslavsky, N.: Deep learning and the information bottleneck principle. In: 2015 IEEE Information Theory Workshop (ITW)
21. Wang, P., Peng, N., Ilievski, F., Szekely, P.A., Ren, X.: Connecting the dots: a knowledgeable path generator for commonsense question answering. In: EMNLP (Findings). Findings of ACL, vol. EMNLP 2020, pp. 4129–4140. ACL (2020)
22. Yasunaga, M., Ren, H., Bosselut, A., Liang, P., Leskovec, J.: QA-GNN: reasoning with language models and knowledge graphs for question answering. In: NAACL-HLT, pp. 535–546. ACL (2021)
23. Yoo, D., Fan, H., Boddeti, V.N., Kitani, K.M.: Efficient k-shot learning with regularized deep networks. In: McIlraith, S.A., Weinberger, K.Q. (eds.) Proceedings of the Thirty-Second (AAAI-18), pp. 4382–4389. AAAI Press (2018)
24. Zbontar, J., Jing, L., Misra, I., LeCun, Y., Deny, S.: Barlow twins: self-supervised learning via redundancy reduction. In: ICML (2021)

Rethinking Unsupervised Domain Adaptation for Nighttime Tracking

Jiaying Chen, Qiyu Sun, Chaoqiang Zhao, Wenqi Ren, and Yang Tang[✉]

Key Laboratory of Smart Manufacturing in Energy Chemical Process, Ministry of Education, East China University of Science and Technology, Shanghai 200237, China
tangtany@gmail.com

Abstract. Despite the considerable progress that has been achieved in visual object tracking, it remains a challenge to track in low-light circumstances. Prior nighttime tracking methods suffer from either weak collaboration of cascade structures or the lack of pseudo supervision, and thus fail to bring out satisfactory results. In this paper, we develop a novel unsupervised domain adaptation framework for nighttime tracking. Specifically, we benefit from the establishment of pseudo supervision in the mean teacher network, and further extend it with three components at the input level and the optimization level. For the unlabeled target domain dataset, we first present an assignment-based object discovery strategy to generate suitable training patches. Additionally, a low-light enhancer is embedded to improve the pseudo labels that facilitate the following consistency learning. Finally, with the aid of better training data and pseudo labels, we replace the common mean square error with two stricter losses, which are entropy-decreasing classification consistency loss and confidence-weighted regression consistency loss, for better convergence. Experiments demonstrate that our proposed method achieves significant performance gains on multiple nighttime tracking benchmarks, and even brings slight enhancement on the source domain.

Keywords: Visual object tracking · Nighttime circumstances · Unsupervised domain adaptation

1 Introduction

Visual Object Tracking (VOT) aims to estimate the location and scale of an arbitrary object in a video sequence. As one of the most important tasks in the computer vision community, VOT has wide applications in many aspects, such as robotics [19], video surveillance [23], and visual localization [24]. Despite the

Supported by National Natural Science Foundation of China (62233005, 62293502), Program of Shanghai Academic Research Leader Under Grant 20XD1401300, Sino-German Center for Research Promotion (Grant M-0066) and Fundamental Research Funds for the Central Universities(222202317006)

B. Luo et al. (Eds.): ICONIP 2023, CCIS 1968, pp. 391–404, 2024.
https://doi.org/10.1007/978-981-99-8181-6_30

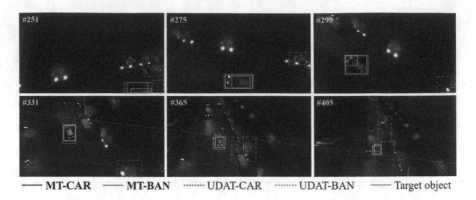

Fig. 1. Visualization comparison of our proposed unsupervised domain adaptive trackers (i.e., MT-CAR and MT-BAN) and their corresponding baselines (i.e., UDAT-CAR and UDAT-BAN) [29]. Our methods outperform state-of-the-arts UDAT in nighttime scenes.

considerable progress in visual tracking, most previous works focus on daytime sequences under favorable illumination circumstances [5,10]. However, a huge domain gap exists between daytime videos and nighttime videos. Compared to daytime tracking datasets [8,12,21], videos captured at night commonly have dim brightness, low contrast, and high noise. Therefore, annotating the indiscernible objects of a large amount of nighttime videos for supervised training is quite time-consuming and expensive [29]. As a result, tracking in low-light conditions remains a challenge and impedes the development of real-world applications like unmanned aerial vehicle (UAV) tracking.

To address this problem, domain adaptation is introduced to nighttime tracking. Commonly-used domain adaptation in computer vision tasks can be divided into three categories: domain translation, domain alignment and self-training [4]. Current works of nighttime tracking are focused on the first two categories. They either build a cascade structure to transfer a nighttime image to a daytime-like one at the input level [9,27,28], or employ a domain discriminator at the feature level [29]. The former way has weak collaboration between different parts, while the latter one lacks pseudo supervision in the target domain. Thus, both of them fail to bring out satisfactory predictions. Compared with domain translation and domain alignment, self-training removes the necessity to train extra style transfer networks or domain discriminators [4]. Moreover, self-training can make good use of both labeled and unlabeled data to build pseudo supervision, which is a superiority that the other two categories are lack of [20]. In view of this, our proposed method, which will be introduced detailedly later, falls into the self-training category, or more specifically, the mean teacher (MT) network [25].

There exist three challenges when applying self-training to unsupervised domain adaptive visual tracking. The first problem is the generation of training patches for unlabeled datasets. Different from other vision tasks that can directly feed original images into the network [11], visual tracking often crops training pairs from raw images according to the bounding boxes of target objects first. When it comes to unlabeled datasets, pseudo bounding boxes are required to generate training patches. Previous unsupervised domain adaptation work [29] regards the object discovery process as a salient object detection with a dynamic programming and linear interpolation problem, neglecting the quality divergence between frames. The unsuitable training patches inevitably impede the tracker training. The second challenge is that the pseudo labels produced in self-training are liable to be noisy, which may mislead the network training [22]. Prior researches [7,20] set a well-designed confidence threshold to filter inferior pseudo boxes at the output level. In this case, the performance of the tracker highly depends on the threshold selection. Thus, this method is lack of flexibility in different scenes. The third challenge is that, even with high-quality training data and pseudo labels, the predictions of trackers is still likely to be inferior if the consistency losses are unbefitting. Original mean teacher [25] and some recent studies [6,33] use mean square error (MSE) as the consistency constraint. However, MSE may lead to poor training convergence in some cases [16], which will limit the final performance.

To solve these three problems, we propose an extended mean teacher network for unsupervised domain adaptive nighttime tracking. First, for the unlabeled dataset, we present a novel object discovery strategy that regards this process as a salient object detection with an assignment problem. In this way, we can discard sequences with low quality and obtain multiple training pairs from one sequence with high quality simultaneously. It is beneficial for the network to learn template matching using our generated training patches. Then, in order to alleviate the effects of noisy pseudo labels, we propose to increase image brightnesses at the input level instead of elaborately selecting filtering thresholds at the output level. Finally, with the aid of preferable training patches and more accurate pseudo labels, we are able to replace MSE with stricter consistency losses. Specifically, We design an entropy-decreasing consistency loss to increase the certainty of classification predictions, and a confidence-weighted consistency loss to make the regression branch more focused on the area that is likely to be the target. Figure 1 is the visualization comparison of our methods and the state-of-the-art (SOTA) trackers [29]. Our methods demonstrate better performance in nighttime tracking.

In summary, the main contributions of this work are as follows:

- We present an effective extended mean-teacher paradigm for unsupervised domain adaptive tracking. Knowledge is transferred from the daytime domain to the nighttime domain to alleviate the effect of low illumination.
- Three key components are proposed to extend mean teacher networks. These enable our model to be trained with more appropriate training pairs, higher-quality pseudo labels, and also stricter consistency losses.
- Our method achieves state-of-the-art results on multiple nighttime tracking benchmarks and meanwhile keeps its superiority on daytime datasets, which demonstrates the effectiveness of our methods in low-light tracking.

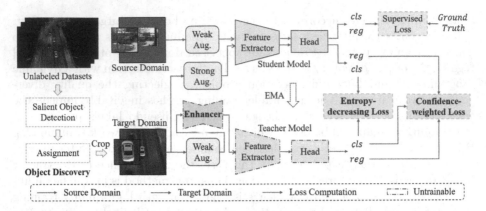

Fig. 2. Illustration of our proposed method. The unlabeled target datasets are first processed with a two-step object discovery strategy to generate training patches. Then, before being fed into the feature extractor, the brightness of target domain images are improved by a light enhancer for the purpose of obtaining better pseudo labels. Finally, the outputs of the classification branch and the regression branch are supervised by the pseudo labels using the entropy-decreasing consistency loss and the confidence-weighted regression loss, respectively. *cls* and *reg* refer to the outputs of the classification and the regression branch. **Bold** represents our proposed three key components.

2   Methods

In this section, we first introduce the object discovery strategy used to generate training patches for unlabeled tracking data (Sect. 2.1). Then, we explain the details of our mean teacher tracking architecture, including the image enhancement for higher-quality pseudo labels (Sect. 2.2) and our designed consistency losses for both the classification and regression branches (Sect. 2.3). The overall framework is illustrated in Fig. 2.

2.1   Object Discovery Strategy

Different from [29] that generates pseudo bounding boxes for each frame by dynamic programming and linear interpolation, we propose an assignment-based object discovery approach. The entire process involves only two stages (see Fig. 3). In the first stage, we aim to identify distinctive objects within the unlabeled images. Due to the inherent scale variation often observed in target sequences, it becomes challenging for conventional salient object detection methods to accurately locate potentially meaningful objects, especially when they are distant. To address this issue, we leverage an edge-guided network called EGNet [32]. EGNet effectively combines rich edge information with location cues to locate salient objects, providing robustness regardless of object sizes. After generating saliency maps, post-processing steps such as closing operations and removal of small objects and holes are applied to obtain the minimum bounding

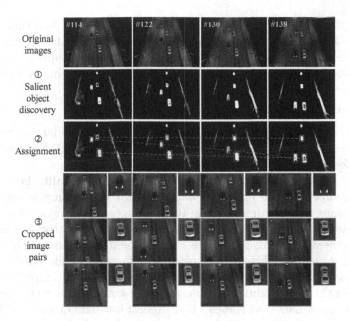

Fig. 3. Illustration of our object discovery strategy. It is composed of image salient object detection and Hungarian assignment. The pseudo boxes generated by this strategy are used to crop training patches. (For better exhibition, we select images that are several frames apart, but they actually should be adjacent images.)

rectangles of the detected salient objects. Consequently, each image generates zero, one, or more candidate bounding boxes.

The second stage involves obtaining sequences of bounding boxes for the same objects based on the candidate boxes. We treat this step as an assignment problem and utilize the Hungarian algorithm to assign objects across consecutive frames. As the sizes of the boxes may change frequently along the sequence dimension after the first stage, the distances between the centers of the boxes provide a more robust measure compared to metrics such as Intersection over Union (IoU) or Convolutional Neural Network (CNN) features. To build the cost matrix C for the assignment problem, we calculate the distances between the central points of the bounding rectangles in adjacent frames. Each element c_{ij} in the i-th row and j-th column of the cost matrix represents the cost of assigning the j-th object in one frame to the i-th object in the adjacent frame:

$$c_{ij} = \sqrt{(x_{1i} - y_{1i})^2 + (x_{2j} - y_{2j})^2},\tag{1}$$

where (x_{1i}, y_{1i}) and (x_{2j}, y_{2j}) refer to the center coordinates of the i-th and j-th object in two adjacent frames, respectively. The optimal assignment is determined by Hungarian algorithm based on the cost matrix. As a result, candidate boxes of the same object in different frames are associated with each other, and thus multiple sequences of potential objects are produced from one video.

Subsequently, we discard the sequences whose lengths are less than the given threshold, since their poor continuity implies unreliability to some extent. Finally, the training patches for the target domain are cropped from original images based on the box sequences generated by the foregoing two steps.

2.2 Mean Teacher with Low-Light Enhancement

In mean teacher network, the outputs of the teacher model are regarded as pseudo labels, which have a non-negligible effect on the model training. Typically, the data fed into the teacher model are often weak-augmented to leave it in its original state as much as possible to generate high-quality labels that are utilized to supervise the unlabeled data through consistency losses. However, when it comes to nighttime, the captured images are always with dim light, low contrast, and high noise, rendering the feature extractor less effective in extracting meaningful image features. Consequently, relying solely on weakly augmented data is insufficient to generate satisfactory pseudo labels.

To alleviate this problem, we tackle the issue from the input perspective. Recognizing that low brightness is a major factor contributing to the degradation of tracking performance, we embed a light-weight low-light enhancer [15] between the weak augmentation step and the teacher model, which enhances images by estimating pixel-wise and high-order curves for nonlinear curve mapping. Thus, the unlabeled target domain datasets are first weakly augmented as usual, then lighted by the image enhancer, and finally fed into the teacher model. In this way, images that are used to generate pseudo labels are in better illumination conditions, decreasing the negative effect of nighttime unsatisfactory brightness.

2.3 Objective Functions

The design of consistency losses is another critical factor in self-training. Visual tracking models typically comprise a classification branch and a regression branch to predict object bounding boxes [1]. Therefore, optimizing the self-training model requires minimizing the consistency losses in both branches. The first part is the classification branch. It entails calculating two probabilities for each position that represents the likelihood of belonging to either background or foreground. Ideally, the outputs are expected to have low-entropy, which indicates higher reliability. For this purpose, we design an entropy-decreasing function $f_{ed}(.)$ on pseudo labels:

$$f_{ed}(y) = \sigma(y/\tau), \tag{2}$$

where τ is a temperature hyperparameter that controls the degree of entropy reduction, and $\sigma(.)$ is the softmax function. The pseudo labels generated by the teacher model are fed into this entropy-decreasing function. Accordingly, pseudo labels with lower entropy are employed to guide the student training, and thus encourage the predictions of the student model to become higher-certainty. We then compute the classification consistency loss $\mathcal{L}_{cons}^{cls}$ as:

$$\mathcal{L}_{cons}^{cls} = f_{ce}(\phi_{stu}^{cls}(x^t), f_{ed}(\phi_{tea}^{cls}(f_{le}(x^t)))), \tag{3}$$

where x^t refers to the target domain images, ϕ represents the tracking model (stu and tea means the student and teacher model, cls means the classification branch), f_{ce} is the standard cross-entropy function and f_{le} is the light enhancement model.

The second component is the regression branch. We propose a confidence-weighted consistency loss for regression. The confidence here is derived from the classification outputs of the teacher model. In other words, the loss weights of the areas where the target is less likely to be in are reduced. In this way, there is no necessity to conduct lots of experiments to obtain an appropriate threshold to filter the positions that could be the background. The confidence-weighted regression consistency loss $\mathcal{L}_{cons}^{reg}$ is based on smooth L1 loss and is defined as:

$$
\begin{aligned}
\mathcal{L}_{cons}^{reg} = &\sigma(\phi_{tea}^{cls}(f_{le}(x^t)))* \\
&f_{smoothl1}(\phi_{stu}^{reg}(x^t) - \phi_{tea}^{reg}(f_{le}(x^t))),
\end{aligned}
\tag{4}
$$

where $\sigma(.)$ is the softmax function.

The overall objective functions $\mathcal{L}$ can be summarized as:

$$
\mathcal{L} = \mathcal{L}_{sup} + \lambda_1 \mathcal{L}_{cons}^{cls} + \lambda_2 \mathcal{L}_{cons}^{reg},
\tag{5}
$$

where $\mathcal{L}_{sup}$ is the supervised loss of the source domain, λ_1 and λ_2 are the weight coefficients.

3 Experiments and Results

3.1 Datasets and Implementation Details

Datasets. During training, two daytime datasets (GOT-10K [12] and VID [21]) are mixed for the source domain, while a nighttime dataset NAT2021 [29] is used for the target domain. During evaluation, we compare our methods with state-of-the-art trackers on three nighttime tracking benchmarks: NAT2021-test [29], UAVDark70 [14], and NAT2021-L-test [29]. Among the three test sets, the first two are short-term tracking datasets, while the last one is a long-term tracking dataset.

Implementation Details.: We conduct our experiments on the same baseline trackers (SiamCAR [10] and SiamBAN [5]) as the SOTA unsupervised domain adaptive tracking work [29], and name them as MT-CAR and MT-BAN, respectively. Our experiments are implemented by Pytorch with 2 NVIDIA A100 GPUs. During the object discovery step, the box sequences that are shorter than 100 are discarded. We sample 20,000 training pairs per epoch and utilize the SGD optimizer to train our model for 20 epochs. The total training costs about 5 h. The initial learning rate is 1.5×10^{-3} and decreases following a logarithmic decay schedule. Weak augmentation applied in our work includes scaling, image shift, and color jitter, while strong augmentation applied before the teacher model additionally contains channel dropout and shuffle, histogram equalization, image blur, and Gaussian noise [2]. The weight coefficients λ_1 and λ_2 in MT-CAR and MT-BAN are set as 2, 1 and 1, 0.1.

Table 1. The performance of our method and other state-of-the-art trackers on NAT2021-test.

Methods	Prec.↑	Norm. Prec.↑	Succ.↑
HiFT [3]	55.2	43.7	36.7
SiamFC++ [26]	60.2	49.5	41.6
Ocean [31]	58.7	46.1	38.3
SiamRPN++ [13]	62.2	49.1	41.0
UpdateNet [30]	58.6	42.2	35.9
D3S [17]	64.0	45.3	39.6
SiamBAN [5]	64.7	50.9	43.7
SiamBAN+DarkLighter [28]	64.8	50.4	43.4
SiamBAN+HighlightNet [9]	60.5	45.9	39.2
UDAT-BAN [29]	69.4	54.6	46.9
MT-BAN (Ours)	**69.9**	**56.2**	**49.4**
SiamCAR [10]	66.3	54.2	45.3
SiamCAR+DarkLighter [28]	64.0	51.7	42.6
SiamCAR+HighlightNet [9]	61.0	48.3	40.3
UDAT-CAR [29]	68.7	56.4	48.3
MT-CAR (Ours)	**72.0**	**59.2**	**50.7**

3.2　State-of-the-Art Comparisons

NAT2021-Test. The evaluation results on NAT2021-test are reported in Table 1. Our proposed trackers, MT-CAR and MT-BAN, achieve the top two rankings among other SOTA trackers. Notably, MT-BAN demonstrates superior performance compared to UDAT-BAN, with improvements of **2.5%**, **1.6%**, and **0.5%** on three metrics. Similarly, MT-CAR surpasses UDAT-CAR by **2.4%**, **2.8%**, and **3.3%**. Interestingly, the performance of domain translation methods is adversely affected due to their cascade structures.

NAT2021-L-Test. Long-term tracking remains a challenge in visual tracking, so we further evaluate our trackers on NAT2021-L-test. The evaluation results are reported in Table 2. Our proposed MT-CAR and MT-BAN outperform other trackers by convincing margins, demonstrating the competitive performance in long-term tracking.

UAVDark70. In addition to evaluating our trackers on datasets with similar data distribution, we also conducted tests on a dataset with significantly different distribution [14]. The performance results are reported in Table 3. Despite the challenging disparity in data distribution, MT-BAN continues to outperform UDAT-BAN [29] in such circumstances. However, we observed that MT-CAR exhibits relatively unsatisfactory performance. We suspect that this discrepancy could be attributed to the divergence between the NAT dataset and the UAVDark70 dataset, as they were captured using different types of cameras.

Table 2. The performance of our method and other state-of-the-art trackers on NAT2021-L-test.

Methods	Prec.↑	Norm. Prec.↑	Succ.↑
HiFT [3]	43.3	31.6	28.7
SiamFC++ [26]	42.5	34.4	29.7
Ocean [31]	45.4	37.0	31.5
SiamRPN++ [13]	43.1	34.2	29.9
UpdateNet [30]	43.2	31.4	27.5
D3S [17]	49.2	36.4	33.2
SiamBAN [5]	46.4	36.6	31.6
SiamBAN+DarkLighter [28]	42.7	32.9	28.4
SiamBAN+HighlightNet [9]	42.7	32.9	28.4
UDAT-BAN [29]	49.6	40.6	35.2
MT-BAN (Ours)	**55.6**	**45.5**	**39.9**
SiamCAR [10]	47.7	37.5	33.0
SiamCAR+DarkLighter [28]	44.8	36.3	31.5
SiamCAR+HighlightNet [9]	43.3	33.8	30.5
UDAT-CAR [29]	50.6	41.3	37.6
MT-CAR (Ours)	**54.3**	**44.2**	**39.0**

Table 3. The performance of our method and other state-of-the-art trackers on UAVDark70.

Methods	Prec.↑	Norm. Prec.↑	Succ.↑
HiFT [3]	50.0	42.7	35.6
SiamFC++ [26]	53.7	47.6	38.6
Ocean [31]	46.3	40.2	33.0
SiamRPN++ [13]	49.7	41.6	35.1
UpdateNet [30]	53.2	40.7	37.1
D3S [17]	59.3	50.7	40.9
SiamBAN [5]	67.7	57.0	48.9
SiamBAN+DarkLighter [28]	57.5	49.2	42
SiamBAN+HighlightNet [9]	53.5	46.1	39.4
UDAT-BAN [29]	70.2	59.7	51.0
MT-BAN (Ours)	**71.3**	**61.0**	**52.2**
SiamCAR [10]	66.9	58.0	49.1
SiamCAR+DarkLighter [28]	62.0	53.4	44.5
SiamCAR+HighlightNet [9]	61.6	52.6	43.4
UDAT-CAR [29]	**69.5**	59.2	**51.2**
MT-CAR (Ours)	68.2	**59.5**	50.1

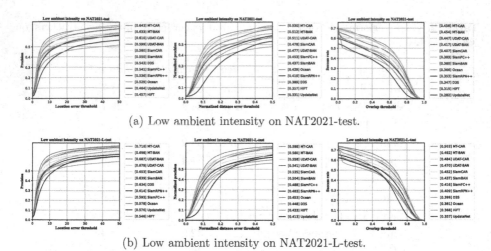

(a) Low ambient intensity on NAT2021-test.

(b) Low ambient intensity on NAT2021-L-test.

Fig. 4. Precision plots, Normalized precision plots and success plots of low ambient intensity attribute on NAT2021-test and NAT2021-L-test.

Tracking Speed. Since no additional modules are required during test, the tracking speed of our proposed methods remains the same as the base trackers they are built upon (SiamCAR and MT-CAR: 78 FPS, SiamBAN and MT-BAN: 81 FPS in our experimental environment). In comparison, UDAT [29] adds an extra bridging layer within the tracking models to narrow the domain gap, so the speeds drop to 54 FPS (UDAT-CAR) and 52 FPS (UDAT-BAN), respectively.

3.3 Illumination-Related Evaluation

Due to the significant impact of low illumination intensity on the performance of generic trackers, we conducted further evaluations of our methods specifically on the low ambient intensity (LAI) attribute. The evaluation results are shown in Fig. 4. Notably, our MT-CAR and MT-BAN exhibit improvements of approximately 3% over UDAT-CAR [29] and UDAT-BAN [29] in most cases. These results clearly indicate that current trackers face considerable challenges posed by adverse illumination conditions. However, our proposed extended mean teacher methods for nighttime tracking effectively alleviate this effect to a significant extent.

3.4 Source Domain Evaluation

Apart from excelling in low-light conditions, it is crucial for trackers to maintain their accuracy during daytime as well. To assess this aspect, we conducted evaluations on a benchmark dataset for normal-light tracking [18] and the results are summarized in Table 4. Experiments show that our methods not only yield performance improvements in the target domain but also in the source domain, showcasing their effectiveness across different lighting conditions.

Table 4. Evaluation on the source domain.

Trackers	SiamCAR	UDAT-CAR	MT-CAR	SiamBAN	UDAT-BAN	MT-BAN
Prec.	80.0	78.6	81.2	78.1	78.2	78.7
Norm. Prec.	70.2	66.8	70.4	68.1	68.2	68.8
Succ.	60.9	58.8	61.5	59.7	59.0	60.2

Table 5. Ablation study of our proposed method on NAT2021-test. LE, CL, and OD denote light enhancer, consistency loss, and object discovery, respectively.

LE	CL		OD	Prec.↑	Norm. Prec.↑	Succ.↑
	cls	reg				
				68.9	56.6	49.1
✓				69.2	56.8	49.3
✓	✓			68.7	56.6	48.7
✓		✓		69.9	57.0	49.3
✓	✓	✓		71.7	58.2	49.9
✓	✓	✓	✓	72.0	59.2	50.7

3.5 Ablation Study

In this section, we investigate the impact of various components in our proposed method, including the light-enhancer module before the teacher model, the consistency losses for classification and regression branches, as well as our object discovery strategy. We conduct experiments on the NAT-test dataset [29], using SiamCAR [10] as the base tracker. The performance improvements achieved by each component are summarized in Table 5. In the LE column, a blank entry represents the absence of a light enhancer before the teacher model. In the CL column, a blank entry indicates the use of mean square error for consistency losses. In the DP column, a blank entry signifies the adoption of the object discovery strategy proposed in UDAT [29]. From the results presented in Table 5, we observe that all three components contribute to the overall performance improvement, indicating their effectiveness in our method.

4 Conclusion

In this paper, we propose a novel unsupervised domain adaptation framework for nighttime visual tracking based on the mean teacher network. Compared to other nighttime tracking methods, no extra modules like style transfer networks or bridging layers are needed during test. We first present an assignment-based object discovery strategy for the unlabeled target domain to make the generated training patches more befitting to the tracker learning. Meanwhile, we incorporate a light enhancer to further improve the pseudo labels of these training samples. Finally, more accurate pseudo labels allow us to design stricter consistency losses instead of common mean square error for both classification and regression branches to get better model convergence. Extensive experiments on multiple nighttime tracking datasets demonstrate the effectiveness and domain transferability of our model. Future work will extend to more diverse and challenging scenarios, such as space debris tracking, to improve the autonomy and accuracy of space debris observation systems.

References

1. Bertinetto, L., Valmadre, J., Henriques, J.F., Vedaldi, A., Torr, P.H.S.: Fully-convolutional Siamese networks for object tracking. In: Hua, G., Jégou, H. (eds.) ECCV 2016. LNCS, vol. 9914, pp. 850–865. Springer, Cham (2016). https://doi.org/10.1007/978-3-319-48881-3_56
2. Buslaev, A., Iglovikov, V.I., Khvedchenya, E., Parinov, A., Druzhinin, M., Kalinin, A.A.: Albumentations: fast and flexible image augmentations. Information **11**(2), 125 (2020)
3. Cao, Z., Fu, C., Ye, J., Li, B., Li, Y.: HiFT: hierarchical feature transformer for aerial tracking. In: Proceedings of the IEEE/CVF International Conference on Computer Vision, pp. 15457–15466 (2021)
4. Chen, M., et al.: Learning domain adaptive object detection with probabilistic teacher. In: International Conference on Machine Learning, pp. 3040–3055. PMLR (2022)
5. Chen, Z., Zhong, B., Li, G., Zhang, S., Ji, R.: Siamese box adaptive network for visual tracking. In: Proceedings of the IEEE/CVF Conference on Computer Vision and Pattern Recognition, pp. 6668–6677 (2020)
6. Chen, Z., Zhu, L., Wan, L., Wang, S., Feng, W., Heng, P.A.: A multi-task mean teacher for semi-supervised shadow detection. In: Proceedings of the IEEE/CVF Conference on Computer Vision and Pattern Recognition, pp. 5611–5620 (2020)
7. Deng, J., Li, W., Chen, Y., Duan, L.: Unbiased mean teacher for cross-domain object detection. In: Proceedings of the IEEE/CVF Conference on Computer Vision and Pattern Recognition, pp. 4091–4101 (2021)
8. Fan, H., et al.: LaSOT: a high-quality benchmark for large-scale single object tracking. In: Proceedings of the IEEE/CVF Conference on Computer Vision and Pattern Recognition, pp. 5374–5383 (2019)
9. Fu, C., Dong, H., Ye, J., Zheng, G., Li, S., Zhao, J.: HighlightNet: highlighting low-light potential features for real-time UAV tracking. In: 2022 IEEE/RSJ International Conference on Intelligent Robots and Systems, pp. 12146–12153. IEEE (2022)
10. Guo, D., Wang, J., Cui, Y., Wang, Z., Chen, S.: SiamCAR: Siamese fully convolutional classification and regression for visual tracking. In: Proceedings of the IEEE/CVF Conference on Computer Vision and Pattern Recognition, pp. 6269–6277 (2020)
11. Hoyer, L., Dai, D., Van Gool, L.: DAFormer: improving network architectures and training strategies for domain-adaptive semantic segmentation. In: Proceedings of the IEEE/CVF Conference on Computer Vision and Pattern Recognition, pp. 9924–9935 (2022)

12. Huang, L., Zhao, X., Huang, K.: GOT-10k: a large high-diversity benchmark for generic object tracking in the wild. IEEE Trans. Pattern Anal. Mach. Intell. **43**(5), 1562–1577 (2019)
13. Li, B., Wu, W., Wang, Q., Zhang, F., Xing, J., Yan, J.: SiamRPN++: evolution of siamese visual tracking with very deep networks. In: Proceedings of the IEEE/CVF Conference on Computer Vision and Pattern Recognition, pp. 4282–4291 (2019)
14. Li, B., Fu, C., Ding, F., Ye, J., Lin, F.: ADTrack: target-aware dual filter learning for real-time anti-dark UAV tracking. In: 2021 IEEE International Conference on Robotics and Automation, pp. 496–502. IEEE (2021)
15. Li, C., Guo, C., Chen, C.: Learning to enhance low-light image via zero-reference deep curve estimation. IEEE Trans. Pattern Anal. Mach. Intell. **44**, 4225–38 (2021)
16. Liu, Y., Tian, Y., Chen, Y., Liu, F., Belagiannis, V., Carneiro, G.: Perturbed and strict mean teachers for semi-supervised semantic segmentation. In: Proceedings of the IEEE/CVF Conference on Computer Vision and Pattern Recognition, pp. 4258–4267 (2022)
17. Lukezic, A., Matas, J., Kristan, M.: D3S-a discriminative single shot segmentation tracker. In: Proceedings of the IEEE/CVF Conference on Computer Vision and Pattern Recognition, pp. 7133–7142 (2020)
18. Mueller, M., Smith, N., Ghanem, B.: A benchmark and simulator for UAV tracking. In: Leibe, B., Matas, J., Sebe, N., Welling, M. (eds.) ECCV 2016. LNCS, vol. 9905, pp. 445–461. Springer, Cham (2016). https://doi.org/10.1007/978-3-319-46448-0_27
19. Qiao, H., Zhong, S., Chen, Z., Wang, H.: Improving performance of robots using human-inspired approaches: a survey. Sci. China Inf. Sci. **65**(12), 221201 (2022)
20. Ramamonjison, R., Banitalebi-Dehkordi, A., Kang, X., Bai, X., Zhang, Y.: SimROD: a simple adaptation method for robust object detection. In: Proceedings of the IEEE/CVF International Conference on Computer Vision, pp. 3570–3579 (2021)
21. Russakovsky, O., et al.: ImageNet large scale visual recognition challenge. Int. J. Comput. Vision **115**(3), 211–252 (2015)
22. Sun, Q., Zhao, C., Tang, Y., Qian, F.: A survey on unsupervised domain adaptation in computer vision tasks. Scientia Sinica (Technologica) **52**(1), 26–54 (2022)
23. Tang, S., Andriluka, M., Andres, B., Schiele, B.: Multiple people tracking by lifted multicut and person re-identification. In: Proceedings of the IEEE Conference on Computer Vision and Pattern Recognition, pp. 3539–3548 (2017)
24. Tang, Y., et al.: Perception and navigation in autonomous systems in the era of learning: a survey. IEEE Trans. Neural Netw. Learn. Syst. (2022)
25. Tarvainen, A., Valpola, H.: Mean teachers are better role models: weight-averaged consistency targets improve semi-supervised deep learning results. In: Advances in Neural Information Processing Systems, vol. 30 (2017)
26. Xu, Y., Wang, Z., Li, Z., Yuan, Y., Yu, G.: SiamFC++: towards robust and accurate visual tracking with target estimation guidelines. In: Proceedings of the AAAI Conference on Artificial Intelligence, vol. 34, pp. 12549–12556 (2020)
27. Ye, J., Fu, C., Cao, Z., An, S., Zheng, G., Li, B.: Tracker meets night: a transformer enhancer for UAV tracking. IEEE Robot. Autom. Lett. **7**(2), 3866–3873 (2022)
28. Ye, J., Fu, C., Zheng, G., Cao, Z., Li, B.: DarkLighter: light up the darkness for UAV tracking. In: 2021 IEEE/RSJ International Conference on Intelligent Robots and Systems, pp. 3079–3085. IEEE (2021)
29. Ye, J., Fu, C., Zheng, G., Paudel, D.P., Chen, G.: Unsupervised domain adaptation for nighttime aerial tracking. In: Proceedings of the IEEE/CVF Conference on Computer Vision and Pattern Recognition, pp. 8896–8905 (2022)

30. Zhang, L., Gonzalez-Garcia, A., Weijer, J.V.D., Danelljan, M., Khan, F.S.: Learning the model update for Siamese trackers. In: Proceedings of the IEEE/CVF International Conference on Computer Vision, pp. 4010–4019 (2019)
31. Zhang, Z., Peng, H., Fu, J., Li, B., Hu, W.: Ocean: object-aware anchor-free tracking. In: Vedaldi, A., Bischof, H., Brox, T., Frahm, J.-M. (eds.) ECCV 2020. LNCS, vol. 12366, pp. 771–787. Springer, Cham (2020). https://doi.org/10.1007/978-3-030-58589-1_46
32. Zhao, J.X., Liu, J.J., Fan, D.P., Cao, Y., Yang, J., Cheng, M.M.: EGNet: edge guidance network for salient object detection. In: Proceedings of the IEEE/CVF International Conference on Computer Vision, pp. 8779–8788 (2019)
33. Zhou, H., Jiang, F., Lu, H.: SSDA-YOLO: semi-supervised domain adaptive yolo for cross-domain object detection. arXiv preprint: arXiv:2211.02213 (2022)

A Bi-directional Optimization Network for De-obscured 3D High-Fidelity Face Reconstruction

Xitie Zhang, Suping Wu$^{(\boxtimes)}$, Zhixiang Yuan, Xinyu Li, Kehua Ma,
Leyang Yang, and Zhiyuan Zhou

NingXia University, NingXia 750000, China
pswuu@nxu.edu.cn

Abstract. 3D detailed face reconstruction based on monocular images aims to reconstruct a 3D face from a single image with rich face detail. The existing methods have achieved significant results, but still suffer from inaccurate face geometry reconstruction and artifacts caused by mistaking hair for wrinkle information. To address these problems, we propose a bi-directional optimization network for de-obscured 3D high-fidelity surface reconstruction. Specifically, our network is divided into two stages: face geometry fitting and face detail optimization. In the first stage, we design a global and local bi-directional optimized feature extraction network that uses both local and global information to jointly constrain the face geometry and ultimately achieve an accurate 3D face geometry reconstruction. In the second stage, we decouple the hair and the face using a segmentation network and use the distribution of depth values in the facial region as a prior for the hair part, after which the FPU-net detail extraction network we designed is able to reconstruct finer 3D face details while removing the hair occlusion problem. With only a small number of training samples, extensive experimental results on multiple evaluation datasets show that our method achieves competitive performance and significant improvements over state-of-the-art methods.

1 Introduction

3D face reconstruction is one of the most popular and well-researched areas at the intersection of computer graphics, computer vision, and machine learning. Currently, 3D face reconstruction is widely used in the fields of face recognition, animation, facial puppetry, and virtual reality. Existing approaches typically use a parametric face modeling framework [7,14], the most representative one of which is the 3D deformable model (3DMM) proposed by Volker Blanz and

This work was supported by National Natural Science Foundation of China under Grant 62062056, in part by the Ningxia Natural Science Foundation under Grant 2022AAC03327, and in part by the Ningxia Graduate Education and Teaching Reform Research and Practice Project 2021.

Thomas Vetter in 1999 [1]. With the rise of neural networks, recent research has shown that regression of 3DMM coefficients based on CNN networks is feasible and can achieve very good results [4, 8, 13, 16, 22].

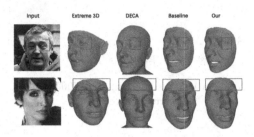

Fig. 1. Details of the face reconstructed by existing methods and our method. The first row shows that our method reconstructs more and more realistic details of the face than the other methods. The second row shows the effectiveness of our method for resolving hair occlusion.

Typically, regression of 3DMM parameters based on CNN networks requires a large amount of data. For supervised learning-based methods [8, 13], a large amount of data with sticky notes is required, while 3D face labels are obtained by scanning the face with a scanner and then manually calibrating it. In contrast, self-supervised learning-based methods [4, 11] learn directly from unlabelled face images.

For example, MoFA [13] learns to regress 3DMM parameters by constraining the rendered image after 3D reconstruction to have similar pixel values in the face region as the input image. Genova et al. [4] renders an image of a face from multiple views and uses a trained face recognition network to measure the perceived similarity between the input face and the rendered face. Chen [2] et al. propose a two-stage neural network framework to learn information such as the shape expression of a face as well as face wrinkle information respectively. These methods can reconstruct more accurate facial geometry, but they suffer from detail loss when reconstructing facial wrinkles and require large amounts of face data for training. Also a two-stage framework, DECA [3] proposes an end-to-end self-supervised network training framework, but the use of a generator to generate details in the second stage of the network resulted in less realistic reconstructed face detail information. Although these methods are effective, they still suffer from inaccurate face geometry and loss of detail when reconstructing facial wrinkle details, and are unable to remove hair scattered in the face region.

To address the above issues, a two-stage learning framework is based. We propose a global and local bi-directional optimized 3D face geometry fitting method and a face detail reconstruction method for hair occlusion repair. Specifically, Fig. 2 illustrates the workflow of our proposed GLIN. In the first stage of the network, we propose a global and local bi-directional optimized 3D face geometry fitting method. In the second stage of the network, based on the U-Net,

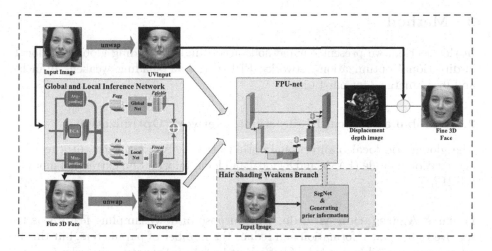

Fig. 2. Workflow of the 3D detail face reconstruction network.

we introduce an FPN structure that can fully capture the finer details of the face by adaptively fusing the feature information after deconvolution and bilinear difference. In addition, to address the effect of hair occlusion on face detail reconstruction, we introduce a semantic segmentation network that feeds the average of the visible part of the face as the a priori information of the hair occlusion region into the second stage of the face detail network for extraction, which can achieve the effect of repairing the hair occlusion region. The core contributions of our work are summarised as follows:

- We design a global and local bi-directional optimized feature extraction network that makes full use of both global and local features to jointly extract face geometry information while reducing the loss of information sampled under the network.
- We design a face detail extraction network with U-Net and FPN, which can further capture finer details in the face by adaptively weighting the features obtained by deconvolution as well as bilinear difference, while ensuring smooth detail edges.
- For some face regions that will be mixed with hair information, we introduce a semantic segmentation network to segment the hair and face regions, using the average depth information value of the visible regions of the face as the prior for the hair part, which reduces the influence of hair on the face detail reconstruction to a certain extent.
- The experimental results show that our method can effectively reconstruct 3D faces with details such as wrinkles and bags under the eyes by only training on small sample data. It is also able to guarantee a high reconstruction accuracy. The qualitative results are shown in Fig. 1.

2   Method

In this section, we present our methods accordingly, including global and local bi-directional optimization networks, FPU-net, hair shading weakens branch, and loss functions for our tasks.

2.1   Global and Local Bi-directional Network Optimization

The global and local bi-directional optimization network consists of a feature aggregation module (FAM) and a global and local information extraction module (GLIEM).

Feature Aggregation Module. The purpose of downsampling features is to remove redundant information and reduce the number of parameters, but this is often accompanied by the loss of important feature information. In order to minimize the loss of important information, we design a feature aggregation module (FAM) to extract and retain the features at different degrees using different downsampling methods. Moreover, we adopt a dynamic aggregation function to assign weights to the features obtained by different downsampling methods, so that the relatively important information is retained as much as possible. Specifically, as shown in Fig. 3, the features F_{init} goes through three branches. The first is average pooling: after this branch, the average state information of the features in the local range is obtained F_{avg}. In the second branch, the features are firstly convolved by $3*3$ convolution and $1*1$ convolution to obtain local spatial information, and then after channel attention ECA, the channel information features are reweighted. The features are then dimensionally reduced after $3*3$ convolution so that the active information is retained and feature F_{eca} [17] is obtained from both spatial and channel dimensions. The third is max-pooling:

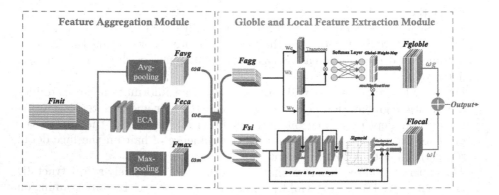

Fig. 3. The detailed structure of our proposed GLIN: the initial feature Finit is dynamically aggregated by three different downsampling, respectively, and then enters the global and local feature extraction network to finally obtain features with global and local information.

after max-pooling, we obtain the most active features F_{max} in the current local region. Finally, the normalized dynamic weighting $\omega_a, \omega_e, \omega_m$ for F_{avg}, F_{eca}, F_{max} yields F_{agg}. The process of this module can be described as:

$$F_{avg} = AvgP\left(F_{init}\right) \tag{1}$$

$$F_{eca} = C_{3\times3}\left(ECA\left(C_{1\times1}\left(C_{3\times3}\left(F_{init}\right)\right)\right)\right) \tag{2}$$

$$F_{max} = MaxP\left(F_{init}\right) \tag{3}$$

$$F_{agg} = \omega_a F_{avg} + \omega_e F_{eca} + \omega_m F_{max} \tag{4}$$

$$\omega_\alpha = \frac{e^{\lambda_\alpha}}{\sum e^{\lambda_\alpha}}, \alpha \in \{a, e, m\} \tag{5}$$

where $AvgP\left(\cdot\right)$ is average pooling, $C_{n\times n}$ is $n \times n$ convolution, ECA is channel attention, $MaxP\left(\cdot\right)$ is max pooling, $\omega_\alpha, \alpha \in \{a, e, m\}$ are learnable weights and λ is the hyperparameter.

Global and Local Information Extraction Module. To further improve the extraction of face geometry information by the network, we propose a global and local information extraction module(GLIEM) that aims to extract global and local scale geometric features of faces. Specifically, our information extraction module is a two-stream parallel network, one branch of which is a global information extraction network, where global features F_{global} can be enhanced by means of a global self-attentive network, which in turn allows for a weighted fusion of global features, effectively avoiding interference due to elements such as local occlusions in the image. The final output of this branch is

$$F_{global}(x) = x + \lambda \cdot \text{Attention}\left(Q, K, V\right), \tag{6}$$

where λ is the learnable hyperparameter, x is the input feature F_{agg}, and *Attention* is the feature after the attention network.

Another branch is the local feature focus network, which first divides the features F_{agg} into four equal parts as $F_{si}, i \in \{1, 4\}$ from the channel dimension, then each part F_{si} is filtered by $3 * 3$ convolution on the local information separately, then the single spatial feature information is weighted and learned by $1 * 1$ convolution, and finally the four weighted features are stitched together in the channel dimension. This will result in a greater expressiveness of the active information. Note that the reason why the local feature aggregation network first slices the features by channel dimension is that as the network deepens, the channel dimension keeps increasing, making the feature information in a given channel less weighted. Without the segmentation, the presence of moderately active information in some channels may be reduced due to the shared weight of the convolution. After the segmentation into four blocks of features, each F_{si} can be learned separately, preserving the moderately active feature information to a certain extent. This process can be described as:

$$F_{local}(x_i) = [1 + \eta \cdot \omega(\mathbf{x}_i)] \odot \mathbf{x}_i \tag{7}$$

where x_i is one of the input features $F_{si}, i \in \{1, 4\}$, ω is the weight of the feature after the local feature focused network, and η is the learnable hyperparameter.

Finally, we obtain the fused feature F_{fusion} by weighting global feature $F_{global}(x)$ and local feature $F_{local}(x_i)$ together, which can be described as:

$$F_{fusion} = \omega_1 \cdot F_{global} + \omega_2 \cdot F_{local} \tag{8}$$

where ω_i is the weight of the feature (Fig. 3).

2.2 Feature Pyramid U-Net: FPU-Net

The learning of face detail network generally adopts the network of U-net architecture, the commonly used u-net decoder up-samples the encoded features by deconvolution to obtain the decoded features. This non-linear operation can make the result smoother, but the detailed information in the features will be lost, although the shallow information in the same size transverse encoded features can enrich the decoding information but still cannot compensate for the lack of detailed information. Therefore, we introduced the SPP Module [6]to integrate the information under different sensory fields. In addition, inspired by the FPN structure [9], we introduce bilinear interpolation for upsampling in the original U-Net, which can compensate for the loss of information caused by the original non-linear transformation. However, the use of bilinear interpolation for

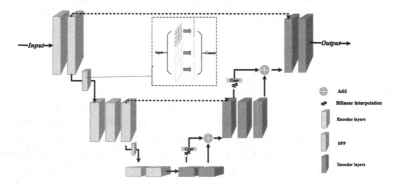

Fig. 4. The detailed structure of our proposed FPU-net: features are convolved in a number of columns as well as downsampled by SPP and then upsampled by means of a dynamic fusion of bilinear interpolation and deconvolution.

upsampling introduces some additional noise information, and to reduce this information generation, we filter it by $1 * 1$ convolution to obtain the detail features. The final features F'_{fi} are then obtained by dynamically weighting the detail features, the decoded features and the same-layer coded features by normalization. With our proposed FPU-net, as much detail information as possible such as wrinkles is extracted while ensuring smooth facial details. The whole process can be described by the following equation:

$$F_{f0} = F_{fusion} \tag{9}$$

$$F_{fi} = SPP(F_{f(i-1)}) \tag{10}$$

$$F'_{fI} = F_{fI} \tag{11}$$

$$F'_{fi} = BiConv(F'_{f(i+1)}) + Uconv(F'_{f(i+1)}) + F_{fi} \tag{12}$$

$$F'_{f0} = BiConv(F'_{f(1)}) + Uconv(F'_{f(1)}) + F_{f0} \tag{13}$$

$$F_{output} = F'_{f0} \tag{14}$$

2.3   Hair Shading Weakens Branches

Since the hair of some images obscures the face region, existing methods mistake the hair as part of the face and reconstruct the hair in the second stage of face detail reconstruction. Since the distribution of the data in the hair region and the face region is different. Therefore, in order to decouple these two parts, we introduce a semantic segmentation network to segment the visible face region and the hair region, and use the distribution of the depth values of the face region as a prior for the depth values of the hair region, and then enter the FPU-net network for face detail extraction, which can finally reconstruct a 3D detailed face without the influence of hair. The whole process is shown in Fig. 5.

2.4   Objective Function

Coarse 3D Face Reconstruction. The total objective function is denoted as:

$$\mathcal{L} = w_1\mathcal{L}_p + w_2\mathcal{L}_{lm} + w_3\mathcal{L}_{id} + w_4\mathcal{R}_{param} \tag{15}$$

where $\mathcal{L}_p$ is the pixel loss, which calculates the average $L_{2,1}$ distance between each pixel value of the input image and the roughly rendered 2D image of the 3D face. $\mathcal{L}_{lm}$ is the landmark consistency loss, which measures the average L_2 distance between the GT 68 2D landmarks of the input image and the 68 landmarks of the reconstructed 3D face. The $\mathcal{L}_{id}$ is the perceptual loss of identity and reflects the perceived similarity between the two images. We sent both the

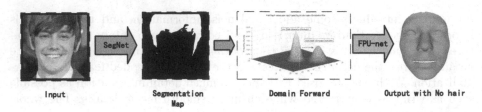

| Input | Segmentation Map | Domain Forward | Output with No hair |

Fig. 5. The workflow of weakening branches in hair shading begins by segmenting the face and hair parts of the input image using a segmentation network. Next, the depth value distribution of the face region is utilized as a prior for the hair region. Finally, the FPU-net is employed to obtain a final 3D detailed face without hair occlusion.

input image and a rough 3D face rendering to VGG-16 to extract features, and then calculated the L_2-distance between the two extracted feature vectors. In addition, $\mathcal{R}_{param}$ represents the regularisation term of the 3DMM parameters estimated by VGG-16. The weights w_1, w_2, w_3, and w_4 are constants to balance the effect of each loss term.

Detailed 3D Face Reconstruction. We adopt the same objective functions to train RI2ITN in both small and large pose reconstructions. The total objective function is denoted as:

$$\mathcal{L} = a_1 \mathcal{L}_p + a_2 \mathcal{L}_s + a_3 \mathcal{R}_{disp} \tag{16}$$

where $\mathcal{L}_p$ is the pixel loss and it calculates the difference in the value of each pixel of the input image and 2D image which is rendered by the detailed 3D face. $\mathcal{L}_s$ is the smoothness loss that is employed on both the UV displacement normal map and facial displacement depth map. In addition, $\mathcal{R}_{disp}$ is a regularization term for smoothness loss to reduce severe depth changes, which may introduce distortion in the face on the 3D mesh.

3 Experiments

In this section, we evaluate the performance of our proposed method on a 3D face reconstruction task. We first present the dataset used in our experiments. We then compare our results quantitatively and qualitatively with several other state-of-the-art methods.

3.1 Datasets

The CelebA dataset [10], contains 202,599 face images of 10,177 celebrity identities, each image is well- tagged with features, including face bbox annotation boxes, 5 face feature point coordinates and 40 attribute tags.

The NoW dataset [12] contains 5 2D images of 20 subjects taken with the iPhone X, as well as individual 3D head scans of each subject. This head scan was the base point for the assessment.

3.2   Experimental Details

We obtained 1000 random images in the CelebA dataset [10] as the training dataset. The weighting parameters $w_1, w_2, w_3, and\, w_4$ are set as 1.0, 1.5, 0.8, 1.0, and the weighting parameters $w_i, w_s, and\, w_d$ are set as 3.0, 10.0, and 1.0. The learning rate is set to 0.00001, decaying every 5000 steps at a rate of 0.98 and the batch size is set to be 20. The input image resolution is 300×300. We adopt Adam optimizer for training on NVIDIA Tesla V100 for over 50,000 steps. The data augmentation strategy is then continued in 300 steps by generating different angular renderings for retraining on top of the currently trained model.

3.3   Evaluation Metrics

We use the 3D root mean square error (3D-NME) between the reconstructed 3D face mesh and the ground truth 3D face scan as an evaluation metric. The predicted mesh and the ground truth mesh were first rigidly aligned (by translation, rotation, and scaling) based on seven points: the inner and outer corners of the eyes, the base of the nose and the corners of the mouth. The area where the face reconstruction was evaluated was then determined with a 95mm radius centered on the coordinates of the base of the nose tip. The points in the 3D face scan that are the distance between all points in the region and the reconstructed mesh is then calculated as the corresponding points, and the difference between all points and points in the region is calculated. Finally, the median and variance of the difference between all points of all reconstructed meshes are calculated as per the above method.

3.4   Results and Analysis

Quantitative Evaluation. To validate the effectiveness of these methods for small and large poses, we divided each dataset into two metrics for evaluation: small and large poses. We use the evaluation protocol proposed by ESRC [19] to calculate the normalized mean error (NME), which refers to the normalized mean of the distances between all reference scan vertices to the nearest point on the surface of the reconstructed mesh for rigidly aligned scan groudtruth and reconstructed meshes. Table 1 shows the quantitative results of our method and existing crude 3D face reconstruction methods. Extreme 3D, baseline and DECA are methods that use 2D keypoints as labels, with DECA being the most advanced method. GCN [20] and RANDA [21] are methods that use 3D keypoint labels. Our method, which like the baseline uses 2D keypoints as labels, achieves an overall improvement in 4 metrics compared to our baseline. This is because we design global and local optimization networks that use both local and global information to jointly constrain the face geometry, ultimately achieving accurate 3D face geometry reconstruction. Compared to DECA, our method achieves comparable performance and outperforms on the now validated metric and the MICC validated metric of the large pose. As the FLAME used by DECA has more expressive parameters than the BFM model we used, it is logical that

Table 1. Coarse 3D Face Reconstruction Quantitative results in two poses on two datasets. Our training dataset is 1000 images, while the other methods are 160,000 and 200,000 respectively.

Method/NME	NOW Large Pose	NOW Small Pose	MICC Large Pose	MICC Small Pose	Training Methods
Extreme 3D [15]	3.685 ± 3.406	3.437 ± 3.361	3.407 ± 3.275	3.219 ± 3.261	Unsupervised
Baseline [2]	3.559 ± 2.928	3.049 ± 2.901	3.177 ± 2.618	2.498 ± 2.167	Unsupervised
GCN [20]	2.598 ± 2.240	2.527 ± 2.321	2.387 ± 2.080	2.347 ± 2.045	Supervised
RADAN [21]	2.591 ± 2.205	2.506 ± 2.309	2.292 ± 2.061	2.325 ± 2.059	Supervised
DECA [3]	2.357 ± 2.128	2.148 ± 1.938	2.401 ± 2.200	2.294 ± 2.212	Unsupervised
Our	2.341 ± 1.968	2.207 ± 1.805	2.602 ± 2.206	2.470 ± 2.058	Unsupervised

Table 2. Coarse 3D Face Reconstruction Quantitative results compared with the DECA method trained on the small-scale training set.

Method/NME	NOW Small Pose	NOW Large Pose	MICC Small Pose	MICC Large Pose
DECA [3]	2.402 ± 2.334	2.628 ± 2.507	2.478 ± 2.374	2.587 ± 2.425
Our	**2.207±1.805**	**2.341±1.968**	**2.470±2.058**	**2.602±2.206**

the same FLAME-based method reconstructs a 3D face more accurately than when it is based on BFM, so part of the gap between our method and DECA is caused by the model. Note that our training set has only 1,000 images, while DECA has 2,000,000 images.

To further validate the effectiveness of our method, a comparison between our method and Deca's method for the same 1000 image dataset is shown in Table 2 and it can be seen that our method is significantly better than DECA.

For detailed 3D face reconstruction, we chose detailed 3D face reconstruction methods of the same topological model for a fair quantitative comparison, as shown in Table 3. After the baseline, to better reconstruct facial details, we used 52 clear facial images with small poses cropped from the MICC as input images. We cut the reconstructed 3D face and the ground truth 3D scanned image to 95 mm centered on the tip of the nose, then performed a rough rigid alignment of the two based on seven key points, then ran an iterative closest point algorithm for exact rigid alignment, and finally calculated the point-to-point distance error between the two. Our method achieves low mean errors and standard deviations, indicating that our method can reconstruct the detailed 3D faces more accurately and consistently.

Table 3. Detailed 3D Face Reconstruction Quantitative results in MICC.

Method	Point-to-point Distances Error
Extreme 3D [15]	1.985 ± 1.523
Baseline [2]	1.802 ± 1.369
Our	**1.653±1.243**

Qualitative Experiments. As shown in Fig. 6, we show the performance of existing methods for face detail reconstruction. In all figures, Extreme 3D is able to reconstruct most of the details of the face, but these details contain many pseudo shadows and are not very realistic; DECA performs poorly in reconstructing details, and many details of the input face image are not reconstructed, and even the reconstructed parts are not faithful to the input face image, such as the forehead part of the face in the last row; The baseline is able to reconstruct the details of the face more realistically, but they are not very detailed. Compared to the other methods, our method gives a very fine reconstruction, especially of the details in the red box. The reason for this is that our face detail network incorporates multi-scale features through the SPP [5] module, and the bilinear difference and inverse convolution up-sampling methods allow the reconstructed details to be preserved as fine as possible while maintaining a smooth and realistic appearance.

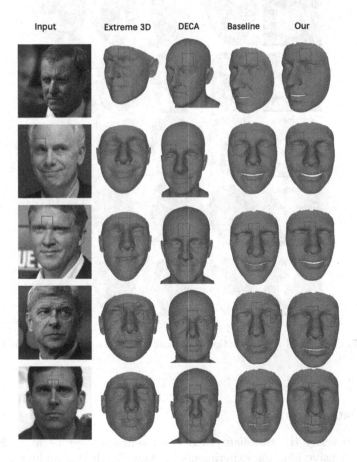

Fig. 6. Results of existing methods in face detail reconstruction

Figure 7 shows the results of face detail reconstruction with hair occlusion. Extreme 3D and baseline are particularly affected by hair occlusion, although they are able to reconstruct fine face detail. DECA generates face detail generatively and is somewhat robust to hair occlusion, but face detail is not obvious. Our method is not only able to reconstruct fine faces, but it also solves the problem of hair occlusion affecting face detail reconstruction. The reason for this is that we separate the face from the hair through the semantic segmentation network, and then use the average of the face replacement map as the prior for the hair occlusion part, and then refine the face reconstruction by FPU-net, so that the face details can be reconstructed without the influence of hair occlusion.

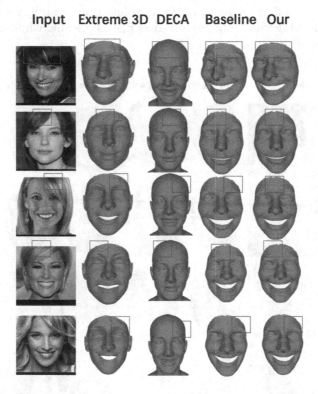

Fig. 7. Face detail reconstruction with hair obscuration in existing methods

3.5   Ablation Study

In order to verify the rationality and effectiveness of our network design, we conduct extensive ablation experiments on Now Small Pose. The results of the ablation study are shown in Table 4. It can be clearly seen that the first row is the accuracy of the baseline, the second row is the accuracy of the baseline with our proposed GLIEM module, and it can be seen that the index has been greatly improved, the third row is the accuracy of the baseline with the FAM module, and the fourth row is the accuracy of the baseline with the DES data augmentation strategy. The fourth behavior has some improvement after applying DES data enhancement strategy based on our proposed method. Combining the above metrics, we can find that our proposed GLIN network can capture accurate face geometry under the joint constraints of global and local information.

Table 4. Ablation Study on NOW Small Pose

Method	NME (mm)
Baseline	3.049 ± 2.901
Baseline + GLIEM	2.405 ± 2.044
Baseline + GLIEM + FAM	2.276 ± 1.911
Baseline + GLIEM + FAM + DES	$\mathbf{2.207 \pm 1.805}$

3.6   Generalisation Experiments

To further validate the generality of our approach, we conducted qualitative experiments on two other datasets, ESRC [19] and FaceScape [18]. As shown in Fig. 8, For the part of the face that is obscured by hair, the baseline may reconstruct the hair as part of the face, as in the first and second columns of the figure. The baseline may also reconstruct pseudo shadows, as in the third, fourth, and fifth columns. While our method is not affected by hair occlusion and can reconstruct a fine 3D face.

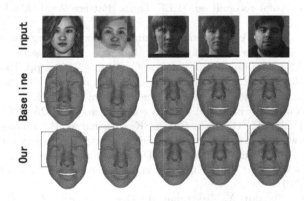

Fig. 8. Qualitative experiments on other data sets

4   Conclusion

In this paper, we present a. In the face geometry fitting phase, we propose global and local optimization networks, including a feature aggregation module that can aggregate feature information and a feature extraction network with local and global bi-directional optimization, which can use both local and global information to infer and fit the face geometry. The face detail extraction network FPU-net, which we have then designed, can effectively improve the sensitivity of the network to details such as facial wrinkles. In particular, if there is hair in the face region, a face prior will be added to the hair part after first separating the face from the hair through the segmentation network. With the network we have designed, a 3D face can eventually be reconstructed with rich details and without the interference of hair.

References

1. Blanz, V., Vetter, T.: A morphable model for the synthesis of 3d faces. In: Proceedings of the 26th Annual Conference on Computer Graphics and Interactive Techniques, pp. 187–194 (1999)
2. Chen, Y., Wu, F., Wang, Z., Song, Y., Ling, Y., Bao, L.: Self-supervised learning of detailed 3d face reconstruction. IEEE Trans. Image Process. **29**, 8696–8705 (2020)
3. Feng, Y., Feng, H., Black, M.J., Bolkart, T.: Learning an animatable detailed 3d face model from in-the-wild images (2020)
4. Genova, K., Cole, F., Maschinot, A., Sarna, A., Vlasic, D., Freeman, W.T.: Unsupervised training for 3d morphable model regression. In: Proceedings of the IEEE Conference on Computer Vision and Pattern Recognition,. pp. 8377–8386 (2018)
5. He, K., Zhang, X., Ren, S., Sun, J.: Spatial pyramid pooling in deep convolutional networks for visual recognition. IEEE Trans. Pattern Anal. Mach. Intell. **37**(9), 1904–16 (2014)
6. He, K., Zhang, X., Ren, S., Sun, J.: Spatial pyramid pooling in deep convolutional networks for visual recognition. IEEE Trans. Pattern Anal. Mach. Intell. **37**(9), 1904–1916 (2015)
7. Jackson, A.S., Bulat, A., Argyriou, V., Tzimiropoulos, G.: Large pose 3d face reconstruction from a single image via direct volumetric cnn regression. In: Proceedings of the IEEE International Conference on Computer Vision, pp. 1031–1039 (2017)
8. Kim, H., Zollhöfer, M., Tewari, A., Thies, J., Richardt, C., Theobalt, C.: Inverse-facenet: Deep monocular inverse face rendering. In: Proceedings of the IEEE conference on computer vision and pattern recognition. pp. 4625–4634 (2018)
9. Lin, T.Y., Dollar, P., Girshick, R., He, K., Hariharan, B., Belongie, S.: Feature pyramid networks for object detection. IEEE Computer Society (2017)
10. Liu, Z., Luo, P., Wang, X., Tang, X.: Deep learning face attributes in the wild. In: Proceedings of the IEEE International Conference on Computer Vision, pp. 3730–3738 (2015)
11. Parkhi, O.M., Vedaldi, A., Zisserman, A.: Deep face recognition (2015)
12. Sanyal, S., Bolkart, T., Feng, H., Black, M.J.: Learning to regress 3d face shape and expression from an image without 3d supervision. In: Proceedings of the IEEE/CVF Conference on Computer Vision and Pattern Recognition, pp. 7763–7772 (2019)

13. Tewari, A., Zollhofer, M., Kim, H., Garrido, P., Bernard, F., Perez, P., Theobalt, C.: Mofa: model-based deep convolutional face autoencoder for unsupervised monocular reconstruction. In: Proceedings of the IEEE International Conference on Computer Vision Workshops, pp. 1274–1283 (2017)
14. Tran, L., Liu, X.: Nonlinear 3d face morphable model. In: Proceedings of the IEEE Conference on Computer Vision and Pattern Recognition, pp. 7346–7355 (2018)
15. Trn, A.T., Hassner, T., Masi, I., Paz, E., Nirkin, Y., Medioni, G.: Extreme 3d face reconstruction: Seeing through occlusions. In: Proceedings of the IEEE Conference on Computer Vision and Pattern Recognition, pp. 3935–3944 (2018)
16. Tuan Tran, A., Hassner, T., Masi, I., Medioni, G.: Regressing robust and discriminative 3d morphable models with a very deep neural network. In: Proceedings of the IEEE Conference on Computer Vision and Pattern Recognition, pp. 5163–5172 (2017)
17. Wang, Q., Wu, B., Zhu, P., Li, P., Hu, Q.: Eca-net: efficient channel attention for deep convolutional neural networks. In: 2020 IEEE/CVF Conference on Computer Vision and Pattern Recognition (CVPR) (2020)
18. Yang, H., Zhu, H., Wang, Y., Huang, M., Cao, X.: Facescape: a large-scale high quality 3d face dataset and detailed riggable 3d face prediction. IEEE (2020)
19. Yin, L., Wei, X., Yi, S., Wang, J., Rosato, M.J.: A 3d facial expression database for facial behavior research. IEEE (2006)
20. Zheng, X., Cao, Y., Li, L., Zhou, Z., Jia, M., Wu, S.: Gnc: Geometry normal consistency loss for 3d face reconstruction and dense alignment. In: 2022 IEEE International Conference on Multimedia and Expo (ICME), pp. 1–6. IEEE (2022)
21. Zhou, Z., Li, L., Wu, S., Li, X., Ma, K., Zhang, X.: Replay attention and data augmentation network for 3d face and object reconstruction. IEEE Trans. Biometrics Behav. Identity Sci. (2023)
22. Zhu, X., Lei, Z., Liu, X., Shi, H., Li, S.Z.: Face alignment across large poses: a 3d solution. In: Proceedings of the IEEE Conference on Computer Vision and Pattern Recognition, pp. 146–155 (2016)

Jointly Extractive and Abstractive Training Paradigm for Text Summarization

Yang Gao, Shasha Li[(✉)], Pancheng Wang, and Ting Wang

National University of Defense Technology, Changsha 410073, Hunan,
People's Republic of China
{gy,shashali,wangpancheng13,tingwang}@nudt.edu.cn

Abstract. Text summarization is a classical task in natural language
generation, which aims to generate concise summary of the original article. Neural networks based on the Encoder-Decoder architecture have
made great progress in recent years in generating abstractive summaries
with high fluency. However, due to the randomness of the abstractive
model during generation, the summaries risk missing important information in the articles. To address this challenge, this paper proposes
a jointly trained text summarization model that combines abstractive
and extractive summarization. On the one hand, extractive models have
higher ROUGE scores but poorer readability on the other hand, abstractive models can produce a more fluent summary but suffer from the problem of omitting important information in the original text. Therefore,
We share the encoder of both models and jointly train both models to
obtain a text representation that benefits from regularisation. We also
add document level information obtained from an extractive model to
the decoder of the abstractive model to improve abstractive summary.
Experiments on CNN/Daily Mail dataset, Pubmed dataset and Arxiv
dataset demonstrate the effectiveness of the proposed model.

Keywords: Encoder · Decoder · Jointly training · Abstractive
Summarization · Extractive Summarization

1 Introduction

Automatic text summarization task is designed to extract key information from
a text to produce a shorter summary. With the advent of information explosion, high quality text summaries can effectively help people to save time on
information retrieval. There are two main genres of text summarization methods, extractive methods and abstractive methods. Extractive methods collect
sumaries directly from the source text, usually selecting one complete sentence
at a time. In contrast, abstractive methods can generate new sentences and
expressions without being limited by the sentence structure and wording of the
original text. Hence, Abstractive summaries can be more coherent and concise
than extractive summaries [7,22].

© The Author(s), under exclusive license to Springer Nature Singapore Pte Ltd. 2024
B. Luo et al. (Eds.): ICONIP 2023, CCIS 1968, pp. 420–433, 2024.
https://doi.org/10.1007/978-981-99-8181-6_32

The extractive methods are usually simpler. It calculates the probability of each sentence output and takes the set of sentences with the highest probability as the summary. Many earlier works on summarization focus on extractive summarization. Among them, Nallapati et al. (2017) [15] uses a heuristic approach to capture features in the text to extract the important sentences of the text. On the other hand, abstractive methods are more complex and it typically uses a sequence-based generation approach, where a word or phrase is generated at each time step and repeat the process until a complete summary is generated. For example, Nallapati et al. [16] proposed an encoder-decoder model capable of generating abstractive summaries. Later, See, Abigail and Liu [23] proposed a pointer generator network with the ability to copy words from the source text as well as generate unseen words. Since the process of abstractive methods involve a great deal of randomness and uncertainty, if we do not control its generation process, it often leads to the omission of important information in the summaries [13].

To solve this problem, we combine the extractive and abstractive summarization models and introduce a jointly training paradigm. Firstly, we share the encoder of the abstractive and the extractive model. Secondly, we use the extractive model to obtain document-level information and add it to the decoder of the abstractive model, thus limiting the randomness of the generation and enabling the summary to contain more important information in the original text. Finally, we train the two models jointly. To the best of our knowledge, the technical contribution of this paper could be concluded as follows:

- We propose a combined model which share the encoder of the abstractive and extractive models so that they get a better representation of the text.
- We propose a new decoder-side combination method that adds the document-level information obtained from an extractive model to the decoder of abstractive model to control the randomness in generation process and thus improve the recall of important content in the original text.
- We propose a jointly training process combining extractive and abstractive methods and demonstrate the effectiveness of it on Pubmed dataset, CNN/Daily Mail dataset and Arxiv dataset.

2   Related Work

Text summarization has been extensively researched in recent years, and below we present related work on neural network-based extractive and abstractive summarization.

2.1   Extractive Summarization

Yin and Pei (2015) [25] modelled extractive summaries as a sequence annotation task, extracting sentences as summaries based on their vector representations. Cheng and Lapata (2016) [3], Nallapat i et al. (2016a) [17] and Nallapati et al.

(2017) [15] use recurrent neural networks to obtain chapter-level text representations and use some features, such as information content, salience and novelty, to calculate a score for each sentence and select the highest scoring few. Liu, Yang (2019) [14] inserted a [CLS] token before each sentence and a [SEP] token after each sentence, and then fed into the BERT model to obtain the sentence vector corresponding to each [CLS] position.

2.2 Abstractive Summarization

Nallapati et al. (2016b) [16] used a sequence-to-sequence model based on recurrent neural networks to generate abstract summaries. Later, See, Abigail and Liu (2017) [23] used a pointer generation network to solve the OOV problem for abstract summarization. The latest abstract summarization is based on Transformer, a network model that uses multi-headed attention to process complex information and can operate in parallel. There are a large number of Transformers currently in use in abstract summarization models, such as Google's BERT [6], PEGASUS [26], T5 [20] and Facebook's BART [12] and Open AI's Generative Pre-trained Transformer (GPT) [18], GPT-2 [19] and GPT-3 [2]. An extension of these models is the so-called Longformer [1], which is an improved Transformer architecture with a self-attention operation that scales linearly with sequence length, making it suitable for processing long documents. For example, LongT5 [10] integrates attention mechanisms from long-input transformers (ETC) and pre-training strategies from summarization pre-training (PEGASUS). It also introduces a new attention mechanism called Transient Global (TGlobal), which mimics the local/global attention of ETC, but without requiring additional side-inputs.

3   Our Model

We propose a jointly trained model that combines the advantages of abstractive and extractive summarization model in the Encoder and Decoder of abstractive summarization model, respectively. Before describing our model in detail, we define the abstractive summarization and extractive summarization (Fig. 1).

3.1   Abstractive Summarization

A model for abstractive summarization can be described based on conditional probabilities. Assuming that the input text is x, the reference summary is y, and the generated summary is $\hat{y}$, a conditional probability model can be defined as follows:

$$p(\hat{y}|x) = \sum_{t=1}^{T} p(\hat{y}_t|\hat{y}_{<t}, x) \tag{1}$$

where T is the length of the generated summary and $\hat{y}_{<t}$ denotes the summary generated in the first $t-1$ time steps. The generation probability $p(\hat{y}_t|\hat{y}_{<t}, x)$ for

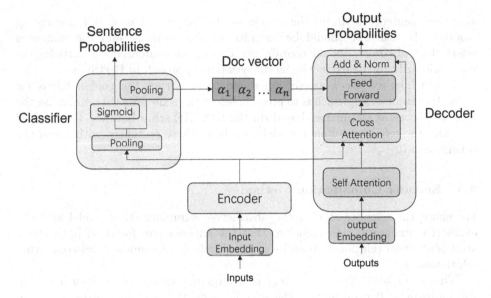

Fig. 1. Jointly trained text summarization model. We used LongT5 pretraing model as our encoder and decoder. In the middle is the shared Encoder layer, on the left is the classification layer of the extractive summarization model, and on the right is the decoder layer of the abstractive summarization model.

each time step in this model is based on the input text x, and the conditional probability of the previous summary $\hat{y}_{<t}$ that has been generated.

We process the input and output using an encoder-decoder structure, which is used to generate concise and accurate summaries from a given input text. The model consists of two main components: encoder and decoder.

Encoder is a neural network that maps the input text x to a fixed length vector representation, often called a context vector. This vector contains key information about the input text and can be used by decoder to generate a summary.

Decoder is another neural network that receives a context vector and a special start token as input and generates an output sequence that progressively builds the generated summary. Each time step of decoder uses the output of the previous time step and the context vector as input, while guiding the output content by cross-attention.

We use Transformer structures to implement the abstractive summarization model.

3.2   Extractive Summarization

In this paper, we take the extractive model as a sequence annotation problem. Suppose that d represents an article and that d contains many sentences $[sent_1, sent_2, ...sent_m]$, where $sent_i$ denotes the i-th sentence in the article. We give each sentence $sent_i$ in the article a label y_i, where $y_i \in 0, 1$, indicating whether the sentence should be included in the abstract, and the sentences selected as abstracts should contain the important content of the article. We refer primarily to the extractive summarization approach in BertSum.

Meanwhile, we use a greedy algorithm to construct the golden labels for extractive summaries, which is to greedily select the sentences that maximise the ROUGE score as summaries based on the ROUGE score between the original text and the reference summary, and then label these sentences with 1 and the others with 0.

3.3   Encoder Combination Method

We share the encoder side of the abstractive summarization model and the extractive summarization model so that the encoder can focus on both word-level contextual relationships and sentence-level and document-level contextual relationships.

The input text $X = [x_1, ...x_n]$ is a sequence of tokens of length n. We use multi-layer transformer as the encoder, and the encoding process can be expressed as following:

$$h_j = LayerNorm(x_j + MultiHeadAttention(x_j, x, x)) \tag{2}$$

$$h_j = LayerNorm(h_j + FeedForward(h_j)) \tag{3}$$

We end up with a sequence of hidden states for the input text:

$$H = \{h_1, h_2, ..., h_n\} \tag{4}$$

3.4   Decoder Combination Method

The abstractive model requires constant decisions to be made when generating the next token, and these decisions are influenced by several factors, making the generation process full of randomness and uncertainty, which leads to the generation of summaries that omit important information in the original text. To solve this problem, we first obtain a document-level vector representation of the original text through an extractive method, and then add the document-level vector to the decoder of the abstractive method to reduce the randomness of the summary generation.

Specifically, the set of word-level hidden states on the encoder side is obtained as following. Then based on the hidden states of the words we obtain the set of sentence-level vectors:

$$H_w = \{h_1, h_2, \ldots, h_n\} \tag{5}$$

$$s_i = Pooling\,(h_{i,1}, h_{i,2}, \ldots, h_{i,j}) \tag{6}$$

$$H_s = \{s_1, s_2, \ldots, s_m\} \tag{7}$$

where H_w is the set of word vectors, H_s is the set of sentence vectors. s is the sentence-level vector, m denotes the number of sentences of the text, i denotes the current sentence serial number and j denotes the current sentence length.

Finally we get the document-level vector, which we add to the decoder of the abstractive model to limit the randomness in the generation of the summary. As the decoder is made up of multiple layers of transformers, we need to add this vector at each layer, and the process of obtaining the document vector and introducing the document vector into the decoder is as follows:

$$h_d = Pooling\,(s_1, s_2, \ldots, s_m) \tag{8}$$

$$H_d = \{h_d, .., h_d\} \tag{9}$$

$$X_{hidden} = Relu[LayerNorm\,(X_{attention})] + X_{attention} + \delta Hd \tag{10}$$

$$X_{hidden} \in \mathbb{R}^{seq\_len \times hidden\_dim} \tag{11}$$

where $X_{attention}$ is the set of output vectors of the cross-attention layer, h_d is the document vector, H_d is the set of document vectors, X_{hidden} is the output vector of each decoder layer, and δ is a constant which serves to is to scale the document vector to the same order of magnitude as the decoder hidden vector.

3.5   Training Procedure

Algorithm 1: Jointly training procedure

---

1 **for** *epoch in* $1, 2, \ldots, epoch_{max}$ **do**
2   **for** *batch in Datasets* **do**
3   1.Compute loss: $\mathcal{L}(\theta)$
4         $\mathcal{L}(\Theta)_{ext}$ =Equation12 for extractive model
5         $\mathcal{L}(\Theta)_{abs}$ =Equation13 for abstractive model
6   2.Compute gradient:$\nabla(\Theta)$
7   3.Update Model:
8         $\Theta_{ext} = \Theta_{ext} - \epsilon\nabla(\Theta_{ext})$
9         $\Theta_{abs} = \Theta_{abs} - \epsilon\nabla(\Theta_{abs})$
10   **end**
11 **end**

We use a joint training approach to train the abstractive and extractive models. The training process is divided into three steps. In the first step, we first perform forward propagation, which means that during each iteration step, the loss of the extracted model is calculated and a document-level vector is obtained, then this vector is added to the decoder of the abstractive model, and the loss is calculated by forward propagation of the it. In the second step the gradient of the model is calculated; in the third step, the parameters of the two models are back-propagated separately.

For the extractive model we use binary crossentropy loss as the objective:

$$\mathcal{L}_{ext} = -\sum_{i=1}^{N}[y_i log((x_i)) + (1-y_i)log((1-x_i))] \tag{12}$$

where y_i is the true label, which can only take 0 or 1, indicating whether the sentence or phrase belongs to the summary; x_i is the model output, indicating the score of the sentence or phrase; and $\sigma(x)$ is the sigmoid function that maps x to the $(0,1)$ interval, indicating the probability that the sentence or phrase belongs to the summary.

For the abstractive model, we use the cross-entropy loss function as the objective function:

$$\mathcal{L}_{abs} = -\sum_{i=1}^{N}[y_i log(\hat{y}_i)] \tag{13}$$

where y_i is the true label, representing the probability distribution of the i-th word in the summary; $\hat{y}_i$ is the model output, representing the probability distribution of the summary word corresponding to the i-th word in the original text.

4   Experiments

In this section we will introduce the datasets and evaluation metrics we used, as well as the baselines and implementation details of our experiments.

4.1   Datasets

In this section we will introduce the datasets used and their characteristics (Table1).

Table 1. The datasets we used and its characteristics

Dataset	train	val	test	text length	summary length
CNN/Daily Mail	287k	13k	11k	781	56
Pubmed	120k	6.63k	6.66k	3043	215
Arxiv	203k	6.43k	6.44	6038	299

CNN/Daily Mail [17]. This dataset is an English language dataset containing over 300k news articles written by journalists from CNN and the Daily Mail. Each article in the dataset is paired with a hand-written summary.

Pubmed [5]. This dataset is made up of abstracts and articles from scientific and technical literature, mainly from the Pubmed database, in the fields of medicine, biology and chemistry. The purpose of this dataset is to train and evaluate models for text abstracts, especially in the scientific domain.

Arxiv [4]. The Arxiv dataset for the Text Summarization Task is a number of papers and data collected for text summarization research on the Arxiv website, which is a collection of preprints of papers in physics, mathematics, computer science, biology and mathematical economics.

4.2 Baselines

Pointer-Generator [23] is an traditional abstractive model with attention and copy mechanism.

SummaRuNNer [15] is a RNN based sequence model for extractive summarization of documents.

UnifiedSum [11] is an unified model that combines extractive and abstractive summarization methods. The model uses sentence-level attention to guide word-level attention, and a new inconsistency loss to align them.

BertSumExt [14] is an extractive summarization model built on the top of a novel document level encoder. Specifically, the document-level encoder is designed based on BERT.

LongT5 [10] is a model that combines long-input transformers and summarization pre-training with the T5 architecture. The model uses a new attention mechanism called Transient Global, which does not need extra inputs. The model achieves state-of-the-art results on summarization and question answering tasks.

PEGASUS [26] model is a pre-training model specifically designed for summary generation.

BART [12] is a abstractive pre-training model, the basic idea of which is to corrupt the original text using multiple noises in the pre-training phase and then reconstruct the original text using a seq2seq model.

HAT-BART [21] proposes a Hierarchical Attention Transformer-based architecture (HAT) for long sequence to sequence tasks.

DANCER PEGASUS [9] introduces a divide-and-conquer method for neural summarization of long documents based on discourse structure and sentence similarity.

FactorSum [8] factorizes summarization into two steps: generating abstractive summary views and combining them into a final summary.

ExtSum-LG [24] is an extractive summarization model which reduces the redundancy in the sentence scoring phase.

4.3 Implementation Details

Our implementation of the model is based on the PyTorch implementation of LongT5. We used Adam as our optimizer with a learning rate of 1e-5. The maximum number of epochs was set to 5. To avoid the exploding gradient problem, we clipped the gradient norm within 1. For the CNN/Daily Mail dataset we truncated the length of the input text to 512 tokens; for the Pubmed dataset and Arxiv dataset we truncated the input text to 16384 tokens.

5 Result

In this section, we report the experimental results of our model and the baseline model on the CNN/DM, Pubmed and Arxiv datasets.

5.1 Experimental Result

Table 2 shows the results of the extractive methods, which was constructed based on the LongT5 pretrained model, and in general worked better than the Bert-SumExt approach based on the Bert pretrained model, which have benefited from a better pretrained model. Table 2 also shows the results of the abstractive methods, in general, our approach significantly improves the informativeness of the abstractive summary, specifically in that our approach is modified based on LongT5, while the ROUGE-1 scores are better than the most basic LongT5 model on all datasets, with the best on the Pubmed dataset, outperforming all abstractive models, and the The improvement on ROUGE-1 reaches 1.45. indicating that our model is better able to capture the important information in the original text.

Table 2. Abstractive and Extractive Method results, ROUGE-1/2/L F1 scores on CNN/DM, PubMed, and arXiv datasets

	CNN/DM			Pubmed			Arxiv		
Extractive Method	R-1	R-2	R-L	R-1	R-2	R-L	R-1	R-2	R-L
SummaRuNNer	39.60	16.20	35.30	-	-	-	-	-	-
BertSumExt	43.23	20.22	39.60	41.92	15.98	38.10	-	-	-
ExtSum-LG	-	-	-	44.81	19.74	-	43.58	17.37	-
Ours(Ext)	**43.51**	**20.47**	**40.14**	**45.87**	**20.11**	**41.48**	-	-	-
Abstractive Method	R-1	R-2	R-L	R-1	R-2	R-L	R-1	R-2	R-L
Pointer-Generator	39.53	17.28	36.38	35.86	-	-	32.06	-	-
UnifiedSum	40.68	17.97	37.13	-	-	-	-	-	-
PEGASUS	44.17	**21.47**	41.11	45.09	-	-	44.67	-	-
BART	44.16	21.28	40.90	-	-	-	-	-	-
FactorSum	-	-	-	47.5	20.33	43.76	49.32	20.27	**44.76**
DANCER PEGASUS	-	-	-	46.34	19.97	42.42	45.01	17.6	40.56
HAT-BART	44.48	21.31	**41.52**	48.36	21.43	37.00	46.74	19.19	42.2
LongT5-large	42.49	20.51	40.18	49.11	23.66	**45.87**	48.28	**21.63**	44.11
Ours(Abs)	**44.64**	20.01	40.59	**50.56**	**24.07**	45.55	**49.27**	21.17	43.85

5.2 Informativeness Evaluation

In the previous section we calculated the ROUGE score between the joint training summary and the reference summary and demonstrated its validity. To further analyse the role played by our model, in this section we calculate the recall of the original text of the jointly training model in the generated summaries. Recall reflects how much of the original text appears in the generated summaries. The document level vector is added to ensure that the summaries do not miss important content in the original text. Calculating the recall of the original text verifies that our summary captures more of the information in the original text. The Table 3 shows that joint training of the decoders improves the recall of the abstracts by a large margin, and the recall of ROUGE-1 improves the most by 9.36.

Table 3. Informative Study. ROUGE-1/2/L recall scores on PubMed

Method	ROUGE-1	ROUGE-2	ROUGE-L
LongT5-Large	15.36	5.23	13.19
+enc	14.04	4.49	13.11
+dec	20.63	**8.64**	**18.70**
+enc+dec	**22.59**	8.51	18.25

5.3 Human Evaluation

We randomly selected 20 articles from the Pubmed dataset and generated summaries, which included reference summary, abstractive summary, extractive summary, and jointly training summary. We gave the original articles and summaries to three graduate students who were proficient in reading English literature for evaluation. The summaries were assessed manually in three aspects: (1) Informativeness: Informativity: how well does the summary capture the important parts of the article? (2) Readability: how well-written (fluent and grammatical) the summary is? (3) Logicality: Whether the sentences in the article are logical to each other? For a fair comparison among them, model names were not shown to the annotators.

The results are displayed in Table 4. Our model received the highest score for informativeness, indicating that our model was able to capture more of the important information in the text, and our model scored second only to the reference summary in the readability and logicality score. At the same time, we can see that the extractive model received the lowest scores for both readability and logic, which validates the limitations of extractive summaries.

Table 4. Human Evaluation Result and the score is out of five.

Method	Informativeness	Readability	Logicality
Golden Summary	3.6	**4.1**	**3.9**
Extractive Summary	3.6	3.7	3.3
Abstractive Summary	3.2	4.0	3.7
Jointly training Summary(Ours)	**3.7**	3.9	3.8

5.4 Ablation Study

To evaluate the contribution of each design of our model, we conducted an ablation study in the Pubmed dataset. Specifically we benchmarked three simplified versions of the model: a simple LongT5-Large, a model with only Encoder joint training, and a model with only Decoder joint training. As can be seen in Table 5, there is a small improvement in ROUGE scores when only the Encoder side is jointly trained, which indicating that both models benefit from sharing the text representation of Encoder to obtain a better vector space representation of the input text. In contrast, the ROUGE scores of the Decoder-only joint training model improved more, mainly for ROUGE-1, and ROUGE-2, indicating that the addition of the doc-vector allows the model to generate more relevant content to the original text.

Table 5. Ablation experiment. ROUGE-1/2/L F1 scores on the test set of PubMed

Method	ROUGE-1	ROUGE-2	ROUGE-L
LongT5-Large	49.11	23.66	45.87
+enc	49.37	23.60	45.40
+dec	50.43	23.70	45.16
+enc+dec	**50.56**	**24.07**	**45.55**

5.5 Case Study

In this section, we compare the abstractive and extractive summaries with our jointly training summaries to demonstrate the effectiveness of our model from a qualitative perspective. Table 2 shows the original Pubmed document, the reference summary, the extractive summary generated by BertSumExt, the abstractive summary generated by LongT5 and the summary generated by our jointly training model. We can find the article on 'the author's association of the word quantitation with the contrast between two 19th-century German scientists'. The summary generated by the classical abstractive model LongT5 does not include the article's subject word 'quantitation' and assumes that the two 19th-century

---

Pubmed Document:
When i hear the word quantitation , what comes to mind is the contrast between two
famous 19th - century german scientists : Alexander von humboldt and carl gauss ,as
described in measuring the world . Von humboldt was a great prussian explorer who
made many discoveries , including electric eels , in his expeditions to south and central
america , which included the casiquiare canal , which con-nects the orinoco and amazon
rivers ,a man of extraordinary energy . He was the consummate quantitater He measured
the heights of mountains and the numbers of lice on the heads of natives .Gauss , as
surveyor and astronomer for the king-dom of hannover . He was also a quantifier . But
with a different style ,charged with devising a practical method to survey on a spherical
earth , he laid the foun-dation for a school of geometry that culminated in the work of
riemann and the notion of curved space - time in einstein 's general theory of relativity .

Reference Summary:
The article presents the author's association of the word quantitation with the contrast
between two 19th-century German scientists: the explorer Alexander von Humboldt and
the mathematician Carl Gauss. The author describes their different styles and
achievements in measuring the world.

Abstractive Method:
Alexander von humboldt and Carl Gauss were two famous 19th century German
scientists. They were both great explorers and scientists. Alexander von Hum-boldt
made many discoveries, including electric eels, in his expeditions to south and central
America, which included the casiquiare canal. Carl gauss was a surveyor and astronomer
for the kingdom of Hannover.

Extractive Method:
When i hear the word quantitation , what comes to mind is the contrast between two
famous 19th - century german scientists : Alexander von humboldt and carl gauss ,as
described in measuring world .He was the consummate quantitater . He laid the
foundation for a school of geometry that culminated in the work of riemann and the
notion of curved space - time in einstein 's general theory of relativity .

Jointly Training Method(Ours):
In this essay, i present the history of quantitation from 19th century german scientists ,
these two famous 19th - century german scientists were found to have different styles of
quantitation.Alexander von humboldt was a great explorer who made many discoveries.
He measured the heights of mountains . Gauss was also a quantifier . But with a different
style ,charged with devising a practical method to survey on a spherical earth.

Fig. 2. Sample generated summaries by our model.

German scientists are both explorers. While, Alexander von Humboldt was an
explorer and Carl gauss was not an explorer. In contrast, the extracted summary
have ambiguous reference in the extractive sentence 'He was the consummate
quantitater', which is not conducive to reader understanding. For our model, we
can see that by adding the doc-vector, the summary includes a discussion of the
subject word 'quantitation' and explains that the subsequent discussion of the
two scientists is intended to explain the concept of quantitation.

6 Conclusion

We propose a jointly trained summarization model that combines the advantages of abstractive and extractive summarizations, enabling the abstractive method to generate better summaries by adding the document vectors obtained from the extractive method to the Decoder of the abstractive method. And the effectiveness of our method is demonstrated by experiments on Transformer structures.

References

1. Beltagy, I., Peters, M.E., Cohan, A.: Longformer: the long-document transformer. arXiv preprint arXiv:2004.05150 (2020)
2. Brown, T., et al.: Language models are few-shot learners. Adv. Neural. Inf. Process. Syst. **33**, 1877–1901 (2020)
3. Cheng, J., Lapata, M.: Neural summarization by extracting sentences and words. arXiv preprint arXiv:1603.07252 (2016)
4. Clement, C.B., Bierbaum, M., O'Keeffe, K.P., Alemi, A.A.: On the use of arxiv as a dataset. arXiv preprint arXiv:1905.00075 (2019)
5. Dernoncourt, F., Lee, J.Y.: Pubmed 200k rct: a dataset for sequential sentence classification in medical abstracts. arXiv preprint arXiv:1710.06071 (2017)
6. Devlin, J., Chang, M.W., Lee, K., Toutanova, K.: BERT: Pre-training of deep bidirectional transformers for language understanding. In: Proceedings of the 2019 Conference of the North American Chapter of the Association for Computational Linguistics: Human Language Technologies, Volume 1 (Long and Short Papers), pp. 4171–4186. Association for Computational Linguistics, Minneapolis, Minnesota, June 2019. https://doi.org/10.18653/v1/N19-1423. https://aclanthology.org/N19-1423
7. El-Kassas, W.S., Salama, C.R., Rafea, A.A., Mohamed, H.K.: Automatic text summarization: a comprehensive survey. Expert Syst. Appl. **165**, 113679 (2021). https://doi.org/10.1016/j.eswa.2020.113679. https://www.sciencedirect.com/science/article/pii/S0957417420305030
8. Fonseca, M., Ziser, Y., Cohen, S.B.: Factorizing content and budget decisions in abstractive summarization of long documents. In: Proceedings of the 2022 Conference on Empirical Methods in Natural Language Processing, pp. 6341–6364 (2022)
9. Gidiotis, A., Tsoumakas, G.: A divide-and-conquer approach to the summarization of long documents. IEEE/ACM Trans. Audio Speech Lang. Process. **28**, 3029–3040 (2020)
10. Guo, M., et al.: LongT5: efficient text-to-text transformer for long sequences. In: Findings of the Association for Computational Linguistics: NAACL 2022, pp. 724–736. Association for Computational Linguistics, Seattle, United States, July 2022. https://doi.org/10.18653/v1/2022.findings-naacl.55. https://aclanthology.org/2022.findings-naacl.55
11. Hsu, W.T., Lin, C.K., Lee, M.Y., Min, K., Tang, J., Sun, M.: A unified model for extractive and abstractive summarization using inconsistency loss. arXiv preprint arXiv:1805.06266 (2018)
12. Lewis, M., et al.: BART: denoising sequence-to-sequence pre-training for natural language generation, translation, and comprehension. In: Proceedings of the 58th Annual Meeting of the Association for Computational Linguistics, pp. 7871–7880. Association for Computational Linguistics, Online, July 2020. https://doi.org/10.18653/v1/2020.acl-main.703, https://aclanthology.org/2020.acl-main.703

13. Lin, D., Tang, J., Li, X., Pang, K., Li, S., Wang, T.: Bert-smap: paying attention to essential terms in passage ranking beyond bert. Inform. Process. Manage. **59**(2), 102788 (2022)
14. Liu, Y., Lapata, M.: Text summarization with pretrained encoders. In: Proceedings of the 2019 Conference on Empirical Methods in Natural Language Processing and the 9th International Joint Conference on Natural Language Processing (EMNLP-IJCNLP), pp. 3730–3740. Association for Computational Linguistics, Hong Kong, China, November 2019. https://doi.org/10.18653/v1/D19-1387, https://aclanthology.org/D19-1387
15. Nallapati, R., Zhai, F., Zhou, B.: Summarunner: a recurrent neural network based sequence model for extractive summarization of documents. In: Proceedings of the AAAI Conference on Artificial Intelligence, vol. 31 (2017)
16. Nallapati, R., Zhou, B., Gulcehre, C., Xiang, B., et al.: Abstractive text summarization using sequence-to-sequence rnns and beyond. arXiv preprint arXiv:1602.06023 (2016)
17. Nallapati, R., Zhou, B., Ma, M.: Classify or select: Neural architectures for extractive document summarization. arXiv preprint arXiv:1611.04244 (2016)
18. Radford, A., Narasimhan, K., Salimans, T., Sutskever, I., et al.: Improving language understanding by generative pre-training (2018)
19. Radford, A., Wu, J., Child, R., Luan, D., Amodei, D., Sutskever, I., et al.: Language models are unsupervised multitask learners. OpenAI blog **1**(8), 9 (2019)
20. Raffel, C., et al.: Exploring the limits of transfer learning with a unified text-to-text transformer. J. Mach. Learn. Res. **21**(1), 5485–5551 (2020)
21. Rohde, T., Wu, X., Liu, Y.: Hierarchical learning for generation with long source sequences. arXiv preprint arXiv:2104.07545 (2021)
22. Ru, C., Tang, J., Li, S., Xie, S., Wang, T.: Using semantic similarity to reduce wrong labels in distant supervision for relation extraction. Inf. Process. Manage. **54**(4), 593–608 (2018)
23. See, A., Liu, P.J., Manning, C.D.: Get to the point: Summarization with pointer-generator networks. In: Proceedings of the 55th Annual Meeting of the Association for Computational Linguistics (Volume 1: Long Papers), pp. 1073–1083. Association for Computational Linguistics, Vancouver, Canada (Jul 2017). https://doi.org/10.18653/v1/P17-1099. https://aclanthology.org/P17-1099
24. Xiao, W., Carenini, G.: Extractive summarization of long documents by combining global and local context. arXiv preprint arXiv:1909.08089 (2019)
25. Yin, W., Pei, Y.: Optimizing sentence modeling and selection for document summarization. In: Twenty-Fourth International Joint Conference on Artificial Intelligence (2015)
26. Zhang, J., Zhao, Y., Saleh, M., Liu, P.: Pegasus: pre-training with extracted gap-sentences for abstractive summarization. In: International Conference on Machine Learning, pp. 11328–11339. PMLR (2020)

A Three-Stage Framework for Event-Event Relation Extraction with Large Language Model

Feng Huang[1], Qiang Huang[1], YueTong Zhao[2], ZhiXiao Qi[3],
BingKun Wang[5（✉）], YongFeng Huang[3（✉）], and SongBin Li[4]

[1] College of Information Science and Engineering, Xinjiang University, Urumqi,
China
[2] International College, Beijing University of Posts and Telecommunications, Beijing,
China
[3] Department of Electronic Engineering, Tsinghua University, Beijing, China
yfhuang@mail.tsinghua.edu.cn
[4] Haikou Lab, Institute of Acoustics, Chinese Academy of Sciences, Haikou 570105,
China
[5] College of Information Engineering, Zhengzhou University of Industrial Technology,
Xinzheng, China
bkwang77@pdsu.edu.cn

Abstract. Expanding the parameter count of a large language model
(LLM) alone is insufficient to achieve satisfactory outcomes in natural language processing tasks, specifically event extraction (EE), event
temporal relation extraction (ETRE), and event causal relation extraction (ECRE). To tackle these challenges, we propose a novel three-stage
extraction framework (ThreeEERE) that integrates an improved automatic chain of thought prompting (Auto-CoT) with LLM and is tailored based on a golden rule to maximize event and relation extraction
precision. The three stages include constructing examples in each category, federating local knowledge to extract relationships between events,
and selecting the best answer. By following these stages, we can achieve
our objective. Although supervised models dominate for these tasks, our
experiments on three types of extraction tasks demonstrate that utilizing
these three stages approach yields significant results in event extraction
and event relation extraction, even surpassing some supervised model
methods in the extraction task.

Keywords: Auto-CoT · Event-Event Relation Extraction · Large
Language Model

1 Introduction

Extracting events and event relations is a critical subtask of natural language
processing. However, most state-of-the-art relation extractors are supervised
models that heavily rely on annotated training documents to extract events

B. Luo et al. (Eds.): ICONIP 2023, CCIS 1968, pp. 434–446, 2024.
https://doi.org/10.1007/978-981-99-8181-6_33

and their relations from test sentences. Annotating event relations in training documents requires significant domain expertise and effort [9], resulting in high costs. Although Manual-CoT approaches [19] have been proposed as an alternative to supervised models, they significantly increase the requirement for manual labor and are not universally effective.

Zero-shot learning (ZSL) [21] is a method that aims to train models capable of generalizing to unseen data without annotated training data, which is distinct from supervised learning methods. ZSL has garnered considerable attention in recent years due to its potential to reduce the need for expensive and time-consuming annotation efforts. Large language models (LLMs), such as ChatGPT [1]which is also the LLM used in our experiment, have recently demonstrated significant zero-shot learning capabilities across several natural language processing (NLP) and healthcare tasks. These include information extraction [20], machine translation [3], summary evaluation [6], and mental health detection [22]. However, the performance of LLMs in detecting temporal and causal relationships between events has yet to be explored. Thus, it remains a crucial and pressing question whether LLMs, guided by the improved Auto-CoT prompting strategy [24], can effectively complete zero-shot extraction tasks with multiple events and event relations. Moreover, this also raises the possibility of LLMs becoming a new paradigm for temporal and causal extraction, which warrants further investigation. Based on LLM's improved Auto-CoT [24] prompting mechanism, the extraction task aims to extract all events and the relationships between two event triggers from the non-annotated text. This approach enables the construction of a temporal sequence of events from the given text, along with the causal relations between them.

In this research paper, we investigate the performance of improved Auto-CoT+LLMs on extraction tasks with multiple events and event relations. Specifically, we propose three stages of prompting learning strategies to interact with LLMs. We begin by employing Zero-Shot-CoT to construct our samples and establish the golden rules for our extraction tasks. Then, we proceed with extracting event and event relation according to the demo. and, finally, set a threshold to select the optimal results.

Our experimental results demonstrate that our framework significantly improves the precision of event extraction, outperforming some traditional supervised learning methods. Furthermore, it reduces manual annotation costs and enhances the stability of LLM. These findings also highlight the importance of the improved Auto-CoT prompting strategy in better harnessing the potential of LLM for future work.

2 Related Work

2.1 Automatic Chain of Thought Prompting

Auto-CoT prompts are a technique that induces large language models (LLMs) to generate intermediate reasoning steps leading to the final answer without the need for gradient information. A formal study by [19] examined CoT prompt

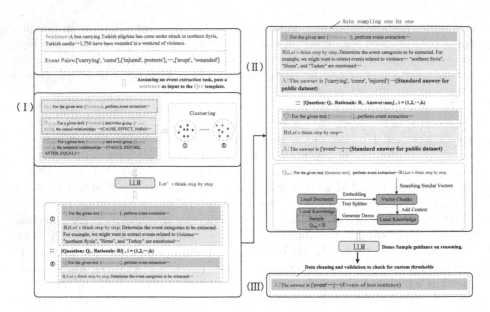

Fig. 1. Illustration for the framework.

themes in language models. The technique prompts LLMs to generate a sequence of coherent intermediate reasoning steps to arrive at the final answer to a question. The research demonstrated that LLMs could perform CoT reasoning through zero-shot prompts (Zero-Shot-CoT) [4], manually crafted few-shot demos (Manual-CoT) [19], or Auto-CoT [24]. It is important to note that Auto-CoT, proposed by Zhang et al. [24], is an encouraging mechanism for large language models. They have demonstrated that this prompting method greatly enhances LLMs' reasoning ability using eight datasets and automatically builds a "Let's think step by step" prompt one after another (total: k) for LLMs.

2.2 Extraction of Event and Event Relation

Event Extraction (EE) is a fundamental information extraction task in the field of natural language processing(NLP), with widespread applications in intelligence work across industries such as business and the military. Various approaches have been developed for extracting events and event relations, including (1) pattern-based EE [17], (2) deep learning-based approaches [23] and (3) joint learning-based approaches. The first method extracts events by applying pre-defined rules to new texts based on patterns discovered from corpora. And the latter two methods have shown promise in overcoming the limited learning ability of traditional machine learning methods by increasing the knowledge gained from continuous data incrementation and addressing dependency issues. Despite their potential, most of these methods rely on supervised learning, which incurs high training costs. As a result, the event extraction and event relation

Extraction of zero-shot LLM has emerged as a noteworthy alternative worth considering.

3   Methodology for Event-Event Relation Extraction

3.1   Challenge of Auto-CoT for Event-Event Relation Extraction

To evaluate the effectiveness of integrating the improved Auto-CoT prompt strategy with LLM for event extraction and event relation extraction, we conducted a thorough validation experiment using two well-established sampling techniques: Random-Q-CoT and Retrieval-Q-CoT.

(1) Retrieval-Q-CoT: For each test question Q^{Test} in the test dataset, we sample k questions for each class of questions Q_i^{demo} ($i = 1, ...$) based on cosine similarity search of the top$-k$ (e.g., $k = 5$) similar questions.
(2) Random-Q-CoT: It randomly selects k instances of questions from the question set Q^{Test}.

Both Random-Q-CoT and Retrieval-Q-CoT invoke zero-shot-COT to generate the reasoning chains R_i^{demo} (rationale and answer) for each sampling question Q_i^{demo}.

Interestingly, our experiment demonstrated that Random-Q-CoT outperformed Retrieval-Q-CoT in terms of F1 score after event extraction on the EventStoryLine [2] dataset. This finding raises questions about why a Retrieval-based approach yielded inferior results compared to Random-Q-CoT. One possible explanation is that Retrieval-Q-CoT retrieves similar questions to the initial one in an attempt to generate a chain of thoughts. However, if the initial chain of thoughts is incorrect, the machine will learn an erroneous thought pattern.

To validate this hypothesis, we conducted Retrieval-Q-CoT experiments on two additional event temporal and event data sets, namely MATRES [12] and DuEE1.0 [5]. Fifty samples were randomly sampled from each of the two datasets, and the reasoning chains of these samples were manually annotated. In the annotated reasoning chain setting, Retrieval-Q-CoT outperformed Random-Q-CoT. The result demonstrates that Retrieval-Q-CoT is effective when manual annotation is available (see Table 1).

Table 1. F1 score (%) of different sampling methods.

	EventStoryLine	MATRES	DuEE1.0
Random-Q-CoT	**58.2**	69.7	68
Retrieval-Q-CoT	50.1	**70.5**	**70.3**

Although manual annotation is helpful, this manual labour is costly. However, automatically generating reasoning chains using zero-shot-CoT is inferior to Manual-CoT, particularly without solving the challenge of problem sampling. We need a better sampling solution to design a more effective Auto-CoT for the extraction of event and event relations.

3.2 Framework for Event-Event Relation Extraction

Based on the challenges discussed earlier with Auto-CoT, we propose an event and event relation extraction system consisting of three stages. The first stage involves constructing suitable demonstration examples from the test dataset. In the second stage, we extract events and their relations by combining the test data with the answers, resulting in the formation of $[Q: q_j^{(i)}, R{:}R_j^{(i)}, A{:}ans_j^{(i)}]$. Finally, in the third stage, we select the best answer using custom thresholds. An overview of our framework is shown in Fig. 1. In the following sections, we will provide a detailed introduction to these three stages, using event extraction as an example.

Stage I. we extract sentences from publicly available datasets and feed them into our predefined event extraction problem template (Q_{EE}). Next, we group all questions into K categories using clustering algorithms and select the question that is closest to the cluster center from each category. To prevent LLM from learning erroneous patterns, we employ the cluster-Q-CoT method, which utilizes sampling diversity to improve Retrieval-Q-CoT, as discussed later in Sect. 3.3. We generate rationale for each question using Zero-Shot-CoT+LLM, resulting in the formation of $[Q: q_j^{(i)}, R{:}R_j^{(i)}]$. By leveraging the diversity of categories, we construct demonstration examples for each category.

Stage II. As outlined in Sect. 3.3 of our paper, the formation of a chain-like structure for every category in the first stage can be observed. Subsequently, we replace $ans_j^{(i)}$ automatically generated by improved Auto-CoT+LLM in this chain-like structure with the standard answers obtained from the question set Q^{test}. And we need to generate a demo that includes local knowledge, sample, and Q^{test}, as discussed later in Sect. 3.4. This post-processing step allows us to generate high-quality extraction examples that act as guidelines for event extraction from new sentences. Finally, we utilize these guidelines to infer the presence of events in new sentences.

Stage III. Results verification stage. We set our own threshold, which is the average F1 score of 50 randomly sampled examples using the model, with an average of 60.1%. If the score exceeds the threshold, it meets the requirements; otherwise, we resample. After three attempts, we select the best extraction result. Finally, we obtain the best extraction results for events and event relationships. The reason for setting a threshold is to filter the best results generated by the model. During the testing process, the model may exhibit randomness, where the extraction performance may be poor in the first few attempts, but improve in subsequent attempts. By setting a threshold, we aim to encourage the model to generate the optimal solution as much as possible.

3.3 Question Clustering and Selecting Optimal Sampling

Due to the potential reduction of misleading similarities by diversity-based clustering, we perform a clustering analysis on the given set of questions Q. We first compute the vector representations of each question in Q using Sentence-BERT [16]. Contextualized vectors are averaged to form a fixed-size question representation. Then, we process the question representations using k-means clustering algorithm to generate k clusters of questions. For each cluster of questions i, questions within it are sorted into a list $Q^{(i)}=[Q_1^{(i)}, Q_2^{(i)}, ...]$ in ascending order of their distances to the center point of the cluster i. This algorithm summarizes the clustering stage of this problem.

In the first stage, we select the optimal example $S^{(i)}$ for each cluster i ($i = 1$, ..., k), which consists of a question and a reasoning process. We iterate through $[Q_1^{(i)}, Q_2^{(i)}, ...]$ in cluster i using the above clustering algorithm until satisfying our selection criteria. In other words, a question that is closer to the center of cluster i is considered first. Assuming the closest j-th question $q_j^{(i)}$ is being considered, we formulate an input prompt as: $[Q: q_j^{(i)}, R:[R_j^{(i)}]]$, where $[R_j^{(i)}]$ is a prompt "Let's think step to step" and a series of thinking processes. This input is then fed into an LLM using Zero-Shot-CoT to output the reasoning chain consisting of the rationale $R_j^{(i)}$ and the extracted answer $ans_j^{(i)}$. A candidate demonstration $S_j^{(i)}$ for the i-th cluster is constructed by concatenating the question, rationale, and answer: $[Q: q_j^{(i)}, R:R_j^{(i)}, A:ans_j^{(i)}]$.

3.4 Constructing Demo

First, we encode local documents in the same way as clustering between them. These local documents do not contain data from public datasets. Then these text vectors are segmented into vector blocks. In this way, the test problem can be retrieved in the vector block to find a vector similar to the Q^{test}, then add the context of this block to form local knowledge, and finally generate a Demo, including local knowledge, previously constructed $[Q: q_j^{(i)}, R:R_j^{(i)}, A:ans_j^{(i)}]$, and Q^{test}.

4 Experiments

4.1 Datasets

EE. DuEE1.0 [5]is a Chinese event extraction dataset released by Baidu, which includes 65 event types. The ACE05[1]corpus provides document and sentence-level event annotations from various domains (such as news agencies and online forums).

---

[1] https://catalog.ldc.upenn.edu/LDC2006T06.

ETRE. MATRES [12] adopts a novel annotation scheme that focuses on the main timeline, where the temporal relations between events are determined by their endpoints, resulting in a consistent inter-annotator agreement (IAA) in event annotation. TCR [10] follows the same annotation scheme, but has many fewer pairs of event relations compared to MATRES.

ECRE. The EventStoryLine [2] corpus (version 0.9) annotates stories with the PLOT-LINK label to express explanatory relations, which provide information about the structural relationship of a story to help readers understand it. there are 258 documents, 22 topics, 4316 sentences, 5334 event mentions, and 1770 of 7805 event mention pairs with causal relation in a sentence. Causal-TimeBank [8] proposes a more widely covering linguistic approach to enrich the TimeML corpus by including causality and trigger words, where events are annotated based on linguistic features. there are 184 documents, 6813 event mentions, and 318 of 7608 event mention pairs annotated with causal relation.

4.2   Evaluation Metrics

EE. Regarding event extraction, we used different evaluation metrics for the DuEE1.0 and ACE05 datasets. For the DuEE1.0 dataset, we evaluated based on F-measure (F14) score with word-level matching. For the ACE05 dataset, predicted entity results were matched with manually labeled entity results at the entity level, and evaluated with micro F1.

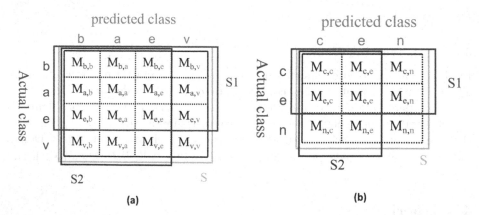

Fig. 2. For this confusion matrix (a), there are four types of classifications, namely before(b), after(a), equal(e), and vague(v). It is worth noting that there are also three variables S, S1, and S2, which represent the sum of all elements in the corresponding areas. For this confusion matrix (b), there are three types of classifications, namely cause(c), effect(e), and NoRel(n). Meanwhile, there are also three variables S, S1, and S2, which represent the sum of all elements in the corresponding areas.

ETRE. The temporal relationships between events are a multi-classification problem in computer NLP, with four common relationships: <before, after, equal, vague (NoRel)>. In event relationship extraction, "vague" is treated as "NoRel", meaning negative examples of no relationship. Therefore, the confusion matrix is presented in Fig. 2.

$$Precision = \frac{M_{b,b} + M_{a,a} + M_{e,e}}{S_1}, Recall = \frac{M_{b,b} + M_{a,a} + M_{e,e}}{S_2}, F1 = \frac{2 * P * R}{P + R} \quad (1)$$

ECRE. The causal relationships between events are also a multi-classification problem, generally including three relationships <Cause, Effect, NoRel>, where NoRel is a negative example of no relationship. Therefore, the confusion matrix is shown in Fig. 2.

$$Precision = \frac{M_{c,c} + M_{e,e}}{S_1}, Recall = \frac{M_{c,c} + M_{e,e}}{S_2}, F1 = \frac{2 * P * R}{P + R} \quad (2)$$

4.3 Results

For event extraction, Fig. 3 shows that using our framework significantly improves the F1 value compared to some supervised models and standard prompt mechanisms. As the number of extracted samples increases, the Large Language Model (LLM) can still maintain the stability of the extraction, even surpassing the supervised model in the DuEE1.0 dataset.

Regarding event temporal relationships, ChatGPT performs worse than state-of-the-art supervised models such as Poincaré Event Embeddings [18] and LSTM + knowledge [11]. However, Table 3 demonstrates that utilizing Three-EERE outperforms traditional supervised models (CogCompTime [13]) and also

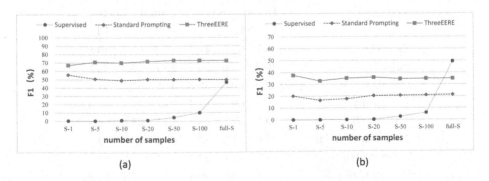

(a)   (b)

Fig. 3. The figure compares the F1 values extracted by events on two datasets in three extraction methods, supervised Text2Event, standard prompt strategy which is a strategy without CoT prompt, and our framework(ThreeEERE). The charts (a) and (b) are on the dataset DuEE1.0 and ACE05 respectively.

beats some unsupervised Zero-Shot-CoT [4], Few-Shot and Manual-CoT [19], and Auto-CoT [24] strategies on LLM in the same dataset. The aforementioned experiments indicate that the model performs better in ambiguous event relations extraction after integrating an external knowledge base, surpassing the performance of the original Auto-CoT approach.

Table 2. Model Performance of event temporal extraction on two datasets.

Model	MATRES			TCR		
	P	R	F1	P	R	F1
CogCompTime [13]	61.6	72.5	66.6	-	-	70.7
Poincaré Event Embeddings [18]	74.1	84.3	78.9	85.0	86.0	85.5
LSTM + knowledge [11]	71.3	82.1	76.3	72.2	86.2	78.6
Zero-Shot-CoT [4]	48.0	57.7	52.4	51.2	60.0	55.3
Few-Shot [19]	49.6	58.0	53.5	54.6	61.3	57.8
Manual-CoT [19]	67.1	70.8	68.9	71.5	72.2	71.8
Auto-CoT [24]	69.0	73.7	71.3	72.6	74.2	73.4
Zero-Shot without CoT	24.5	35.3	28.9	30.1	37.5	33.4
ThreeEERE	**70.8**	**74.0**	**72.4**	**73.0**	**74.8**	**74.0**

Finally, we compare our model's ability in event causal relationship extraction with supervised BERT-based methods. Our model outperforms these specific task supervised models, even improving the F1 score on the Causal-TimeBank by more than 10%, And can surpass these unsupervised prompt strategies (see Table 3).

Table 3. Model Performance of event causality extraction on two datasets.

Model	EventStoryLine			Causal-TimeBank		
	P	R	F1	P	R	F1
KnowDis [26]	39.7	66.5	49.7	42.3	60.5	49.8
RichGCN [14]	49.2	63.0	55.2	39.7	56.5	46.7
LearnDA [25]	42.2	69.8	52.6	41.9	67.7	51.8
T5 Classify [15]	39.1	69.5	47.7	39.1	67.7	48.3
GenECI [7]	59.5	57.1	58.8	60.1	53.3	56.5
Zero-Shot-CoT [4]	42.6	44.5	43.5	40.2	45.0	42.5
Few-Shot [19]	44.5	41.7	43.1	46.1	40.3	43.0
Manual-CoT [19]	56.1	63.0	59.4	52.0	56.7	54.2
Zero-Shot without CoT	20.5	24.1	22.2	29.2	24.0	26.3
ThreeEERE	**58.2**	**64.3**	**61.1**	**59.0**	**61.2**	**60.1**

5   Ablation Study

To demonstrate the effectiveness of CoT (Context of Thinking) and the local knowledge base in the domain of event and event relation extraction, Table 2 shows that adding the augmentation to the CoT prompting mechanism boosts the performance from 28.9% to 52.4% (+23.5%). Additionally, Auto-CoT improves the F1 score from 52.4% to 71.3% (+18.9%). Surprisingly, we found that incorporating a local knowledge base, our architecture, can further increase the F1 score to 72.4% (+1.1%). In Table 3, the addition of the CoT component to LLM (Language Model) increases the F1 score from 22.2% to 43.5% (+21.3%). However, the threshold settings are primarily aimed at selecting the best answer from three trial results. This experiment demonstrates that the inclusion of CoT and the local knowledge base can enhance the extraction performance to varying degrees across different datasets.

6   Conclusion

This paper proposes and implements a three-stage framework using the improved Auto-CoT prompt mechanism, which significantly improves the extraction performance of LLM (Language Model). In the first stage, sample diversity is used to prevent LLM from learning incorrect reasoning processes caused by the same problem category, avoiding reasoning errors for one problem and creating a category reasoning error. In the second stage, a Demo is generated by combining local knowledge to prevent large language models from extracting ambiguous event relationships. This is achieved by incorporating relevant contextual relationships into the prompt as reference examples. In the third stage, the extraction stability of LLM is controlled by setting the extraction threshold. Extensive experiments demonstrate that the proposed strategy enhances LLM's ability in event extraction, event temporal relationship extraction, and event causality relationship, surpassing some supervised learning methods and ordinary prompt strategies, such as Zero-Shot-CoT and Few-Shot, on LLM. These experimental results also indicate that this strategy can significantly impact on natural language processing applications.

Acknowledgements. This work was supported in part by the Important Science and Technology Project of Hainan Province under Grant ZDKJ2020010.

References

1. Bubeck, S., et al.: Sparks of artificial general intelligence: Early experiments with GPT-4. CoRR abs/2303.12712 (2023). https://doi.org/10.48550/arXiv.2303.12712
2. Caselli, T., Vossen, P.: The event storyline corpus: A new benchmark for causal and temporal relation extraction. In: Caselli, T., et al. (eds.) Proceedings of the Events and Stories in the News Workshop@ACL 2017, Vancouver, Canada, August 4, 2017, pp. 77–86. Association for Computational Linguistics (2017). https://doi.org/10.18653/v1/w17-2711. https://doi.org/10.18653/v1/w17-2711

3. Jiao, W., Wang, W., Huang, J., Wang, X., Tu, Z.: Is chatgpt A good translator? A preliminary study. CoRR abs/2301.08745 (2023). https://doi.org/10.48550/arXiv. 2301.08745

4. Kojima, T., Gu, S.S., Reid, M., Matsuo, Y., Iwasawa, Y.: Large language models are zero-shot reasoners. In: NeurIPS (2022). http://papers.nips.cc/paper_files/ paper/2022/hash/8bb0d291acd4acf06ef112099c16f326-Abstract-Conference.html

5. Li, X., et al.: DuEE: a large-scale dataset for Chinese event extraction in real-world scenarios. In: Zhu, X., Zhang, M., Hong, Yu., He, R. (eds.) NLPCC 2020. LNCS (LNAI), vol. 12431, pp. 534–545. Springer, Cham (2020). https://doi.org/10.1007/ 978-3-030-60457-8_44

6. Luo, Z., Xie, Q., Ananiadou, S.: Chatgpt as a factual inconsistency evaluator for abstractive text summarization. CoRR abs/2303.15621 (2023). https://doi.org/10. 48550/arXiv.2303.15621

7. Man, H., Nguyen, M., Nguyen, T.: Event causality identification via generation of important context words. In: Nastase, V., Pavlick, E., Pilehvar, M.T., Camacho-Collados, J., Raganato, A. (eds.) Proceedings of the 11th Joint Conference on Lexical and Computational Semantics, *SEM@NAACL-HLT 2022, Seattle, WA, USA, July 14–15, 2022, pp. 323–330. Association for Computational, Linguistics (2022). https://doi.org/10.18653/v1/2022.starsem-1.28

8. Mirza, P.: Extracting temporal and causal relations between events. In: Proceedings of the 52nd Annual Meeting of the Association for Computational Linguistics, ACL 2014, June 22–27, 2014, Baltimore, MD, USA, Student Research Workshop, pp. 10–17. The Association for Computer Linguistics (2014). https://doi.org/10.3115/ v1/p14-3002

9. Naik, A., Breitfeller, L., Rosé, C.P.: Tddiscourse: A dataset for discourse-level temporal ordering of events. In: Nakamura, S., Gasic, M., Zuckerman, I., Skantze, G., Nakano, M., Papangelis, A., Ultes, S., Yoshino, K. (eds.) Proceedings of the 20th Annual SIGdial Meeting on Discourse and Dialogue, SIGdial 2019, Stockholm, Sweden, September 11–13 2019, pp. 239–249. Association for Computational Linguistics (2019). https://doi.org/10.18653/v1/W19-5929

10. Ning, Q., Feng, Z., Wu, H., Roth, D.: Joint reasoning for temporal and causal relations. In: Gurevych, I., Miyao, Y. (eds.) Proceedings of the 56th Annual Meeting of the Association for Computational Linguistics, ACL 2018, Melbourne, Australia, July 15–20, 2018, Volume 1: Long Papers, pp. 2278–2288. Association for Computational Linguistics (2018). https://doi.org/10.18653/v1/P18-1212

11. Ning, Q., Subramanian, S., Roth, D.: An improved neural baseline for temporal relation extraction. In: Inui, K., Jiang, J., Ng, V., Wan, X. (eds.) Proceedings of the 2019 Conference on Empirical Methods in Natural Language Processing and the 9th International Joint Conference on Natural Language Processing, EMNLP-IJCNLP 2019, Hong Kong, China, November 3–7, 2019, pp. 6202–6208. Association for Computational Linguistics (2019). https://doi.org/10.18653/v1/D19-1642

12. Ning, Q., Wu, H., Roth, D.: A multi-axis annotation scheme for event temporal relations. ArXiv abs/1804.07828 (2018)

13. Ning, Q., Zhou, B., Feng, Z., Peng, H., Roth, D.: Cogcomptime: a tool for understanding time in natural language. In: Blanco, E., Lu, W. (eds.) Proceedings of the 2018 Conference on Empirical Methods in Natural Language Processing, EMNLP 2018: System Demonstrations, Brussels, Belgium, October 31 - November 4, 2018, pp. 72–77. Association for Computational Linguistics (2018). https://doi.org/10. 18653/v1/d18-2013

14. Phu, M.T., Nguyen, M.V., Nguyen, T.H.: Fine-grained temporal relation extraction with ordered-neuron LSTM and graph convolutional networks. In: Xu, W., Ritter, A., Baldwin, T., Rahimi, A. (eds.) Proceedings of the Seventh Workshop on Noisy User-generated Text, W-NUT 2021, Online, November 11, 2021, pp. 35–45. Association for Computational Linguistics (2021). https://doi.org/10.18653/v1/2021.wnut-1.5

15. Raffel, C., et al.: Exploring the limits of transfer learning with a unified text-to-text transformer. J. Mach. Learn. Res. **21**, 140:1–140:67 (2020). http://jmlr.org/papers/v21/20-074.html

16. Reimers, N., Gurevych, I.: Sentence-bert: sentence embeddings using siamese bert-networks. In: Inui, K., Jiang, J., Ng, V., Wan, X. (eds.) Proceedings of the 2019 Conference on Empirical Methods in Natural Language Processing and the 9th International Joint Conference on Natural Language Processing, EMNLP-IJCNLP 2019, Hong Kong, China, November 3–7, 2019, pp. 3980–3990. Association for Computational Linguistics (2019). https://doi.org/10.18653/v1/D19-1410

17. Song, S., Gao, Y., Wang, C., Zhu, X., Wang, J., Yu, P.S.: Matching heterogeneous events with patterns. IEEE Trans. Knowl. Data Eng. **29**(8), 1695–1708 (2017). https://doi.org/10.1109/TKDE.2017.2690912

18. Tan, X., Pergola, G., He, Y.: Extracting event temporal relations via hyperbolic geometry. In: Moens, M., Huang, X., Specia, L., Yih, S.W. (eds.) Proceedings of the 2021 Conference on Empirical Methods in Natural Language Processing, EMNLP 2021, Virtual Event/Punta Cana, Dominican Republic, 7–11 November, 2021, pp. 8065–8077. Association for Computational Linguistics (2021). https://doi.org/10.18653/v1/2021.emnlp-main.636

19. Wei, J., et al.: Chain-of-thought prompting elicits reasoning in large language models. In: NeurIPS (2022). http://papers.nips.cc/paper_files/paper/2022/hash/9d5609613524ecf4f15af0f7b31abca4-Abstract-Conference.html

20. Wei, X., et al.: Zero-shot information extraction via chatting with chatgpt. CoRR abs/2302.10205 (2023). https://doi.org/10.48550/arXiv.2302.10205

21. Xian, Y., Schiele, B., Akata, Z.: Zero-shot learning - the good, the bad and the ugly. In: 2017 IEEE Conference on Computer Vision and Pattern Recognition, CVPR 2017, Honolulu, HI, USA, July 21–26, 2017, pp. 3077–3086. IEEE Computer Society (2017). https://doi.org/10.1109/CVPR.2017.328

22. Yang, K., Ji, S., Zhang, T., Xie, Q., Ananiadou, S.: On the evaluations of chatgpt and emotion-enhanced prompting for mental health analysis. CoRR abs/2304.03347 (2023). https://doi.org/10.48550/arXiv.2304.03347

23. Yang, S., Feng, D., Qiao, L., Kan, Z., Li, D.: Exploring pre-trained language models for event extraction and generation. In: Korhonen, A., Traum, D.R., Màrquez, L. (eds.) Proceedings of the 57th Conference of the Association for Computational Linguistics, ACL 2019, Florence, Italy, July 28- August 2, 2019, Volume 1: Long Papers. pp. 5284–5294. Association for Computational Linguistics (2019). https://doi.org/10.18653/v1/p19-1522, https://doi.org/10.18653/v1/p19-1522

24. Zhang, Z., Zhang, A., Li, M., Smola, A.: Automatic chain of thought prompting in large language models. CoRR abs/2210.03493 (2022). https://doi.org/10.48550/arXiv.2210.03493

25. Zuo, X., Cao, P., Chen, Y., Liu, K., Zhao, J., Peng, W., Chen, Y.: Learnda: Learnable knowledge-guided data augmentation for event causality identification. In: Zong, C., Xia, F., Li, W., Navigli, R. (eds.) Proceedings of the 59th Annual Meeting of the Association for Computational Linguistics and the 11th International Joint Conference on Natural Language Processing, ACL/IJCNLP 2021, (Volume

1: Long Papers), Virtual Event, August 1–6, 2021, pp. 3558–3571. Association for Computational Linguistics (2021). https://doi.org/10.18653/v1/2021.acl-long.276

26. Zuo, X., Chen, Y., Liu, K., Zhao, J.: Knowdis: Knowledge enhanced data augmentation for event causality detection via distant supervision. In: Scott, D., Bel, N., Zong, C. (eds.) Proceedings of the 28th International Conference on Computational Linguistics, COLING 2020, Barcelona, Spain (Online), December 8–13, 2020, pp. 1544–1550. International Committee on Computational Linguistics (2020). https://doi.org/10.18653/v1/2020.coling-main.135

MEFaceNets: Muti-scale Efficient CNNs for Real-Time Face Recognition on Embedded Devices

Jiazhi Li[ID], Degui Xiao[⊠][ID], Tao Lu[ID], and Shiping Dong[ID]

College of Computer Science and Electronic Engineering, Hunan University,
Changsha, China
dgxiao@hnu.edu.cn

Abstract. The growing trend of applying face recognition technology on terminals and embedded devices highlights the critical need to strike a balance between recognition accuracy and real-time inference latency. In response to this challenge, we propose an efficient bottleneck named MEBottleneck, which utilizes convolution kernels of different sizes on two parallel branches to capture multi-scale features in the bottleneck, followed by a 1×1 expansion layer to fuse multi-scale features, thereby improving the representation ability. Then, to balance the trade-off between accuracy and latency, we design a family of lightweight models with MEBottleneck, specifically tailored for face recognition and named MEFaceNets. Large kernels are used for depthwise convolutions in shallow layers, leading to notable improvements in accuracy. We evaluate the proposed models on several popular face recognition benchmarks. Our primary model achieves 99.80% face verification accuracy on LFW and exhibits excellent performance on the larger and more challenging benchmarks, including MegaFace Challenge 1, IJB-B and IJB-C. Furthermore, our proposed models have demonstrated impressive real-time performance on both the CPU and GPU of embedded devices.

Keywords: Deep learning · Face recognition · Lightweight CNNs · Multi-sacle Features · Real-time Response

1 Introduction

Face recognition [1,2] is an important biometric technology for identity authentication. Deriving high performance of face recognition on terminals and embedded platforms with limited computing resources is widely regarded as a major challenge in this field. Figure 1 shows several faces captured in unconstrained environment, including various poses, illuminations, occlusions, and ages.

To address the challenge of achieving high-precision and low-latency face recognition on embedded devices, Wu et al. [3] introduced a series of LightCNN, aiming to learn a compact embedding using a large-scale noisy labeled dataset. For real-time face verification on embedded terminals, Chen et al. [4] introduced

© The Author(s), under exclusive license to Springer Nature Singapore Pte Ltd. 2024
B. Luo et al. (Eds.): ICONIP 2023, CCIS 1968, pp. 447–461, 2024.
https://doi.org/10.1007/978-981-99-8181-6_34

Fig. 1. Faces captured in unconstrained environment.

an improved efficient lightweight model known as MobileFaceNets. However, in comparison to state-of-the-art (SOTA) methods [5–8], these lightweight models still exhibit a significant gap in face recognition accuracy. These SOTA methods typically employ ResNet networks with depths ranging from 50 to 100 layers.

To bridge the gap in accuracy between lightweight and large models in the field of face recognition, in this paper, we propose an efficient bottleneck called MEBottleneck and subsequently design a family of lightweight models for face recognition deployed on the embedded device. To enhance the model's performance, we employ convolution kernels of varying sizes simultaneously within two parallel branches integrated into the MEBottleneck unit. This utilization of different kernel sizes results in distinct receptive fields, facilitating the capture of multi-scale features within the bottleneck. Following this, we apply a 1×1 convolution layer for feature fusion and channel expansion, thereby enhancing the bottleneck's representational capabilities. The primary model demonstrates impressive real-time performance on the embedded device and also exhibits remarkable performance on several popular benchmarks, including LFW [9], MegaFace Challenge 1 [10], IJB-B [11], and IJB-C [12]. Extensive results affirm the effectiveness of the proposed method. To further enhance the performance of real-time face recognition on embedded devices, in this paper, we propose an efficient bottleneck called MEBottleneck and subsequently design a family of lightweight models for face recognition deployed on the embedded device. To enhance the model's performance, we employ convolution kernels of varying sizes simultaneously within two parallel branches integrated into the MEBottleneck unit. This utilization of different kernel sizes results in distinct receptive fields, facilitating the capture of multi-scale features within the bottleneck. Following this, we apply a 1×1 convolution layer for feature fusion and channel expansion, thereby enhancing the bottleneck's representational capabilities. The primary model demonstrates impressive real-time performance on the embedded device and also exhibits remarkable performance on several popular benchmarks, including LFW [9], MegaFace Challenge 1 [10], IJB-B [11], and IJB-C [12]. Extensive results affirm the effectiveness of the proposed method.

Our main contributions are presented as follows:

- We propose an efficient bottleneck called MEBottleneck, which employs convolution kernels of various sizes within two parallel depthwise convolution branches to capture multi-scale features within the inverted bottleneck. Furthermore, through experimentation, we observed that replacing ReLU with GELU can significantly enhance accuracy.
- We design a family of lightweight models with MEBottleneck blocks specifically for face recognition, named MEFaceNets. We introduce large kernels for depthwise convolutions in shallow layers to acquire large receptive field. We provide users with different model choices according to their own needs and configurations.
- Our proposed MEFaceNets strike a balance between accuracy and latency. MEFaceNets achieve impressive real-time performance on the embedded device and achieve high-accuracy on several popular benchmarks.

2   Related Work

2.1   Face Recognition Loss Functions

Softmax loss has been widely used in face recognition and is capable of learning separable feature representation. However, this representation is not discriminative enough for face recognition. Schroff et al. [2] proposed a new triplet loss, and the CNN architecture of the model was based on GoogLeNet Inception [13]. The model was trained on a large-scale private dataset and achieved 99.63% verification. Liu et al. [14] proposed a generalized large-margin softmax (L-Softmax) loss, which explicitly encouraged intra-class compactness and inter-class separability between learned features. Consequently, the decision margin between classes could be well controlled, and the intra-class features could be compact through the margin parameter m. Sphereface [5] introduces the angular softmax (A-Softmax) loss, which enables CNNs to learn angularly discriminative features. CosFace [6,15] and ArcFace [7] introduce cosine/angular margin in an additive manner, achieving state-of-the-art performance on several popular face recognition benchmarks. However, these methods adopt a unified margin, which limits the recognition performance of the model to some extent. CurricularFace [8] introduces curriculum learning strategies into the loss function and improves the performance of face recognition with new feature learning methods and measurement learning. To further improve the discriminative face representation, IHEM Loss [16] introduces a hard example selection approach to efficiently identify hard positive examples, and then penalizes the cosine distance between hard examples and their class centers during model training, achieving state-of-the-art (SOTA) performance on several popular face recognition benchmarks.

2.2   Lightweight CNNs

In recent years, researchers have conducted extensive experiments to strike an optimal balance between accuracy and inference performance. To address this challenge, there is a growing focus on the design of lightweight models. MobileNet [17] leverages depthwise separable convolution to replace the conventional standard convolution, achieving a substantial reduction in the number of parameters. Depthwise separable convolution factorizes a standard convolution into two layers, including depthwise and pointwise convolution layers. More importantly, computation and model size are drastically reduced by factorization. Thus, depthwise separable convolution has been extensively used in lightweight CNNs after the proposition of MobileNet [17]. MobileNetV2 [18] introduces the concept of inverted residuals, enabling efficient information flow throughout the network. Since then, lightweight models have become popular, and many excellent lightweight CNN models have emerged [19–21]. In contrast to very deep CNNs, an excellent lightweight CNN model not only consumes less memory and incurs lower CPU latency but also achieves sufficiently high accuracy for practical applications. Additionally, lightweight CNNs are relatively easy to train.

2.3 Lightweight CNNs for Face Recognition

At present, mainstream lightweight CNNs are mainly designed for common recognition tasks. However, the direct use of existing common lightweight CNNs in face recognition tasks may not yield satisfactory performance. In 2018, Wu et al. [3] developed a light CNN framework to learn a compact embedding on a large-scale noisy labeled dataset. Light CNN-29 model occupies 12 M parameters and extracts features with approximately 121 ms latency on a single core i7-4790. This model achieved state-of-the-art results on five face benchmarks. Chen et al. [4] proposed a class of efficient lightweight models called MobileFaceNetsbased, based on MobileNet-V2 and specifically optimized for real-time face verification on mobile and embedded platforms. MobileFaceNets have fewer than 1 million parameters, and the primary model obtained 99.55% face verification accuracy on LFW and 92.59% TAR (@FAR = 1e−6) on MegaFace Challenge 1. The performance and accuracy of the model exceeded those of existing state-of-the-art lightweight models for face verification.

Compared to existing methods [3,4,7,8], our proposed approach not only achieves a high level of facial recognition accuracy but also demonstrates excellent real-time inference performance on embedded devices.

3   MEFaceNets

In this section, we first describe our motivation and then introduce a novel multi-scale efficient bottleneck called MEBottleneck, which utilizes convolution kernels with different sizes for depthwise convolution to acquire features in different scales, aiming to enhance feature representational power. Then, we provide a detailed description of the proposed CNN structure for face recognition, called MEFaceNets. Figure 2 exhibits the framework of our proposed MEFaceNets.

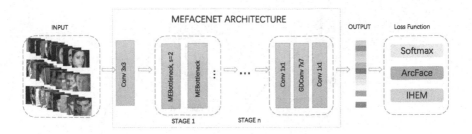

Fig. 2. The framework of MEFaceNets.

3.1 Motivation

The inverted bottleneck [18] improves the model efficiency through depthwise separation convolutions, and compensates for the loss of accuracy through channel expansion in the bottleneck block. Previous studies [21–23] have demonstrated the benefits of using convolution kernels of different sizes at various depths of the network's bottleneck. This approach aids in learning discriminative representations and improving model accuracy. In contrast to these previous works, our approach introduces convolution kernels of different sizes in two parallel branches of the bottleneck for deep convolutions. This enables us to capture local multi-scale features and subsequently fuse these multi-scale features through a 1×1 expansion layer. The goal is to further enhance the discriminative power of the learning features.

3.2 MEBottleneck: An Muti-scale Efficient Bottleneck

In convolutional neural networks, the bottleneck structure is a fundamental unit commonly used to construct CNN models. Figure 3 illustrates three different bottleneck structures. M. Sandler [18] introduced an inverted residual with a linear bottleneck, as depicted in Fig. 3(a). This structure incorporates an expansion layer in the bottleneck and filters features using depthwise convolutions to introduce non-linearity. In the inverted residual bottleneck, low-dimensional input is expanded to high-dimension and filtered features via a depthwise convolution layer. Subsequently, features are mapped to a low-dimensional space with a linear 1×1 convolution. Inspired by the MSA block design in Transformers [24], ConvNeXt bottleneck [25] utilizes a large convolution kernel for depthwise convolution and moves up the depthwise convolution layer, as shown in Fig. 3(b). The proposed MEBottleneck is as shown in Fig. 3(c). To further improve the performance of inverted bottleneck, we introduce kernels with different sizes for depthwise convolutions on two branches. This allows it to capture features at multi-scales within the inverted bottleneck unit. These two branches are concatenated and then followed with 1×1 convolution to fuse multi-scale features. To enhance the model capacity, the 1×1 convolution layer expands the features to a high dimension. Finally, these expanded features are subsequently compressed back to low-dimension using a linear 1×1 convolution.

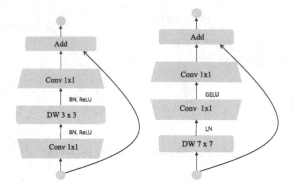

(a) Inverted bottleneck (b) ConvNeXt bottleneck

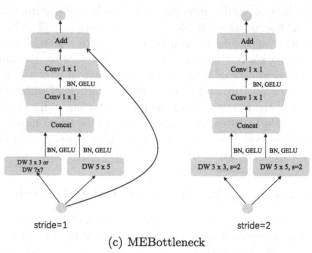

(c) MEBottleneck

Fig. 3. Comparison of different bottlenecks.

The convolution cost of the standard inverted bottleneck is compose of depthwise convolution cost and 1×1 pointwise convolution cost, the total computational cost is:

$$M_s = h \cdot w \cdot t \cdot c \cdot k \cdot k + h \cdot w \cdot c \cdot t \cdot c + h \cdot w \cdot t \cdot c \cdot c' \qquad (1)$$

where the input shape is $h \cdot w \cdot c$, the output channel is c', t represents the
expansion factor, the kernel size for depthwise convolution is $k \times k$.

For the proposed MEBottleneck, the implementation structure is as shown
in Table 1, the total number of multiply-adds (MAdds) computation can be formulated as:

$$M = h \cdot w \cdot c \cdot (k \cdot k + k' \cdot k') + h \cdot w \cdot 2 \cdot c \cdot t \cdot c + h \cdot w \cdot t \cdot c \cdot c' \qquad (2)$$

Table 1. The structure of the MEBottleneck.

Input	Operator	Output
$h \times w \times c$	$k \times k$ dw, s $= s'$, PReLU	$\frac{h}{s} \times \frac{w}{s} \times c$
$h \times w \times c$	$k' \times k'$ dw, s $= s'$, PReLU	$\frac{h}{s} \times \frac{w}{s} \times c$
$\frac{h}{s} \times \frac{w}{s} \times 2 \times c$	1×1 conv2d, PReLU	$\frac{h}{s} \times \frac{w}{s} \times t \times c$
$\frac{h}{s} \times \frac{w}{s} \times t \times c$	Linear 1×1 conv2d	$\frac{h}{s} \times \frac{w}{s} \times c'$

The two kernel sizes are $k \times k$ and $k' \times k'$. Compared with the standard inverted bottleneck, the increase of MAdds is:

$$M_\triangle = h \cdot w \cdot c \cdot k' \cdot k' + h \cdot w \cdot c \cdot t \cdot c \qquad (3)$$

As can be seen from the Eq. (3), the amount of MAdds increased by the bottleneck can be adjusted by the hyperparameter t. Using a small t results in a small increase in FLOPs.

Given the excellent performance of GELU in recent studies [25–27], we replace ReLU [28] with it in the bottleneck. We find in the experiment that the utilization of GELU is very important for lightweight model performance improvement in face recognition.

3.3   MEFaceNet Architecture

Table 2. Architecture of MEFaceNet.

Stage	Input	Operator	t	c	n	s
	$112^2 \times 3$	conv 3×3	-	64	1	2
Stage 1	$56^2 \times 64$	dwconv 7×7	-	64	-	1
	$56^2 \times 64$	MEBottleneck($5 \times 5, 3 \times 3$)	2	64	1	2
Stage 2	$56^2 \times 64$	MEBottleneck($5 \times 5, 3 \times 3$)	2	64	2	1
	$28^2 \times 64$	MEBottleneck($5 \times 5, 3 \times 3$)	4	128	1	2
Stage 3	$14^2 \times 128$	MEBottleneck($5 \times 5, 3 \times 3$)	2	128	6	1
	$14^2 \times 128$	MEBottleneck($5 \times 5, 3 \times 3$)	4	128	1	2
Stage 4	$7^2 \times 128$	MEBottleneck($5 \times 5, 3 \times 3$)	2	128	2	1
	$7^2 \times 128$	conv 1×1	-	512	1	1
	$7^2 \times 512$	Linear GDConv 7×7	-	512	1	1
	$1^2 \times 512$	Linear conv 1×1	-	256	1	1

Table 2 presents the primary architecture of MEFaceNet. Parameter t represents the expansion factor in the bottleneck, c refers to the number of output channels, n indicates the number of times the bottleneck is repeated, and s denotes the stride size. We categorize the bottlenecks into four stages based on the output resolution of these bottlenecks within the MEFaceNet. To optimize computational efficiency, we employ depthwise convolution layers in Stage 1, which can be substituted with MEBottleneck for larger models. MEBottleneck is used instead of the original inverted residual block in MEFaceNet. Recent studies have shown that using larger kernels is more effective in achieving a larger receptive field than stacking numerous smaller kernels [27]. In addition, it is effective to avoid optimization problems caused by increased depth. Inspired by these findings, we utilize 7×7 kernels for depthwise convolution in the shallower layers and 5×5 and 3×3 kernels for depthwise convolution in the deeper layers within the MEBottleneck block. Compared to MobileFaceNet, we reduced the number of repetitions in Stage 2 to offset the increase in FLOPs caused by the use of larger convolution kernels and changes in the bottleneck structure. In pursuit of higher recognition accuracy, we extend the model to 1G FLOPs, named MEFaceNet-2X, which replaces 7×7 depthwise convolution layer with MEBottleneck (5×5, 7×7) in Stage 1 and doubles the number of bottlenecks from Stage 2 to Stage 4.

4   Experiments

In this section, we first investigate the effection of different convolution kernels in MEFaceNet. Then, we evaluate our models on several popular benchmarks, including LFW [9], CFP-FP [29], AgeDB [30], MegaFace [10], IJB-B [11] and IJB-C [12]. Experimental results indicate that our method is competitive with existing approaches.

4.1   Data Set

Table 3. Datasets for evaluation.

Date Set	Identity	Image
LFW [9]	5,749	13,233
CFP-FP [29]	500	7,000
AgeDB [30]	568	16,488
MegaFace [10]	530(p)	1M(g)
IJB-B [11]	1,845	76,824
IJB-C [12]	3,531	148,876

We employ CASIA-Webface [31] and MS1MV3 [32] as our training data. CASIA-Webface is the first widely available public large-scale face dataset, consisting of 10k subjects and 500k images, collected from the Internet in a semi-automatic manner. MS1MV3 [32] is a refined version of MS-Celeb-1M [33], containing 91k ids and 5.1M images.

We evaluate the performance of our proposed MEFaceNets on several popular face test benchmarks, including LFW [9], CFP-FP [29], AgeDB [30], MegaFace [10], IJB-B [11] and IJB-C [12]. The detail of the test benchmarks is as shown in Table 3.

4.2 Experimental Settings

We generate 112×112 face crops with five landmarks [7], then use MTCNN [34] for face detection and align the faces with the landmarks. Our models are trained on four GPUs (24 GB) and the total batch size is set as 512. For the training on CASIA-WebFace, we initially set the learning rate as 0.1 and then divide it by 10 at 26k, 40k iterations. While on MS1MV3, we initialize the learning rate to 0.1 but divide it by 10 at 100k, 150k, and 180k iterations.

To evaluate inference latency, we deployed models on RK3399 and NVIDIA Jetson NANO B01, which are two popular embedded development boards for applications. We used RK3399 to evaluate the model's CPU performance with two threads running on two big cores, and Jetson NANO B01 to evaluate its GPU performance by running 200 threads loops 10,000 times.

4.3 Effection of Different Convolution Kernels in MEFaceNet

To investigate the effect of different convolution kernels in MEFaceNet, we progressively replace 3×3 convolution kernels with 7×7 kernels for depthwise convolutions from Stage 1 to Stage 4 to obtain a series of models. We train these models on CASIA-Webface [31] with Softmax loss and evaluate them on LFW, CFP-FP and AgeDB. Table 4 shows the evaluation results of these models. "S1", "S2", "S3" and "S4" refer to "Stage 1", "Stage 2", "Stage 3", and "Stage 4", respectively. "Latency" refers to the inference latency evaluated on RK3399 with two threads.

Table 4. Effection of different Convolution Kernels in MEFaceNet.

Model	S1	S2	S3	S4	LFW	CFP-FP	AgeDB	FLOPs	Latency (ms)
MEFaceNet-0	3×3	3×3 5×5	3×3 5×5	3×3 5×5	97.05	89.60	84.26	428M	32.08
MEFaceNet-1	7×7	3×3 5×5	3×3 5×5	3×3 5×5	**97.93**	**89.83**	84.40	444M	36.32
MEFaceNet-2	7×7	5×5 7×7	3×3 5×5	3×3 5×5	97.73	89.65	**84.46**	452M	39.42
MEFaceNet-3	7×7	5×5 7×7	5×5 7×7	3×3 5×5	97.53	89.07	83.42	464M	42.05
MEFaceNet-4	7×7	5×5 7×7	5×5 7×7	5×5 7×7	97.65	89.45	82.73	465M	42.40

From Table 4, we can see that MEFaceNet-1 achieves the best results on both LFW and CFP-FP. Compared to MEFaceNet-0, MEFaceNet-1 gains 0.88%,

0.23% and 0.14% on LFW, CFP-FP and AgeDB, respectively, and increases inference latency by 4 ms. MEFaceNet-2 uses 7×7 convolution kernels in the first two stages, achieving the best accuracy on AgeDB. It can be observed that as the 7×7 convolution kernels increase, the FLOPs and inference latency increase accordingly, but the accuracy on the evaluation data sets does not continue to increase. Experimental results exhibited that using 7×7 convolution kernels for depthwise convolutions in shallow layers is effective to improve the model performance.

4.4 Effection of GELU

To investigate the influence of GELU, we conducted a comparative analysis of experimental results involving three distinct activation functions. Our model was trained using the CASIA-Webface dataset [31], with MEFaceNet employed as the primary network architecture. Table 5 presents a comparison of MEFaceNets utilizing various activation functions. It is evident that GELU exhibited superior performance on LFW and CFP-FP, resulting in the highest average accuracy. Compared to ReLU, it improved LFW and CFP-FP by 0.72% and 0.52%, respectively, leading to an overall average accuracy improvement of 0.41%.

Table 5. Comparison of MEFaceNets with different activation functions. 'AVG' refers to the average accuracy over the three test benchmarks.

ACTIVATION	LFW	CFP-FP	AgeDB	AVG
ReLU	97.21	89.30	84.43	90.31
PRelu	97.35	89.26	**84.70**	90.44
GELU	**97.93**	**89.82**	84.40	**90.72**

4.5 Model Evaluation

In this section, we evaluate the proposed MEFaceNets based on several challenging testing benchmarks, including LFW [9], MegaFace [10], IJB-B [11] and IJB-C [12] benchmarks.

Evaluation on LFW. LFW [9] is a widely used benchmark for unconstrained face recognition. It contains 5,749 different identities and 13,233 images collected from the web, under arbitrary poses, expression and illuminations.

We evaluate the our method on LFW [9]. Table 6 shows the remarkable performance of MEFaceNets compared with other competitors. Compared with MobileFaceNet [4], the accuracy of the proposed MEFaceNet-(ArcFace) improved by 0.10%, at the cost of an additional 2 ms delay. The accuracy

Table 6. Comparison of MEFaceNets and other competitors on LFW.

Methods	LFW	Size (MB)	FLOPs	Latency (ms)	
				CPU	GPU
FaceNet [2]	99.63	-	30G	-	-
Light CNN-29 [3]	99.33	50	-	-	-
R64, SphereFace [5]	99.27	-	-	-	-
R100, ArcFace [7]	99.83	248	27G	595.18	143.86
R50, ArcFace [7]	99.80	166	12.6G	358.45	90.32
R100, CurricularFace [8]	99.80	400	27G	595.18	143.86
Mobilefacenet [4]	99.55	4	440M	34.01	17.55
EfficientNet_B0, ArcFace [7,21]	99.65	14.5	1G	155.12	40.32
MNasNet, ArcFace [7,20]	99.70	12.6	598M	58.42	23.90
MEFaceNet, ArcFace (ours)	99.68	5.4	444M	36.32	16.40
MEFaceNet, IHEM (ours)	99.70	5.4	444M	36.32	16.40
MEFaceNet-2X, ArcFace(ours)	99.73	9.4	1G	90.19	32.08
MEFaceNet-2X, IHEM (ours)	99.80	9.4	1G	90.19	32.08

of MEFaceNet-2X(IHEM) on LFW achieves 99.80%, which is equal to Arc-Face(R50) and CurricularFace(R100). However, the model size of MEFaceNet-2X is only 9.4 MB, with 1G FLOPs. Compared to ArcFace(R50), the inference latency of MEFaceNet-2X is 3x faster on CPU and 2x faster on GPU. Compared to CurricularFace(R100), the inference latency of MEFaceNet-2X is 5.6x faster on CPU and 3.4x faster on GPU. Experiment results exhibit that our proposed method achieves a good balance between accuracy and latency on the embedded device.

Evaluation on MegaFace Challenge 1. The MegaFace dataset [10] is a challenging test benchmark, which is the largest public face recognition dataset that contains 1 million faces of 690K individuals. We use the Facescrub [35] dataset, which contains 100K photos of 530 unique individuals, as the probe set to evaluate the performance of MEFaceNets on MegaFace Challenge 1.

Table 7 shows the results of identification and verification accuracy under the large protocol. "Rank-1" denotes the rank-1 face identification accuracy with 1 million distractors, and "VR" denotes the face verification TAR (@FAR = $1e-6$). Our proposed models are trained on MS1MV3 [32]. As shown in Table 7, the proposed MEFaceNet-2X(IHEM) achieves the best accuracy on the face verification TAR(@FAR = $1e-6$). Compared to MobileFaceNet, the proposed MEFaceNet demonstrates a 0.90% improvement in face verification TAR(@FAR = $1e-6$). Compare to CurricularFace(R100), the proposed MEFaceNet-2X(IHEM) gains on the face verification TAR(@FAR=$1e-6$) by 0.18%, but drops 1.4% on Rank-1

Table 7. Comparison performance of different methods on MegaFace Challenge 1.

Methods	Rank-1	VR	FLOPs	Latency(ms)	
				CPU	GPU
FaceNet	70.49	86.47	-	-	-
Light CNN-29 [3]	73.49	84.73	-	-	-
R100, ArcFace [7]	81.03	96.98	27G	595.18	143.86
R100,CurricularFace [8]	**81.26**	97.26	27G	595.18	143.86
MobileFaceNet [4]	-	90.16	440M	34.01	17.55
MEFaceNet,IHEM (ours)	76.28	91.06	444M	36.32	16.40
MEFaceNet-2X,ArcFace (ours)	79.53	97.07	1G	90.19	32.08
MEFaceNet-2X,IHEM (ours)	79.86	**97.44**	1G	90.19	32.08

accuracy. However, the FLOPs of MEFaceNet-2X(IHEM) decreases by 26x, and the CPU and GPU inference performance is 5.6x and 3.4x faster, respectively.

Experimental results exhibit that our proposed method achieves competitive face recognition performance at low inference latency.

Table 8. Evaluation on 1:1 verification protocol on IJB-B and IJB-C dataset.

Methods	IJB-B	IJB-C	FLOPs	Latency(ms)	
				CPU	GPU
ResNet50+DCN(Kpts) [36]	85.0	86.7	12.6G	358.45	-
SENet50+DCN(Divs) [36]	84.9	88.5	-	-	-
R50, ArcFace [7]	89.8	92.1	12.6G	358.45	90.32
R100, ArcFace [7]	94.2	95.6	27G	595.18	143.86
R100, CurricularFace [8]	94.80	96.10	27G	595.18	143.86
mNasNet, ArcFace [7,20]	90.81	91.49	598M	58.42	23.90
EfficientNet_B0, ArcFace [7,21]	91.74	91.89	1G	156.18	40.32
MEFaceNet-2X,ArcFace(ours)	92.58	93.90	1G	90.19	32.08
MEFaceNet-2X,IHEM(ours)	94.05	95.65	1G	90.19	32.08

Evaluation on IJB-B and IJB-C. The IARPA Janus Benchmark-B (IJB-B) [11] dataset, which includes 1,845 K subjects with 21.8 K still images and 55 K frames from 7,011 videos, is a widely used evaluation benchmark for face recognition. The IJB-C dataset [12] is an extension of IJB-B, which encompasses 3,531 identities and 148,876 faces.

We evaluate the proposed models on the standard protocols for IJB-B [11] and IJB-C [12] on the 1:1 verification protocol. Table 8 compares the 1:1 verification TAR (@FAR = 1e−4) results with previous methods on IJB-B. Our proposed MEFaceNet-2X(IHEM) exhibited very high performance on these two aforementioned benchmarks.

Compared to EfficientNet_B0(Arcface) [7,21], the performance of proposed MEFaceNet-2X(Arcface) improves by 0.84% and 2.01% on the TAR (@FAR = 1e−4) on IJB-B and IJB-C, respectively. The latencies of MEFaceNet-2X on CPU and GPU are 73% and 25% faster than EfficientNet_B0. Compared with ArcFace(R100), the proposed MEFaceNet-2X(IHEM) gains 0.05% on IJB-C and drops on 0.15% on IJB-B, with 5.6x and 3.4x faster inference speed on CPU and GPU, respectively. Figure 4 shows the ROC curves of 1:1 verification protocol on IJB-B and IJB-C datasets.

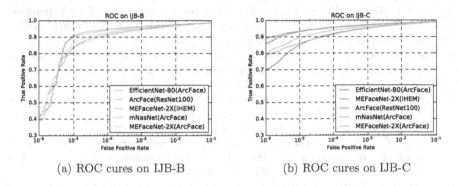

(a) ROC cures on IJB-B   (b) ROC cures on IJB-C

Fig. 4. ROC cures on IJB-B and IJB-C.

5   Conclusion

In this study, we propose an efficient MEBottleneck, which utilizes convolution kernels of different sizes to capture multi-scale features in the inverted bottleneck. Furthermore, we propose a family of efficient and lightweight CNNs, named MEFaceNets, which are specially tailored for real-time face recognition on embedded devices. We evaluate MEFaceNets on several popular face recognition benchmarks. Extensive experimental evaluations demonstrate that our proposed method achieves comparable accuracy to state-of-the-art large model approaches on several popular test benchmarks. Meanwhile, the proposed models achieve impressive real-time performance on both the CPU and GPU of embedded devices. In future work, we will validate the effectiveness of the proposed method in visual tasks such as image classification and object detection.

Acknowledgements. This work was supported in part by the National Natural Science Foundation of China (Grant No. 62172150).

References

1. Turk, M., Pentland, A.: Eigenfaces for recognition. J. Cogn. Neurosci. **3**(1), 71–86 (1991)
2. Schroff, F., Kalenichenko, D., Philbin, J.: Facenet: a unified embedding for face recognition and clustering. In: Proceedings of the IEEE Conference on Computer Vision and Pattern Recognition, pp. 815–823 (2015)
3. Wu, X., He, R., Sun, Z., Tan, T.: A light CNN for deep face representation with noisy labels. IEEE Trans. Inf. Forensics Secur. **13**(11), 2884–2896 (2018)
4. Chen, S., Liu, Y., Gao, X., Han, Z.: Mobilefacenets: Efficient cnns for accurate real-time face verification on mobile devices. In: Chinese Conference on Biometric Recognition, pp. 428–438. Springer (2018)
5. Liu, W., Wen, Y., Yu, Z., Li, M., Raj, B., Song, L.: Sphereface: deep hypersphere embedding for face recognition. In: Proceedings of the IEEE Conference on Computer Vision and Pattern Recognition, pp. 212–220 (2017)
6. Wang, H., Wang, Y., Zhou, Z., Ji, X., Gong, D., Zhou, J., Li, Z., Liu, W.: Cosface: large margin cosine loss for deep face recognition. In: Proceedings of the IEEE Conference on Computer Vision and Pattern Recognition, pp. 5265–5274 (2018)
7. Deng, J., Guo, J., Xue, N., Zafeiriou, S.: Arcface: additive angular margin loss for deep face recognition. In: Proceedings of the IEEE/CVF Conference on Computer Vision and Pattern Recognition, pp. 4690–4699 (2019)
8. Huang, Y., et al.: Curricularface: adaptive curriculum learning loss for deep face recognition. In: Proceedings of the IEEE/CVF Conference on Computer Vision and Pattern Recognition, pp. 5901–5910 (2020)
9. Huang, G.B., Mattar, M., Berg, T., Learned-Miller, E.: Labeled Faces in the Wild: A Database for Studying Face Recognition in Unconstrained Environments. In: Workshop on Faces in 'Real-Life' Images: Detection, Alignment, and Recognition. Erik Learned-Miller and Andras Ferencz and Frédéric Jurie, Marseille, France, October 2008. https://hal.inria.fr/inria-00321923
10. Kemelmacher-Shlizerman, I., Seitz, S.M., Miller, D., Brossard, E.: The megaface benchmark: 1 million faces for recognition at scale. In: Proceedings of the IEEE Conference on Computer Vision and Pattern Recognition, pp. 4873–4882 (2016)
11. Whitelam, C., et al.: Iarpa janus benchmark-b face dataset. In: Proceedings of the IEEE Conference on Computer Vision and Pattern Recognition Workshops, pp. 90–98 (2017)
12. Maze, B., Adams, J., Duncan, J.A., Kalka, N., Grother, P.: Iarpa janus benchmark - c: Face dataset and protocol, pp. 158–165 (2018)
13. Szegedy, C., et al.: Going deeper with convolutions. In: Proceedings of the IEEE Conference on Computer Vision and Pattern Recognition, pp. 1–9 (2015)
14. Liu, W., Wen, Y., Yu, Z., Yang, M.: Large-margin softmax loss for convolutional neural networks. In: ICML, vol. 2, p. 7 (2016)
15. Wang, F., Cheng, J., Liu, W., Liu, H.: Additive margin softmax for face verification. IEEE Signal Process. Lett. **25**(7), 926–930 (2018)
16. Xiao, D., Li, J., Li, J., Dong, S., Lu, T.: Ihem loss: intra-class hard example mining loss for robust face recognition. IEEE Trans. Circuits Syst. Video Technol. (2022)
17. Howard, A.G., et al.: Mobilenets: efficient convolutional neural networks for mobile vision applications. arXiv preprint arXiv:1704.04861 (2017)
18. Sandler, M., Howard, A., Zhu, M., Zhmoginov, A., Chen, L.C.: Mobilenetv 2: inverted residuals and linear bottlenecks. In: Proceedings of the IEEE Conference on Computer Vision and Pattern Recognition, pp. 4510–4520 (2018)

19. Zhang, X., Zhou, X., Lin, M., Sun, J.: Shufflenet: an extremely efficient convolutional neural network for mobile devices. In: Proceedings of the IEEE Conference on Computer Vision and Pattern Recognition, pp. 6848–6856 (2018)
20. Tan, M., et al.: Mnasnet: platform-aware neural architecture search for mobile. In: Proceedings of the IEEE/CVF Conference on Computer Vision and Pattern Recognition, pp. 2820–2828 (2019)
21. Tan, M., Le, Q.V.: Efficientnet: rethinking model scaling for convolutional neural networks, pp. 6105–6114 (2019)
22. Tan, M., Chen, B., Pang, R., Vasudevan, V., Le, Q.V.: Mnasnet: platform-aware neural architecture search for mobile. arXiv preprint arXiv:1807.11626 (2018)
23. Tan, M., Pang, R., Le, Q.V.: Efficientdet: scalable and efficient object detection. In: Proceedings of the IEEE/CVF Conference on Computer Vision and Pattern Recognition, pp. 10781–10790 (2020)
24. Liu, Z., et al.: Swin transformer: hierarchical vision transformer using shifted windows. In: Proceedings of the IEEE/CVF International Conference on Computer Vision, pp. 10012–10022 (2021)
25. Liu, Z., Mao, H., Wu, C.Y., Feichtenhofer, C., Darrell, T., Xie, S.: A convnet for the 2020s. In: Proceedings of the IEEE/CVF Conference on Computer Vision and Pattern Recognition, pp. 11976–11986 (2022)
26. Hendrycks, D., Gimpel, K.: Gaussian error linear units (gelus). arXiv preprint arXiv:1606.08415 (2016)
27. Ding, X., Zhang, X., Han, J., Ding, G.: Scaling up your kernels to 31x31: revisiting large kernel design in CNNs. In: Proceedings of the IEEE/CVF Conference on Computer Vision and Pattern Recognition, pp. 11963–11975 (2022)
28. Nair, V., Hinton, G.E.: Rectified linear units improve restricted Boltzmann machines. In: Proceedings of the 27th International Conference on Machine Learning (ICML-10), pp. 807–814 (2010)
29. Sengupta, S., Chen, J.C., Castillo, C., Patel, V.M., Chellappa, R., Jacobs, D.W.: Frontal to profile face verification in the wild. In: 2016 IEEE Winter Conference on Applications of Computer Vision (WACV), pp. 1–9. IEEE (2016)
30. Moschoglou, S., Papaioannou, A., Sagonas, C., Deng, J., Kotsia, I., Zafeiriou, S.: Agedb: the first manually collected, in-the-wild age database. In: Proceedings of the IEEE Conference on Computer Vision and Pattern Recognition Workshops, pp. 51–59 (2017)
31. Yi, D., Lei, Z., Liao, S., Li, S.Z.: Learning face representation from scratch. arXiv preprint arXiv:1411.7923 (2014)
32. Deng, J., Guo, J., Zhang, D., Deng, Y., Lu, X., Shi, S.: Lightweight face recognition challenge. In: Proceedings of the IEEE/CVF International Conference on Computer Vision Workshops (2019)
33. Guo, Y., Zhang, L., Hu, Y., He, X., Gao, J.: Ms-celeb-1m: a dataset and benchmark for large-scale face recognition. In: European Conference on Computer Vision, pp. 87–102. Springer (2016)
34. Zhang, K., Zhang, Z., Li, Z., Qiao, Y.: Joint face detection and alignment using multitask cascaded convolutional networks. IEEE Signal Process. Lett. **23**(10), 1499–1503 (2016)
35. Ng, H.W., Winkler, S.: A data-driven approach to cleaning large face datasets. In: 2014 IEEE International Conference on Image Processing (ICIP), pp. 343–347. IEEE (2014)
36. Xie, W., Li, S., Zisserman, A.: Comparator networks (2018)

Optimal Low-Rank QR Decomposition with an Application on RP-TSOD

Haiyan Yu[1,3], Jianfeng Ren[1,2(✉)], Ruibin Bai[1,2], and Linlin Shen[3]

[1] School of Computer Science, University of Nottingham Ningbo China, Ningbo, China
[2] Nottingham Ningbo China Beacons of Excellence Research and Innovation Institute, University of Nottingham Ningbo China, Ningbo, China
jianfeng.ren@nottingham.edu.cn
[3] Computer Vision Institute, School of Computer Science and Software Enginering, Shenzhen University, Shenzhen, China

Abstract. Low-rank matrix approximation has many applications, *e.g.*, denoising, recommender systems and image reconstruction. Recently, a Randomized Pivoted Two-Sided Orthogonal Decomposition (RP-TSOD) was developed to exploit the randomization in approximating a high-dimensional matrix using QR decomposition. Instead of random projection, we propose to optimize the projection matrix for low-rank QR decomposition with the target of minimizing the approximation error. A method based on gradient descent is developed to derive optimal projections. The developed techniques can be used in not only RP-TSOD, but also other decompositions. Experimental results on both synthetic data and real data show that the proposed method could more accurately approximate a high-dimensional matrix than RP-TSOD.

Keywords: Low-rank matrix approximation · optimal projection · QR decomposition · RP-TSOD

1 Introduction

A tremendous amount of high-dimensional data has been generated with the increase of computational power [9, 26, 30, 33, 35], which is hard to be processed using conventional software tools [13]. Low-rank matrix approximation [11, 22] plays an important role in modern data analysis, with representative applications on face recognition [17, 25], denoising [32], recommender systems [5, 29] and medical image reconstruction [6].

Many matrix decomposition methods have been designed for low rank approximation [21], *e.g.*, eigen-decomposition in Principal Component Analysis (PCA) [2] for dimension reduction [10], Singular Value Decomposition (SVD)

This work was supported in part by the National Natural Science Foundation of China under Grant 91959108 and 72071116, and in part by the Ningbo Municipal Bureau Science and Technology under Grant 2019B10026 and 2022Z173.

[19] for decomposing non-square matrices, Cholesky decomposition [24] for positive definite matrices, and QR decomposition [31]. Among these methods, SVD is most popular, but it has a relatively high computational cost as it needs to determine the singular values of the matrix [15]. In contrast, Two-Sided Orthogonal Decomposition (TSOD) [16] only requires QR factorization to derive the orthonormal row bases and column bases at the same time. Furthermore, as discussed in [15], TSOD could well exploit the parallel architecture to accelerate the decomposition by using QR decomposition while SVD could not.

One of important applications of low rank approximation is to approximate streaming data such as time series and streaming videos [12,28], where the whole data is not available to derive the projection matrix. In these applications, a proper projection matrix needs to be learned from "historical" data to optimally decompose the newly arrived data matrix to facilitate fast processing or efficient compression of newly arrived data [12,18,27,28,34]. However, the aforementioned methods often assume that all the data is available to derive the optimal decomposition. Many randomized algorithms have been developed [8,14,15,34], which randomly project the matrix into a subspace before further decomposition, e.g., Randomized SVD [15], Improved HMT [23], Tropp's method [28], and Randomized Pivoted Two-Sided Orthogonal Decomposition (RP-TSOD) [15]. These methods are computationally efficient, accurate and easy to implement. But there is no guarantee that these random projection methods will lead to the optimal decomposition, as demonstrated later in the experiments.

In this paper, we propose an optimization framework based on QR decomposition for such a low-rank approximation. More specifically, instead of using random projections in RP-TSOD, we define a new objective function to optimize the projection matrix from "historical data" in such a way that the approximate orthonormal bases capture most information of the original matrices, so that the matrix reconstruction error is minimized. We develop a solution method based on gradient descent to derive the optimal projection matrix. The proposed method is named as Optimized Pivoted Two-Sided Orthogonal Decomposition (OP-TSOD). We conduct experiments on both synthetic data and real data. Experimental results show that the proposed OP-TSOD can derive orthonormal bases that retain more information of original matrices than RP-TSOD and other compared methods, in terms of both reconstruction error and visual inspection of reconstructed images.

Our contributions are three-fold: 1) The proposed OP-TSOD could derive the optimal projection matrix for the low-rank approximation problem. 2) A solution based on gradient descent is developed to solve the optimization problem. 3) The proposed OP-TSOD can better approximate both synthetic data and real data than other matrix decomposition methods.

2   Proposed OP-TSOD

2.1   Preliminary of Randomized QR Decomposition

The low-rank approximation can be formulated as the following minimization problem with the Frobenius Norm,

$$\min \left\| \boldsymbol{A} - \hat{\boldsymbol{A}} \right\|_F, \text{subject to:} f_r(\hat{\boldsymbol{A}}) \leq f_r(\boldsymbol{A}), \tag{1}$$

where $\hat{\boldsymbol{A}}$ is a low-rank approximation of matrix $\boldsymbol{A}$, $\|\cdot\|_F$ represents Frobenius Norm and $f_r(\boldsymbol{A})$ returns the rank of matrix $\boldsymbol{A}$. Typical rank-revealing methods include Column-Pivoted QR (CPQR) [14], TSOD [21] and SVD [19]. While SVD requires the calculation of determinant, TSOD only requires QR decomposition, which is much simple to implement. Before introducing RP-TSOD [15], we first briefly describe the QR decomposition. Formally, given a matrix $\boldsymbol{A} \in \mathbb{R}^{n_1 \times n_2}$, QR decomposition decomposes matrix $\boldsymbol{A}^\top$ as:[1]

$$\boldsymbol{A}^\top = \boldsymbol{Q}\boldsymbol{R}, \tag{2}$$

where $\boldsymbol{Q} \in \mathbb{R}^{n_2 \times n_1}$ is a unitary matrix and $\boldsymbol{R} \in \mathbb{R}^{n_1 \times n_1}$ is an upper triangular matrix. When using the Gram-Schmidt process for QR decomposition on $\boldsymbol{A}$ with dependent columns, a Column-Pivoted QR (CPQR) can decompose $\boldsymbol{A}$ as:

$$\boldsymbol{A}\boldsymbol{P} = \boldsymbol{Q}_C \boldsymbol{R}_C, \tag{3}$$

where $\boldsymbol{P} \in \mathbb{R}^{n_2 \times n_2}$ is a permutation matrix to preserve the most important column vectors [15], $\boldsymbol{Q}_C \in \mathbb{R}^{n_1 \times n_2}$ and $\boldsymbol{R}_C \in \mathbb{R}^{n_2 \times n_2}$ are the respective QR decomposition. One of the challenges in QR decomposition is its high computational complexity. Randomized QR decomposition has been developed to reduce the complexity,

$$\boldsymbol{A}^\top \boldsymbol{G} = \boldsymbol{Q}_T \boldsymbol{R}_T, \tag{4}$$

where $\boldsymbol{G} \in \mathbb{R}^{n_1 \times d}$ is a random projection matrix, and $d \ll \min\{n_1, n_2\}$. $\boldsymbol{Q}_T \in \mathbb{R}^{n_2 \times d}$ and $\boldsymbol{R}_T \in \mathbb{R}^{d \times d}$ are the unitary matrix and the upper triangular matrix for $\boldsymbol{A}^\top \boldsymbol{G}$, respectively. As the dimensionality of $\boldsymbol{A}^\top \boldsymbol{G}$ is much smaller, the computational load is greatly reduced.

2.2   Problem Analysis of RP-TSOD

Before introducing RP-TSOD [15], we first briefly review TSOD. The main procedures of TSOD are summarized in Algorithm 1. TSOD is a rank-revealing low-rank approximation method. It decomposes the matrix $\boldsymbol{A} \in \mathbb{R}^{n_1 \times n_2}$ into the orthonormal column basis $\boldsymbol{U} \in \mathbb{R}^{n_1 \times n_1}$, the lower triangular matrix $\boldsymbol{L} \in$

---

[1] We decompose $\boldsymbol{A}^\top$ instead of $\boldsymbol{A}$ for better understanding RP-TSOD.

Algorithm 1. TSOD decomposition.

Input: $A \in \mathbb{R}^{n_1 \times n_2}$;
Output: $U \in \mathbb{R}^{n_1 \times n_2}$, $L \in \mathbb{R}^{n_2 \times n_2}$, and $V \in \mathbb{R}^{n_2 \times n_2}$.
1: Decompose A using CPQR [15] as: $AP_1 = Q_1R_1$, $Q_1 \in \mathbb{R}^{n_1 \times n_2}$, $R_1 \in \mathbb{R}^{n_2 \times n_2}$, $P_1 \in \mathbb{R}^{n_2 \times n_2}$ is a permutation matrix;
2: Decompose $R_1^\top$ using CPQR [15] as: $R_1^\top P_2 = Q_2 L_2^\top$, where $Q_2 \in \mathbb{R}^{n_2 \times n_2}$, $L_2 \in \mathbb{R}^{n_2 \times n_2}$, $P_2 \in \mathbb{R}^{n_2 \times n_2}$ is a permutation matrix;
3: TSOD decomposition on A as: $A = Q_1 P_2 L_2 Q_2^\top P_1^\top = ULV^\top$, where $U = Q_1 P_2$, $L = L_2$, $V = P_1 Q_2$.

$\mathbb{R}^{n_1 \times n_2}$ approximating the singular values of A, and the orthonormal row basis $V \in \mathbb{R}^{n_2 \times n_2}$, where $A = ULV^\top$. Note that L is a lower triangular matrix that reveals the rank of A, which is different from the diagonal matrix in SVD that stores singular values.

TSOD is time-consuming when A is large. RP-TSOD [15] hence projects $A^\top$ to a subspace as $A^\top G$ by a low-dimensional random matrix $G \in \mathbb{R}^{n_1 \times d}$. QR decomposition is then applied on $A^\top G$ to obtain the approximate orthonormal bases Q_T of the row space of A. A is projected into the subspace spanned by Q_T as $Y = AQ_T$. TSOD decomposition is then applied on Y as $Y = U_Y L_Y V_Y^\top$, where $U_Y \in \mathbb{R}^{n_2 \times d}$, $L_Y \in \mathbb{R}^{d \times d}$, and $V_Y \in \mathbb{R}^{d \times d}$ are the orthonormal row bases, the diagonal matrix and the orthonormal row bases for Y, respectively. Finally, A is approximated as $A \approx YQ_T^\top$. RP-TSOD algorithm is summarized in Algo. 2.

Algorithm 2. RP-TSOD decomposition.

Input: $A \in \mathbb{R}^{n_1 \times n_2}$, Random matrix $G \in \mathbb{R}^{n_1 \times d}$;
Output: $U_T \in \mathbb{R}^{n_1 \times d}$, $L_T \in \mathbb{R}^{d \times d}$, and $V_T \in \mathbb{R}^{d \times d}$.
1: QR decomposition on $A^\top G$ as: $A^\top G = Q_T R_T$;
2: Project A into its subspace as: $Y = AQ_T$;
3: TSOD decomposition on Y as: $Y = U_Y L_Y V_Y^\top$;
4: $A \approx YQ_T^\top = U_T L_T V_T^\top$, where $U_T = U_Y$, $L_T = L_Y$, $V_T = Q_T V_Y$.

It is difficult to find the random projection matrix G tightly approximating A by $YQ_T^\top$. As $Y = AQ_T$, we can have $AQ_T Q_T^\top = YQ_T^\top$. Only if $Q_T Q_T^\top \approx I$, where I is an identity matrix, we have $A \approx YQ_T^\top$. The random projection matrix G can't guarantee $Q_T Q_T^\top \approx I$. Therefore, our goal is to learn a projection matrix G so that the approximate subspace spanned by Q_T captures most information of A,

$$\hat{A} = AQ_T Q_T^\top, \text{subject to: } Q_T^\top Q_T = I, \tag{5}$$

which minimizes the approximation loss defined in Eq. (1).

2.3   Objective Function for OP-TSOD

Consider a set of historical matrices $\{A_1, A_2, \ldots, A_N \in \mathbb{R}^{n_1 \times n_2}\}$ sampled from a certain distribution $\mathcal{D}$, our goal is to learn a projection matrix $G \in \mathbb{R}^{n_1 \times d}$ that preserves the column spaces of $A_i^\top$ once projecting them in the form of $A_i^\top G$. To show that the optimization variable is the matrix G, we denote Q_T by $Q_T = g_Q(A^\top G)$, where $g_Q(\cdot)$ denotes the QR decomposition that returns Q_T. The approximation $\hat{A}$ defined in Eq. (5) can be re-written as,

$$\hat{A}^\top = g_Q(A^\top G) g_Q^\top (A^\top G) A^\top. \tag{6}$$

The objective function defined in Eq. (1) can be revised as,

$$\min_{G \in \mathbb{G}} \frac{1}{N} \sum_{i=1}^{N} \underbrace{\frac{1}{2} \left\| (g_Q(A_i^\top G) g_Q^\top (A_i^\top G) A_i^\top - A_i^\top \right\| F^2}_{f(G;A_i)}, \tag{7}$$

where $\mathbb{G} \subset \mathbb{R}^{n_1 \times d}$ is the set of targeted projection matrices. Note that the term $\left(I - g_Q(A^\top G) g_Q^\top(A^\top G)\right) A_i^\top$ represents the projection of $A_i^\top$ onto the orthogonal complement of the column space of Q_T.

2.4   Optimizing Projection Matrix

Gradient-based methods [20] have been widely used to solve the optimization problem defined in Eq. (7). However, it is not straightforward to derive the partial derivative $\frac{\partial g_Q(A^\top G)}{\partial G}$. Towards that end, we establish the gradient of the loss function in (7) in the following theorem.

Theorem 1. *Given $A \in \mathbb{R}^{n_1 \times n_2}$, $G \in \mathbb{R}^{n_1 \times d}$, let $M = A^\top G = Q_T R_T$ be the QR decomposition of $A^\top G$, where $Q_T \in \mathbb{R}^{n_2 \times d}$ is an unitary matrix and is calculated by implementing Gram-Schmidt process on matrix $M = A^\top G$, and $R_T \in \mathbb{R}^{d \times d}$ is an upper triangle matrix, $u_i \in \mathbb{R}^{n_2}$ and $w_i \in \mathbb{R}^{n_2}$ denotes the i-th column vector of M and Q_T respectively. Then the gradient of $f(G;A)$ in (7) is given by*

$$\nabla f(G; A) = ADW, \tag{8}$$

where

$$D = 2(HA + A^\top H^\top) Q_T, D \in \mathbb{R}^{n_2 \times d}; \tag{9}$$

$$H = Q_T Q_T^\top A^\top - A^\top, H \in \mathbb{R}^{n_2 \times n_1}; \tag{10}$$

$$W = \{\frac{\partial w_i}{\partial u_j}\}_{1 \leq i \leq d; 1 \leq j \leq d}, W \in \mathbb{R}^{d \times d}; \tag{11}$$

$$\frac{\partial w_i}{\partial u_j} = \begin{cases} 0, & i < j, \\ \frac{1}{\|u_i - \sum_{k=1}^{i-1}(u_i, w_k)w_k\|}, & i = j, \\ -\frac{\sum_{k=j+1}^{i}(u_i, w_{k-1})\frac{\partial w_{k-1}}{\partial u_j}}{\|u_i - \sum_{k=1}^{i}(u_i, w_k)w_k\|}, & i > j. \end{cases} \tag{12}$$

We then outline the key steps for a sketch proof of Theorem 1. The gradient $\nabla f(G; A) = \frac{\partial f(G;A)}{\partial G}$ of the objective function $f(G; A)$ w.r.t. G can be calculated by using the chain rule,

$$\frac{\partial f}{\partial G} = \frac{\partial f}{\partial Q_T} \frac{\partial Q_T}{\partial M} \frac{\partial M}{\partial G}, \tag{13}$$

where $M = A^\top G$. The next step is to calculate $\frac{\partial f}{\partial Q_T}$, $\frac{\partial Q_T}{\partial M}$ and $\frac{\partial M}{\partial G}$, respectively. Note that $\frac{\partial f}{\partial Q_T}$ will produce a matrix, while $\frac{\partial Q_T}{\partial M}$ and $\frac{\partial M}{\partial G}$ will produce a tensor. Following the trace trick in [3,7], we can derive $\frac{\partial f}{\partial Q_T} = D \in \mathbb{R}^{n_2 \times d}$, where D is defined in Eqn. (9). $\frac{\partial Q_T}{\partial M}$ and $\frac{\partial M}{\partial G}$ are matrix-matrix derivatives, which can be calculated by using the vectorization approach [3,7]. More specifically, $\frac{\partial Q_T}{\partial M} = I_{n_2} \bigotimes W \in \mathbb{R}^{n_2 d \times n_2 d}$, where $I_{n_2} \in \mathbb{R}^{n_2 \times n_2}$ is an identity matrix, $\bigotimes$ is the Kronecker product and W is defined in Eqn. (11). $\frac{\partial M}{\partial G} = I_d \bigotimes A \in \mathbb{R}^{n_1 d \times n_2 d}$ as given in [7], where $I_d \in \mathbb{R}^{d \times d}$ is an identity matrix. Finally, $\frac{\partial f(G;A)}{\partial vec(G)} = (1 \bigotimes D)(I_{n_2} \bigotimes W)(I_d \bigotimes A) \in \mathbb{R}^{1 \times n_1 d}$ following [1,7], where $vec(G)$ is the vector form of the G. Unstacking the vector form $vec(G)$ back to matrix form will result in $\frac{\partial f(G;A)}{\partial G} = ADW$. The detailed proof is beyond the scope of this paper. Readers can refer to here for more details.

Then, gradient descent [4] is applied to solve the problem. The procedures to derive the optimal projection matrix G^* of proposed OP-TSOD are summarized in Algorithm 3. After obtaining the optimal projection matrix G^*, we can replace G by the learned G^* in OP-TSOD in Algorithm 2 for low-rank matrix approximation. Specifically, we first project the original matrix A by using G^* as $A^\top G^*$, and then implement QR decomposition on $A^\top G^*$ to obtain the orthogonal projection matrix Q^*. Finally, we apply TSOD decomposition on AQ^* for low-rank matrix factorization.

Algorithm 3. Learning projection matrix for OP-TSOD.

Input: N given matrix $\{A_i \in R^{n_1 \times n_2}\}_{i=1}^N$, learning rate η.
Output: Learned projection matrix $G^* \in R^{n_2 \times d}$.
  1: Initialize G randomly.
  2: **for** m iterations **do**
  3:     **for** i $= \{1, 2, \ldots N\}$ **do**
  4:         Calculate $f(G, A_i)$ defined in Eqn. (7);
  5:     **end for**
  6:     Update G using gradient descent method as: $G_{j+1} = G_j - \frac{\eta}{N} \sum_{i=1}^N \nabla f(G, A_i)$, where $\nabla f(G, A_i)$ is defined in Theorem 1.
  7: **end for**

---

3   Experimental Results

3.1   Experimental Settings

In this section, we compare the proposed OP-TSOD with RP-TSOD [15] and other methods on both the synthetic dataset and the real dataset. RP-TSOD [15] is the baseline method, which utilizes a random sketch matrix for dimension reduction. Tropp's method has been utilized for matrix approximation in video streaming, which improves the bilateral randomized algorithm by a two-side linear update method [28]. Indyk's approach is a learning-based method that derives the projection matrix by utilizing power iteration method during training, which reduces the reconstruction error at an extra computational cost [12]. All algorithms are implemented in MATLAB R2022a.

3.2   Experimental Results on Synthetic Data

We generate two synthetic datasets consisting of N training matrices and M testing matrices of size $n_1 \times n_2$ using standard normal distribution, to simulate the scenario where training data and test data come from the same distribution. **Setup 1**: We set $N = 500$, $M = 200$, $n_1 = n_2 = 100$, $d = 5, 7, 9, 11$, and $\eta = 0.1$. **Setup 2**: We set $N = 500$, $M = 200$, $n_1 = 720, n_2 = 1280$, $d = 20, 25, 30, 35$, and $\eta = 10$.

For each setup, we randomly initialize $G_0 \in \mathbb{R}^{n_1 \times d}$ from standard normal distribution. The maximum number of iterations is set to 2000. The following relative Mean Squared Error (MSE) is used to evaluate the reconstruction error.

$$\epsilon = \frac{1}{2N} \sum_{i=1}^{N} \frac{\left\| \left(g_Q(A_i^\top G) g_Q^\top(A_i^\top G) - I \right) A_i^\top \right\| F^2}{\left\| A_i^\top \right\| F^2}. \tag{14}$$

We first plot the convergence of the proposed OP-TSOD in Fig. 1(a) and Fig. 1(b) for two setups, respectively. The proposed OP-TSOD converges fast in both setups. We then compare the proposed OP-TSOD with RP-TSOD in terms of MSE defined in Eq. (14). As shown in Fig. 1(c) and Fig. 1(d), the proposed OP-TSOD achieves much better performance than RP-TSOD for all the choices of d.

3.3   Experimental Results on Real Data

Following the setup in [34], we conduct experiments on 720×1280 frames extracted from five different videos, "Logo"[2], "Friends" [12][3], "Eagle"[4], "Sheldon"[5], "Big Bang Theory"[6], downloaded from Youtube. For each video, we use

---

[2] http://youtu.be/L5HQoFIaT4I.
[3] http://youtu.be/xmLZsEfXEgE.
[4] http://youtu.be/ufnf_q_3Ofg.
[5] https://www.youtube.com/watch?v=8t_qFivFGaM.
[6] https://www.youtube.com/watch?v=61FasQ6KQCI.

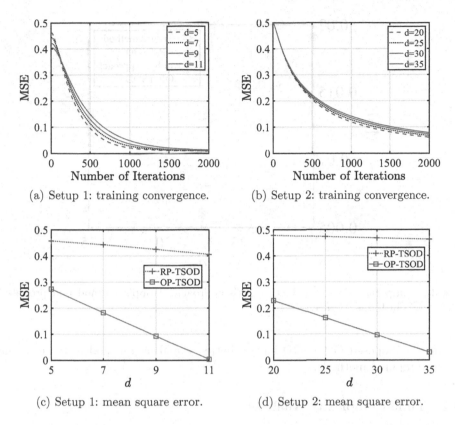

(a) Setup 1: training convergence.   (b) Setup 2: training convergence.

(c) Setup 1: mean square error.   (d) Setup 2: mean square error.

Fig. 1. Comparisons of training convergence and MSE between RP-TSOD and the proposed OP-TSOD.

the first 400 frames for training, and the next 100 frames for testing. In total, there are 2000 training frames and 500 test frames. Besides RP-TSOD [15], we compare the proposed OP-TSOD with another two matrix approximation methods, Indyk's approach [12] and Tropp's method [28].

We first compare the proposed OP-TSOD with these three methods in terms of MSE defined in Eq. (14). Figure 2 shows that the proposed OP-TSOD significantly outperforms RP-TSOD and the other two compared methods. Indyk's approach [12] is a previous learning-based method, and it performs the second best among all the compared methods. The proposed OP-TSOD significantly and consistently outperforms Indyk's approach [12] for various d. The experimental results demonstrate that the optimal matrix G^* learned by the proposed framework could better capture the information than other compared methods.

Then, we compare the visualization of original images, the approximated images using Tropp's method [28], RP-TSOD [15], Indyk's approach [12] and the proposed OP-TSOD in Fig. 3. We can observe that the reconstructed images

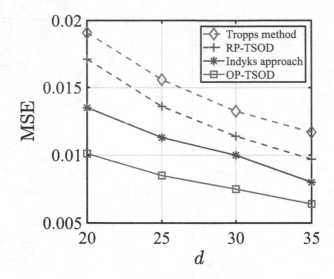

Fig. 2. Comparison of RP-TSOD, Indyk's approach, Tropp's method and proposed OP-TSOD in terms of MSE.

using the proposed OP-TSOD have a better visualization quality than all the three compared methods.

3.4 Time Complexity Analysis

The time complexity and run time per matrix of all the compared methods are summarized in Table 1, where m is the number of iterations. OP-TSOD requires additional one-time training compared with RP-TSOD, but the test time is the same as RP-TSOD. Compared with Tropp's method, the proposed OP-TSOD requires more time for training and testing, but as shown in Fig. 2, the proposed OP-TSOD has a much lower MSE than Tropp's method.

Table 1. Comparisons of time complexity and run time.

Method	Time Complexity		Running Time	
	Train	Test	Train	Test
RP-TSOD [15]	-	$\mathcal{O}(Mn_1n_2d)$	-	0.061 s
Indyk's approach [12]	$\mathcal{O}(Nmn_1^2n_2)$	$\mathcal{O}(Mn_1n_2d)$	0.465 s	0.061 s
Tropp's method [28]	$\mathcal{O}(Nmn_1n_2d)$	$\mathcal{O}(Mn_1n_2d)$	0.036 s	0.015 s
OP-TSOD	$\mathcal{O}(Nn_1n_2^2 + Nmn_2n_1d)$	$\mathcal{O}(Mn_1n_2d)$	0.318 s	0.061 s

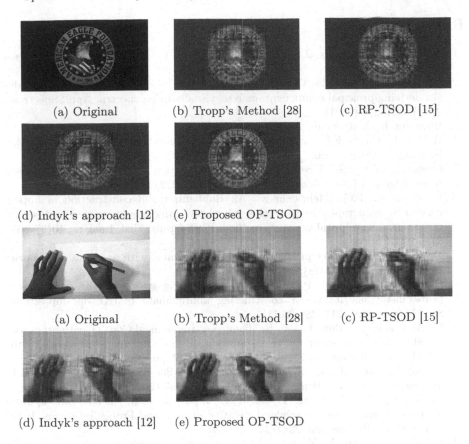

(a) Original   (b) Tropp's Method [28]   (c) RP-TSOD [15]

(d) Indyk's approach [12]   (e) Proposed OP-TSOD

(a) Original   (b) Tropp's Method [28]   (c) RP-TSOD [15]

(d) Indyk's approach [12]   (e) Proposed OP-TSOD

Fig. 3. Visualization of reconstructed images using first 10 dimensions with $d = 20$.

4   Conclusion

In this paper, we propose a learning framework to derive an optimal projection matrix from historical data so that the derived subspace captures most of the information to facilitate fast processing or efficient compression of newly arrived data. Randomized methods such as RP-TSOD can't guarantee the minimal reconstruction error. Instead of random projection, we propose to optimize the projection matrix by directly minimizing the reconstruction error. We derive the gradient of the objective function and obtain the optimal projection matrix using the gradient descent method. Experimental results on both synthetic and real data demonstrate that the orthonormal basis obtained by the proposed OP-TSOD can effectively achieve low-rank approximation.

References

1. Abadir, K.M., Magnus, J.R.: Matrix algebra, vol. 1. Cambridge University Press (2005)
2. Beattie, J.R., Esmonde-White, F.W.: Exploration of principal component analysis: deriving principal component analysis visually using spectra. Appl. Spectrosc. **75**(4), 361–375 (2021)
3. Bodewig, E.: Matrix calculus. Elsevier (2014)
4. Bottou, L., Curtis, F.E., Nocedal, J.: Optimization methods for large-scale machine learning. SIAM Rev. **60**(2), 223–311 (2018)
5. Chen, Z., Wang, S.: A review on matrix completion for recommender systems. Knowledge and Information Systems, pp. 1–34 (2022)
6. Daneshmand, P.G., Mehridehnavi, A., Rabbani, H.: Reconstruction of optical coherence tomography images using mixed low rank approximation and second order tensor based total variation method. IEEE Trans. Med. Imaging **40**(3), 865–878 (2020)
7. Graham, A.: Kronecker products and matrix calculus with applications. Courier Dover Publications (2018)
8. Halko, N., Martinsson, P.G., Tropp, J.A.: Finding structure with randomness: probabilistic algorithms for constructing approximate matrix decompositions. SIAM Rev. **53**(2), 217–288 (2011)
9. He, W., Zhang, J., Ren, J., Bai, R., Jiang, X.: Hierarchical ConViT with attention-based relational reasoner for visual analogical reasoning. In: Proceedings of the AAAI Conference on Artificial Intelligence, vol. 37, pp. 22–30 (2023)
10. Huang, D., Jiang, F., Li, K., Tong, G., Zhou, G.: Scaled pca: a new approach to dimension reduction. Manage. Sci. **68**(3), 1678–1695 (2022)
11. Huang, Y., Liao, G., Xiang, Y., Zhang, L., Li, J., Nehorai, A.: Low-rank approximation via generalized reweighted iterative nuclear and Frobenius norms. IEEE Trans. Image Process. **29**, 2244–2257 (2020)
12. Indyk, P., Vakilian, A., Yuan, Y.: Learning-based low-rank approximations. arXiv preprint arXiv:1910.13984 (2019)
13. Jiang, P., Sinha, S., Aldape, K., Hannenhalli, S., Sahinalp, C., Ruppin, E.: Big data in basic and translational cancer research. Nature Reviews Cancer, pp. 1–15 (2022)
14. Kaloorazi, M.F., Chen, J.: Projection-based QLP algorithm for efficiently computing low-rank approximation of matrices. IEEE Trans. Signal Process. **69**, 2218–2232 (2021)
15. Kaloorazi, M.F., Chen, J.: Low-rank approximation of matrices via a rank-revealing factorization with randomization. In: IEEE International Conference on Acoustics, Speech and Signal Processing, pp. 5815–5819. IEEE (2020)
16. Kishore Kumar, N., Schneider, J.: Literature survey on low rank approximation of matrices. Linear Multilinear Algebra **65**(11), 2212–2244 (2017)
17. Lai, J., Jiang, X.: Classwise sparse and collaborative patch representation for face recognition. IEEE Trans. Image Process. **25**(7), 3261–3272 (2016)
18. LE Thanh, T., Abed-Meriam, K., Nguyen, T.L., Hafiane, A.: Tracking online low-rank approximations of higher-order incomplete streaming tensors. Karim and Nguyen, Trung Linh and Hafiane, Adel, Tracking Online Low-Rank Approximations of Higher-Order Incomplete Streaming Tensors (2022)
19. Li, P., Wang, H., Li, X., Zhang, C.: An image denoising algorithm based on adaptive clustering and singular value decomposition. IET Image Proc. **15**(3), 598–614 (2021)

20. Long, B., Zhu, Z., Yang, W., Chong, K.T., Rodríguez, J., Guerrero, J.M.: Gradient descent optimization based parameter identification for FCS-MPC control of LCL-type grid connected converter. IEEE Trans. Industr. Electron. **69**(3), 2631–2643 (2021)
21. Lu, J.: Numerical matrix decomposition and its modern applications: a rigorous first course. arXiv preprint arXiv:2107.02579 (2022)
22. Markovsky, I.: Low Rank Approximation: Algorithms, Implementation, Applications, vol. 906. Springer, London (2012). https://doi.org/10.1007/978-1-4471-2227-2
23. Nakatsukasa, Y.: Fast and stable randomized low-rank matrix approximation. arXiv preprint arXiv:2009.11392 (2020)
24. Nottoli, T., Gauss, J., Lipparini, F.: Second-order CASSCF algorithm with the Cholesky decomposition of the two-electron integrals. J. Chem. Theory Comput. **17**(11), 6819–6831 (2021)
25. Shakeel, M.S., Lam, K.M.: Deep low-rank feature learning and encoding for cross-age face recognition. J. Vis. Commun. Image Represent. **82**, 103423 (2022)
26. Song, X., Jin, J., Yao, C., Wang, S., Ren, J., Bai, R.: Siamese-discriminant deep reinforcement learning for solving jigsaw puzzles with large eroded gaps. In: Proceedings of the AAAI Conference on Artificial Intelligence, vol. 37, pp. 2303–2311 (2023)
27. Tropp, J.A., Yurtsever, A., Udell, M., Cevher, V.: Fixed-rank approximation of a positive-semidefinite matrix from streaming data. In: Advances in Neural Information Processing Systems 30 (2017)
28. Tropp, J.A., Yurtsever, A., Udell, M., Cevher, V.: Streaming low-rank matrix approximation with an application to scientific simulation. SIAM J. Sci. Comput. **41**(4), A2430–A2463 (2019)
29. Wang, J., Zhu, L., Dai, T., Xu, Q., Gao, T.: Low-rank and sparse matrix factorization with prior relations for recommender systems. Appl. Intell. **51**(6), 3435–3449 (2021)
30. Wang, S., Ren, J., Bai, R.: A semi-supervised adaptive discriminative discretization method improving discrimination power of regularized naive Bayes. Expert Syst. Appl. **225**, 120094 (2023)
31. Wu, F., Li, Y., Li, C., Wu, Y.: A fast tensor completion method based on tensor QR decomposition and tensor nuclear norm minimization. IEEE Trans. Comput. Imaging **7**, 1267–1277 (2021)
32. Xie, T., Li, S., Sun, B.: Hyperspectral images denoising via nonconvex regularized low-rank and sparse matrix decomposition. IEEE Trans. Image Process. **29**, 44–56 (2020)
33. Yao, C., Ren, J., Bai, R., Du, H., Liu, J., Jiang, X.: Mask attack detection using vascular-weighted motion-robust rPPG signals. IEEE Trans. Inf. Forensics Secur. **18**, 4313–4328 (2023)
34. Yu, H., Qin, Z., Zhu, Z.: Learning approach for fast approximate matrix factorizations. In: IEEE International Conference on Acoustics, Speech and Signal Processing, pp. 5408–5412. IEEE (2022)
35. Zhang, J., Ren, J., Zhang, Q., Liu, J., Jiang, X.: Spatial context-aware object-attentional network for multi-label image classification. IEEE Trans. Image Process. **32**, 3000–3012 (2023)

EDDVPL: A Web Attribute Extraction Method with Prompt Learning

Yuling Yang[1,2], Jiali Feng[1,2], Baoke Li[1,2], Fangfang Yuan[1(✉)], Cong Cao[1],
and Yanbing Liu[1,2]

[1] Institute of Information Engineering, Chinese Academy of Sciences, Beijing, China
{yangyuling,fengjiali,libaoke,yuanfangfang,caocong,
liuyanbing}@iie.ac.cn
[2] School of Cyber Security, University of Chinese Academy of Sciences, Beijing, China

Abstract. Since labeling web pages requires a lot of human resources and time, web attribute extraction methods based on few-shot learning have gained the attention of researchers. However, these methods still rely heavily on sufficient labeled data of several seed websites. In order to effectively alleviate the lack of domain information, we design a web attribute extraction model based on dual-view prompt learning named EDDVPL, achieving page-level few-shot learning which uses only a small number of labeled web pages for training. Specifically, we first retrieve semantic prompt information of DOM tree view by a simplified algorithm to stimulate domain-related knowledge of the pre-trained language model. Then, we introduce task prompt information of template view by constructing a template indicating the extraction target, which can help the pre-trained language model quickly understand the task of web attribute extraction. Finally, we integrate the dual-view prompt information by template filling to jointly guide the training of the pre-trained language model at semantic and task levels. Extensive experimental results on the public SWDE dataset show that EDDVPL performs the best results compared to the baselines.

Keywords: Web Information Extraction · Attribute Extraction · Prompt Learning

1 Introduction

Web attribute extraction is to extract target attributes from web pages and form them into structured data. Compared to common web pages, the content of semi-structured web pages (such as web pages of encyclopedia websites and official websites) is often carefully edited or reviewed, and thus contains more comprehensive, real-time, and high-quality information [1]. Therefore, extracting attributes from large-scale semi-structured web pages (as shown in Fig. 1) is an important problem worthy to be studied.

Most of the existing methods formulate web attribute extraction as a multi-class classification task of DOM tree nodes, training models based on sufficient labeled data. However, the process of labeling web pages often requires a lot of human resources and

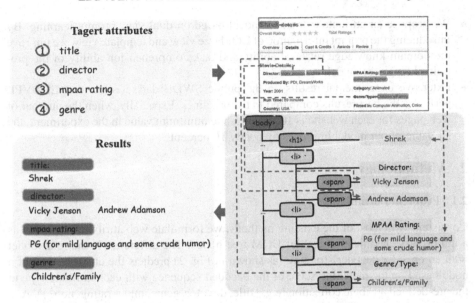

Fig. 1. An example of extracting target attributes from a semi-structured web page

takes a long time for new domains. In order to reduce annotation work, Lin et al. [2] and Zhou et al. [3] propose to build learning-based models that can extract attributes from unseen websites by training on a small number of seed websites. Although only a few seed websites are needed, the number of web pages contained in the seed websites is still large. Therefore, we propose to reduce the dependence on the amount of training data at the web page level. Generally, the pre-trained language model that contains rich prior knowledge can alleviate the problem of lacking training data for downstream tasks, and prompt learning can help the pre-trained language model retrieve existing knowledge to further understand downstream tasks. Thus, we mine prompt information to guide the pre-trained language model to extract attributes from semi-structured web pages.

In this paper, we propose a web attribute extraction model based on dual-view prompt learning, called EDDVPL. Firstly, in order to enhance the domain knowledge retrieval ability of the pre-trained language model, we retrieve semantic prompt information for candidate nodes from the DOM tree according to the simplification algorithm. Then, to stimulate the task comprehension ability of the pre-trained language model, we construct a template indicating the extraction target to obtain task prompt information. Finally, we integrate the prompt information of DOM tree view and template view through template filling, thus jointly guiding the training of pre-trained language model at the semantic and task levels. In addition, to deal with the variation of attribute label length, we use an encoder-decoder based pre-trained language model and utilize the generated results to predict the attribute types of DOM tree nodes.

In summary, the main contributions of our paper are as follows:

- To our knowledge, we are the first to introduce the prompt learning paradigm to benefit the task of web attribute extraction.

- We propose an attribute extraction model based on dual-view prompt learning. By introducing the prompt information of DOM tree view and template view, we improve the domain knowledge retrieval ability and task comprehension ability of the pre-trained language model.
- Extensive experimental results on the public SWDE dataset show that EDDVPL performs the best results compared to the baselines. Especially, when the number of web pages for each website is 10, which is the minimum value in the experiment, the F1 value of our model improves by nearly 10 percent.

2   Methodology

2.1   Problem Formulation

Consistent with most of the existing methods, we formulate web attribute extraction as a multi-class classification task of DOM tree nodes. The difference is that our model with an encoder-decoder structure (as shown in Fig. 2) predicts the attribute type of a node based on the matching score of the decoded sequence with each attribute type in the pre-defined attribute collection (e.g. {title, director, genre, mpaa-rating, none}). And we follow the convention that one node can correspond to at most one attribute type [4].

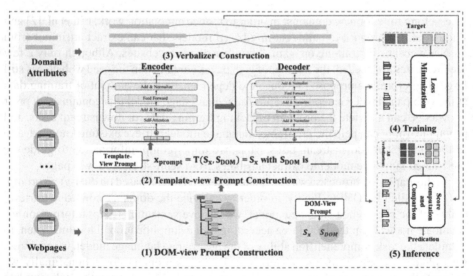

Fig. 2. The overall architecture of EDDVPL. (1) DOM-view Prompt Construction, constructing semantic-level DOM-view prompt learning based on DOM tree structure. (2) Template-view Prompt Construction, integrating the prompt information of DOM-view and template-view through template filling. (3) Verbalizer Construction, the pre-trained language model predicts masked text in template whitespace. (4) Model Training and (5) Inference, using the attribute type with the highest score as the prediction result

Note that we only need to classify nodes with textual content. Textual nodes in the DOM tree can be divided into fixed nodes N_f and variable nodes N_v. Fixed nodes remain consistent across different web pages in the same website, while variable nodes often change in contents. Since attribute values usually vary across different web pages, we further narrow the range of nodes to be classified into variable nodes.

2.2 DOM-View Prompt Construction

A DOM tree is a hierarchical structure composed of a series of nodes. Nodes in the DOM tree have contextual relationships, and different contexts contribute differently to helping the model understand the content of the current candidate node. Among them, the context before the candidate node is usually descriptive text, which clearly and intuitively indicates what information the candidate node contains at the semantic-level and has important reference significance. Therefore, in this module, our goal is to retrieve the descriptive context for each variable node x as its semantic prompt information of DOM tree view.

Specifically, we first define the set of contexts for each variable node x as $X_n \subseteq N_f \cup N_v$, where the distances from x and $x_n \in X_n$ to their lowest common ancestor don't exceed constant D. And the prompt information of DOM tree view exists in a fixed node set $X_{DOM} \subseteq X_n$ that contains at most one node. This fixed node set satisfies that $x_{DOM} \in X_{DOM}$ is the only fixed node that originates from the lowest common ancestor of x_{DOM} and x in the DOM tree.

2.3 Template-View Prompt Construction

For prompt learning, a template $T(\cdot)$ is used to guide the model on what to do next by adding task descriptions to the original input text. With such task-level prompts, the model can quickly retrieve and fully utilize task-relevant knowledge learned during pre-training.

For each variable node $x \in N_v$ in a web page, we denote its textual content as S_x and its semantic prompt of DOM tree view as S_{DOM}. In this module, we first design a template that indicates the target of extraction task, and then rewrite both S_x and S_{DOM} based on the template to achieve the integration of the semantic and task prompts. In this way, the dual-view prompt input x_{prompt} can be represented as:

$$x_{prompt} = T(S_x, S_{DOM}) \tag{1}$$

Specifically, the template designed by this model is:

$$T(S_x, S_{DOM}) = "S_x \text{ with } S_{DOM} \text{ is} \underline{\qquad}" \tag{2}$$

The goal of the model is to fill in the blank at decoder when the input to encoder is x_{prompt}.

2.4 Verbalizer Construction

After obtaining the dual-view prompts, the pre-trained language model needs to predict the text z that is masked in the template. This requires a one-to-one verbalizer:

$$\phi : Y \rightarrow Z, z = \phi(y) \tag{3}$$

The above equation indicates that the function ϕ can map each attribute $y \in Y$ to $z \in Z$ consisting of words in the pre-trained language model vocabulary. Specifically, the label mapping relationships in various domains are shown in Table 1.

Table 1. The mapping relationships of labels in all domains

Domains	Mapping relationships of labels
Auto	model → model, price → price, engine → engine, fuel_economy → fuel economy
Book	title → title, author → author, isbn_13 → ISBN 13, publisher → publisher, publication_data → publication data
Camera	model → model, price → price, manufacturer → manufacturer
Job	title → title, company → company, loction → location, date_posted → post date
Movie	title → title, director → director, genre → genre, mpaa_rating → the motion picture association of America rating
Nbaplayer	name → current player name, team → current NBA team, height → height, weight → weight
Restaurant	name → name, address → address, phone → phone, cuisine → cuisine
University	name → name, phone → phone, website → website, type → type

For classification tasks, prompt learning based on the auto-encoding pre-trained language model needs to mask the mapped label in the template, and then get the final result by predicting each token at each mask position. However, the length of the mapped label is unknown during inference, which makes it impossible for the model to determine the number of masks.

To solve this problem, existing methods [5, 6] attempt to fix the length of each mapped label. However, this method may cause the loss of semantic information [7]. Therefore, we adopt a pre-trained language model with an encoder-decoder structure. After a mask is given as the label placeholder in the template, the model can learn both the content and length information of the mask position during training, so as to generate prediction results of arbitrary lengths at the decoder.

2.5 Model Training and Inference

When the pre-trained language model is denoted as $\mathcal{M}$, given the dual-view prompt input $\mathcal{T}(\mathcal{S}_x, \mathcal{S}_{DOM})$ and the target result $\phi(y)$, we denote the output sequence at decoder as o. The matching probability between the output sequence and the target is calculated as follows:

$$\prod_{t=1}^{|\phi(y)|} P(o_t = \phi_t(y)|o_{<t}, \mathcal{T}(\mathcal{S}_x, \mathcal{S}_{DOM})) \tag{4}$$

where o_t and $\phi_t(y)$ denote the t-th word of the output sequence o and the target result $\phi(y)$, $o_{<t}$ denotes the output sequence to the left of the t-th word. The objective of

the model is to minimize the negative log-likelihood of all variable nodes, and the loss function is formulated as follows:

$$L = -\frac{1}{|N_v|} \sum_{x \in N_v} \sum_{t=1}^{|\phi(y)|} \log P(o_t = \phi_t(y)|o_{<t}, T(S_x, S_{DOM})) \tag{5}$$

When predicting the attribute type of a node, the model first obtains the normalized probability output at each position of the decoder. Next, it computes the matching score of the output sequence with each attribute type $y \in Y$ as follows:

$$score_y = \frac{1}{|\phi(y)|} \sum_{t=1}^{|\phi(y)|} P(o_t = \phi_t(y)) \tag{6}$$

here $P(o_t = \phi_t(y))$ is the probability value of the word $\phi_t(y)$ in the normalized probability output of decoder at time step t. When calculating scores, in order to avoid favoring the prediction results to attribute types with longer text [8], we normalized the scores using the length of the mapped label, and finally selected the attribute type with the highest score as the prediction result.

3 Experiments

3.1 Dataset

We evaluate EDDVPL on the public SWDE [4] dataset. This dataset contains 8 domains, each of which consists of 10 websites and has 3 to 5 attributes. For each domain, k (k can be set as 10, 50, and 100) pages from each website are used to construct the training set, and the remaining pages are used to construct the test set. The statistics of the training and test sets for each domain when k is set as different values are shown in Table 2.

Table 2. The statistics of training and test sets in all domains

Domains	$k = 10$		$k = 50$		$k = 100$	
	train	test	train	test	train	Test
Auto	100	17,823	500	17,423	1,000	16,923
Book	100	19,900	500	19,500	1,000	19,000
Carmera	100	5,158	500	4,758	1,000	4,258
Job	100	19,900	500	19,500	1,000	19,000
Movie	100	19,900	500	19,500	1,000	19,000
Nbaplayer	100	4,305	500	3,905	1,000	3,405
Restaurant	100	19,900	500	19,500	1,000	19,000
University	100	16,605	500	16,205	1,000	15,705

3.2 Evaluation Metrics

Following the evaluation metrics of [2–4], we calculate the page-level F1 score to evaluate the performance of EDDVPL. The page-level F1 score is the harmonic mean of precision and recall. Specifically, for each attribute, the *precision* of a method is the number of pages for which the ground-truth attribute values are correctly extracted divided by the number of pages from which the method extracts values. *Recall* is the number of pages for which the ground-truth attribute values are correctly extracted divided by the number of pages containing ground-truth attribute values. For each domain, we calculate the average page-level F1 score for all attributes as the final experimental result.

3.3 Implementation Details

In the data preprocessing stage, we first use the open-source LXML library to process each HTML source code to get DOM tree structures. Then, we use a simple heuristic algorithm [2] to distinguish fixed nodes and variable nodes in web pages. In addition, we set the maximum distance D as 2.

In the model training stage, we use the $T5base$ [9] provided by the Transformers library of Hugging Face as the pre-trained language model for EDDVPL and set the training batch size as 16, the learning rate as 0.0002, and the number of training iterations as 20.

3.4 Baseline Models

We compare the proposed EDDVPL with existing methods including SimpDOM and DOM2R-Graph.

- SimpDOM [3]: SimpDOM uses the DOM tree structure to avoid using rendering features. It enhances node representations by retrieving context and capturing discrete features for DOM tree nodes. Since the extracted features have consistency across websites, SimpDOM can perform extraction on unseen websites after training on a few seed websites.
- DOM2R-Graph [10]: DOM2R-Graph simplifies and models the DOM tree as a heterogeneous graph. It captures the influence of context structural relations on semantic interactions in the graph to obtain fine-grained representations of nodes. Since context structural relations are generalizable features, DOM2R-Graph can achieve excellent extraction performance.

3.5 Overall Performance

Table 3 reports the comparison results between EDDVPL and baselines on different sizes of training sets.

It can be seen that SimpDOM outperforms DOM2R-Graph when $k = 10$. This is because that the rich discrete features of SimpDOM can provide better reasoning basis compared to DOM2R-Graph which only focuses on complex features such as semantics and structure in the case of few labeled data. We can also see that EDDVPL significantly outperforms all the above methods when $k = 10$. This is because SimpDOM and

DOM2R-Graph rely on enough labeled web pages to learn complex semantic knowledge or web page structure that is consistent across websites. And the extremely limited training data cannot provide sufficient learning patterns for them. In contrast, for EDDVPL, the following two factors determine its superior performance:

(1) The pre-trained language model contains rich prior knowledge, which can provide a good foundation for the model to infer node attribute types with little training data.

(2) EDDVPL guides the pre-trained language model to understand "what needs to be done" by constructing the task template, and assists the model in stimulating domain knowledge and making judgments at the semantic-level by introducing DOM tree view prompts. EDDVPL realizes an effective combination of the pre-trained language model with task target and domain information, thus making full use of the limited domain-labeled data.

Table 3. Comparison results of EDDVPL with baseline methods

k	10			50			100		
Models	SimpDOM	DOM2R-Graph	EDDVPL	SimpDOM	DOM2R-Graph	EDDVPL	SimpDOM	DOM2R-Graph	EDDVPL
Auto	92.50	92.16	97.20	93.75	97.94	97.27	96.00	98.66	97.57
Book	83.60	69.94	97.37	89.80	91.85	98.13	92.40	94.42	98.19
Camera	89.00	92.64	96.85	96.67	97.51	98.17	95.33	98.47	99.04
Job	90.75	52.81	95.76	95.50	89.74	99.42	97.50	95.35	99.39
Movie	77.00	75.01	93.69	91.00	95.27	96.77	93.75	96.16	96.52
Nbaplayer	68.50	88.41	82.11	84.25	98.92	90.72	90.25	99.01	92.00
Restaurant	86.50	83.67	95.87	96.00	96.07	98.48	95.00	96.39	98.87
University	92.75	85.23	98.65	98.00	97.21	99.80	98.00	97.80	99.80
Average	85.08	79.98	**94.69**	93.12	95.56	**97.35**	94.78	97.03	**97.67**

With the increase of k, the existing models can learn enough types of feature information that is more targeted to the domain and task. The generalization ability of them is further improved. However, our model still achieves the best results. It can be seen that our model is not sensitive to the amount of data which has stronger robustness.

3.6   Ablation Study

To verify the effectiveness of each design of EDDVPL, we conduct ablation studies on the public SWDE [4] dataset. Specifically, two variant models are designed to demonstrate the contributions of different view prompts, which are described as follows:

(1) Template-view: To demonstrate the importance of the semantic prompt in DOM tree view, we remove the semantic prompt and only use the text of the nodes themselves for template filling.
(2) DOM-view: To demonstrate the effectiveness of the task prompt of template view, we remove the template and input the nodes' text and DOM tree view prompt as a continuous sequence to the model.

When taking 10 web pages per website as the training set (i.e. $k = 10$), the results of ablation experiments in each domain are shown in Fig. 3. It shows that the effectiveness

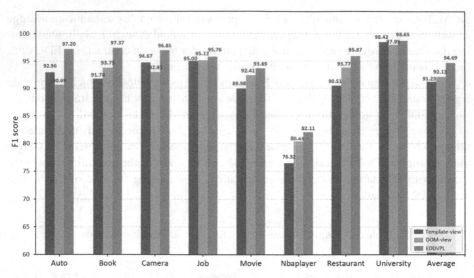

Fig. 3. Ablation results of EDDVPL

of both Template-view and DOM-view decreases compared to the complete model. This is due to the following two reasons:

(1) When removing the semantic prompt of DOM tree view, relevant knowledge cannot be accurately retrieved due to the lack of domain information.
(2) When removing the task prompt of template view, the model only has domain-related semantic information, and the task target is still ambiguous.

The above experimental results prove that the dual-view prompt paradigm plays a positive role in learning better model discrimination ability. The results also show that the pre-trained language model can maximize the utilization of existing knowledge and limited labeled data by closely integrating with task and domain.

4 Related Work

4.1 Web Information Extraction

Web information extraction methods are divided into three main categories: (1) Wrapper-based methods [11–13], which construct extraction rules or templates based on shallow rules of target data in web pages. (2) Visual feature-based methods [14, 15], which consider visual features such as the distribution of content blocks, the size and proportion of text blocks, and layout relevance according to the characteristics of web page layout. And (3) DOM feature-based methods [16–18], which focus on the underlying DOM tree features of web pages. However, the labeling process for web pages is time-consuming and labor-intensive, and the scarcity of labeled data can limit the learning ability of existing models. Therefore, existing research has implemented few-shot extraction at the site-level, but a significant amount of labeling work is still required when new domains appear.

4.2 Prompt Learning

In NLP tasks, few-shot learning is typically achieved through pre-training and fine-tuning, utilizing the prior knowledge and semantic representation of pre-trained language models to assist downstream tasks [19–21]. On the other hand, prompt learning [22] can help pre-trained language models retrieve existing knowledge to further understand downstream tasks. For example, Ding et al. [23] proposed a prompt-learning method on fine-grained entity typing, which extracts entity types without overfitting by performing distribution-level optimization. Hambardzumyan et al. [24] proposed an automatic prompt generation method that transfers knowledge from large pre-trained language models to downstream tasks by appending embeddings to the input text, which significantly outperforms the baseline in few-shot settings. To our knowledge, prompt learning has not been used for web attribute extraction. Therefore, inspired by the above methods, we propose the paradigm of few-shot learning to assist with web attribute extraction tasks.

5 Conclusion

In this paper, we propose an encoder-decoder structure web attribute extraction model based on dual-view prompt learning named EDDVPL. EDDVPL first retrieves descriptive context for DOM tree nodes to retain important semantic prompt information of DOM tree view. Then, by observing the characteristics of extraction task, EDDVPL constructs a template to provide intuitive and effective task prompt information for the pre-training language model. In addition, the effective integration of dual-view prompts is achieved by template filling and is used to guide the training of the pre-trained language model. Extensive experiments demonstrate the effectiveness of EDDVPL in extracting attributes from semi-structured web pages when there is a small number of labeled web pages.

Acknowledgments. This work is supported by Key National Defense Science and Technology Innovation Program of the Chinese Academy of Sciences (No. KGFZD-145–22-16), Strategic Priority Research Program of the Chinese Academy of Sciences (No.XDC02030400).

References

1. Lockard, C., Dong, X.L., Einolghozati, A., Shiralkar, P.: CERES: distantly supervised relation extraction from the semi-structured web. arXiv preprint arXiv:1804.04635 (2018)
2. Lin, B.Y., Sheng, Y., Vo, N., Tata, S.: FreeDOM: a transferable neural architecture for structured information extraction on web documents. In: Proceedings of the 26th ACM SIGKDD International Conference on Knowledge Discovery & Data Mining, pp. 1092–1102 (2020)
3. Zhou, Y., Sheng, Y., Vo, N., Edmonds, N., Tata, S.: Simplified DOM trees for transferable attribute extraction from the web. arXiv preprint arXiv:2101.02415 (2021)
4. Hao, Q., Cai, R., Pang, Y., Zhang, L.: From one tree to a forest: a unified solution for structured web data extraction. In: Proceedings of the 34th international ACM SIGIR Conference on Research and Development in Information Retrieval, pp. 775–784 (2011)

5. Schick, T., Schütze, H.: Exploiting cloze questions for few shot text classification and natural language inference. arXiv preprint arXiv:2001.07676 (2020)
6. Han, X., Zhao, W., Ding, N., Liu, Z., Sun, M.: PTR: prompt tuning with rules for text classification. arXiv preprint arXiv:2105.11259 (2021)
7. Han, J., Zhao, S., Cheng, B., Ma, S., Lu, W.: Generative prompt tuning for relation classification. arXiv preprint arXiv:2210.12435 (2022)
8. Chen, Y., Harbecke, D., Hennig, L.: Multilingual relation classification via efficient and effective prompting. arXiv preprint arXiv:2210.13838 (2022)
9. Raffel, C., et al.: Exploring the limits of transfer learning with a unified text-to-text transformer. J. Mach. Learn. Res. **21**(1), 5485–5551 (2020)
10. Feng, J., et al.: DOM2R-Graph: a web attribute extraction architecture with relation-aware heterogeneous graph transformer. In: Neural Information Processing: 29th International Conference, ICONIP 2022, Virtual Event, November 22–26, 2022, Proceedings, Part I, pp. 468–479. Springer, Cham (2023) https://doi.org/10.1007/978-3-031-30105-6_39
11. Sahuguet, A., Azavant, F.: Building intelligent web applications using lightweight wrappers. Data Knowl. Eng. **36**(3), 283–316 (2001)
12. Chang, C.H., Hsu, C.N., Lui, S.C.: Automatic information extraction from semistructured web pages by pattern discovery. Decis. Support. Syst. **35**(1), 129–147 (2003)
13. Yu, H.T., Guo, J.Y., Yu, Z.T., Xian, Y.T., Yan, X.: A novel method for extracting entity data from deep web precisely. In: The 26th Chinese Control and Decision Conference (2014 CCDC), pp. 5049–5053. IEEE (2014)
14. Wang, J., Chen, Q., Wang, X., Guo, H.: Basic semantic units based web page content extraction. In: 2008 IEEE International Conference on Systems, Man and Cybernetics, pp. 1489–1494. IEEE (2008)
15. Anderson, N., Hong, J.: Visually extracting data records from the deep web. In: Proceedings of the 22nd International Conference on World Wide Web, pp. 1233–1238 (2013)
16. Wang, H., Zhang, F., Zhao, M., Li, W., Xie, X., Guo, M.: Multi-task feature learning for knowledge graph enhanced recommendation. In: The World Wide Web Conference, pp. 2000–2010 (2019)
17. Wang, X., et al.: Heterogeneous graph attention network. In: The World Wide Web Conference, pp. 2022–2032 (2019)
18. Wu, S., et al.: Fonduer: knowledge base construction from richly formatted data. In: Proceedings of the 2018 international conference on management of data, pp. 1301–1316 (2018)
19. Alt, C., Hübner, M., Hennig, L.: Fine-tuning pre-trained transformer language models to distantly supervised relation extraction. arXiv preprint arXiv:1906.08646 (2019)
20. Liu, Y.: Fine-tune BERT for extractive summarization. arXiv preprint arXiv:1903.10318 (2019)
21. Brown, T., et al.: Language models are few-shot learners. Adv. Neural. Inf. Process. Syst. **33**, 1877–1901 (2020)
22. Liu, P., Yuan, W., Fu, J., Jiang, Z., Hayashi, H., Neubig, G.: Pre-train, prompt, and predict: a systematic survey of prompting methods in natural language processing. ACM Comput. Surv. **55**(9), 1–35 (2023)
23. Ding, N., et al.: Prompt-learning for fine-grained entity typing. arXiv preprint arXiv:2108.10604 (2021)
24. Hambardzumyan, K., Khachatrian, H., May, J.: WARP: word-level adversarial reprogramming. arXiv preprint arXiv:2101.00121 (2021)

CACL:Commonsense-Aware Contrastive Learning for Knowledge Graph Completion

Chuanhao Dong, Fuyong Xu, Yuanying Wang, Peiyu Liu[✉], and Liancheng Xu[✉]

Shandong Normal University, Jinan, China
{liupy,lcxu}@sdnu.edu.cn

Abstract. Most knowledge graphs (KGs) are incomplete in the real world, so knowledge graph completion (KGC) is widely investigated to predict the most credible missing facts from given knowledge. However, existing KGC methods heavily rely on the given facts to predict missing relations between entities, ignoring the value of external knowledge. In addition, previous knowledge representation methods ignore the multi-perspective characteristics of cognate knowledge, which leds to the inability to obtain high-level semantic representation of knowledge. To alleviate the above issues, this paper proposes a Commonsense Aware Contrastive Learning (CACL) framework, which extracts relevant knowledge triples from existing commonsense knowledge base to assist in the KGC. Moreover, our method employs knowledge contrast representation learning method to acquire the higher-order representation from multiple perspectives. Experiments show that our method improves the performance of basic knowledge graph embedding (KGE) models. Our method also could be easily adapted to various KGE models.

Keywords: Knowledge Graph · Knowledge Graph Completion · Commonsense · Contrastive Learning

1 Introduction

Knowledge Graph (KG) is a relational graph network that links all kinds of information and stores a large amount of knowledge in form of the triple (head entity, relationship, tail entity). In recent years, KGs have played a very important role in various applications, such as question answering [1,2], dialogue systems [3,4], and recommendation systems [5,6]. However, most KGs are sparse, the missing information in KGs needs to be predicted by knowledge graph completion (KGC).

Existing works indicate that fusing external knowledge can effectively predict missing information in KG [7]. Commonsense knowledge is widely used in various Natural Language Processing (NLP) tasks, such as commonsense quiz, commonsense dialogue, and commonsense textual reasoning. Commonsense knowledge is

© The Author(s), under exclusive license to Springer Nature Singapore Pte Ltd. 2024
B. Luo et al. (Eds.): ICONIP 2023, CCIS 1968, pp. 485–496, 2024.
https://doi.org/10.1007/978-981-99-8181-6_37

more effective for KGC, Niu et al. [8] extract commonsense knowledge from fact triples with entity concepts for KGC. In addition, the KGE approach also needs to be paid attention in KGC tasks, and an effective KGE approach can also assist in improving the performance of KGC. Zeb et al. [9] generate two rich and high-quality embedding vectors for each node by aggregating information from neighboring nodes, and then accurately model the KG triples in the form of tensor representations. However, the multi-view higher-order representation features of knowledge and relations in KG are ignored.

At present, there are few models extracting relevant knowledge triples from the existing commonsense knowledge base, and the examples of using external knowledge to supplement the knowledge graph are very rare, and the acquired knowledge are all low-order representations, which cannot deepen the model's understanding and perception ability. In KG, the higher-order representation with multiple views can better acquire the characteristics of knowledge and relationships, improve the model's understanding and perception of relevant knowledge, and thus improve the accuracy of KGC. Therefore, we replace the entities in each fact triple with the corresponding commonsense knowledge concepts to obtain higher-order representations of the same source knowledge through comparative learning, and achieve a complementary knowledge graph through an effective KGE approach.

In summary, existing KGC models still have some issues: (1) The existing models can not fully and effectively merge multiple commonsense to predict information in KG. (2) The inability of KGE methods to characterize knowledge from multiple perspectives, resulting in the performance of KGC being limited by the performance of KGE. To alleviate the above issues, we propose CACL, a contrastive learning framework for incorporating commonsense knowledge. In addition, higher-order representation features of knowledge are acquired by comparing different knowledge perspectives through a Contrastive learning approach. In summary, the contributions are as follows:

- We propose CACL, a commonsense-aware contrastive fusion framework for KGC, which can effectively discover the implicit complex relations in commonsense.
- We design a Contrastive learning schema to compare different perspective features of the knowledge to obtain multi-view representation in KG.
- Experiments show that our method can be easily adopted into basic KGE models, and CACL can also improve the effectiveness of KGE models.

2   Related Work

2.1   Reasoning About Commonsense Knowledge Graphs

The commonsense triple, unlike the fact triple, can inject some abstract knowledge into the knowledge graph to complement the missing parts of the knowledge graph. In the field of NLP, some studies have constructed commonsense knowledge graphs by inference to complement the knowledge graph. Yu et al. [10] propose MoKGE, a novel method which improves the generation quality of commonsense reasoning tasks by diversifying formative reasoning through mixed expert (MoE) reasoning. Hao et al. [11] propose an attention guided commonsense reasoning network (ACENet) that continuously reinforces correct choice information through multiple layers of interaction of answer choices and guides GNN messaging, exhibiting superior performance in commonsense inference. Kim et al. [12] extend the zero-shot transfer learning scheme to a multi-source setting. Different graphs can be utilized synergistically, by developing a modular knowledge aggregation variant. As a new framework for piecewise commonsense reasoning, the loss of knowledge due to interference between different knowledge sources is mitigated. Previous studies have proposed training language models that target existing relations in the commonsense KG and are used for commonsense reasoning. However, these commonsense graphs contain only concepts without links to corresponding entities, making them inappropriate for KGC tasks. In our work, we focus on addressing the missing relations in the commonsense KG for better commonsense reasoning.

2.2 Contrastive Learning

Contrastive learning is a machine learning technique that learns general features of unlabeled data sets by training models to discriminate between data points that are similar or different. Contrastive learning is also widely used in natural language processing, where models learn to predict positive or negative pairs to obtain unsupervised representations [13]. Contrastive learning is an effective representation learning method that learns positive pairs in contrast to negative pairs [14,15]. The key idea of contrast learning is to pull semantically tight pairs together and push negative pairs apart. The unsupervised contrastive learning framework [14] will use data augmentation to construct positive pairs in order to compute contrast loss. The supervised contrastive learning framework [15] computes the contrast loss for all positive instances within the same small batch. Inspired by these studies, we propose a triadic contrast loss to mitigate the sparsity of the knowledge graph and perform KGC using a commonsense KG, which improves the higher-order representation of triples and the accuracy of KGC, and ultimately improves the prediction performance of sparse entities.

2.3 KGE Model

Knowledge Graph Embedding (KGE) learns the embedding representation of entities and relations in a knowledge base, and is the basic research for many applications such as semantic retrieval, knowledge question answer, and recommendation. We know that KGE models are embedding representations of entities and relations in knowledge base, and the existing KGC models are as follows: (1) translation models, which are to treat relations as translations between head and tail entities, including models such as TransE [16]. (2) Bilinear models, which calculate the plausibility of potential semantics of entities and relations in vector space, including models such as RESCAL [17] and ComplEx [18]. (3) Neural network models, which are models that use neural networks to solve KGE problems, including ConvE [19], etc. Compared with other methods, KGE methods achieve higher efficiency and better performance on KGC. However, the uncertainty of embedding limits the accuracy of KGC relying only on facts. More specifically, KGE models usually require a negative triple acquisition procedure to randomly or purposefully train some triples that are not observed in KG as negative triples.

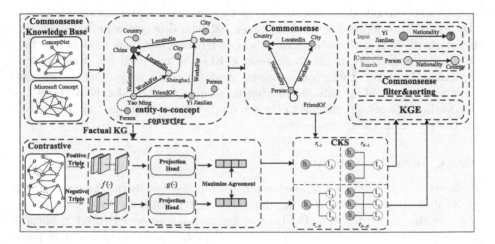

Fig. 1. Overview of the CACL framework.

3   Methodology

In this section, we introduce a simple and effective CACL framework to predict the missing facts by a given fact for KGC. As shown in Fig. 1, firstly, the factual entities are converted into conceptual entities by means of an entity-to-concept converter. Secondly, commonsense knowledge is extracted from fact triples with entity concepts, and the generated commonsense knowledge is used to generate high-quality negative triples by characterizing complex relationships, and the relevant knowledge triples are extracted from the commonsense knowledge base using the contrastive learning. Finally, the scores of candidate triples are ranked according to the learned KG embeddings, so as to infer the missing entities or relations in the triples.

3.1   Commonsense Knowledge Translator

Knowledge Graph (KG) is expressed as triplet set $\mathcal{G} = \{(h_i, r_j, t_k)\}$, h_i, r_j and t_k denote the head entity, relation and tail entity, respectively. $i \leq |\mathcal{V}_E|$, $j \leq |\mathcal{V}_R|$,

$k \leq |\mathcal{V}_E|$. $\mathcal{V}_E$ and $\mathcal{V}_R$ denote the entity set and relation set. Commonsense Knowledge Translator (CKT) aims to convert factual triplets (h_i, r_j, t_k) into concept-level triplets $\{(c_h, r_c, c_t)\}$. Specifically, we first build Commonsense Knowledge Base (CKB) $\mathcal{C}$ from existing CKBs, ConceptNet [20] and Microsoft Concept [21] Graph. $\mathcal{C}$ is the sets of triplets while each triplet is constituted of a head entity c_h and a tail entity c_t associated with their relation r_c. Besides, we also change the order of head entity and tail entity to augment the CKB, and the final CKB can be formulated as:

$$\mathcal{C} = \{(c_h, r_c, c_t)\} \cup \{(c_t, r_c, c_h)\}. \tag{1}$$

Then, we convert the factual triplets to concept-level triplets with CKB $\mathcal{C}$. For example shown in Fig. 1, the factual triplet (Yao Ming, Nationality, China) is convert to concept-level triplet (Person, Nationality, Country).

3.2 Contrastive Learning

The unsupervised contrastive learning framework [22] use data augmentation to calculate contrast loss. Here, we employ contrasitve learning schema to improve the representation ability of triplets in Knowledge Graph (KG). The supervised contrastive learning framework [15] computes the contrast loss for all positive instances within the same small batch. Our negative triple sampling is to select a target triple in a batch as a positive triple and other triples as negative triples. Knowledge can be obtained from multiple angles, enabling the model to better understand and perceive the knowledge. Contrastive loss is calculated as:

$$f_{\text{CL}}(F_i, Z_i) = \frac{-1}{|P(i)|} \log \left(\sum_{Z_i \in P(i), F_j \in N(i)} \frac{e^{\text{sim}(F_i, Z_i)/\tau}}{e^{\text{sim}(F_i, F_j)/\tau}} \right), \tag{2}$$

where F_i and Z_i are the representations of negative and positive examples. $P(i)$ is the set of all positive triples in the batch, and $N(i)$ is the set of all negative triples in the batch, such as the triples in the red and blue circles Fig. 1. F_j is the negative triple in the set of negative triples of the batch. $\text{sim}(\cdot)$ is the distance measurement function, and it can be defined as:

$$\text{sim}(F_i, Z_i) = \frac{F_i \cdot Z_i}{\| Fi \|_2 \| Z_i \|_2}. \tag{3}$$

3.3 KGE Score Function

Our proposed framework is independent of Knowledge Graph Embedding (KGE) models, any KGE models can be utilized to learn entity and relation embeddings. Therefore, it is essential to rule the KGE score function $E(e_h, e_r, e_t)$, which is used to evaluate the plausibility of the triple (h, r, t). More specifically, three most typical scoring function patterns are as follows: **(a)** Translation-based scoring functions such as TransE [16]:

$$E(e_h, e_r, e_t) = \|e_h + e_r - e_t\|, \tag{4}$$

where e_h, e_r and e_t denote the embeddings of head entity h, relation r and tail entity t, respectively. (b) Rotation-based scoring functions, such as RotatE [23]:

$$E(e_h, e_r, e_t) = \|e_h \odot e_r - e_t\|, \tag{5}$$

where $\odot$ indicates the hardmard product. (c) Scoring functions based on tensor decomposition, such as DistMult [24]:

$$E(e_h, e_r, e_t) = e_h^T \operatorname{diag}(M_{e_r}) e_t, \tag{6}$$

$\operatorname{diag}(M_{e_r})$ represents the diagonal matrix of the relation r.

3.4 Commonsense Knowledge Generation

Our approach can theoretically generate commonsense knowledge automatically. Specifically, the entities in fact triples can be converted into commonsense knowledge triples by replacing the entities in each fact triple with the corresponding commonsense knowledge entities in the generated commonsense knowledge through commonsense knowledge converters. In particular, the commonsense knowledge in C is achieved by removing duplicate commonsense knowledge entities, and then, the generated commonsense knowledge entities are combined into a set to construct a commonsense knowledge triple in the set form C'

To reduce false negative triples in commonsense sampling, we negatively sample entities using features of complex relations defined in TransH (i.e., 1-1, 1-N, N-1, and N-N), where 1 indicates that an entity is unique when given a relation and another entity, and conversely, N indicates that multiple entities may exist in this case. For a better understanding, the Commonsense Knowledge Sampling (CKS) process is shown in Fig. 2. The whole process can be divided into two main steps. Step 1: Select candidate head concepts with commonsense knowledge bases C. The candidate head concept "city" is determined based on commonsense C and non-unique sampling. Furthermore, based on the unique sampling strategy, candidate tail concepts are selected as the same concept province as "Jiangsu". Step 2: Concept to entity conversion. To reduce the appearance of false triples while ensuring the high quality of negative triples, corrupt entities belonging to candidate concepts are sampled from the following distribution:

$$
\begin{aligned}
\omega\left(h_i', r, t\right) &= 1 - p\left(\left(h_i', r, t\right) \mid \left\{\left(h_i, r_j, t_k\right)\right\}\right) \\
&= 1 - e^{\alpha E\left(h_i', r, t\right)} \Big/ \sum\nolimits_i e^{\alpha E\left(h_i', r, t\right)} \\
\omega\left(h, r, t_k'\right) &= p\left(\left(h, r, t_k'\right) \mid \left\{\left(h_i, r_j, t_k\right)\right\}\right) \\
&= e^{\alpha E\left(h, r, t_k\right)} \Big/ \sum\nolimits_i e^{\alpha E\left(h, r, t_k\right)}
\end{aligned} \tag{7}
$$

where h_i' and t_k' are the damaged head and tail entities obtained by sampling. ω and p denote the weights and probabilities of the negative triples, respectively. α is the temperature of sampling motivated by the self adversarial sampling [23].

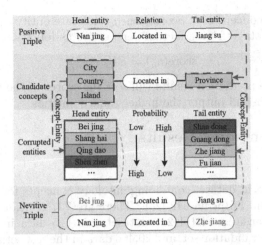

Fig. 2. Sampling process from commonsense knowledge base.

3.5 Train the Model

Positive and negative triples are used for contrastive learning so that the KGE models can acquire higher-order knowledge representations from multiple perspectives for training through transformation. The training loss function is as follows:

$$L = \sum_{h \in \mathcal{V}_E + \mathcal{V}_E^C} \sum_{t \in \mathcal{V}_E + \mathcal{V}_E^C} -[log\sigma(\gamma - E(e_h, e_r, e_t))$$

$$+ \frac{1}{2} \sum_i^n \omega(h_i^{'}, r, t) log\sigma(E(h_i^{'}, r, t) - \gamma) \tag{8}$$

$$+ \frac{1}{2} \sum_i^n \omega(h, r, t_i^{'}) log\sigma(E(h, r, t_i^{'}) - \gamma)]$$

$$Loss = (L + f_{CL}(F_i, Z_i))/2. \tag{9}$$

where γ is the margin and σ is the sigmoid function. L is the loss between positive and negative triples, and $f_{CL}(F_i, Z_i)$ is the loss after contrastive learning.

3.6 Link Prediction

This paper uses a head entity/tail entity and the correlate relation to predict a tail entity/head entity. Specifically, taking the query $(h, r, ?)$ as an example, commonsense $\mathcal{C}$ is used to filter the concepts of objective tail entities, and the candidate concepts of objective entities are defined as:

$$C^t = \{e_k^t | (e_i^h, r)\}, \tag{10}$$

where e_i^h is the i concept of h. Scoring each candidate entity e_k^t derived from the rough prediction stage from a factual perspective as follows:

$$\text{score}(e_k^t) = E(h, r, e_i^h). \tag{11}$$

Subsequently, the prediction results will sort the scores of the candidate entities in ascending order and output the higher-ranked entities.

4   Experiments and Results

4.1   Dataset

We used a real-world dataset containing ontology concepts for our experiments, FB15K237 [25] and DBpedia-242 [26] . There are 237 relations, 14,505 entities, and 89 commonsense in FB15K237 dataset; 272,115 data in the training set, 17,535 data in the validation set, and 20,466 data in the test set,and 298 relations, 99744 entities and 242 commonsense in the DBpedia-242 dataset, 592,654 data in the training set, 35,851 data in the validation set and 30,000 data in the test set. It is noteworthy that the entities in FB15K237 always belong to multiple concepts, and in the other datasets there is only one concept per entity, DBpedia-242 has only one concept.

4.2   Experiment Settings

To verify the effectiveness and efficiency of CACL, we compare it with the currently widely used KGC models, including CANS, MVLP and CAKE. In addition, we also compare different KGE models(TransE [16], RotatE [23], HAKE [27]) based on the above KGC methods. Specific experimental setup, the embedding size is 1,024 and the batch size is 512. The negative sample size is set to 128 for all models. Learning rate 5e-4. The margins are adjusted to {9,12,18,24,30}. Sampling temperature 1.0. All experiments are performed on 3090 GPU.

4.3   Automatic Metrics

MR is the Mean Rank, the smaller the MR value the better. MRR is the Mean Reciprocal Ranking, the larger the MRR value the better, the specific algorithm is as follows:

$$\mathbf{MR} = \frac{1}{|S|} \sum_{i=1}^{|S|} \mathbf{rank}_i,$$

$$\mathbf{MRR} = \frac{1}{|S|} \sum_{i=1}^{|S|} \frac{1}{\mathbf{rank}_i}. \tag{12}$$

Hit@N is the average percentage of triples ranked less than or equal to n in the link prediction, according to the following algorithm:

$$\mathbf{Hits@N} = \frac{1}{|S|} \sum_{i=1}^{|S|} \mathrm{II}(\mathbf{rank}_i \le n). \tag{13}$$

S is the set of triples, $|S|$ is the number of sets of triples, and $rank_i$ is the link prediction rank of the i triple.

4.4 Experiment Results

We experiment with the CACL model on three basic KGC models, the experimental results are shown in Table 1, 2 and 3, we can observe that the contrast fusion module effectively improves the performance of the three basic KGE models.

Furthermore, compared to each individual module, the overall model framework achieves a significant performance boost and outperforms all baselines significantly, demonstrating the superiority and effectiveness of combining commonsense, contrastive fusion with the original KGE model. On the FB15K237 dataset,compared with RotatE+CAKE, the MR, MRR, Hits@10, Hits@3, and Hits@1 indicators have improved significantly. The experiment was particularly strong on MR, improving 8 rankings, 9 rankings, and 78 rankings, respectively.On the DBpedia-242 dataset, there was a significant improvement in all metrics compared to RotatE+CAKE and HAKE+CAKE. The experiments were particularly strong on MR, MRR, Hits@10, Hits@3 and Hits@1.

Table 1. Prediction results on TransE KGE models

Models	FB15K-237					DBpedia-242				
	MR	MRR	Hits@10	Hits@3	Hits@1	MR	MRR	Hits@10	Hits@3	Hits@1
TransE	195	0.268	0.454	0.298	0.176	2733	0.242	0.468	0.344	0.100
TransE+CANS	175	0.298	0.49	0.333	0.203	1889	0.287	0.575	0.427	0.103
TransE+MVLP	181	0.29	0.476	0.323	0.186	881	0.322	0.585	0.450	0.152
TransE+CAKE	175	**0.301**	0.493	0.335	0.206	881	0.330	0.595	0.458	0.160
TransE+CACL	**167**	0.298	**0.52**	**0.348**	**0.298**	**872**	**0.345**	**0.602**	**0.463**	**0.164**

Table 2. Prediction results on RotatE KGE models

Models	FB15K-237					DBpedia-242				
	MR	MRR	Hits@10	Hits@3	Hits@1	MR	MRR	Hits@10	Hits@3	Hits@1
RotatE	204	0.269	0.452	0.298	0.179	1950	0.374	0.582	0.457	0.249
RotatE+CANS	182	0.296	0.486	0.329	0.202	1063	0.407	0.593	0.476	0.300
RotatE+MVLP	188	0.308	0.493	0.34	0.217	983	0.393	0.594	0.474	0.273
RotatE+CAKE	181	0.318	0.511	0.354	0.223	1027	0.423	0.603	0.486	0.320
RotatE+CACL	**172**	**0.322**	**0.522**	**0.363**	**0.305**	**953**	**0.442**	**0.612**	**0.489**	**0.331**

4.5 Ablation Study

We verify the effectiveness of the model by conducting ablation experiments on three bases of KGE models. As shown in Table 4, by removing the comparative

Table 3. Prediction results on HAKE KGE models

Models	FB15K-237					DBpedia-242				
	MR	MRR	Hits@10	Hits@3	Hits@1	MR	MRR	Hits@10	Hits@3	Hits@1
HAKE	176	0.306	0.486	0.337	0.216	1757	0.408	0.579	0.463	0.312
HAKE+CANS	174	0.315	0.501	0.344	0.221	1147	0.427	0.587	0.472	0.341
HAKE+MVLP	172	0.32	0.508	0.352	0.227	1083	0.411	0.580	0.463	0.319
HAKE+CAKE	170	**0.321**	0.515	0.355	0.226	931	0.437	0.593	0.481	0.353
HAKE+CACL	**91.7**	0.304	**0.572**	**0.373**	**0.245**	**912**	**0.457**	**0.603**	**0.497**	**0.372**

learning module experiment, we can draw the following conclusions: **(a)** On the three basic KGE models, the CACL indicators (MR, MRR, Hits@10, Hits@3, Hits@1) are all the best; **(b)** On the TransE model, the performance of MR and Hits@10 is improved most significantly through comparative learning; **(c)** On the RotatE model, the improved performance of MRR, Hits@3, and Hits@1 is average. **(d)** On the HAKE model, Hits@3 is similar to the other two models, and MR, MRR, Hits@10, Hits@3, Hits@1 are all better than the ablation experimental results.

Through ablation experiments, it is shown that contrastive learning brings triple embeddings from the same entity closer, and pushes triple embeddings from different entities further, which better demonstrates the model performance of CACL. At the same time, introducing features of commonsense and complex relations in the contrastive loss process helps to generate more efficient triples.

Table 4. Ablation experiment of integrating CL model and various KGE models on FB15K-237 dataset

Models	FB15K-237				
	MR	MRR	Hits@10	Hits@3	Hits@1
TransE-CACL	**167**	**0.298**	**0.52**	**0.348**	**0.298**
W/O CL	234	0.241	0.43	0.323	0.292
RotatE-CACL	**172**	**0.322**	**0.522**	**0.363**	**0.305**
W/O CL	215	0.263	0.489	0.302	0.261
HAKE-CACL	**91.7**	**0.304**	**0.572**	**0.373**	**0.245**
W/O CL	147.8	0.271	0.312	0.324	0.212

4.6   Case Study

As shown in Fig. 2, given a missing trailing entity ($Nanjing, Locatedin, ?$) of the query, our model can output the answer entity and provide the corresponding entity concept as well as the commonsense specific to the query. We can observe that all entities, including the correct entity, belong to the notion of conforming to commonsense. More interestingly, the commonsense and entity concepts can explain the reasonableness of the predicted answer entities to improve the user's confidence in the answers.

5   Conclusion

This paper proposes CACL to alleviate the ubiquitous missing issue in KGC. First, CACL can extract commonsense concepts from fact triples for KGC task. Second, we propose a triple contrastive loss to alleviate the absentness of the KG, and increase the higher-order representation of triples and the accuracy of KGC, thereby improving the prediction performance of missing entities. Extensive experiments on one dataset and three KGE models show that our proposed framework can outperform baseline methods, demonstrating its good effectiveness and scalability.

References

1. Cao, X., Liu, Y.: ReLMKG: reasoning with pre-trained language models and knowledge graphs for complex question answering. Appl. Intell. **53**(10), 12032–12046 (2023). https://doi.org/10.1007/s10489-022-04123-w
2. Cheng, Z., et al.: HiTab: a hierarchical table dataset for question answering and natural language generation. In: ACL, pp. 1094–1110 (2022)
3. Sheu, J.-S., Siang-Ru, W., Wen-Hung, W.: Performance improvement on traditional Chinese task-oriented dialogue systems with reinforcement learning and regularized dropout technique. IEEE **11**, 19849–19862 (2023)
4. Ji, T., Graham, Y., Jones, G.J.F., Lyu, C., Liu, Q.: Achieving reliable human assessment of open-domain dialogue systems. In: ACL, pp. 6416–6437 (2022)
5. Ma, X., Dong, L., Wang, Y., Li, Y., Zhang, H.: AKUPP: attention-enhanced joint propagation of knowledge and user preference for recommendation systems. Knowl. Inf. Syst. **65**(1), 163–182 (2023)
6. Bi, Q., Li, J., Shang, L., Jiang, X., Liu, Q., Yang, H.: MTRec: multi-task learning over BERT for news recommendation. In: ACL, pp. 2663–2669 (2022)
7. Li, M., Sun, Z., Zhang, W., Liu, W.: Leveraging semantic property for temporal knowledge graph completion. Appl. Intell. **53**(8), 9247–9260 (2023)
8. Niu, G., Li, B., Zhang, Y., Pu, S.: CAKE: a scalable commonsense-aware framework for multi-view knowledge graph completion. In: ACL, pp. 2867–2877 (2022)
9. Zeb, A., Ul Haq, A., Zhang, D., Chen, J., Gong, Z.: KGEL: a novel end-to-end embedding learning framework for knowledge graph completion. Expert Syst. Appl. **167**, 114164 (2021)
10. Yu, W., Zhu, C., Qin, L., Zhang, Z., Zhao, T., Jiang, M.: Diversifying content generation for commonsense reasoning with mixture of knowledge graph experts. In: ACL, pp. 1896–1906 (2022)
11. Hao, C., Xie, M., Zhang, P.: ACENet: attention guided commonsense reasoning on hybrid knowledge graph. In: EMNLP, pp. 8461–8471 (2022)
12. Kim, Y.J., Kwak, B.-W., Kim, Y., Amplayo, R.K., Hwang, S.-W., Yeo, J.: Modularized transfer learning with multiple knowledge graphs for zero-shot commonsense reasoning. In: NAACL, pp. 2244–2257 (2022)
13. Zhang, J., et al.: JointContrast: skeleton-based interaction recognition with new representation and contrastive learning. Algorithms **16**(4), 190 (2023)

14. Chen, T., Kornblith, S., Norouzi, M., Hinton, G.E.: A simple framework for contrastive learning of visual representations. In: ICML, volume 119 of Proceedings of Machine Learning Research, pp. 1597–1607 (2020)
15. Khosla, P., et al.: Supervised contrastive learning. In: Advances in Neural Information Processing Systems 33: Annual Conference on Neural Information Processing Systems 2020, NeurIPS 2020, December 6–12, 2020, virtual (2020)
16. Bordes, A., Usunier, N., García-Durán, A., Weston, J., Yakhnenko, O.: Translating embeddings for modeling multi-relational data. In: Advances in Neural Information Processing Systems 26: 27th Annual Conference on Neural Information Processing Systems 2013. Proceedings of a meeting held December 5–8, 2013, Lake Tahoe, Nevada, United States, pp. 2787–2795 (2013)
17. Nickel, M., Tresp, V., Kriegel, H.-P.: A three-way model for collective learning on multi-relational data. In: ICML, pp. 809–816 (2011)
18. Trouillon, T., Welbl, J., Riedel, S., Gaussier, É., Bouchard, G.: Complex embeddings for simple link prediction. In: ICML, volume 48 of JMLR Workshop and Conference Proceedings, pp. 2071–2080 (2016)
19. Dettmers, T., Minervini, P., Stenetorp, P., Riedel, S.: Convolutional 2D knowledge graph embeddings. In: AAAI, pp. 1811–1818 (2018)
20. Speer, R., Chin, J., Havasi, C.: ConceptNet 5.5: an open multilingual graph of general knowledge. In: AAAI, pp. 4444–4451 (2017)
21. Ji, L., Wang, Y., Shi, B., Zhang, D., Wang, Z., Yan, J.: Microsoft concept graph: mining semantic concepts for short text understanding. Data Intell. 1(3), 238–270 (2019)
22. Li, J., Shang, J., McAuley, J.J.: UCTopic: unsupervised contrastive learning for phrase representations and topic mining. In: ACL, pp. 6159–6169 (2022)
23. Sun, Z., Deng, Z.-H., Nie, J.-Y., Tang, J.: Knowledge graph embedding by relational rotation in complex space. In: ICLR, Rotate (2019)
24. Yang, B., Yih, W.-T., He, X., Gao, J., Deng, L.: Embedding entities and relations for learning and inference in knowledge bases. In: ICLR (2015)
25. Toutanova, K., Chen, D.: Observed versus latent features for knowledge base and text inference. In: CVSC, pp. 57–66 (2015)
26. Lehmann, J., et al.: Dbpedia - a large-scale, multilingual knowledge base extracted from wikipedia, pp. 167–195 (2015)
27. Zhang, Z., Cai, J., Zhang, Y., Wang, J.: Learning hierarchy-aware knowledge graph embeddings for link prediction. In: AAAI, pp. 3065–3072 (2020)

Graph Attention Hashing via Contrastive Learning for Unsupervised Cross-Modal Retrieval

Chen Yang[1], Shuyan Ding[2]($\boxtimes$), Lunbo Li[1], and Jianhui Guo[1]

[1] School of Computer Science and Technology, Nanjing University of Science and Technology, Nanjing, China
{kogenta,lunboli,guojianhui}@njust.edu.cn
[2] School of Electronic and Optical Engineering, Nanjing University of Science and Technology, Nanjing, China
shuyanding@njust.edu.cn

Abstract. Hashing-based cross-modal retrieval maps multi-modal features into binary codes into a common Hamming space. Due to its small storage consumption and high efficiency, hashing has received extensive attention in recent years. However, the current researches have difficulty in constructing a well-defined joint semantic space and conduct more detailed and in-depth learning guidance. In this paper, Graph Attention Hashing via Contrastive Learning (GAHCL) is proposed to address these issues. First, we use the idea of contrastive learning to generate positive samples, and propose a novel contrastive adjacency matrix through a graph attention network. Specifically, this matrix assigns higher weights to node pairs whose source is the same sample, and assigns lower weights to node pairs that do not match each other. The key semantic features can be captured more carefully and accurately under the influence of attention weights. In addition, the contrastive loss function is constructed by taking the output features of different modalities in an instance and its generated positive sample features as a positive sample pair. Extensive experiments on two datasets show that the proposed method can significantly outperform existing competitors.

Keywords: Unsupervised hashing · Cross-modal retrieval · Graph attention networks · Contrastive learning

1 Introduction

With the rapid growth of social networking platforms such as Tiktok and Twitter in recent years, there has been an exponential increase in the volume and variety of multimedia content, including images, texts, and videos. As a result, the effective organization and retrieval of this vast amount of information has become a pressing issue that needs to be addressed. The essence of the cross-modal retrieval task lies in inputting one type of data and retrieving another type of data, ensuring their semantic similarity to the greatest extent possible. Traditional cross-modal retrieval (CMR) [1] methods map heterogeneous data into a common space. While the computational constraints make them unsuitable for

© The Author(s), under exclusive license to Springer Nature Singapore Pte Ltd. 2024
B. Luo et al. (Eds.): ICONIP 2023, CCIS 1968, pp. 497–509, 2024.
https://doi.org/10.1007/978-981-99-8181-6_38

large-scale retrieval work. In this case, cross-modal hashing(CMH) [7] techniques have received extensive attention. CMH methods usually map multi-modal data into a unified binary Hamming space. Since different modality instances are heterogeneous in terms of their feature representation and distribution, it is necessary for CMH methods to explore appropriate reliable learning guidance to mitigate the cross-modal discrepancy.

Existing CMH techniques can be broadly categorized into supervised methods and unsupervised methods [16]. The supervised ones rely on semantic labels, which are usually unavailable. In contrast, unsupervised CMH (UCMH) methods usually utilize global-level intra-modality and inter-modality similarities to learn the hashing functions, which is more suitable for real-life scenarios.

However, for many complex data, barely generating hash codes through cross-modal features is insufficient to describe the deep relationship between them. When two data have an underlying semantic correlation but different feature representations, suboptimal hash codes are generated. Thus it creates a major challenge. Many recent studies have tried to use the idea of graphs in the process of feature extraction. They often use the semantic relevance of graphs to obtain better learning guidance [11]. AGCH [18] is the first application of GCN [5] to unsupervised hashing learning, using various distance-based similarity measures to build affinity graphs that aggregate information from different modalities. However, GCN is a full-graph computation method, which is difficult to handle inductive tasks [2]. SGEH [19] proposes dense hybrid semantic graph which restricts the process of hash code learning to learn more discriminative results. Graph Convolutional Networks for Hyperspectral Image Classification [3] designs a miniGCNs to find a better and more robust local optimal solution, but that is still not enough.

In light of these issues, we propose the Graph Attention Hashing via Contrastive Learning (GAHCL) model. On the one hand, we use the graph attention network [14] and propose a contrastive adjacency matrix as the graph structure to better exploit the correlation between different vertex features. The attention mechanism is combined with our proposed contrastive graph structure and the semantic information of different views between instances can be better captured. On the other hand, we use the idea of contrastive learning to create new positive samples from the original data set. The similarity between these positive samples and the original data set is calculated, and it is intervened to make the distance between samples with the same semantic relationships closer. The pre-trained CLIP [10] model is selected for training, which allows us to introduce a hashing model that consists of only a single hashing network and thus naturally mitigates the difficulty in bridging different modalities.

GAHCL has the following main contributions:

- To the best of our knowledge, GAHCL is the first work using contrastive learning in the field of unsupervised cross-modal hashing. New positive samples are created from the original datasets by data augmentation method. A contrast loss function is selected to shorten the distance of positive samples in the projection space and widen the distance of negative samples in the space.

- A novel graph structure is designed based on contrast. The intervention of the graph structure makes the model focus more on learning the potential connections between positive samples, and the similarity between negative samples gradually decreases.
- Extensive experiments demonstrate the superiority of GAHCL over existing competitors on two public datasets, MIRFlickr-25k and NUS-WIDE.

2 Method

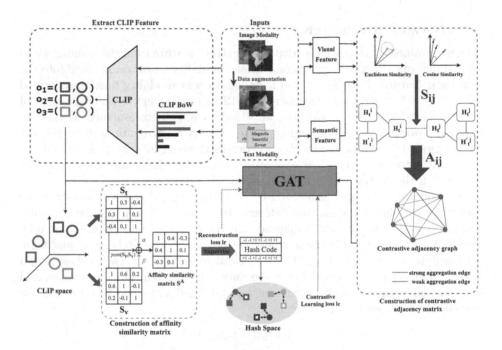

Fig. 1. The framework of GAHCL

The procedure of GAHCL is illustrated in Fig. 1. Initially, GAHCL conducts data augmentation on all input images and extracts both image and text features. Subsequently, the similarity between any two nodes is evaluated using Euclidean and cosine similarity measures, generating a contrastive adjacency matrix. This matrix is then fed into the GAT module, along with the pre-processed features extracted from the CLIP model. The resulting output is transformed into hash codes. Finally, GAHCL is trained using the contrastive loss and reconstruction loss functions.

2.1 Notation and Problem Definition

As we use the method of batch training, the samples of each batch can be expressed as $\mathcal{O} = \{o_1, o_2, \ldots, o_m\}$, where the m variable denotes the batch size.

Each instance consists of a co-occurred image-text pair description $o_k = (v_k, t_k)$, where $v_k \in \mathbb{R}^{d_v}$ and $t_k \in \mathbb{R}^{d_t}$ respectively represent the d_v-dimensional image feature and the d_t-dimensional text feature in the k-th instance.

Given the training data, GAHCL aims to learn projection functions $f_v(v, \theta_v)$ and $f_t(t, \theta_t)$ for images and texts, where θ_v and θ_t are parameters in our network, and generates hash code $\mathbf{B} = [b_1, b_2, \ldots, b_m] \in \mathbb{R}^{m \times c}$, where c is the code length we preset and b_k is the final binary code of the k-th instance in a training batch. Such binary codes are eventually generated for each modality to represent similar semantic features of the original data.

2.2   Network Architecture

The pre-trained CLIP model is adopted to extract feature from the training set in our model. In each epoch, instances in image modality v are transformed into the nonlinear features $\mathbf{F}_v \in \mathbb{R}^{m \times Z}$ and instances in text modality t are transformed into $\mathbf{F}_t \in \mathbb{R}^{m \times Z}$ respectively, where Z (512 in our paper) is the dimension of the extracted feature. In addition, we use data augmentation(such as random flipping, rotating, cropping, adding noise, etc.)as a surrogate task to convert image modality instances v to v', and transform them into $\mathbf{F}'_v \in \mathbb{R}^{m \times Z}$ either. The contrastive features $\mathbf{F}'_v$ is employed to broaden the scope of comparisons and enrich the diversity of positive samples.

Although the features generated by the CLIP model can well preserve the similarity correlations of original datasets [10], they need more similar semantic guidance. $\mathbf{F}_v$, $\mathbf{F}_t$ and $\mathbf{F}'_v$ with adjacency matrix $\mathbf{A}$ are fed into the GAT module, where $\mathbf{A}$ will be defined in the next subsection. More concretely, $\mathbf{v}$, $\mathbf{v}'$ and $\mathbf{t}$ are all graph nodes in the module. The structural relationship between these graph nodes enables them to learn deeper semantic information. For any two related nodes i and j, the similarity coefficient e_{ij} is defined as follows to represent the correlation between feature points:

$$e_{ij} = a(Wh_i \| Wh_j), j \in \mathcal{N}_i, \tag{1}$$

where W serves as the convolutional filters and h_k is the input feature of node k. $\mathcal{N}_i$ represents the set of neighbor nodes of i. $a(\cdot)$ is the activation function. $(\|)$ stands for concatenation operation. Then we use the softmax function to obtain the attention coefficient between two nodes as follows:

$$\alpha_{ij} = \frac{exp(LeakyReLU(\mathbf{A}_{ij}e_{ij}))}{\sum_{k \in \mathcal{N}_i} exp(LeakyReLU(\mathbf{A}_{ik}e_{ik}))}. \tag{2}$$

To better integrate features and avoid overfitting, several independent attention layers are used to generate attention weights in parallel, where each attention layer focuses on a different part of the input information. The outputs of multiple focused attentions are stitched together and transformed by another linear projection. Then the final outputs are achieved through the above procedure. The output of each node is defined as follows:

$$h'_i(K) = \|_{k=1}^{K} \sigma\left(\sum_{j \in \mathcal{N}_i} \alpha_{ij}^k W^k h_j\right), \tag{3}$$

where K represents the number of attention heads.

With GAT modules, GAHCL can mine deeper structural semantic information and the binary codes are generated by the following formulation:

$$\mathbf{B}_k = sign(f(\mathbf{F}_k; \theta_k)),$$
$$\textbf{s.t. } k \in \{v, v', t\}, \tag{4}$$

where θ_v, θ'_v and θ_t are weight parameters of GAT modules. The projection function f is mentioned in Sect. 2.1. The sign activation function $sign(\mathbf{x}) = \begin{cases} 1, & \mathbf{x} > 0 \\ -1, & \mathbf{x} \leq 0 \end{cases}$ is used to binarize the c-dimensional continues representation $\mathbf{F}$ into c-bit hash codes $\mathbf{B} \in \{-1, +1\}^{m \times c}$. However, according to the latest researches, in the backward propagation, the sign function may lead to the gradient vanishing problem, which has adverse effects on the training of parameters. Following [8,13], we use $\mathbf{B} = \tanh(\zeta \mathbf{F}) \in [-1, +1]^{m \times c}$ instead of the symbolic function during training. As the number of training epochs increases, the value of ζ becomes progressively larger to approximate the sign function.

2.3   Construction of Contrastive Adjacency Matrix

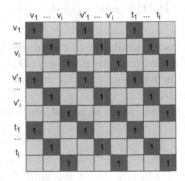

Fig. 2. Contrastive Adjacency Matrix

As shown in Fig. 2, the contrastive adjacency matrix describes the degree of semantic relatedness between $\mathbf{v} \in \mathbb{R}^{m \times d_v}$, $\mathbf{v}' \in \mathbb{R}^{m \times d'_v}$ and $\mathbf{t} \in \mathbb{R}^{m \times d_t}$ in each batch. On one hand, the red parts represent the different modal features of the same sample or its positive sample features. These two nodes have the same semantic relationships, and the degree of correlation is very high. Therefore the edge between them is called the strong aggregation edge, which is given the greatest weight. For example, $\mathbf{v}_1$, $\mathbf{v}'_1$ and $\mathbf{t}_1$ are a set of strong aggregation edges, where $\mathbf{v}_1$ and $\mathbf{t}_1$ are a set of image-text pairs, $\mathbf{v}'_1$ is a positive contrast sample of $\mathbf{v}_1$. On the other hand, the yellow parts in the matrix represent mismatched instances. Although the two nodes do not correspond to each other, there is still an underlying neighborhood structure worth noticing during training, so a small weight is assigned and they are represented as weak aggregation edges. The weak aggregation edges use the similarity affinity matrix $\mathbf{S}$ (mentioned below) to judge the connection between two nodes, and through the attention weights, the model focuses more on amplifying the related semantic information, while the irrelevant information is suppressed or ignored.

Recent studies usually utilize nonlinear features extracted from pre-trained deep network models to create affinity matrices and achieve good results. Therefore, this paper selects the pre-trained CLIP model to extract the original image features $\mathbf{H_v} \in \mathbb{R}^{m \times Z}$, text features $\mathbf{H'_t} \in \mathbb{R}^{m \times Z}$ and features of contrasting images $\mathbf{H'_v} \in \mathbb{R}^{m \times Z}$. Since the features of two modalities are mapped into the same encoding space, the heterogeneity between images and text can be well alleviated.

We combine the three classes of features into a total feature matrix $\mathbf{H} = \begin{bmatrix} \mathbf{H}_t \\ \mathbf{H}_v \\ \mathbf{H}'_v \end{bmatrix} \in \mathbb{R}^{3m \times Z}$. Following [18], we evaluate similarity information from different perspectives and combine them as follows to capture potential similarity relationships:

$$\mathbf{S}_{ij} = \mathbf{C}_{ij} \cdot \mathbf{D}_{ij}$$
$$= (\hat{\mathbf{H}}_i \hat{\mathbf{H}}_j^T) \cdot exp(-\sqrt{\left\| \hat{\mathbf{H}}_i - \hat{\mathbf{H}}_j \right\|_2}/\rho), \tag{5}$$

where $\mathbf{C_{ij}}$ represents the cosine similarity of the normalized features $\hat{\mathbf{H}}_i$ and $\hat{\mathbf{H}}_j$, which are the features of node i and j respectively. $\mathbf{D}$ stands for similarity based on Euclidean distance, which is more suitable for the similarity of image data. And ρ is a scaling parameter.

After combining strong and weak aggregation edges, the contrastive adjacency matrix is as follows:

$$\mathbf{A}_{ij} = \begin{cases} 1, & |i - j| = km \\ \eta S_{ij}, & |i - j| \neq km \ and \ S_{ij} > \varepsilon, \\ 0, & others \end{cases} \tag{6}$$

$$\text{s.t. } k \in \{0, 1, 2\}.$$

The red parts are given the greatest weight, and the yellow parts are given a smaller weight $\eta \mathbf{S}$. When the value of the affinity matrix $\mathbf{S}_{ij}$ is too small (such as $\varepsilon = 0.4$), we can directly assume that there is no relationship between nodes i and j, which not only saves computing resources, but also avoids the overfitting problem.

Essentially, the construction of adjacency matrix interferes with the learning direction of the input features in GAT, which will eventually make the features of the strong aggregation edges in the output closer, and the weak aggregation edges are relatively far away. It is consistent with the idea of contrastive learning.

2.4 Loss Function

In order to ensure the model's discrimination between positive and negative samples, we use the infoNCE loss function [9] to widen the distance between negative samples and narrow the distance between positive samples. We combine the three sets of binary codes obtained in each batch into $\mathbf{B} = \begin{bmatrix} \mathbf{B}_t \\ \mathbf{B}_v \\ \mathbf{B}'_v \end{bmatrix} \in \mathbb{R}^{3m \times c}$.

The objective function is as follows:

$$\min_{\mathbf{B}} \mathcal{L}_c = -\sum_{i=0}^{m-1} \log \frac{\sum_{k=1}^{2} exp(\mathbf{B}_i \mathbf{B}_{i+km}^{T}/\tau)}{\sum_{j=0}^{3m-1} exp(\mathbf{B}_i \mathbf{B}_{j}^{T}/\tau)}, \tag{7}$$

where τ represents temperature coefficient, which regulates the discrimination ability of GAHCL against negative samples.

In fact, even if only Eq. (7) is served as the final loss function, the presented accuracy is already quite satisfactory. But we realize that the above formula focuses more on the relative positions of positive and negative samples in the hash space, so using an affinity similarity matrix $\mathbf{S}^A$ to reconstruct $\mathbf{B}_t$ and $\mathbf{B}_v$ can strengthen the absolute positions of these samples. Following [13], we define the cosine similarity matrices $\mathbf{S}_t = \hat{\mathbf{F}}_t \hat{\mathbf{F}}_t^T \in [-1, +1]^{m \times m}$ and $\mathbf{S}_v = \hat{\mathbf{F}}_v \hat{\mathbf{F}}_v^T \in [-1, +1]^{m \times m}$ to describe the original neighborhood structure for the input texts and images features respectively. They are then fused into a unified affinity similarity matrix as follows:

$$\mathbf{S}^A = \alpha \mathbf{S}_t + \beta \mathbf{S}_v + \gamma \cos(\mathbf{S}_t, \mathbf{S}_v)$$
$$\text{s.t. } \alpha, \beta, \gamma \geq 0, \alpha + \beta + \gamma = 1, \tag{8}$$

where $\cos(\mathbf{S}_t, \mathbf{S}_v) = \frac{\mathbf{S}_t \mathbf{S}_v^T}{\|\mathbf{S}_t\|_2 \|\mathbf{S}_v\|_2} \in [-1, +1]$.

Considering both intra- and inter-modal training in cross-modal network training has been proven to be effective in improving retrieval performance [13], the modal reconstruction is supplemented in the loss function. The reconstruction loss function is:

$$\min_{\mathbf{B}_t, \mathbf{B}_v} \mathcal{L}_r = \|\mu \mathbf{S}^A - \cos(\mathbf{B}_t, \mathbf{B}_v)\|_F^2$$
$$+ \lambda_1 \|\mu \mathbf{S}^A - \cos(\mathbf{B}_t, \mathbf{B}_t)\|_F^2 \tag{9}$$
$$+ \lambda_2 \|\mu \mathbf{S}^A - \cos(\mathbf{B}_v, \mathbf{B}_v)\|_F^2,$$

where λ_1 and λ_2 are the trade-off parameters to balance the intra-modal and inter-modal reconstruction. μ is a scale parameter used to adjust the degree of reconstruction. The final loss function the proposed GAHCL is the combination of the above two functions as:

$$\min_{\mathbf{B}_t, \mathbf{B}_v} \mathcal{L} = \mathcal{L}_c + \delta \mathcal{L}_r. \tag{10}$$

3   Experiments

3.1   Datasets

MIRFlickr-25K: it is a dataset consisting of 25,000 image-text pairs with 24 labels. Each image and text is represented by a 100-dimensional SIFT descriptor and a 500-dimensional label vector respectively.

Table 1. The mAP Result of GAHCL and all baselines on MIRFlickr-25K at four code length. The best performances are shown in boldface.

Task	Method	16bits	32bit	64bits	128bits
I → T	DJSRH [13]	0.659	0.661	0.675	0.684
	JDSH [8]	0.663	0.666	0.676	0.691
	UGACH [17]	0.676	0.693	0.702	0.706
	UKD-SS [4]	0.700	0.706	0.709	0.707
	DSAH [15]	0.701	0.712	0.722	0.726
	VCGH [12]	0.703	0.714	0.717	0.724
	KDCMH [6]	0.713	0.716	0.724	0.728
	GAHCL	**0.734**	**0.745**	**0.758**	**0.774**
T → I	DJSRH [13]	0.655	0.671	0.673	0.685
	JDSH [8]	0.657	0.672	0.673	0.688
	UGACH [17]	0.676	0.692	0.703	0.707
	UKD-SS [4]	0.704	0.705	0.714	0.712
	DSAH [15]	0.707	0.713	0.728	0.730
	VCGH [12]	0.706	0.713	0.725	0.734
	KDCMH [6]	0.711	0.715	0.731	0.733
	GAHCL	**0.730**	**0.744**	**0.756**	**0.773**

NUS-WIDE: it contains 269,648 multi-modal instances. Among all 81 types of labels, the most frequently occurring instances of 10 types of labels are selected, about 186577 instances.

For a fair comparison, 5000 instances are selected as the training set and 2000 as the test set in both datasets. The combination of MIRFlickr-25K and NUS-WIDE in our experiments ensures that we explore cross-modal retrieval on datasets of varying scales and complexities, enhancing the reliability and generalizability of our research findings.

3.2   Baseline Methods and Evaluation Metrics

We selected several competitive unsupervised methods in recent years to compare with our proposed GAHCL. They are DJSRH [13], JDSH [8], UGACH [17], UKD-SS [4], DSAH [15], VCGH [12] and KDCMH [6].

Mean Average Precision (mAP) is adopted to evaluate the effectiveness of GAHCL with other models. The mAP method is a commonly used evaluation tool for information retrieval, which is computed as the mean of average precision (AP) of the retrieval results.

Moreover, we use the top-N precision curves as important evaluation criteria in our experiments.

3.3   Implementation Details

The mini-batch size m in our model is 32. The mini-batch SGD optimizer is selected with momentum 0.9 and 0.0005 weight decay to optimize CLIP module (with image network and text network inside). In addition, the Adam optimizer is used in the GAT module with 0.0005 weight decay.

Table 2. The mAP Result of GAHCL and all baselines on NUS-WIDE at four code length. The best performances are shown in boldface.

Task	Method	16bits	32bit	64bits	128bits
I → T	DJSRH [13]	0.503	0.517	0.528	0.544
	JDSH [8]	0.543	0.551	0.568	0.593
	UGACH [17]	0.613	0.623	0.628	0.631
	UKD-SS [4]	0.584	0.578	0.586	0.613
	DSAH [15]	0.602	0.612	0.632	0.637
	VCGH [12]	0.615	0.628	0.631	0.639
	KDCMH [6]	0.615	0.628	0.637	0.642
	GAHCL	**0.629**	**0.642**	**0.649**	**0.655**
T → I	DJSRH [13]	0.526	0.541	0.539	0.570
	JDSH [8]	0.560	0.576	0.584	0.599
	UGACH [17]	0.603	0.614	0.640	0.641
	UKD-SS [4]	0.587	0.599	0.599	0.615
	DSAH [15]	0.621	0.632	0.646	0.649
	VCGH [12]	0.617	0.628	0.635	0.648
	KDCMH [6]	0.623	0.636	0.647	0.651
	GAHCL	**0.627**	**0.640**	**0.648**	**0.653**

In the graph attention module, a dropout module is introduced multiple times to randomly disable some hidden layer nodes. The purpose of it is to prevent the features from being overweighted and reduce the overfitting of the model. Specifically, we impose dropout on the attention weights (obtained by Eq. (2)). Experimental results show that the dropout module greatly improves the generalization ability of the model in the attention module. In the experiments, when we set the number of heads to 8, the hidden layer dimension of each head to 16, and the dropout rate to 0.6 (60% of nodes fail).

With regard to the settings for other parameters, we set $\alpha = 0.4, \beta = 0.3, \gamma = 0.3$, $\mu = 1.5, \lambda_1 = 0.3, \lambda_2 = 0.3, \delta = 0.5, \eta = 0.7, \epsilon = 0.4$ for NUS-WIDE. For MIRFlickr-25K, $\alpha = 0.5, \beta = 0.1, \gamma = 0.4, \mu = 1.47, \lambda_1 = 0.1, \lambda_2 = 0.1, \delta = 0.5, \eta = 0.7, \epsilon = 0.4$.

3.4   Performance

Table 1 and 2 show the MAP scores from 16 to 128 hash bits on the MIRFlickr dataset and NUS-WIDE dataset. From these tables, it is obvious that GAHCL significantly outperforms the other seven state-of-the-art methods. Specifically, on the MIRFlickr dataset, GAHCL outperformed the second-best competitor by 6.3% on the image←text 128-bit task and by 5.4% on the text←image 128-bit task. On the NUS-WIDE dataset, although the performance is not as good as on the MIRFlick-25K dataset, it still achieves excellent results. GAHCL obtains improvements of 2.0% and 0.3% on the image←text and text←image 128-bit task, respectively.

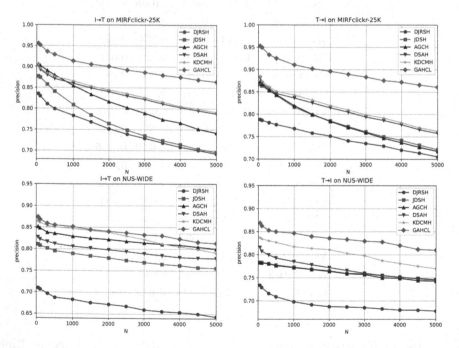

Fig. 3. The mAP curves of several methods with different number of top returned points on MIRFlickr-25K and NUS-WIDE. The code length is 128.

Table 3. The ablation comparison among different variants of GAHCL on MIRFlickr-25K. The best performances for each category are shown in boldface.

Task	Method	16bits	32bit	64bits	128bits
I → T	GAHCL	**0.734**	**0.745**	**0.758**	**0.774**
	GAHCL-1	0.690	0.696	0.705	0.710
	GAHCL-2	0.710	0.719	0.729	0.745
	GAHCL-3	0.689	0.703	0.716	0.723
	GAHCL-4	0.680	0.697	0.708	0.717
T → I	GAHCL	**0.730**	**0.744**	**0.756**	**0.773**
	GAHCL-1	0.691	0.699	0.708	0.713
	GAHCL-2	0.713	0.718	0.729	0.745
	GAHCL-3	0.686	0.703	0.714	0.720
	GAHCL-4	0.682	0.701	0.709	0.723

The mAP curves are plotted Fig. 3. It is observed that our method overpasses other methods in all cases. Furthermore, it is worth noting that as the value of N decreases, the plotted curve exhibits a discernible upward inclination. This phenomenon suggests that a reduction in the retrieval quantity within the retrieval task yields improved outcomes. These improvements demonstrate the ability of GAHCL for UCMH.

3.5 Ablation Studies

The slight modifications are made to the original model from different angles to confirm the importance of each component.

First and foremost, the importance of the contrastive Adjacency Matrix in Fig. 2 should be examined. So in GAHCL-1 we replaced A_{ij} with S_{ij}. GAHCL-2 means to remove the reconstruction function part $\mathcal{L}_r$ in the original loss function, leaving only the contrastive loss function $\mathcal{L}_c$. In GAHCL-3, no data enhancement method is used to create positive samples. Finally, in GAHCL-4 we remove the GAT modules in our framework.

From the result in Table 3, it has been shown that each module plays an important role. Specifically, the construction of contrastive adjacency matrix plays a crucial role in the impact on performance. When using S_{ij} instead of A_{ij}, the performance is even worse than not using GAT (GAHCL-4) due to the excessive number of computing nodes and the features of the same samples are not aggregated by intervention. The contrastive loss function strengthens the deficiencies of the original reconstruction function, and the performance is improved.

4 Conclusion and Future Work

In this paper, we propose the Graph Attention Hashing via Contrastive Learning (GAHCL) approach for UCMH. GAHCL is an end-to-end deep hashing learning framework, based on the Graph Convolutional Networks. We use the data augmentation method on the original samples to obtain corresponding positive samples, and propose a novel contrastive adjacency matrix to better aggregate the information of the positive samples. Finally, the parameters is trained by the contrastive loss function and the reconstruction function. Extensive experimental results show that GAHCL has excellent performance in both datasets.

In the future, other forms of contrastive learning methods in hashing cross-modal retrieval will be further explored. In addition, we will further investigate better data enhancement schemes and contrast loss functions to bridge the gap between modalities.

References

1. Feng, F., Wang, X., Li, R.: Cross-modal retrieval with correspondence autoencoder. In: Proceedings of the 22nd ACM International Conference on Multimedia, pp. 7–16 (2014)
2. Hamilton, W., Ying, Z., Leskovec, J.: Inductive representation learning on large graphs. Adv. Neural Inf. Process. Syst. **30** (2017)
3. Hong, D., Gao, L., Yao, J., Zhang, B., Plaza, A., Chanussot, J.: Graph convolutional networks for hyperspectral image classification. IEEE Trans. Geosci. Remote Sens. **59**(7), 5966–5978 (2021). https://doi.org/10.1109/TGRS.2020.3015157
4. Hu, H., Xie, L., Hong, R., Tian, Q.: Creating something from nothing: unsupervised knowledge distillation for cross-modal hashing. In: Proceedings of the IEEE/CVF Conference on Computer Vision and Pattern Recognition, pp. 3123–3132 (2020)
5. Kipf, T.N., Welling, M.: Semi-supervised classification with graph convolutional networks. arXiv preprint arXiv:1609.02907 (2016)
6. Li, M., Wang, H.: Unsupervised deep cross-modal hashing by knowledge distillation for large-scale cross-modal retrieval. In: Proceedings of the 2021 International Conference on Multimedia Retrieval, pp. 183–191 (2021)
7. Liu, H., Ji, R., Wu, Y., Huang, F., Zhang, B.: Cross-modality binary code learning via fusion similarity hashing. In: Proceedings of the IEEE Conference on Computer Vision and Pattern Recognition, pp. 7380–7388 (2017)
8. Liu, S., Qian, S., Guan, Y., Zhan, J., Ying, L.: Joint-modal distribution-based similarity hashing for large-scale unsupervised deep cross-modal retrieval. In: Proceedings of the 43rd International ACM SIGIR conference on research and development in Information Retrieval, pp. 1379–1388 (2020)
9. Oord, A.V.D., Li, Y., Vinyals, O.: Representation learning with contrastive predictive coding. arXiv preprint arXiv:1807.03748 (2018)
10. Radford, A., et al.: Learning transferable visual models from natural language supervision. In: International Conference on Machine Learning, pp. 8748–8763. PMLR (2021)
11. Shen, F., Shen, C., Shi, Q., Van Den Hengel, A., Tang, Z.: Inductive hashing on manifolds. In: Proceedings of the IEEE Conference on Computer Vision and Pattern Recognition, pp. 1562–1569 (2013)

12. Shi, G., Li, F., Wu, L., Chen, Y.: Object-level visual-text correlation graph hashing for unsupervised cross-modal retrieval. Sensors **22**(8), 2921 (2022)
13. Su, S., Zhong, Z., Zhang, C.: Deep joint-semantics reconstructing hashing for large-scale unsupervised cross-modal retrieval. In: Proceedings of the IEEE/CVF International Conference on Computer Vision, pp. 3027–3035 (2019)
14. Velickovic, P., Cucurull, G., Casanova, A., Romero, A., Lio, P., Bengio, Y.: Graph attention networks. STAT **1050**, 20 (2017)
15. Yang, D., Wu, D., Zhang, W., Zhang, H., Li, B., Wang, W.: Deep semantic-alignment hashing for unsupervised cross-modal retrieval. In: Proceedings of the 2020 International Conference on Multimedia Retrieval, pp. 44–52 (2020)
16. Yu, J., Zhou, H., Zhan, Y., Tao, D.: Deep graph-neighbor coherence preserving network for unsupervised cross-modal hashing. In: Proceedings of the AAAI Conference on Artificial Intelligence, vol. 35, pp. 4626–4634 (2021)
17. Zhang, J., Peng, Y., Yuan, M.: Unsupervised generative adversarial cross-modal hashing. In: Proceedings of the AAAI Conference on Artificial Intelligence, vol. 32 (2018)
18. Zhang, P.F., Li, Y., Huang, Z., Xu, X.S.: Aggregation-based graph convolutional hashing for unsupervised cross-modal retrieval. IEEE Trans. Multimedia **24**, 466–479 (2021)
19. Zhao, Y., Yu, J., Liao, S., Zhang, Z., Zhang, H.: From sparse to dense: semantic graph evolutionary hashing for unsupervised cross-modal retrieval. In: Wang, L., Gall, J., Chin, T., Sato, I., Chellappa, R. (eds.) Computer Vision - ACCV 2022–16th Asian Conference on Computer Vision, Macao, China, 4–8 December 2022, Proceedings, Part IV. Lecture Notes in Computer Science, vol. 13844, pp. 521–536. Springer, Heidelbe (2022). https://doi.org/10.1007/978-3-031-26316-3_31

A Two-Stage Active Learning Algorithm for NLP Based on Feature Mixing

Jielin Zeng[1,2], Jiaqi Liang[1(✉)], Xiaoxuan Wang[1], Linjing Li[1,2],
and Daniel Zeng[1,2]

[1] State Key Laboratory of Multimodal Artificial Intelligence Systems,
Institute of Automation, Chinese Academy of Sciences, Beijing, China
{zengjielin2020,liangjiaqi2014,xiaoxuan.wang,
linjing.li,dajun.zeng}@ia.ac.cn
[2] School of Artificial Intelligence, University of Chinese Academy of Sciences,
Beijing, China

Abstract. Active learning (AL) aims to improve the model performance with minimal data annotation. While recent AL studies have utilized feature mixing to identify unlabeled instances with novel features, applying it to natural language processing (NLP) tasks has been challenged due to the discrete nature of text tokens and the limited contribution of some novel features. To address these issues, we propose a two-stage acquisition method based on feature mixing for NLP tasks. We first create a mixed feature for both labeled and unlabeled instances to identify the features in the unlabeled instances that the model cannot recognize. Next, we evaluate the contribution of these novel features to the model using the entropy of the nearest labeled neighbors. The proposed method enables the model to select the most informative samples in the unlabeled sample pool. Experiments on sentiment analysis, topic classification, and natural language inference validated that our method not only outperforms other AL approaches but improves the efficiency of batch data acquisition.

Keywords: active learning · feature mixing · text classification

1 Introduction

Active learning (AL) [2] is a machine learning paradigm that enables efficient training of models with limited labeled data. In natural language processing (NLP), AL algorithms intelligently select informative instances from an unlabeled dataset for labeling by a human expert. These labeled instances are used to train a model, followed by selecting additional informative instances, then continuing the cycle of AL until a desired level of performance is achieved. AL has been successfully applied in various NLP tasks, such as named entity recognition [24], machine translation [11], and text classification [14], among others. Its effectiveness in reducing the labeling cost while maintaining high accuracy has made it a popular research subject in NLP.

© The Author(s), under exclusive license to Springer Nature Singapore Pte Ltd. 2024
B. Luo et al. (Eds.): ICONIP 2023, CCIS 1968, pp. 510–521, 2024.
https://doi.org/10.1007/978-981-99-8181-6_39

Data acquisition is responsible for selecting the most informative unlabeled data points to be labeled, which is a critical function in AL. The goal is to choose data points likely to provide the maximum information to improve the model's performance. Several different acquisition methods are available, and they can be broadly classified into three categories: uncertainty-based, diversity-based, and hybrid sampling [12]. Examples of uncertainty-based acquisition methods include entropy-based sampling, margin sampling, and variance reduction. Examples of diversity-based acquisition methods include cluster-based sampling, representative sampling, and uncertainty sampling based on ensemble models. Choosing an appropriate acquisition method is essential as it may significantly affect the model's performance and the labeling process's efficiency.

With the development of deep learning (DL) technology, the combination of DL and AL has excellent potential. However, the relatively high computational cost and large label data requirements of the two batch AL method based on feature mixture [10] has been proposed. Unlike traditional acquisition functions, this method aims to identify unlabeled instances with novel features by interpolating their features with those of previously labeled instances. In this way, it is able to identify the instances with sufficiently distinct features and improve the model's overall performance. However, we found that although the method can identify samples containing new features, it does not consider the problem that some of these features have little contribution to the model updating. As a result, the method performed poorly on many benchmark datasets.

The main challenge is how to estimate the samples' contribution to the model updating with new features. In addition, the acquisition function selects a fixed quantity of samples in each iteration, but the acquisition function based on feature mixing always selects more than it. To address these issues, we propose a two-stage acquisition function. Inspired by Margatina et al. [9], in the second stage, we employ the K-nearest neighbor search technique to iteratively estimate how the features contained in each candidate sample in the pool contribute to the model classification.

In this paper, we propose a two-stage AL algorithm based on feature mixing, and an acquisition function for deep learning models that constructs constraints on newly discovered features to address these challenges. Our method consisted of two stages: feature recalling and feature ranking. In the feature recalling stage, the features of the center point of each labeled data type are interpolated with those of the unlabeled candidate instance to identify the innovative features. In the feature ranking stage, the entropy of the nearest labeled neighbors of the novel feature is used to determine its importance. We demonstrate the effectiveness of our method in our annotation system for several NLP tasks. In addition, we use two-stage acquisition function to improve batch data acquisition efficiency in realistic annotation tasks.

The proposed method is evaluated on SST-2, AG_NEWS, and QNLI, and the experiments illustrate a significant improvement over the baseline. The main contributions of this paper can be summarized as follows:

- We propose a novel two-stage AL method based on feature mixing that selects the most informative unlabeled data points to be labeled.
- This paper introduces a feature ranking stage and integrates it into the acquisition function, which greatly facilitates the task.
- The proposed method improves significantly the performance of text classification and natural language inference.

2 Related Work

AL strategies can be broadly classified into three types: uncertainty-based, diversity-based, and hybrid sampling, based on the nature of their acquisition function [12].

Uncertainty-based AL strategies aim to select the most informative samples that the current model is uncertain about [8,19]. These sampling methods estimate the model's uncertainty on each unlabelled sample and select the ones with the highest uncertainty for labeling [8]. Common measures of uncertainty include entropy-based and margin-based [17]. Specifically, entropy-based sampling uses the entropy of the predictive distribution or the variance between predicted probabilities within the ensemble as a measure of uncertainty [21], and margin-based sampling uses the difference between the top two predicted class probabilities as a measure of uncertainty [13]. Although these methods prioritize points close to the decision boundary, they rely solely on predicted class likelihoods and overlook the inherent value of the feature representation.

Diversity-based AL strategies aim to select the most representative samples from the unlabelled dataset to cover the underlying variability of the dataset. These methods prioritize selecting samples that are dissimilar to those already selected for labeling [20]. One common approach is to cluster the unlabelled samples based on their feature representations and then select the most representative samples from each cluster [7]. Another approach is to construct a core-set of the unlabelled dataset based on their latent features, and then select the most diverse samples from the core-set [15]. The two approaches are orthogonal to each other, as uncertainty sampling typically relies on the model's output, and diversity utilizes information from the feature space.

Hybrid sampling in AL strategies aims to balance the exploration-exploitation trade-off by combining both uncertainty-based and diversity-based methods to select the most informative samples. By selecting samples that are both diverse and uncertain, these methods seek to maximize the model's learning from the labeled data while minimizing the amount of data required for training [4,6,23]. Hsu et al. [18] makes the machines adaptively "learn" from the performance of a set of given strategies on a particular dataset, Agarwal et al. [1] exploit the predicted probabilities in images to select samples from diverse contexts. Recently, a notable work proposed clustering the gradients of the final output layer of the target model as the features of the unlabeled samples, which implicitly encompass uncertainty information [3]. Additionally, the idea of contrastive learning has been introduced in AL [9], where data points are selected based on similar model encodings while their model's predictions are significantly different. The closest method to which we propose in this paper is a novel method for batch AL (ALFA-Mix) approach [10], identifying unlabeled instances with novel features by seeking inconsistencies in predictions resulting from mixed representation. Despite their state-of-the-art results on some image datasets, their methods are not scalable to some NLP tasks.

In this paper, the standard ALFA-Mix is modified according to the task specificity and introduced in our method, which outperforms their methods by a large margin in different settings.

3 Methodology

This section introduces the task definition of AL at the beginning. Then we provide an overview of the proposed model. Finally, we explain how our two-stage acquisition function selects samples.

3.1 Task Definition

Assuming a supervised multi-class classification problem with C classes, a model undergoes active training through iterations of interactions with Oracle annotators. At each iteration, we train a model on a set of labeled data $D_l = \{(x_i, y_i)\}_{i=0}^m$ where x_i, y_i and m represent the input text, associated class label and the number of labeled data sets, respectively. Then we use this model and the proposed acquisition function to acquire a batch $\{x_j\}_{j=0}^b$ from a pool of unlabeled data D_u. Examples in this batch are then labeled by an oracle and removed from the pool D_u. Meanwhile, they are added to the labeled set D_l and will be used for the training model in the next iteration. The model's performance is evaluated on the test dataset in each iteration.

Our experiment uses a pre-trained BERT model as our model f, which we fine-tune at each iteration using the current D_l. Φ is the BERT model's encoder which encodes input text to a 512-dimensional representation in latent space, $i.e. z = \Phi(x)$. In our method, we use the encoder Φ to acquire the representation set Z_l for all the labeled data and $Z_u = \{z = \Phi(x), \forall x \in D_u\}$ for all examples in the unlabeled data pool.

3.2 System Overview

Figure 1 illustrates the functional architecture of the annotation system based on our method. The database component stores two pools of data: labeled data, denoted as D_l, and unlabeled data, denoted as D_u. In each iteration, the AL method selects a batch of instances from D_u to be labeled by the oracle. The newly labeled instances are appended to the pool of labeled data, which is then used to retrain and update the deep learning model. This process is repeated until the model's accuracy meets the expected criteria. The AL acquisition strategy consists of two stages: feature recalling and feature ranking. In the next section, we will delve into the details.

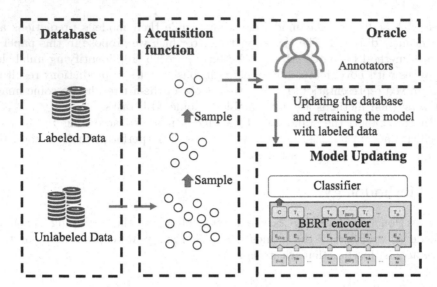

Fig. 1. System Overview. The annotation system based on the proposed method includes four main modules: Database, Acquisition function, Oracle, and Model Updating. In each iteration, the system uses Acquisition function to select samples from Database, which are then given to Oracle to annotate. Then we update the Database, retrain and evaluate the model.

3.3 Two-Stage Acquisition Function

Different from traditional acquisition functions, our method has two acquisition stages: Feature Recalling and Feature Ranking (Fig. 2).

Feature Recalling. The feature recalling stage focuses on selecting unlabeled instances from D_u that contain the novel features. Inspired by the work of Parvaneh *et al.* [10], the representation features of each unlabeled instance are combined with the semantic representation z^* of each labeled data type to see if it has any new hidden features. Specifically, we compute the semantic representation of each class as:

$$z_c^* = \sum_{i=1}^{n_c} z_i | z_i \in Z_l \tag{1}$$

where n_c represents the number of samples in class C. The representations are obtained from a deep neural network $\Phi(\cdot)$. In this paper, we use the Transformer encoder $\Phi(\cdot)$ to extract the feature representation z^* of labeled input instance x^*. Similarly, we can extract the feature representation z_u of the unlabeled candidate instance from D_u. Figure 2 shows how we combine z^u and z^* in a linear way. Through the model of current iteration, the interpolated features $\tilde{z} = \alpha z^* + (1-\alpha)z_u$ are passed to get the prediction labels. It is demonstrated by [10] that if the predicted labels do not match up, the unlabeled instance may have new features. We put such unlabeled instances into a candidate sample set D_{can}. In the following feature ranking stage, we will select the instances from D_{can} for the oracle to annotate.

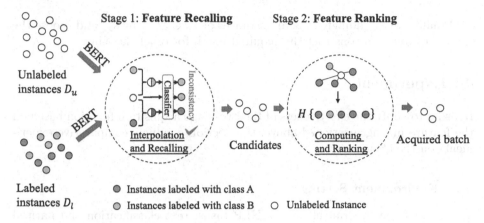

Fig. 2. Acquisition Function. The proposed two-stage acquisition function has two components: Feature Recalling and Feature Ranking. Feature Recalling stage aims to acquire candidates with novel features, while Feature Ranking stage aims to acquire the batch whose novel features contribute more to model updating.

Feature Ranking. After the feature recalling stage, each unlabeled instance in the candidate sample set D_{can} contains novel features. In the feature ranking stage, we rank these novel features and select the critical ones. The proposed feature ranking method is based on the observation that instances containing a critical feature will have the same label. Thus, we select the features with the highest consistency in label prediction across instances.

Based on this, we evaluate the importance of x_c for the model using the entropy of x_c's nearest K neighbors from the labeled instances set D_l. The lower entropy indicates that the candidate's novel feature may be important to the classifier. Specifically, for each $x_c \in D_{can}$, we first use the KNN algorithm to find the top K nearest neighbor instances x_i from D_l. The selected top K nearest neighbor instances are represented as a neighborhood set N_{x_c}, defined as below:

$$N_{x_c} = \{(x_1^l, y_1^l), \dots, (x_k^l, y_k^l) | (x_i^l, y_i^l) \in D_l, d(z_i, z_j^l) < \delta\} \qquad (2)$$

where $d(.)$ is the Euclidean Distance and δ is a pre-determined small value. After obtaining the neighborhood set of the sample x_c, collect the labels of these k labeled samples, and calculate the probability distribution of the labels for the k samples. For a certain category i:

$$p(y_i|x) = \frac{\sum_{j=1}^{k} I(y_j = y_i)}{k} \qquad (3)$$

Then we calculate the entropy value of this probability distribution:

$$H(x_c) = -\sum_{i=1}^{c} p(y_i|x)\mathrm{log}p(y_i|x) \tag{4}$$

Finally, we ascendingly rank all instances of $x_c \in D_{can}^u$ by s_{x_c} and choose the top N instances to construct the acquired batch for oracle labeling.

4　Experiment

In order to verify the effectiveness of the two-stage acquisition function based on the feature mixing mentioned above, this section conducts comparative experiments on the algorithm.

4.1　Environment Setting

We conduct the experiment on two NLP tasks: text classification and natural language inference. Both tasks use models that sample the BERT encoder, a base-size 12-layer Transformer with a hidden size of 768 and additional fully connected layers. The algorithm's effectiveness is verified under the annotation system described in this paper. Both tasks only involve fine-tuning the model on downstream tasks, with a training learning rate of $2e^{-5}$ and a batch size of 256. The batch size used for inference during sampling is also 256. To reduce the impact of randomness, we also repeat all experiments with five different random seeds, and the average of the results from 5 experiments is taken as the presented data. The evaluation metrics used in the experiment include precision, recall, and F1 score. We budget 11% of the training set, 1% of the initial training set, and 2% of the acquisition size for all tasks. Each experiment is run on a single NVIDIA Tesla A100 GPU.

4.2　Dataset and Baselines

We conduct our validation experiments on three different datasets, including text sentiment analysis dataset SST-2 [16] , news topic classification dataset AG_NEWS [25] , and natural language inference dataset QNLI [16] which requires pairs of input sequences. These tasks involve downstream tasks in sentiment classification, topic classification, and natural language inference, details for the datasets are summaried in Table 1.

We compare our method with the following AL baselines:

- **Random**: A simple baseline which selects a random set of a fixed number of samples from the pool of unlabeled data at each round.
- **Alpha-Mix**: A feature mixing-based batch AL method which aims to identify unlabeled instances with novel features by interpolating their features with those of previously labeled instances.

Table 1. Details of the dataset used in the experiment

Dataset	Task	Domain	Train	Val	Test
SST-2	Sentiment classification	Movie reviews	67250	873	1821
AG_NEWS	Topic classification	News	114000	6000	7600
QNLI	Natural language inference	Wikipedia	104743	5463	5461

- **BADGE** [3]: An acquisition function combines diversity and uncertainty sampling by computing gradient embeddings for each candidate point in the unlabeled pool.
- **CAL** [9]: An acquisition function which introduces the concept of contrastive learning into the sampling process, and defines contrastive samples as pairs of samples in the feature space that are very close to each other but have significantly different predicted results by the model.

4.3 Results

We present results for test accuracy across three datasets and acquisition functions in Fig. 3. Our method is consistently top-performing, especially in three datasets, indicating that it could operate with fewer annotations.

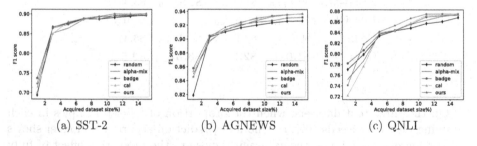

|(a) SST-2|(b) AGNEWS|(c) QNLI|

Fig. 3. Test accuracy during AL iterations for different acquisition functions.

Our method outperforms CAL by a small margin on AGNEWS, and significantly outperforms all baselines on QNLI. CAL ranks second in general, consistently outperforming hybrid baselines (Alpha-Mix, BADGE). This result could be attributed to the fact that BERT can produce accurate uncertainty estimates, making the method which focus on uncertainty a challenging baseline to surpass in AL [5,22]. Our results indict that despite utilizing uncertainty and diversity measures, Alpha-Mix and BADGE performs poorly, suggesting that there may be more effective strategies for data acquisition than clustering the constructed gradient embeddings.

Most methods achieve similar results on SST-2. The informal style of the reviews in SST-2, which consists of noisy social media data, makes the pretrained BERT model a reliable signal for selecting linguistically challenging examples. This selection process may improve the performance of downstream sentiment analysis tasks, ultimately making the differences between various methods insignificant.

In addition, we test our method and baselines on a real business dataset. We collected 5608 speech transcription text from China Mobile, classified into three categories online car appointment scenario, E-commerce scenario, and other scenarios. All texts are transcribed from actual call recordings. In summary, we applied various AL methods to the task of scene classification for transcription text, and the proposed approach outperformed the baselines as shown in Table 2. Compared to random sampling, our method improved of over 7% in average test accuracy in the early AL rounds and 1% in the later rounds. Additionally, our method outperformed all other baselines, demonstrating its effectiveness in reducing labeling costs, which can be a significant obstacle in real-world business tasks.

Table 2. Test accuracy of various AL methods on the actual business dataset. Values on top of each column represent the size of the labeled set.

Method	Iteration			
	500	1000	1500	2000
Random	$75.3_{\pm1.3}$	$79.1_{\pm1.1}$	$81.3_{\pm0.7}$	$82.7_{\pm0.5}$
ALFA-Mix	$80.1_{\pm1.0}$	$81.9_{\pm0.6}$	$82.8_{\pm0.9}$	$83.3_{\pm0.5}$
BADGE	$77.7_{\pm1.4}$	$80.3_{\pm1.0}$	$81.9_{\pm0.8}$	$83.0_{\pm0.6}$
CAL	$80.6_{\pm1.5}$	$82.1_{\pm0.8}$	$82.9_{\pm0.6}$	$83.4_{\pm0.4}$
Ours	$\mathbf{81.0_{\pm1.2}}$	$\mathbf{82.6_{\pm0.8}}$	$\mathbf{83.3_{\pm0.4}}$	$\mathbf{83.5_{\pm0.5}}$

On the three text datasets, when the proportion of labeled samples in each sampling method exceeds 10%, the model's prediction accuracy no longer shows a rapid increase trend, and the accuracy is close to the prediction effect of fully supervised training. In order to further analyze the effect of the AL algorithm proposed in this paper when the proportion of labeled samples reaches a certain threshold, the experiment takes the AG_NEWS dataset as an example, and compares the effect of various sampling methods when the number of labeled samples accounts for 11% of the training set. The results are shown in Table 3.

As seen from Table 3, when the AG_NEWS dataset is labeled with 11%, the dynamic learning algorithm proposed in this paper performs better than other methods in three indicators of model classification precision, recall and F1.

4.4 Ablation Study

In order to verify the effectiveness of the second-stage sampling method based on the K-nearest neighbor search, we also conducte validation experiments on the SST-2, AG_NEWS, and QNLI benchmark text datasets.

Table 3. Experimental results of each algorithm when the proportion of labeled samples is 11%.

Category	Random			BADGE		
	Precision	Recall	F1	Precision	Recall	F1
Sports	93.6	93.7	93.6	93.7	93.8	93.8
World	94.3	85.5	89.9	94.7	86.1	90.3
Business	93.4	93.4	93.4	93.6	93.8	93.7
Sci/Tech	92	92.7	92.3	92.3	92.9	92.6
category	Random			BADGE		
	Precision	Recall	F1	Precision	Recall	F1
Sports	93.9	94.1	94	94.5	94.6	94.6
World	95	86.4	90.7	95.2	87.2	91.2
Business	94.1	94.2	94.1	94.3	94.4	94.4
Sci/Tech	93.1	93.5	93.3	93.3	93.6	93.4

In the ablation experiments, several comparative methods were selected from the same candidate pool for the second-stage sampling while keeping the first-stage sampling process the same. (1) Random: selects a random set of a fixed number of samples from the pool of unlabeled data at each round (2) K-means: perform fixed number clustering centers in the feature space on the samples in the candidate pool, and select the centers as unlabeled samples. The change curve of model prediction accuracy with the change of annotation size on the three datasets is shown in Fig. 4.

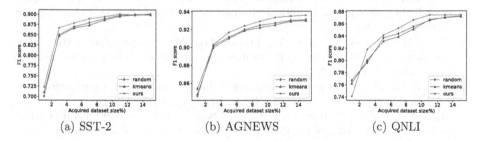

(a) SST-2   (b) AGNEWS   (c) QNLI

Fig. 4. Test accuracy during AL iterations for different second-stage acquisition functions with the same first-stage selection.

These curves illustrate that the proposed method performs best, while the K-means method performs the worst, even worse than the random sampling results in the second stage. The comparison of the results among random sampling experiments indicts the necessity of the second-stage sampling process. The comparison between the two methods verifies the rationality and effectiveness of the second-stage sampling method designed in this paper.

5   Conclusion

This paper proposed a novel active learning acquisition strategy that involves two stages: feature recalling and feature ranking. The second ranking stage is proposed according to the limitations of the first stage, which improves the model performance a lot. The experiments validated that the proposed method outperforms other active learning methods on NLP benchmark datasets and increases data labeling efficiency. In our future work, the two areas of concern will be fused according to the features extracted from the token of words and we will apply the model to more complex environments.

Acknowledgement. This work was supported in part by the Strategic Priority Research Program of Chinese Academy of Sciences under Grant XDA27030100 and the National Natural Science Foundation of China under Grants 72293573 and 72293575.

References

1. Agarwal, S., Arora, H., Anand, S., Arora, C.: Contextual diversity for active learning. In: Proceedings of the European Conference on Computer Vision, pp. 137–153 (2020)
2. Angluin, D.: Queries and concept learning. Mach. Learn. **2**, 319–342 (1988)
3. Ash, J.T., Zhang, C., Krishnamurthy, A., Langford, J., Agarwal, A.: Deep batch active learning by diverse, uncertain gradient lower bounds. arXiv preprint arXiv:1906.03671 (2019)
4. Citovsky, G., et al.: Batch active learning at scale. In: Proceedings of the International Conference on Neural Information Processing Systems, pp. 11933–11944 (2021)
5. Dor, L.E., et al.: Active learning for bert: an empirical study. In: Proceedings of the 2020 Conference on Empirical Methods in Natural Language Processing, pp. 7949–7962 (2020)
6. Fajri, R., Saxena, A., Pei, Y., Pechenizkiy, M.: Fal-cur: fair active learning using uncertainty and representativeness on fair clustering. arXiv preprint arXiv:2209.12756 (2022)
7. Hendrickson, P.: Effect of active learning techniques on student excitement, interest, and self-efficacy. J. Politic. Sci. Educ. **17**(2), 311–325 (2021)
8. Holub, A., Perona, P., Burl, M.C.: Entropy-based active learning for object recognition. In: Proceedings of the Conference of Computer Vision and Pattern Recognition Workshops, pp. 1–8. IEEE (2008)
9. Margatina, K., Vernikos, G., Barrault, L., Aletras, N.: Active learning by acquiring contrastive examples. In: Proceedings of the Conference on Empirical Methods in Natural Language Processing, pp. 650–663 (2021)
10. Parvaneh, A., Abbasnejad, E., Teney, D., Haffari, G.R., Van Den Hengel, A., Shi, J.Q.: Active learning by feature mixing. In: Proceedings of the Conference on Computer Vision and Pattern Recognition, pp. 12237–12246 (2022)
11. Patil, A., Garera, N.: Large-scale machine translation for Indian languages in e-commerce under low resource constraints. In: Proceedings of the Conference on Empirical Methods in Natural Language Processing: Industry Track, pp. 627–634 (2022)

12. Ren, P., et al.: A survey of deep active learning. ACM Comput. Surv. **54**(9), 1–40 (2021)
13. Roth, D., Small, K.: Margin-based active learning for structured output spaces. In: Fürnkranz, J., Scheffer, T., Spiliopoulou, M. (eds.) ECML 2006. LNCS (LNAI), vol. 4212, pp. 413–424. Springer, Heidelberg (2006). https://doi.org/10.1007/11871842_40
14. Schröder, C., Niekler, A.: A survey of active learning for text classification using deep neural networks. arXiv preprint arXiv:2008.07267 (2020)
15. Sener, O., Savarese, S.: Active learning for convolutional neural networks: a core-set approach. arXiv preprint arXiv:1708.00489 (2017)
16. Wang, A., Singh, A., Michael, J., Hill, F., Levy, O., Bowman, S.R.: Glue: A multi-task benchmark and analysis platform for natural language understanding. arXiv preprint arXiv:1804.07461 (2018)
17. Wang, D., Shang, Y.: A new active labeling method for deep learning. In: International Joint Conference on Neural Networks, pp. 112–119. IEEE (2014)
18. Xie, B., Yuan, L., Li, S., Liu, C.H., Cheng, X., Wang, G.: Active learning for domain adaptation: an energy-based approach. In: Proceedings of the AAAI Conference on Artificial Intelligence, vol. 36, pp. 8708–8716 (2022)
19. Yadav, C.S., et al.: Multi-class pixel certainty active learning model for classification of land cover classes using hyperspectral imagery. Electronics **11**(17), 2799 (2022)
20. Yang, Y., Ma, Z., Nie, F., Chang, X., Hauptmann, A.G.: Multi-class active learning by uncertainty sampling with diversity maximization. Int. J. Comput. Vision **113**, 113–127 (2015)
21. Yu, W., Zhu, S., Yang, T., Chen, C.: Consistency-based active learning for object detection. In: Proceedings of the Conference on Computer Vision and Pattern Recognition, pp. 3951–3960 (2022)
22. Yuan, M., Lin, H.T., Boyd-Graber, J.: Cold-start active learning through self-supervised language modeling. In: Proceedings of the Conference on Empirical Methods in Natural Language Processing, pp. 7935–7948 (2020)
23. Yuan, T., et al.: Multiple instance active learning for object detection. In: Proceedings of the Conference on Computer Vision and Pattern Recognition, pp. 5330–5339 (2021)
24. Zhan, X., Wang, Q., Huang, K.H., Xiong, H., Dou, D., Chan, A.B.: A comparative survey of deep active learning. arXiv preprint arXiv:2203.13450 (2022)
25. Zhang, X., Zhao, J., LeCun, Y.: Character-level convolutional networks for text classification. In: Cortes, C., Lawrence, N., Lee, D., Sugiyama, M., Garnett, R. (eds.) Advances in Neural Information Processing Systems, vol. 28. Curran Associates, Inc. (2015)

Botnet Detection Method Based on NSA and DRN

Zhanhong Yin, Renchao Qin[✉], Chengzhuo Ye, Fei He, and Lan Zhang

School of Computer Science and Technology, Southwest University of Science and Technology,
Mianyang 621000, Sichuan, China
qinrenchao@sina.com

Abstract. Botnets are one of the most serious cybersecurity threats facing organizations today. Although the analysis and detection of botnets have achieved a lot of research results, it still has problems such as strong concealment and difficult identification. Therefore, we propose a botnet detection method based on NSA and DRN. This method uses our improved NSA to expand the preprocessed and dimensionally reduced malicious traffic data with fewer samples, and then extracts useful features of network traffic from two dimensions through SENet-based DRN combined with BiGRU. Experimental results based on the CICIDS-2017 and UNSW-NB15 datasets show that our proposed method has a high accuracy for botnet detection and improves the detection accuracy of rare malicious traffic 99.99% and 99.96%. In addition, we further demonstrate the good generalization ability and robustness of our method in botnet detection through an ablation study.

Keywords: Botnet Detection · Traffic Analysis · Negative Selection Algorithm · Dilated Residual Network · Bidirectional Gated Recurrent Unit

1 Introduction

As the global cyber threat situation continues to change, and new cyber malicious methods become more complex and changeable, the task of situational awareness and threat protection for cyberspace security is still very urgent. Alongside common malware such as Trojans and worms, botnets remain one of the most widespread, persistent, and well-hidden security threats. According to the "Acronis Cyber Threat Report 2022" [1] released by Acronis, in 2022, the botnet "Pink" infected more than 1.6 million devices. Another botnet, called "MyKings" has been around for at least five years and has been active. As early as 2017, the botnet infected 500,000 computers to illegally mine and steal Monero coins, making illegal profits of up to 2.3 million US dollars in a single month.

A botnet refers to a group of user terminals whose security protection system has been destroyed and whose control rights have been seized by malicious persons. Infected endpoints are usually controlled by one or more malicious actors [2]. Through botnets, malicious actors can carry out a series of criminal activities, including DDoS, phishing, encrypted extortion, malware distribution, illegal information exchange, etc. With the

B. Luo et al. (Eds.): ICONIP 2023, CCIS 1968, pp. 522–534, 2024.
https://doi.org/10.1007/978-981-99-8181-6_40

continuous evolution of malicious means of botnets, traffic characteristics are becoming more and more difficult to detect. Therefore, it is necessary to develop more novel and effective botnet detection techniques to better prevent the impact of botnets on the real cyberspace environment.

Based on the research on the malicious traffic of botnets, we propose a botnet detection method based on NSA and DRN that can improve the detection ability and generalization ability of the detection model, and accelerate the convergence speed of the model to a certain extent, which is of great significance to botnet detection and cyberspace security management.

The rest of the paper is organized as follows. Section 2 discusses the related work in botnet detection methods. The proposed improved upsampling method and proposed botnet detection model are discussion are in Sect. 3. Section 4 outlines the experimental results and analysis for our method. The conclusion and future work are in Sect. 5.

2 Related Work

In recent years, due to the evolution of botnets, their types have become increasingly diverse and their destructive power has continued to increase, posing a huge threat to cyberspace security. Many methods for detecting botnets have emerged in this field. Among them, the mainstream detection method is the anomaly-based detection method, which can be divided into host-based and network-based detection methods.

The host-based botnet detection method mainly judges whether the terminal is infected by analyzing the process and resource usage. For example, Almutairi et al. [3], proposed a behavior-based host anomaly detection method that analyzes the registry, file system, and the duration of active processes. However, this approach is less accurate at identifying emerging and more stealthy botnets.

Detection methods based on network anomalies perceive malicious traffic by analyzing the behavior, load, response time, and connection characteristics of network traffic generated at different stages of the botnet life cycle. With the development of artificial intelligence, machine learning, and deep learning technologies are gradually applied to botnet detection. VM et al. [4], proposed a method capable of ultra-fast network analysis, based a decision tree algorithm, which increases the speed without significantly reducing the F1-score. However, the generalization ability of this method still needs to be improved, and it cannot achieve good results on all botnets. In addition, Yin et al. [5], proposed an intrusion detection system based on MLP, IGRF-RFE feature selection method. The method uses an information gain algorithm and random forest method to rank the importance of numerical features so that important features are effectively selected. However, there is still room for further improvement in the accuracy and F1-score of the method.

Since the maliciousness in real network traffic is complex and changeable, and the rare maliciousness is rare and hidden, the current public botnet datasets generally have the problem of an unbalanced distribution of dataset samples. To solve this problem, many researchers have proposed some solutions. For example, Yang et al. [6], proposed an intrusion detection system based on NSA, which reduces the dimensionality of features by an unsupervised clustering method and conducts density-based gridding on the feature space to generate detectors. This method effectively reduces the training time, but the classification accuracy still needs to be improved. In addition, Zhang et al. [7] proposed a method based on SMOTE oversampling and GMM clustering under-sampling, resampling all data, and using CNN for feature extraction and classification, which achieved high detection accuracy, but not Considering the temporal dependence between samples, the generalization ability is insufficient. Lee et al. [8], proposed a rare class oversampling method based on the GAN model, using the autoencoder model to process the data features to a lower level. However, the method still has room for improvement in terms of recall and F1-score.

Based on the study of botnet traffic characteristics, we propose a botnet detection method based on NSA and DRN. This method can improve the detection ability and generalization ability of the detection model, which is of great significance in botnet detection. Our main contributions are as follows:

1. We propose an improved upsampling method. First, this method performs PCA dimensionality reduction on the data and then uses the improved NSA to expand the number of malicious traffic samples, which not only reduces the number of parameters but also balances the sample distribution of the dataset.
2. We propose a detection model based on SENet-based DRN combined with BiGRU. Compared with Resnet, this model expands the receptive field of information while reducing calculation parameters. And combined with the BiGRU model, the characteristics of network traffic are extracted from two dimensions of time and space, which effectively expands the feature samples and better captures the dependencies between traffic samples. It not only improves the detection accuracy of the model for rare and unknown botnets but also enhances the generalization ability of the model.

3 The Proposed Method

3.1 Improved Upsampling Method

To address the problem of unbalanced sample distribution in the dataset, we use PCA [9] to reduce the dimensionality of the preprocessed sample data, and then expand the malicious traffic samples after dimensionality reduction through the improved NSA [10], to balance the sample distribution of the dataset.

PCA is a dimensionality reduction algorithm. Its core idea is to map the n-dimensional features of the original data to k-dimensional through the decomposition of the covariance matrix. The k-dimensional features are called principal components, which are orthogonal features reconstructed on the basis of the original n-dimensional features. The working principle of PCA is to sequentially find a set of mutually orthogonal coordinate axes from the original space, and the selection of new coordinate axes is closely related to the data itself. After PCA dimensionality reduction, the data set is easier to train, while reducing the number of parameters. The formula for calculating the covariance is as follows:

$$Con(X, Y) = \frac{1}{n-1} \sum_{i=1}^{n} (x_i - \bar{x})(y_i - \bar{y}). \tag{1}$$

where X and Y are samples, n is the feature dimension of the original data, i is the currently calculated feature dimension, $\bar{x}$ and $\bar{y}$ represents the sample mean.

Inspired by the biological immune system, researchers applied its excellent characteristics to solve various practical problems and formed an artificial immune system [11]. The negative selection algorithm in the artificial immune system has been widely used in the field of information security.

In NSA, the generation mechanism of the detector determines whether the generated detector is effective or not. We improve the generation mechanism of traditional detectors. First, calculate the maximum value, minimum value, and data distribution of each class feature in the autologous collection. If the data distribution of a certain type of feature is relatively discrete, the generated data will be randomly selected from the maximum and minimum values of this type of feature; if the data distribution of a certain type of feature is relatively concentrated or redundant, the generated data will be directly obtained from the existing values is randomly selected from the sample value. After PCA dimensionality reduction samples, unimportant samples and redundant data are removed, so we use the Euclidean distance to calculate the affinity between the generated detector and the autologous collection. The generated detector performs Euclidean distance calculation with each sample in the autologous collection, takes the minimum value as the affinity, and compares it with the threshold. The improved NSA flow chart is shown in Fig. 1.

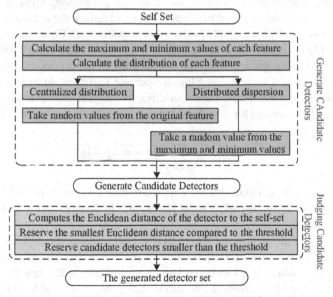

Fig. 1. Improved upsampling method

3.2 Detection Model

We propose a botnet detection model. It is composed of DRN [12], SENet [13] and BiGRU [14]. Its structure is shown in Fig. 2.

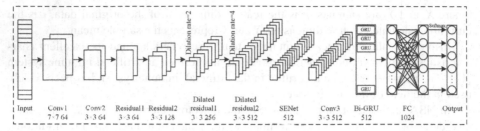

Fig. 2. Model structure

The model first downsamples the input data through two convolutional layers. Next, flow image features are extracted from the spatial dimension using two residual blocks and two dilated residual blocks with dilation rates of 2 and 4, respectively. Then, a convolutional layer with a kernel size of 3 is used instead of the global pooling layer to downsample the sample data to reduce the grid effect of the feature map. Then, SENet is used to model the importance of different feature channels, and the weight of each feature channel is adjusted to improve the classification accuracy. The next step is to use BiGRU to extract traffic statistics from the time series to capture the temporal dependencies between sample data. Finally, the output is passed to Softmax through a fully connected layer for classification.

In the convolutional neural network, the data after the pooling layer downsampling will lose the spatial information of the feature data and reduce the accuracy. The dilated convolution proposed by Yu et al. [15] solves this problem well. Compared with ordinary convolution, dilated convolution introduces a hyper-parameter called "dilation rate", which defines the value spacing when the convolution kernel processes data. Therefore, dilated convolution is a special convolution operator that uses different dilation rates to use the same size convolution kernel in different ranges, effectively expanding the receptive field. The formula for one-dimensional dilated convolution is as follows:

$$(x * dk)[s] = \sum_{p=-q}^{q} x[s - d * p]k[p].$$ (2)

Among them, x is the input, d is the expansion rate, k is the convolution kernel, s is the size of the convolution kernel, p is the lower limit of convolution, q and is the upper limit of convolution.

Compared with the traditional ResNet [16], DRN retains the residual connection mechanism, and at the same time adjusts the expansion rate to achieve feature extraction at different scales, which helps to better handle multi-scale traffic features. It not only reduces the number of parameters and calculations, but also enables the network to obtain a larger receptive field so that it can capture a wider range of network traffic context information, which helps to deal with long-distance dependencies. We adjusted the structure of the residual block and the dilated residual block in DRN. From the original two-layer convolution to one-layer convolution. In this model, the output of the residual block is:

$$y = x + f[f(x, w_1), w_2].$$ (3)

Among them, x represents the original feature of the input, $f(x, w_1)$ represents the output of the first convolutional layer in the residual block, $f[f(x, w_1), w_2]$ represents the output of the second convolutional layer in the residual block, and w_1 and w_2 are the corresponding weight parameters. The output of the dilated residual block is:

$$y = x + D(f(D(f(x, w_1), w_2), w_3), w_4). \tag{4}$$

Compared with the residual block, the dilated residual block replaces the convolutional layer with a dilated convolutional layer. x represents the original feature of the input, $D(f(x, w_1), w_2)$ represents the output of the first dilated convolutional layer in the dilated residual block, $D(f(D(f(x, w_1), w_2), w_3), w_4)$ represents the output of the second dilated convolutional layer, and w_1, w_2, w_3 and w_4 are the corresponding weight parameters.

Compared with RNN [17] and LSTM [18], GRU not only performs well in processing long sequences but also GRU only contains reset gates and update gates, with fewer parameters and higher computational efficiency. BiGRU introduces bidirectionality on the basis of GRU, including two independent GRUs, which can process input data in two sequences at the same time. Therefore, when dealing with dimensionally reduced data, BiGRU can better capture the temporal relationship and bidirectional dependence of sample features. The calculation formulas of the reset gate and the update gate are as follows:

$$R_t = \sigma(W_r X_t + W_r H_{t-1} + b_r), \tag{5}$$

$$Z_t = \sigma(W_z X_t + W_z H_{t-1} + b_z). \tag{6}$$

Among them, σ is the Sigmoid activation function, W_r and W_z are weight parameters, b_r and b_z are deviation parameters. The inputs of both the reset gate and the update gate are the input X_t of the current time step, and the hidden state H_{t-1} of the previous time step.

4 Experimental

4.1 Dataset

At present, commonly used experimental datasets in the field of botnet detection include CTU-13, ISOT, Bot-IoT, NSL-KDD, ISCX2012, CICIDS-2017 [19], and UNSW-NB15 [20]. However, these datasets generally have problems such as unbalanced sample distribution, inability to represent real-world network conditions, and user privacy.

In order to solve the above problems, after evaluating multiple public datasets, we select the CICIDS-2017 dataset and the UNSW-NB15 dataset as benchmark datasets for training and evaluating the effect of the botnet detection model. Among them, the CICIDS-2017 dataset contains 9 different types of traffic, and the UNSW-NB15 dataset contains 10 different types of traffic. By using these datasets, it is possible to better evaluate the performance of the model in real scenarios.

4.2 Data Preprocessing

We preprocess the data through the following three steps:

Traffic Segmentation. For the study of botnet traffic classification, the traffic needs to be segmented into specific traffic units according to a certain granularity. In this paper, we adopt the commonly used flow granularity, i.e., packets with the same five elements (source IP address, source port, destination IP address, destination port, protocol) and arranged them in chronological order to form a flow. Considering multiple protocol layer information, we use complete data information for traffic segmentation processing.

Traffic Cleaning. To protect user privacy and prevent sensitive information leakage, traffic needs to be cleaned before using the data. A common way is to randomize MAC addresses and IP addresses. In this paper, in addition to privacy-preserving processing, duplicate data is cleaned to eliminate redundancy.

Numericalization and Normalization. Different features may have different magnitudes and units, so the feature data need to be enumerated and normalized to eliminate the differences between data types and sizes. In this paper, One-Hot coding is used to numericalize the non-numerical features and convert the extracted features into new feature vectors. Then, the obtained feature vectors are normalized using Z-Score normalization so that the data in all dimensions conform to a standard normal distribution with a mean of 0 and a variance of 1, thus eliminating errors caused by different magnitudes.

4.3 Sample Distribution

We selected 136,568 pieces of traffic data from the CICIDS-2017 dataset for research. These data contain 8 different types of malicious traffic samples, among which DDoS traffic samples have the largest number, with a total of 22,215 pieces of traffic data. However, the least number of Infiltration traffic samples is only 36 traffic data. In the UNSW-NB15 data set, we select 175,340 flow data for research. The data set contains 9 different types of malicious samples, among which Generic traffic has the largest number of samples, with 40,000 pieces of traffic data. However, Worms traffic had the smallest number of samples, with only 130 entries. Due to the large gap between the number of samples of different traffic categories, the sample distribution of the dataset is unbalanced.

Table 1. Sample distribution

Dataset	Categories	Before Upsampling	After Upsampling
CICIDS-2017	Bot	1028	13677
	Infiltration	36	12490
UNSW-NB15	Shellcode	1133	8684
	Worms	130	3382

To solve this problem, we use PCA and improved NSA to effectively expand the sample numbers of Bot and Infiltration traffic in the CICIDS-2017 dataset and the sample numbers of Shellcode and Worms traffic in the UNSW-NB15 dataset. The number of samples before and after upsampling is shown in Table 1.

It can be seen that the number of Bot traffic samples in the CICIDS-2017 dataset has expanded from 1028 to 13677, and the number of Infiltration traffic samples has expanded from 36 to 12490. In the UNSW-NB15 dataset, the number of samples of Shellcode traffic is expanded from 1133 to 8684, and the number of samples of Worms traffic is expanded from 130 to 3382. Compared with before, the sample distribution of the data set expanded by using NSA becomes more balanced to a certain extent.

4.4 Experimental Results

In order to verify the effectiveness of our proposed method for botnet detection, on the CICIDS-2017 dataset and UNSW-NB15 dataset with SVM + GAN-BAL [21], KNN + GAN, Decision Tree + GAN, Random Forest + GAN [22], ANN, Resnet-18, CNN-LSTM, and Dilated Convolution for comparative experiments. The experimental results are shown in Table 2.

Table 2. Comparative experiment

Dataset	Methods	Precision (%)	Recall (%)	F1-Score (%)
CICIDS-2017	SVM + GAN-BAL	99.21	99.96	99.58
	KNN + GAN	98.30	94.05	96.13
	DT + GAN	97.21	88.48	92.64
	RF + GAN	97.03	87.23	91.87
	ANN-GAN	92.27	75.60	83.11
	Resnet-18	95.63	95.25	94.92
	CNN-LSTM	99.42	99.02	99.22
	Dilated Convolution	98.93	98.97	98.93
	Proposed	**99.99**	**99.99**	**99.99**
UNSW-NB15	SVM-GAN	83.09	91.03	86.88
	KNN + GAN	81.81	80.69	81.20
	DT + GAN	87.04	86.80	86.24
	RF + GAN	87.84	87.38	86.67
	ANN-GAN	87.12	92.44	89.70
	Resnet-18	93.00	91.00	89.00
	CNN-LSTM	82.69	78.10	80.33
	Dilated Convolution	98.60	99.50	99.00
	Proposed	**99.96**	**99.96**	**99.96**

It can be seen from the table that compared with other methods, the method proposed in this paper performs better on evaluation indicators such as precision, recall, and F1-score, while the false alarm rate and missing alarm rate are lower. It shows that our proposed method has high accuracy in botnet detection, which verifies the effectiveness of our proposed method.

The reason for the significant improvement in the two datasets is that in this method, PCA dimensionality reduction is first performed on the preprocessed data, which helps to remove unimportant and redundant sample data. Then, the improved NSA is used to generate a small number of rare malicious traffic, which not only expands the total number of samples but also balances the distribution of samples among different traffic categories. Next, we propose a combination model based on SENet-based DRN combined with BiGRU, which extracts traffic statistical features in the time dimension and traffic image features in the spatial dimension, enriching the diversity of features. Therefore, after effectively upsampling the sample data, the model can more easily learn key features from the two dimensions of time and space, thereby improving the detection accuracy of the model and reducing the calculation parameters of the model to a certain extent.

4.5 Ablation Study

In order to further verify the impact of NSA, DRN, SENet, and BiGRU on detection accuracy, we conducted an ablation study. Among them, Baseline is DRN to perform feature extraction and classification on the data after PCA dimension reduction.

Table 3. Ablation study result

Dataset	Methods	Precision (%)	Recall (%)	F1-Score (%)
CICIDS-2017	Baseline	99.02	99.03	98.99
	+SENet	99.27	99.24	99.28
	+SENet + BiGUR	99.98	99.92	99.95
	+SENet + BiGUR + NSA	99.99	99.99	99.99
UNSW-NB15	Baseline	98.36	98.95	98.65
	+SENet	99.60	99.69	99.64
	+SENet + BiGUR	99.89	99.89	99.89
	+SENet + BiGUR + NSA	99.96	99.96	99.96

As can be seen from Table 3, we use the DRN and SENet on the UNSW-NB15 dataset, evaluation indicators have significantly improved, and the precision and F1-score have increased by more than 1%. Combining BiGRU, the precision, recall, and F1-score on the CICIDS-2017 dataset have all been significantly improved by more than 0.6%. After using NSA to expand the dataset, it can be seen that the precision, recall, and F1-score on the two datasets are closer to 100%. It is verified that our proposed method has good generalization ability and robustness.

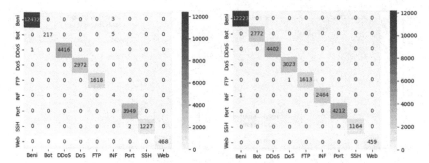

Fig. 3. Confusion matrix on CICIDS-2017

From the confusion matrix in Fig. 3 and Fig. 4, it can be further seen that the classification results of each malicious traffic are in the CICIDS-2017 dataset and UNSW-NB15 dataset. The left picture in Fig. 3 is the classification result of the CICIDS-2017 dataset without using NSA to expand the sample data. It can be seen that the detection accuracy of the model for Bot and Infiltration two kinds of malicious traffic is compared with the detection accuracy of other malicious traffic. There is a high error. Therefore, we use the improved NSA to expand these two types of malicious traffic. From the confusion matrix on the right in Fig. 3, we can see that the false alarm rate and missing alarm rate of Bot and Infiltration malicious traffic detection by the model have been significantly reduced.

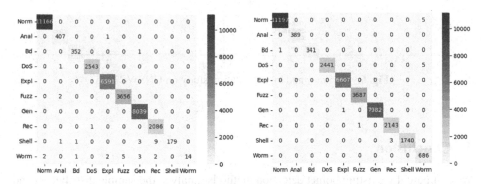

Fig. 4. Confusion matrix on UNSW-NB15

The left picture in Fig. 4 is the classification result of the UNSW-NB15 dataset without using NSA to expand the number of samples. It can be seen that there are many false positives in the malicious traffic of Shellcode and Worms. Among them, 14 pieces of Worms malicious traffic were detected correctly, but 15 pieces were detected as other traffic. The right figure in Fig. 4 is the confusion matrix of the traffic sample detection results after NSA upsampling of Shellcode and Worms malicious traffic. It can be seen that the detection accuracy of Shellcode and Worms malicious traffic has been significantly improved, reducing the rate of missed alarms and false alarms.

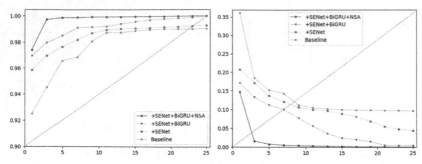

Fig. 5. Accuracy and loss of ablation study on CICIDS-2017

Figures 5 and 6 show the changes in accuracy and loss functions in the training sets of the two datasets, respectively. After combining NSA, SENet, and BiGRU on the baseline, the convergence speed of the model is faster, the accuracy is higher, and it is easier to train.

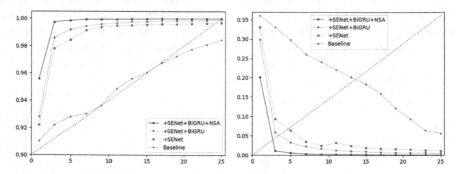

Fig. 6. Accuracy and loss of ablation study on UNSW-NB15

5   Conclusion

We compare the existing botnet detection methods, analyze the sample distribution and traffic characteristics of botnet datasets, and propose a botnet detection method based on NSA and DRN. This method first uses PCA to reduce the dimensionality of all sample data to reduce non-critical and redundant data in the sample. Then, the improved NSA is used to effectively expand the rare malicious traffic with a small number of samples, which balances the distribution of samples among different traffic categories in the dataset and improves the quality of the overall data. Finally, a combination model based on SENet-based DRN combined with BiGRU is used to extract sample features from two dimensions of time and space, which enriches the diversity of extracted traffic features and makes the model easier to train. Experiments were carried out on the CICIDS-2017 dataset and the UNSW-NB15 dataset, and the statistical precision, recall, and F1-score results verified that our proposed method has high precision and robustness in botnet detection. It improves the generalization ability of the model while reducing the false alarm rate, which has certain practical significance and value.

Since the detection model based on deep learning increases the training overhead while improving the detection accuracy, it has high requirements for hardware. In the future, we will study a more lightweight detection model or use a distributed architecture for model training, to improve the detection accuracy without increasing the calculation parameters and training time of the model too much.

Acknowledgement. This research was supported by Sichuan Science and Technology Program (No. 2022YFG0339).

References

1. Alexander, L., Candid, W.: Acronis Cyberthreats Report 2022. Acronis (2022)
2. García, S., Zunino, A., Campo, M.: Survey on network-based botnet detection methods. Secur. Commun. Netw. **7**(5), 878–903 (2014)
3. Almutairi, S., et al.: Hybrid botnet detection based on host and network analysis. J. Comput. Netw. Commun. **2020**, 1–16 (2020)
4. Velasco-Mata, J., et al.: Real-time botnet detection on large network bandwidths using machine learning. Sci. Rep. **13**(1), 4282 (2023)
5. Yin, Y., et al.: IGRF-RFE: a hybrid feature selection method for MLP-based network intrusion detection on UNSW-NB15 dataset. J. Big Data **10**(1), 1–26 (2023)
6. Yang, G., et al.: A modified gray wolf optimizer-based negative selection algorithm for network anomaly detection. Int. J. Intell. Syst. (2023)
7. Zhang, H., et al.: An effective convolutional neural network based on SMOTE and Gaussian mixture model for intrusion detection in imbalanced dataset. Comput. Netw. **177**, 107315 (2020)
8. Liu, X., et al.: A GAN and feature selection-based oversampling technique for intrusion detection. Secur. Commun. Netw. **2021**, 1–15 (2021)
9. Maćkiewicz, A., Ratajczak, W.: Principal components analysis (PCA). Comput. Geosci. **19**(3), 303–342 (1993)
10. Ji, Z., Dasgupta, D.: V-detector: an efficient negative selection algorithm with "probably adequate" detector coverage. Inf. Sci. **179**(10), 1390–1406 (2009)
11. Dasgupta, D.: Advances in artificial immune systems. IEEE Comput. Intell. Mag. **1**(4), 40–49 (2006)
12. Yu, F., Koltun, V., Funkhouser, T.: Dilated residual networks. In: CVPR, pp. 472–480 (2017)
13. Hu, J., Shen, L., Sun, G.: Squeeze-and-excitation networks. In: CVPR, pp. 7132–7141 (2018)
14. Cho, K., et al.: Learning phrase representations using RNN encoder-decoder for statistical machine translation. arXiv preprint arXiv:1406.1078 (2014)
15. Yu, F., Koltun, V.: Multi-scale context aggregation by dilated convolutions. arXiv preprint arXiv:1511.07122 (2015)
16. He, K., et al.: Deep residual learning for image recognition. In: CVPR, pp. 770–778 (2016)
17. Schuster, M., Paliwal, K.K.: Bidirectional recurrent neural networks. IEEE Trans. Signal Process. **45**(11), 2673–2681 (1997)
18. Bengio, Y., Simard, Y., Frasconi, P.: Learning long-term dependencies with gradient descent is difficult. IEEE Trans. Neural Netw. **5**(2), 157–166 (1994)
19. Panigrahi, R., Borah, S.: A detailed analysis of CICIDS2017 dataset for designing intrusion detection systems. Int. J. Eng. Technol. **7**(3.24), 479–482 (2018)

20. Moustafa, N., Slay, J.: UNSW-NB15: a comprehensive data set for network intrusion detection systems (UNSW-NB15 network data set). In: 2015 Military Communications and Information Systems Conference, pp. 1–6. IEEE (2015)
21. Liu, I.H., et al.: Data balanced algorithm based on generative adversarial network. In: 27th International Conference on Artificial Life and Robotics (ICAROB 2022), pp. 645–649. ALife Robotics Corporation Ltd. (2022)
22. Alabrah, A.: A novel study: GAN-based minority class balancing and machine-learning-based network intruder detection using chi-square feature selection. Appl. Sci. 12(22), 11662 (2022)

ASRCD: Adaptive Serial Relation-Based Model for Cognitive Diagnosis

Zhuonan Liang[1], Dongnan Liu[1], Yuqing Yang[2,3], Caiyun Sun[2], Weidong Cai[1], and Peng Fu[4(✉)]

[1] The University of Sydney, Sydney, NSW 2006, Australia
[2] Nanjing University of Aeronautics and Astronautics, Nanjing 210016, China
[3] Nanjing University of Finance and Economics, Nanjing 210046, China
[4] Nanjing University of Science and Technology, Nanjing 210094, China
fupeng@njust.edu.cn

Abstract. Cognitive diagnosis (CD) is a critical task in the education field, aimed at assessing the true concept proficiency of learners. Recent studies have highlighted the significance of concept relations (e.g., concept Addition and concept Multiplication in mathematics) in CD. While advanced research has contributed to concept relation modeling, there remains a gap in automatic building and adaptive integration of relation modeling. To address these challenges, we present an innovative approach called the Adaptive Serial Relation-based model for Cognitive Diagnosis (ASRCD). Our method begins by constructing a Concept Serial Relation Graph (CSRG) to automatically mine concept relations from learner response sequences. Next, a refined graph attention network (GAT) is designed to weight the concept relations for effective aggregation. Finally, we establish a comprehensive CD model that incorporates concept relations, leveraging the extendibility of CSRG, allowing it to be seamlessly integrated into various existing CD methods. To evaluate the performance of our model, we conduct experiments on two real-world datasets from education practice. The experimental results demonstrate that our proposed ASRCD model achieves outstanding accuracy and exhibits excellent extendibility, showcasing its effectiveness and potential in enhancing cognitive diagnosis tasks.

Keywords: Cognitive diagnosis · Graph relation-building · Learner evaluation · Concept serial relation

1 Introduction

With the rapid development of online education, numerous online intelligent education systems have entered the public vision, providing the chance of self-learning to learners [2,18,25]. Cognitive diagnosis provides the foundation for the learner's personalized learning and resource recommendation [3,6]. The goal of cognitive diagnosis is to evaluate learner proficiency by recognizing learner performance in exercise. Specifically, it assumes an association between the exercise and the knowledge concept [4]. The learner's performance on the test is

B. Luo et al. (Eds.): ICONIP 2023, CCIS 1968, pp. 535–551, 2024.
https://doi.org/10.1007/978-981-99-8181-6_41

influenced by the learner's mastery of the knowledge concept [12,13,17,20,28]. To achieve processing massive learning data and automatic diagnosis, previous research has made great contributions including Deterministic Inputs, Noisy And gate (DINA) model [5], Item Response Theory (IRT) [1], Multidimensional IRT (MIRT) [19], and Matrix Factorization (MF) [10]. The above models have achieved considerable efficiency and accuracy. However, they still rely on a lot of labeled intermediate results. To address the above issues, the Neural Cognitive Diagnosis Model (NCD) [26] was proposed. It can effectively analyze cognitive diagnosis adaptively from learners' answers. NCD greatly reduces the participation of artificial intermediate results and improves accuracy and recognition.

There is an objective concept relation graph structure within the knowledge system [9,16,21]. Despite the association of knowledge concepts and exercises, cognitive diagnosis ignoring the concept relation leads to the deviated evaluation of the learners [8]. Advanced research has made great contributions to concept relation modeling [15]. RCD employs a hierarchic map for integrating concept relations [7]. There are still certain problems with the concept relation modeling. One of those is that existing methods suffer from the subjective deviation caused by manually labeled concept relation. The other one is that a learning-based integration method without constraint causes invalid relations. To address these problems, we construct the association mining method to adaptively build the serial relation of knowledge concepts by student behavior recognition.

Serial relation refers to the latent concept association observed within the interaction sequences of learners. In these sequences, related concepts often exhibit statistical synchronism in terms of learner performance. For instance, when exercises involve related concepts, learners tend to show synchronous patterns, such as consistently answering both correctly or incorrectly for these exercises. This observation motivates the construction of serial relations by identifying and recognizing the synchronization patterns within the interaction sequences. Our research makes several key contributions to the field of cognitive diagnosis. Firstly, we introduce a validation method to confirm the concept serial relation, ensuring its accuracy and reliability. This validation method serves as a benchmark, simulating real-world cognitive diagnosis scenarios. Secondly, we enhance the graph attention networks (GAT) by modifying them to effectively weight and integrate the concept relation derived from the Concept Serial Relation Graph (CSRG) [23]. This integration considers the actual significant differences among related concepts, providing a more realistic and robust cognitive diagnosis framework. Lastly, we conduct extensive experiments using real-world datasets to validate the effectiveness of our proposed Adaptive Serial Relation-based model for Cognitive Diagnosis (ASRCD) and the extendibility of CSRG. Our results demonstrate the outstanding performance of ASRCD and the versatility of CSRG, which can be seamlessly incorporated into existing models to enhance their diagnostic capabilities.

2   Related Works

Cognitive diagnosis (CD) models are designed to diagnose subjects' potential abilities through their problem-making history [1,7]. Current research in the education field is primarily based on IRT for its adoptability in massive data process [27]. For education applications, the parameter logistics model (PLM) is widely used [24,29]. It considers the difficulty and discrimination h_disc of the exercise. Additionally, it also concludes the guessing factor in the test. The formula of PLM is defined as:

$$P_{(\theta)} = h_{guess} + \frac{1 - h_{guess}}{1 + \exp{-h_{disc}(\theta - h_{diff})}},$$
(1)

where θ donates the subject factor, or learner factor in educational practice, while we do not introduce the guessing factor in subjective exercises and is set to zero. The educational implementation of CD is to model the learner's knowledge by accurately determining the results of the answer, i.e., their result is the probability that the learner will answer correctly on each question [26]. Thus, we will proceed the $P_{(\theta)}$ after the PLM inspired by NCD. The end of procedure is extracting the learner possibility of correct response.

3   ASRCD Model

In this section, we firstly make a formal introduction to the serial relation cognitive diagnosis task. Then, we formally propose the ASRCD model and provide a general description. In the following description, we present the details of every module of the ASRCD model.

3.1   Task Overview

First, we assume that the relationships of knowledge concepts can be constructed as a concept-aware graph $\mathcal{G} = \mathcal{V} \times \mathcal{E}$. $\mathcal{V}$ presents the nodes of the concept, which are one-to-one relationships. $\mathcal{E}$ presents the edges of nodes which carry out the latent relation between concepts. Specifically, $\mathcal{E}$ can be represented as $\mathcal{E} \subseteq \mathcal{V} \times \mathcal{V}$. Furthermore, the N_i is the neighbor of node i, which presents the relevant concepts of concept i. Then, suppose that a learning system includes N learners, M exercises, and K knowledge concepts which can be represented as: $N = \{u_1, u_2, \cdots, u_n\}$, $M = \{e_1, e_2, \cdots, e_m\}$, and $K = \{s_1, s_2, \cdots, s_k\}$, respectively. In the system, log R of learners which responded to exercises are donated as a set of quartets (u_i, e_j, s_k, r_l) where $u_i \in N, e_j \in M, s_k \in K$ and r is the score of learner s_i in the exercise e_j. As described above, we have the latent graph $\mathcal{G}$. We then extend the concept-exercise relation and concept-learner relation through the aggregation of neighbor concepts which are included in the graph $\mathcal{G}$.

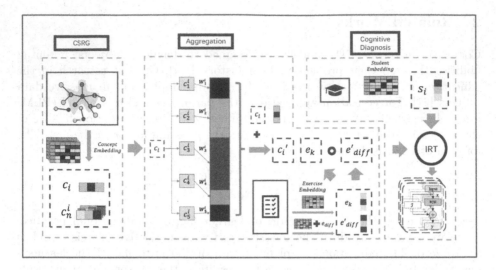

Fig. 1. ASRCD structure. CSRG: Build concept relation. Aggregation: Integrate the related concept into exercise factor. Cognitive Diagnosis: Predict the student performance to diagnose proficiency.

3.2   Model Description

Our ASRCD model can model the concept relation by recognizing the learner response sequences. As shown in Fig. 1, the ASRCD model contains three modules: the CSRG module, the relation aggregation module, and the general cognitive diagnosis module. The CSRG module learns the serial relation of concepts through the exercise responses of learners. In the latent graph module, we implement a concept-aware serial graph structure. Then, the aggregation module associates the learner factor with the relative concepts and extracts a concept-embedded learner factor which will be represented as the updated learner factor in the following cognitive diagnosis process, while the exercise factor is extracted via the aggregation module as the learner factor. We clarify the details in the following description. In the cognitive diagnosis module, the modified learner factor and the exercise factor are sent to the interactive function. Moreover, the ASRCD is state-of-the-art since the concept-aware graph neural network was applied which exposed the inner relationships among concepts and expanded the receptive domain of the diagnosis rationally.

It is the module that analyzes and exposes the relationships among the concepts. To distinguish the different relations, we define the concept relations labeled by experts as the explicit synergy. The implicit synergy is defined as the latent relations exposed via post analysis. The module implements the CSRG structure as shown in Fig. 2. The graph structure can expose the neighbors of the concept nodes. In the following description, we propose a statistics-based latent graph implementation.

Table 1. The frequency matrix of concept response.

	$r_{(i)}^1$	$r_{(i)}^0$	Sum
$r_{(j)}^1$	$\left|r_{(i)}^1 \cap r_{(j)}^1\right|$	$\left|r_{(i)}^1 \cap r_{(j)}^0\right|$	$\left|r_{(j)}^1\right|$
$r_{(j)}^0$	$\left|r_{(i)}^0 \cap r_{(j)}^1\right|$	$\left|r_{(i)}^0 \cap r_{(j)}^0\right|$	$\left|r_{(j)}^0\right|$
Sum	$\left|r_{(i)}^1\right|$	$\left|r_{(i)}^0\right|$	$\sum_{l\in\{i,j\}} \sum_{k\in\{0,1\}} \left|r_l^k\right|$

First, we construct a frequency matrix (Table 1) for each pair of concepts represented as (s_i, s_j): Here is the definition of $r_{(i)}^0$ and $r_{(i)}^1$

$$r_{(i)}^0 = \{r \,|\, r = (u_*, e_*, s_i, 0)\, , r \subseteq R\}, \tag{2}$$

$$r_{(i)}^1 = \{r \,|\, r = (u_*, e_*, s_i, 1)\, , r \subseteq R\}, \tag{3}$$

where R is the log of responses. In the frequency matrix, the continuous score has been mapped to binary value. The frequency matrix provides the intuitive relevant judgement and the construction of following relevance analysis.

Before further relevance investigation, an expectation adequacy test needs to be recruited to guarantee the asymptotic statistical result [11]. Compared with the transition matrix, the frequency matrix can adaptively explore the concept relations to a sequence scale. The adaptive association exploration is more realistic modeling the concept relevant situation of the education system. As to the implantation of relation exploration, we design a composite exploring method based on statistical independence methods. The method concludes two components, expectation test and relevant test.

Firstly, we define an expectation as:

$$e_{(ij)}^{k_1 k_2} = S \times \frac{\left|r_{(i)}^1\right|}{S} \times \frac{\left|r_{(j)}^1\right|}{S} = \frac{\left|r_{(i)}^{k_1}\right| \cdot \left|r_{(j)}^{k_2}\right|}{S}, \tag{4}$$

$$S = \sum_{l\in\{i,j\}} \sum_{k\in\{0,1\}} \left|r_l^k\right| \; k_1, k_2 \in \{0,1\}, \tag{5}$$

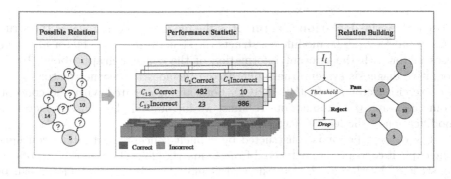

Fig. 2. The process of building concept relation in CSRG.

where $e_{(ij)}^{k_1 k_2}$ represents the response situation expectation of each concept which is the quantitative reliable indicator of frequency table. If any $e_{(ij)}^{k_1 k_2} < 5$, it is considered as the small expectation requires the specific fine relevant test. The Fisher exact test proves that it is more suitable to undertake the relevant test in a fine expectation situation [14]. Otherwise, as the general method, the chi-square test will be applied to relevant tests for the normal expectation. Here we have two formulas to calculate an independent index I of a couple of concepts. In the case of the index satisfying the specific conditions, it reveals that concept i and concept j are relevant:

$$
I = \begin{cases} \sum_{x \in \{0,1\}} \sum_{y \in \{0,1\}} \frac{\left(\left|r_{(i)}^x \cap r_{(j)}^y\right| - e_{(ij)}^{xy}\right)^2}{e_{(ij)}^{xy}} \sim \chi^2 & \min e_{(ij)} > 5 \\ \frac{\sqrt{S}\left(\left|r_{(i)}^1 \cap r_{(j)}^1\right| \cdot \left|r_{(i)}^0 \cap r_{(j)}^0\right| - \left|r_{(i)}^1 \cap r_{(j)}^0\right| \cdot \left|r_{(i)}^0 \cap r_{(j)}^1\right|\right)}{\sqrt{\left|r_{(i)}^1\right|\left|r_{(i)}^0\right|\left|r_{(j)}^1\right|\left|r_{(j)}^0\right|}} \sim N & \min e_{(ij)} \leq 5 \end{cases}, \quad (6)
$$

where χ^2 and N represent chi square contribution and gaussian distribution, respectively. It divides the relevant analysis operation into twin conditions due to the value I obeying different distributions in two conditions. After that, we can calculate the value I and compare it with the corresponding threshold value. When the value I satisfies the threshold, it considers that concept i and concept j are relevant. Therefore, the edge $e_{ij} \in \mathcal{E}$ of concept i and concept j will be built in the graph $\mathcal{G}$. Here is a unit of edge construction in the graph $\mathcal{G}$. Pair set of concepts presents as $\{(c_i, c_j) | c_i \in C, c_j \in C\}$. Any element in that set should be collected and tested when the construction of graph $\mathcal{G}$ is finished. The formula calculates the consistency index of learners' response to concepts in every binary set. It provides solid evidence to demonstrate that the concept relation is valid and reasonable.

CSRG is based on the statistical theory of general association. Moreover, the graph is built via a novel graph construct method. It constrains reasonably the following aggregation module.

Relation Aggregation

Concept Serial Relation Graph. It is the module that aggregates the structural input factor. The module constructs the structural learner factor and exercises factor with the relevant information of the concept graph above. In the cognitive diagnosis system, the input stream includes the learner factor h_s and the exercise factor h_e. In the ASRCD, both are aggregated with the concepts from the graph $\mathcal{G}$ aforementioned. Thus, we introduced the concept factor for modification of the learner factor and the exercise factor.

The concept factor is constructed by explicit concepts and latent concepts. The explicit concepts are extracted from Q-matrix [26]. The latent concepts are constructed by the graph $\mathcal{G}$ in the CSRG module. The concept factor can be represented as a knowledge information vector h_c. The explicit concepts and the latent concepts are represented as $h_{explicit}$ and h_{latent}. The formula can be defined as:

$$h_{explicit} = x_e \times Q, \tag{7}$$

$$h_{latent} = h_{explicit} \times A_{\mathcal{G}} \times Q, \tag{8}$$

$$h'_{latent} = f_{GATs}(h_{latent}, 1^{K \times K} \times Q, 1^{K \times K} \times Q), \tag{9}$$

$$h_c = h_{explicit} + h'_{latent} \times A_{lamb}, \tag{10}$$

where x_e presents the one-hot coded vector of exercise, K is the amount of exercise, A_{lamb} is a trainable matrix, $A_{\mathcal{G}}$ is the neighbor matrix extracted from graph $\mathcal{G}$ and $A_{lamb} \in \mathbb{R}^{K \times K}$, $A_{\mathcal{G}} \in \{0,1\}^{K \times K}$. $f_{GATs}(x, y, z)$ denotes the GAT, and x, y, z represent the $Query, Key, Value$ for self-attention in GAT [22]. After the concept aggregation, the graph relation of concept corresponding x_e is embedded into the updated concept factor h_c.

The learner Factor contains the proficiency of learners and relevant concepts. Each learner can be presented as a knowledge proficiency vector h_s. Specifically, h_s is initiated by:

$$h_s = sigmoid\,(u'_i \times A_u)\,, u_i \in N, \tag{11}$$

where u'_i is the one-hot encoded vector of an element in the aforementioned learner set N, A_s is a trainable matrix and $A_s \in \mathbb{R}^{N \times K}$. For integrating the relevant concepts into the initial learner factor h_s, the following aggregation formula is used:

$$h'_s = h_s \circ h_c, \tag{12}$$

where $\circ$ presents the Hadamard production. In the end of procedure below, h_s has been aggregated with the updated concept factor.

The exercise factor contains the difficulty of the exercise and the relevant concepts. It can be represented as the knowledge difficulty vector. The initiation formula for h_e is:

$$h_e = sigmoid\,(e'_i \times A_e + \lambda \cdot d_i)\,, e_i \in M, \tag{13}$$

where e'_i is the one-hot encoded vector of an element in the exercise set M, A_e and λ are trainable $A_e \subseteq \mathbb{R}^{K \times K}, \lambda \subseteq \mathbb{R}$, and d_i is the correct answer probability of e_i. Then, h_e will be aggregated with the relevant concepts as h_s. The integration formula is defined as:

$$h'_e = h_e \circ h_c. \tag{14}$$

Cognitive Diagnosis. For a cognitive diagnostic system, three elements need to be considered: the learner factor, the exercise factor, and the interaction function among them [6]. In the aggregation module, the learner factor h'_s and the exercise factor h'_e are integrated with graph G. The first layer of this module is inspired by the IRT, which presents the interactive function. It can be formulated as:

$$h_{disc} = sigmoid\,(e_i \times A_{disc})\,, \tag{15}$$

$$x = PLM(h_{disc}, h'_s, h'_e), \tag{16}$$

where A_{disc} is a trainable matrix and $A_{disc} \in \mathbb{R}^{K \times 1}$. Following is a five-layer fully-connected residual neural network with the sigmoid activation function. As for the loss function, we opt for the NLLLoss function, which is represented as:

$$loss_{ASRCD} = \sum_{i=1}^{N} \frac{l_i}{N}, \tag{17}$$

$$l_i = r_i \log y_i + (1 - r_i) \log (1 - y_i), \tag{18}$$

where r_i presents the real label and y_i presents the output of the ASRCD. Specially, the h_s denotes the knowledge proficiency of learners after the proper training procedure is finished.

4   Experiments

4.1   Datasets

We conduct the experiment on two real-world educational datasets, ASSISTments 2017[1] and SPOC 2020[2]. ASSISTments 2017 is a data set of ASSISTments Data Mining Competition 2017 published by the online education service "ASSISTment". Multi-skill collapsing has not been done. We therefore collapse the multi-skill exercises according to the exercise ID. SPOC 2020 is the desensitization learning record for the year 2020 on the "Medical Statistics Program of the National Association of Medical College Graduate Schools SPOC Platform". It tracks the learner's performance of exercises and tests during their studies on the online learning platform. Table 2 presents the basic statistics of the datasets.

Both datasets contain real performance records including learners' test behavior records and exercise-concept relation labeled by experts. In those educational systems, learners are allowed to submit their answers more than once until the assignments are over. To satisfy the requirements of statistical diagnosis, we only collect learners' first submission responses for every exercise. To guarantee the integration of learners' cognitive diagnosis, the learners whose count of test behavior record is less than 15 were filtered out.

4.2   Experiment Settings

Evaluation Metrics Parameter Setting. To measure the performance of the model, we select the Area Under an ROC Curve (AUC) to evaluate whether the model recognizes the concept proficiency of each learner in each exercise. Moreover, we also use the Prediction Accuracy (ACC) and Root Mean Square Error (RMSE) to evaluate.

---

[1] https://sites.google.com/view/assistmentsdatamining/data-mining-competition-2017.

[2] https://www.kaggle.com/ds/3713082.

Table 2. The statistics description of datasets.

Dataset	ASSIST 2017	SPOC 2020
Learners	1702	5254
Knowledge Concepts	102	95
Exercises	3162	798
Interactive Logs	942817	1204026
Knowledge Concepts per exercise	1.22	1.09
Response records per student	553.95	229.16

The dimensions of the full connection layer are 512, 256 and the Sigmoid is used as the activation function. To initiate the model, we select the Xavier initialization and generate the parameter from the gauss distribution. The average value and variance are calculated from the training dataset. The mini-batch size is 128. Adam algorithm is applied for optimization. For the graph construction, we opt 0.005 as the significance level according to statistical practice. To be fair, the embedding dimension of learner, exercise, and concept is set as the amount of concept.

Implementation Details. We implement all the experiments with PyTorch via Python and conduct them on a Linux server with two 2.0 GHz Intel Xeon E5-2683 CPUs and two GeForce 1080Ti GPUs.

It is difficult to evaluate the model performance of learner diagnosis as we cannot know the real concept proficiency of the learner. While the diagnosis results usually are related to the learner performance which is easily observed. Considering that all the interactions are objective records collected via a real education system, we use the evaluation metrics including accuracy, RMSE (root mean square error), and AUC (Area under curve) to evaluate the model performance. The implementation code is available in https://github.com/Hi-FishU/ASRCD-iconip2023.

4.3 Baseline Performance Comparison

Baselines. To verify the performance promotion of the proposed model, we implement the baselines and conduct comparative experiments. We select the baselines from two dimensions. One is the widely applied cognitive diagnosis models, including IRT, MIRT, PMF, NCD, and RCD as follows:

- IRT [1] presents the most common diagnosis approach in practice, which filters the learner and exercise factor with a linear function.
- MIRT [19] presents the dimension extended IRT approach, which models the multiple learner and exercise factor.
- PMF [10] models learners' proficiency through decomposing learner and exercise features with a parametric matrix.

- NCD [26] is an advanced neural cognitive diagnosis method, which captures the complex relation between learner and exercise.
- RCD [8] is the latest cognitive diagnosis model, which incorporates the hierarchic relation among concept-exercise-learner.

The other aspects are about the effective graph construction representation learning methods as follows:

- Dense [15] builds simple relation using the densely connection, which weights nodes equally.
- Transition [15] builds relation through calculating a transition probability matrix, which weights the nodes with the possibility of relation in context response sequence.

Performance. To compare effectiveness of ASRCD with baselines on learner prediction performance, we implement the baseline in cognitive diagnosis fields such as IRT and MIRT. In Table 3, the experimental results of all models on two real datasets are shown. The best performance is marked in bold. It can be observed that our model achieved consistent outstanding performance in the prediction task. The results demonstrate that introducing serial relation in diagnosis benefits the evaluation of learner's proficiency. Additionally, the overload response of the learner might limit the performance of RCD according to the observation. This demonstrates further the adaptability of our proposed ASRCD.

Table 3. Experimental results on learner diagnosis.

	ASSIST 2017			SPOC 2020		
	ACC	RMSE	AUC	ACC	RMSE	AUC
IRT [1]	0.688	0.500	0.684	0.832	0.394	0.660
MIRT [19]	0.681	0.512	0.691	0.824	0.393	0.781
PMF [10]	0.651	0.525	0.630	0.745	0.419	0.735
NCDM [26]	0.840	0.331	0.851	0.844	0.344	0.821
RCD [8]	0.720	0.430	0.789	0.863	0.313	0.892
ASRCD	**0.875**	**0.303**	**0.901**	**0.874**	**0.299**	**0.905**

4.4   Ablation Experiment

The Extendibility of CSRG. In this section, we conduct an ablation study to evaluate the effectiveness of the Concept Serial Relation Graph (CSRG) and its integration with existing methods. The purpose of this experiment is to gain insights into the impact of incorporating CSRG on the diagnostic abilities of current methods and to understand the necessity of considering the concept relation derived from the learner interaction sequence.

By implementing CSRG and integrating it into various existing methods, we aim to assess the performance improvements brought about by the inclusion of concept relations. The results of this integration, as shown in Table 4, provide empirical evidence that demonstrates the positive influence of CSRG on the accuracy of cognitive diagnosis. Moreover, these findings emphasize the significance of capturing and incorporating the concept relation information extracted from the learner interaction sequence.

In analyzing the results, several key observations emerge. Firstly, most methods that integrate CSRG exhibit consistent improvements in diagnostic accuracy, underscoring the exceptional extendibility of CSRG. Secondly, it is important to note that one method, CSRG-PMF, experiences a decline in performance. This can be attributed to the increased complexity introduced by the matrix structure in CSRG, which presents limitations for the Probabilistic Matrix Factorization (PMF) method.

Through this ablation study, we aim to demonstrate the valuable insights gained from reconstructing the concept relation based on the interaction sequence. The findings support the assertion that incorporating CSRG significantly enhances the accuracy of cognitive diagnosis in various scenarios. In the following section, we proceed with additional experimental sequences to further validate and reinforce the superiority of our CSRG constructing method.

Table 4. Results of ablation experiments.

	ASSIST 2017			SPOC 2020		
	ACC	RMSE	AUC	ACC	RMSE	AUC
CSRG-IRT	0.865	0.305	0.893	0.868	0.308	0.890
CSRG-MIRT	0.865	0.304	0.894	0.868	0.309	0.891
CSRG-PMF	0.584	0.496	0.507	0.539	0.309	0.539
CSRG-NCDM	0.847	0.321	0.872	0.869	0.305	0.901
CSRG-RCD	0.722	0.428	0.791	0.865	0.313	0.894
ASRCD	**0.875**	**0.303**	**0.901**	**0.874**	**0.299**	**0.905**

The Building Effectiveness of ASRCD. To further demonstrate that ASRCD is effective, we opt several node graphs to build the concept relation. The results are shown in Table 5. Specifically, we implement the different graph constructing methods, including transition graph and dense graph. Both are proposed for concept relation building in GKT. To show the promotion of weighting in ASRCD, we also experimented with the single CSRG in the binary instead of weights. The experiment shows some interesting results. In the comparison of dense and transition, the relation based on context interaction sequence as transition leads to fair performance with simple dense relation. It reports that improper serial relation might cause side effects in the diagnosis. In CSRG, it builds an adaptive serial relation based on full learning sequence. The result supports that the CSRG method is more suitable to model serial relation. Moreover, the ASRCD, CSRG with weighted relation, shows superior performance in the experiment. It demonstrates that the weighting benefits modeling the real relation.

Table 5. Results of different graph building methods.

	ASSIST 2017			SPOC 2020		
	ACC	RMSE	AUC	ACC	RMSE	AUC
Dense	0.842	0.329	0.868	0.838	0.347	0.814
Transition	0.843	0.326	0.867	0.819	0.363	0.802
Single CSRG	0.859	0.311	0.890	0.868	0.310	0.896
ASRCD	**0.875**	**0.303**	**0.901**	**0.874**	**0.299**	**0.905**

5   Discussions

5.1   CSRG Analysis

To visually demonstrate the CSRG ability of modeling the concept relation, we conduct two interesting experiments on sequence-level and system-level concept relation. We visualize the capacity of exposing the concept relation through interaction sequence and rebuilding the concept system based on prior structure.

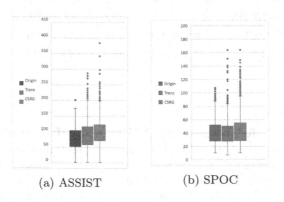

(a) ASSIST   (b) SPOC

Fig. 3. The average exercise distance (AED) comparison among different graph constructions. (a): ASSIST 2017 (b): SPOC 2020

Sequence-Level Concept Relation. The concept relation on the learner interaction sequence reflects the connection of the exercises attached with concept. We measure the situation of related concepts in the exercise sequences through the average exercise distance (AED) of each pair of related concepts. AED here presents the average of the exercise count between two concepts of a related binary in every learner exercise sequence. Figure 3 illustrates the AED result of each graph building method. As shown in the figure, CSRG can discover concept serial relation on longer intervals and keep the ability of mining short interval serial relations. The effect is more significant on datasets with more exercises and concepts for each student. This also fits with the statistical characteristics of CSRG.

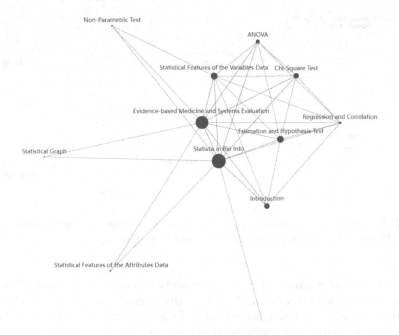

Fig. 4. The unit relation graph based on concept relation.

Unit-Level Concept Relation. For an individual system, the concepts will be grouped by several units in common. Specifically, in the SPOC 2020 dataset, concepts are divided into 12 different topic units. Meanwhile, the units can be properly arranged based on the concept relation among the system. The system-level relation presents the connection of units and discovers a topic framework of the system which is helpful for designing the learning route. We visualize the system-level concept relation in Fig. 4, which presents the unit relations in the system. The size of the unit node depends on the concept relations amount between the target unit and the other units. For more clear representation, we filter out the unit relation with less than 40 concept relations for guaranteeing the relation confidence. There are several obvious central units which connect densely with others. It includes *Statistics in Bio Info, Evidence-based Medicine and Systems Evaluation,* and *Estimation and Hypothesis Test* which are treated as core content of the medical statistics course. In general, the unit relation benefits further course promotion and recommendation.

5.2   Case Study

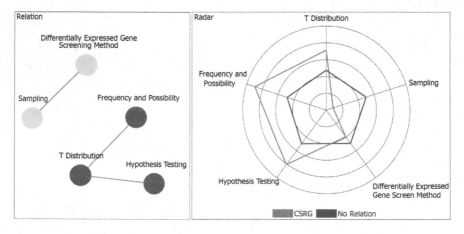

Fig. 5. The diagnosis result comparison. Relation: The subset of concept serial relation graph. Radar: The visualization of learner concept proficiency

The cognitive diagnosis aims to recognize the concept proficiency of learners. In this section, we visualize and analyze the performance of a learner in the education system. Moreover, we evaluate the interpretation of ASRCD through evaluation of whether the diagnosis result of learners is reasonably aligned with the practical education judgement. Specifically, we sample randomly an individual learner from the SPOC educational system and present the concept cognitive state.

Figure 5 (Relation) presents a subset structure of concept relation in SPOC. It contains a triple relation and a binary relation. Figure 5 (Radar) shows the proficiency of sample learner in corresponding concepts. The radar graph in Fig. 5 presents the synchronization of specified concept mastery. The blue line denotes the diagnosis integrated CSRG. The red line denotes the diagnosis without the relation. As to the relation, it is obvious that concept ① (Frequency and probability), concept ⑮ (T distribution), and concept ⑩ (Hypothesis testing) are related. The binary set shows the relation between concept ⑤ (Differentially expressed gene screen method) and concept ⑭ (Sampling). As for the diagnosis, we can observe that the proficiency of concept ① ⑬ ⑩ are equally extremely excellent. The concept mastery ⑤ ⑭ are significantly low. According to the observations, we can obtain that: (1) The related concept shows synchronic learner proficiency. (2) The synchronization of concept proficiency is only effective when diagnosis integrated CSRG. Thus, this demonstrates that CSRG can properly build concept relations. Then, our proposed ASRCD can achieve a better interpretable diagnosis of the learner's proficiency.

6   Conclusion

In this paper, we proposed an extendable relation construct method called Concept Serial Relation Graph (CSRG) and Adaptive Serial Relation-based model for Cognitive Diagnosis (ASRCD), which builds and integrates the concept relation into cognitive diagnosis. Specifically, we first built the concept relation according to the learner response sequence. Then, we constructed a well-defined relation graph to represent the concept relation. After that, we used an improved GNN to integrate and weight the relation. At last, we designed a general cognitive diagnosis module to predict learner performance. We implemented ASRCD on the real-world datasets to demonstrate its effectiveness and extendibility. We will be glad to see this work help further intelligent education studies.

Acknowledgements. This research was in part supported by the High Educational Teaching Reformation Research Project under Grant no. 2021-A-16, and in part of the achievement of National education science "14th Five-Year" planning project in 2021 (Key Project of the Ministry of Education) "Research on the Path of Improving University Teachers' Teaching Force from the Perspective of Digital Empowerment" (DIA210368). The authors would like to thank Mingbo Yang in ITS Lab of Tsinghua Shenzhen International Graduate School for kindly pointing out the issue in a previous iteration of the paper, and Weibo Gao in Anhui Province Key Laboratory of Big Data Analysis and Application for helpful discussions, feedback, and support.

References

1. Akour, M., AL-Omari, H.: Empirical investigation of the stability of IRT item-parameters estimation. Int. Online J. Educ. Sci. **5**(2), 291–301 (2013)
2. Anderson, A., Huttenlocher, D., et al.: Engaging with massive online courses (2014). https://doi.org/10.1145/2566486.2568042
3. Baker, R.S., Inventado, P.S.: Educational data mining and learning analytics. In: Larusson, J.A., White, B. (eds.) Learning Analytics, pp. 61–75. Springer, New York (2014). https://doi.org/10.1007/978-1-4614-3305-7_4
4. Barnes, T.: The Q-matrix method: mining student response data for knowledge. In: American Association for Artificial Intelligence 2005 Educational Data Mining Workshop, Pittsburgh, PA, USA, pp. 1–8. AAAI Press (2005)
5. De La Torre, J.: Dina model and parameter estimation: a didactic. J. Educ. Behav. Stat. **34**(1), 115–130 (2009)
6. DiBello, L.V., Roussos, L.A., Stout, W.: 31A review of cognitively diagnostic assessment and a summary of psychometric models. In: Rao, C., Sinharay, S. (eds.) Handbook of Statistics, vol. 26, pp. 979–1030. Elsevier (2006). https://doi.org/10.1016/S0169-7161(06)26031-0

7. Embretson, S.E., Reise, S.P.: Item Response Theory. Psychology Press (2013)
8. Gao, W., Liu, Q., et al.: RCD: relation map driven cognitive diagnosis for intelligent education systems. In: Proceedings of the 44th International ACM SIGIR Conference on Research and Development in Information Retrieval, pp. 501–510 (2021)
9. Huang, X., Liu, Q., et al.: Constructing educational concept maps with multiple relationships from multi-source data. In: 2019 IEEE International Conference on Data Mining (ICDM), pp. 1108–1113. IEEE (2019)
10. Koren, Y., Bell, R., et al.: Matrix factorization techniques for recommender systems. Computer **42**(8), 30–37 (2009)
11. Larntz, K.: Small-sample comparisons of exact levels for chi-squared goodness-of-fit statistics. J. Am. Stat. Assoc. **73**(362), 253–263 (1978)
12. Leighton, J., Gierl, M.: Cognitive Diagnostic Assessment for Education: Theory and Applications. Cambridge University Press (2007)
13. Liu, Z., Wang, S., Liang, Z., Fu, P.: Concept relative attention based deep knowledge tracing. In: Xie, Q., Zhao, L., Li, K., Yadav, A., Wang, L. (eds.) ICNC-FSKD 2021. LNDECT, vol. 89, pp. 858–865. Springer, Cham (2022). https://doi.org/10.1007/978-3-030-89698-0_88
14. Mehta, C.R., Patel, N.R.: Algorithm 643: FEXACT: a FORTRAN subroutine for fisher's exact test on unordered r × c contingency tables. ACM Trans. Math. Softw. (TOMS) **12**(2), 154–161 (1986)
15. Nakagawa, H., Iwasawa, Y., et al.: Graph-based knowledge tracing: modeling student proficiency using graph neural network. In: IEEE/WIC/ACM International Conference on Web Intelligence, pp. 156–163 (2019)
16. Novak, J.D.: Learning, Creating, and Using Knowledge: Concept Maps as Facilitative Tools in Schools and Corporations. Routledge (2010)
17. Piaget, J., Brown, T., et al.: The Equilibration of Cognitive Structures: The Central Problem of Intellectual Development. University of Chicago Press (1985)
18. Premchaiswadi, W., Porouhan, P.: Process modeling and decision mining in a collaborative distance learning environment. Decis. Anal. **2**(1), 1–34 (2015). https://doi.org/10.1186/s40165-015-0015-5
19. Reckase, M.D.: 18 multidimensional item response theory. In: Handbook of Statistics, vol. 26, pp. 607–642 (2006)
20. Shi, H., Yang, Y., et al.: Dynamic multi-skill knowledge tracing for intelligent educational system. In: Proceedings of the 2022 5th International Conference on Algorithms, Computing and Artificial Intelligence, pp. 1–6 (2022)
21. Tong, S., Liu, Q., et al.: Structure-based knowledge tracing: an influence propagation view. In: 2020 IEEE International Conference on Data Mining (ICDM), pp. 541–550. IEEE (2020)
22. Vaswani, A., Shazeer, N., et al.: Attention is all you need. In: Advances in Neural Information Processing Systems, vol. 30 (2017)
23. Veličković, P., Cucurull, G., et al.: Graph attention networks. arXiv preprint arXiv:1710.10903 (2017)
24. Verhelst, N.D., Glas, C.A.W.: The one parameter logistic model. In: Fischer, G.H., Molenaar, I.W. (eds.) Rasch Models, pp. 215–237. Springer, New York (1995). https://doi.org/10.1007/978-1-4612-4230-7_12
25. Vukicevic, M., Jovanovic, M., et al.: Recommender system for selection of the right study program for higher education students. In: RapidMiner: Data Mining Use Cases and Business Analytics Applications, p. 145 (2013)

26. Wang, F., Liu, Q., et al.: Neural cognitive diagnosis for intelligent education systems. In: Proceedings of the AAAI Conference on Artificial Intelligence, vol. 34-04, pp. 6153–6161 (2020)
27. Wu, R., Liu, Q., et al.: Cognitive modelling for predicting examinee performance. In: Twenty-Fourth International Joint Conference on Artificial Intelligence (2015)
28. Yang, Y., Fu, P., et al.: MOOC learner's final grade prediction based on an improved random forests method. Comput. Mater. Continua **65**(3), 2413–2423 (2020)
29. Yu, X., Li, S., et al.: A three-parameter logistic regression model. Stat. Theor. Relat. Fields **5**(3), 265–274 (2021)

PyraBiNet: A Hybrid Semantic Segmentation Network Combining PVT and BiSeNet for Deformable Objects in Indoor Environments

Zehan Tan[1], Weidong Yang[1,2]([⊠]), and Zhiwei Zhang[3]

[1] School of Computer Science, Fudan University, Shanghai, China
{18110240062,wdyang}@fudan.edu.cn
[2] Zhuhai Fudan Innovation Institute, Hengqin New Area, Zhuhai, Guangdong, China
[3] Gree Electric Appliances, INC. of Zhuhai, Zhuhai, China
zzwyyds0606@gmail.com

Abstract. In this study, we introduce PyraBiNet, an innovative hybrid model optimized for lightweight semantic segmentation tasks. This model ingeniously merges the merits of Convolutional Neural Networks (CNNs) and Transformers. We propose a dual-branch structure that strategically employs the global feature extraction capabilities of the Pyramidal Vision Transformer (PVT) and the local feature extraction proficiency of BiSeNet. Specifically, the global feature branch employs a transformer from PVT to harness high-level patterns from input images, while the local feature branch utilizes a CNN, inspired by BiSeNet, to extract fine-grained details. Comprehensive evaluations conducted on the ADE20K and DOS datasets underscore PyraBiNet's superior performance compared to the existing state-of-the-art methods. With its effective and efficient performance, PyraBiNet proves to be an invaluable asset in the domain of mobile robotics, particularly beneficial for applications such as sweeping robots. The code source and dataset are open at https://github.com/zehantan6970/PyraBiNet.

Keywords: Image processing · Semantic Segmentation · Real-time processing

1 Introduction

Semantic segmentation is a task within the field of computer vision, the goal of which is to classify each pixel in an image, dividing it into distinct semantic categories, thereby enabling a deeper understanding of the image. The challenge of semantic segmentation lies in the precise delineation of object boundaries and assigning them the correct category labels. This necessitates the model to possess substantial perceptual capability, allowing it to comprehend the various objects, colors, textures, and shapes within an image, as well as the relationships among them. Concurrently, the model must be capable of classifying each pixel within

B. Luo et al. (Eds.): ICONIP 2023, CCIS 1968, pp. 552–564, 2024.
https://doi.org/10.1007/978-981-99-8181-6_42

the image since the same object may appear in different locations, sizes, and orientations. With the increasing demand of intelligence, semantic segmentation has become the basic perception component for applications such as autonomous driving [6], medical imaging diagnosis [1] and indoor robot [3,15]. To meet real-time or mobile requirements, researchers have come up with many efficient and effective models in the past for semantic segmentation. The field of lightweight semantic segmentation models has experienced significant evolution, characterized by shifts in underlying network architectures. These transitions can be seen from the initial utilization of Convolutional Neural Networks (CNNs) as typified by Fully Convolutional Networks (FCNs) [21] and extended in BiSeNet series [40,41] and PIDNet [37]. The focus later moved to transformer-based methods, exemplified by LeViT [8] and Pyramid Vision Transformer (PVT) [34]. The latest developments showcase hybrid architectures that combine CNNs with Vision Transformers (ViTs). These include models like the MobileViT series [23,24,33] and Convolutional Vision Transformer (CVT) [36]. Thus, the development of lightweight semantic segmentation models has seen a significant transformation, marked by diverse architectural designs to optimize performance.

By rethinking previous successful lightweight semantic segmentation works with reference to SegNeXt's research [9], we found that these works all face the challenge of how to balance accuracy, parameter scale and inference speed, and improve the fusion of different features. We argue a successful lightweight semantic segmentation model should have the following characteristics:

(i) Feature Extraction: Robust feature extractors not only capture a global features but also discern local detail features. These can acquire features of varying scales.

(ii) Feature Fusion: A rational approach is needed for the integration of local detail features and global features.

(iii) Feature Enhancement: Enhancing the diversity and detailed spatial information of features is essential. Lightweight models have limited capabilities in modeling global relationships, leading to insufficient attention to details in segmentation tasks and often unclear edges.

(iv) Network Architecture Design: The optimization of network structure is necessary, ensuring not only the reasonable utilization of global and local detail features but also control over the number of parameters, while maintaining network inference speed. The key to this network structure is to balance accuracy, parameter scale, and inference speed, while improving the fusion of different features. Given the yearly increase in memory with the widespread use of embedded systems, the size of the parameter scale should be a limiting factor. However, keeping the model size small at the cost of relatively high computation, which also means high latency, is not a sound practice. The parameter volume should not be blindly reduced. Similarly, the network structure should not simply trade accuracy for speed, or vice versa.

Considering the analyses above, we reassess the design of lightweight network architectures for semantic segmentation in this paper. Instead of applying PVT or BiSeNet independently, we propose a novel hybrid architecture, PyraBiNet, which integrates the strengths of both PVT and BiSeNet. The global feature branch of PyraBiNet, powered by a transformer from PVT, extracts the global features from the input images. Concurrently, the local feature branch, inspired by BiSeNet, utilizes a convolutional neural network (CNN) to capture the local detailed features. Subsequently, these two sets of features are fused to generate a final feature map that is utilized for semantic segmentation.

Our primary contributions are:

- We present a novel lightweight network architecture, termed PyraBiNet, which combines the strengths of convolution (inductive bias, translation invariance, exceptional local detail capture ability, and low computational complexity) and Transformers (ability to capture long-range dependencies) in a dual-branch structure optimized for embedded devices, bolstered by an efficiently parametrized Detail Feature Block that adjusts resolution to align with the global feature branch while effectively capturing local spatial information.
- We introduce the Parallel Dual-Feature CBAM (PDF-CBAM) that concurrently applies a Channel Attention Module to the transformer-derived global features and a Spatial Attention Module to the CNN-derived local features, resulting in an enhanced final feature map that effectively integrates detailed spatial information and diversity of features.
- Our experimental results demonstrate that our proposed architecture achieves state-of-the-art (SOTA) on different benchmarks of ADK20K [46] and our proprietary DOS dataset[1].

2   Related Work

The arena of lightweight semantic segmentation [25,32] has witnessed numerous advances over recent years. We primarily focus on three major neural network types in this context: 1) Convolutional Neural Networks (CNNs), 2) Vision transformers (ViTs), and 3) Hybrids of CNNs and ViTs.

2.1   Convolutional Neural Networks (CNNs)

CNN-based models, such as FCNs [21] and MobileNets   [12,13,28], have greatly improved performance by encoding local features, replacing hand-crafted [17,18,29,39] systems. Techniques like channel shuffle, micro-factorized convolution, and dynamic operators help enhance information flow and efficiency. Furthermore, novel methods like DDRNet [26] and BiSeNet [31,40,41] utilize bilateral connections and multi-path frameworks to blend low-level details and high-level semantics. PIDNet [37], one of the latest architectures, is composed of three branches to parse the detailed, context, and boundary information. However, despite these advancements, CNNs still have limitations like high computational time and disregard for global information.

---

[1] https://github.com/zehantan6970/DOS_Dataset.

2.2   Vision Transformers (ViTs)

Drawing from NLP success, transformers have been employed in computer vision tasks [11,14,19], yielding impressive results. Models like ViT [5], DeiT [30], T2T-ViT [44], and Swin Transformer [20] have significantly pushed the boundaries of image classification performance. To create lightweight ViTs, architectures like LeViT [8] and PVT [34] fuse standard convolution layers with improved ViT. PoolFormer [42] replaces the attention module in Transformers with an embarrassingly simple spatial pooling operator to conduct only basic token mixing. Despite these advances, ViTs still face challenges in dealing with visual features of different scales and are often inefficient in terms of memory usage.

2.3   Hybrids of CNNs and ViTs

To capitalize on the strengths of both CNNs and ViTs, hybrid models like Mobile-ViTv3 [33], TopFormer [45], LVT [38], and others have been proposed. These models aim to combine the efficiency of convolution with the global receptive field of Transformers. Other architectures, such as the CeiT [43] and CVT [36], integrate convolutional and self-attention modules in the same architecture. Twins [2] builds upon PVT by substituting its absolute position embedding with relative conditional position embedding and incorporating separable depth-wise convolutions for capturing both local and global image contexts. DFvT [7] opens up the transformer block and enhance it with convolution, both before and after self-attention that tightly integrates transformer and convolution. Despite the efficiency of these hybrid models, they usually come at the cost of performance accuracy.

3   Approach

In this study, we propose PyraBiNet, a novel dual-branch architecture designed to address the challenges of Feature Extraction, Fusion, Enhancement, and Network Architecture Design in lightweight semantic segmentation. PyraBiNet integrates the broad-scale feature extraction capability of PVT with the detailed extraction prowess of BiSeNet. The architecture leverages a transformer from PVT in the global feature branch and a BiSeNet-inspired CNN in the local feature branch for comprehensive Feature Extraction. We utilize a Parallel Dual-Feature Convolutional Block Attention Module for efficient Feature Fusion and a Detail Feature Block for Feature Enhancement. The design of integrating the global feature branch and the local feature branch into a dual branch helps optimize the Network Architecture and provides a robust solution for lightweight semantic segmentation. In this work, the loss function used for training the model is cross-entropy.

Proposed Method: PyraBiNet is a hybrid model that combines the strengths of CNNs and transformers. Figure 1 illustrates the architecture of PyraBiNet, in which the input image is separately processed by the global feature branch and the local feature branch. In our global feature branch, designed following the

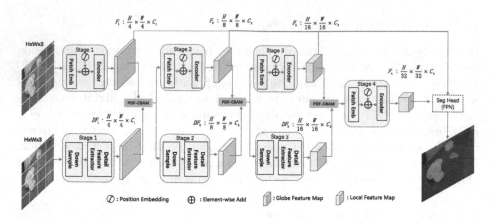

Fig. 1. The pipeline of the proposed PyraBiNet for semantic segmentation. The global feature branch (PVT) contains global branch (up); The local feature branch(reference BiSeNet) contains local branch (down); PyraBiNet contains fusion block and segmentation head.

PVT, the image is processed by a series of self-attention modules. Each of these self-attention modules focuses on different spatial regions of the input image, enabling the PVT to learn global features that are invariant to changes in pose and scale. Following this, the global and local features are fused through a Parallel Dual-Feature Convolutional Block Attention Module (PDF-CBAM), and then separately input into the global feature branch (PVT) and the local feature branch. Notably, in stage 4, only the global feature branch (PVT) is used to generate the final feature map for semantic segmentation. Ultimately, we employ a Semantic FPN [16] as the segmentation head to achieve the final segmentation outcome. Our local feature branch, referenced from BiSeNet, consists of a downsampling module and a Detail Feature Block.

$$\begin{cases} S_i\left(F_g, F_l\right) = PDF\text{-}CBAM_i\left(Attention_i\left(F_g\right), DF_i\left(F_l\right)\right) & ,i = 1,2,3 \\ S_i\left(F_g\right) = Attention_i\left(F_g\right) & ,i = 4 \end{cases} \quad (1)$$

where S_i represents the i-th stage in the architecture of PyraBiNet. PDF-CBAM$_i$ refers to the i-th Parallel Dual-Feature Convolutional Block Attention Module, which is designed for feature fusion. Attention$_i$ symbolizes the self-attention operation implemented at the i-th stage. DF$_i$ denotes the detail feature block deployed at the i-th stage. F_g refers to the global feature map obtained from the global feature branch, and F_l symbolizes the local feature map derived from the local feature branch in our architecture.

Detail Feature Block (DF): The overall framework of the proposed DF is presented in Fig. 2. For each stage, the process begins with a downsampling module to match the resolution of the PVT branch, followed by the use of a detail feature extractor to capture local spatial information. As this involves low-level information, the module requires a substantial channel capacity to encode rich

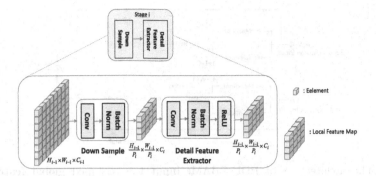

Fig. 2. Detail Feature Block (DF) is meticulously constructed to fulfill two main tasks at each stage: adjusting the resolution to align with the PVT branch, and capturing local spatial information. 'i' represents the corresponding stage. The numbers of Detail Feature Block is set to 1, 2, 2, corresponding to satege1, stage2, stage3.

spatial detail information. More specifically, wide channels and shallow layers are used to process spatial details, following the structural design of BiSeNetV2. Each extractor is composed of 'n' blocks, with each block comprising a convolution, Batch Normalization, and ReLU sequence. Our local feature branch, while inspired by the design philosophy of BiSeNet, diverges significantly from its prototype. BiSeNet consists of two branches, where its global branch employs CNNs with large receptive fields for the implementation. In contrast, our model utilizes PVT for global feature extraction. In terms of the local branch, our design also deviates from BiSeNet in terms of the parameters and quantity of CNN kernels, as well as the overall structure. Additionally, while BiSeNet does not have staged architecture and adopts a pyramid-like method without downsampling, our model is organized into stages, with fusion performed at each stage. Furthermore, the fusion strategy in BiSeNet is achieved by Aggregation Layer before the final feature map, while our approach incorporates a fusion process in each stage.

$$DF_i\left(F_l\right) = Local\_Extractor_i\left(DS_i\left(F_l\right)\right) , i = 1, 2, 3 \qquad (2)$$

where Local_Extractor signifies the mechanism within our model that facilitates the extraction of local features from the input. DS is an acronym for downsampling in each stage. F_l represents the local features that are extracted and processed within our architecture.

Parallel Dual-Feature CBAM (PDF-CBAM): The feature fusion module, an integral part of semantic segmentation, enhances feature representations. However, in our ablation studies (Table 3), we discovered that straightforward strategies such as element-wise summation, multiplication, and concatenation did not yield satisfactory results when fusing local and global features. Considering that VIT has strong attention on space but weak attention on channels, and CNNs, with their local convolution operations, can naturally capture local

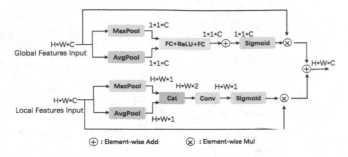

Fig. 3. The architecture of PDF-CBAM. Input the reshaped global feature map (H*W*C) and local feature map (H*W*C) respectively, and the output is a feature map (H*W*C) that combines global features and local features.

details and handle channel-wise information well, but handle space information weak. Notably, the Convolutional Block Attention Module (CBAM) [35] utilizes a sequential combination of the Channel Attention Module and Spatial Attention Module, each processing the input feature layer independently. Aiming for better integration of local detail features with global features and enhancement of the diversity and detailed spatial information of features, we opted for a parallel approach. Our Parallel Dual-Feature CBAM (PDF-CBAM) takes as input local detail features derived from a Convolutional Neural Network (CNN) and global features derived from a transformer. The global features are reshaped into a convolutional feature map (H*W*C), which is then subjected to a Channel Attention Module. Concurrently, a Spatial Attention Module is applied to the local feature map (H*W*C). Finally, the outputs from the Channel Attention Module and Spatial Attention Module are combined via element-wise summation to produce the final feature map, thereby resolving the issue of lack of information screening inherent to simple element-wise summation, multiplication, and concatenation. The architecture of PDF-CBAM is shown in Fig. 3.

$$PDF\text{-}CBAM = (CA\,(F_g) \otimes F_g) \oplus (SA\,(F_l) \otimes F_l) \tag{3}$$

where CA denotes Channel Attention, a CBAM that focuses on the channel-wise information of the input features. SA stands for Spatial Attention, another component of the CBAM, which pays attention to the spatial arrangement of the features. F_g refers to the global features derived from the transformer in the PyraBiNet architecture, whereas F_l represents the local features extracted by the CNN within the same architecture.

Deformable Object Segmentation Dataset for Sweeping Robots (DOS Dataset): We present a novel dataset, designed specifically to serve as a benchmark for semantic segmentation of deformable objects within the context of obstacle avoidance in indoor robotic sweeping scenarios. DOS dataset comprises 3,056 images, We used the open-source LabelMe [27] annotation toolkit, to manually collect the polygon annotations of deformable objects of four types: faeces, socks, plastic bag, and rope. DOS dataset has 7687 annotated object instances.

4   Experiments

PyraBiNet was evaluated on the ADE20K dataset [46] and DOS dataset. ADE20K is a demanding scene parsing dataset designed to benchmark the performance of semantic segmentation. This dataset comprises 150 highly-detailed semantic categories and features 20,210 training images, 2,000 validation images, and 3,352 testing images. DOS dataset comprises 3,056 images, which have been randomly partitioned into training and validation sets at 8:1 ratio. The training set contains 6,800 semantic segmentation labels, while the validation set includes 887 annotated labels. The experiments were performed using an Intel Core i7-10700 CPU, Nvidia V100 16G GPU, and 16 GB memory.

For the quantitative evaluation, we report the performance of baseline methods and the proposed method by three metrics: mean intersection over union ($mIoU$).

Let TP, FP, and FN denote the total number of true positive, false positive, and false negative pixels, respectively. The Intersection over Union (IoU) is calculated as follows:

$$IoU_i = \frac{GT_i \cap \text{Pred}_i}{GT_i \cup \text{Pred}_i} \tag{4}$$

$$mIoU = \frac{1}{n} \sum_{i=1}^{n} IoU_i \tag{5}$$

where GT stands for ground truth, i denotes the semantic categories, and n symbolizes the total number of classes.

Params refers to the number of parameters in the model. FLOPs stands for "Floating Point Operations," and it is used as a measure of computational complexity or the number of calculations the model needs to perform during inference. GFLOPs represents for "Giga Floating Point Operations," equivalent to one billion FLOPs.

4.1   Semantic Segmentation on ADE20K and DOS

The experiments are carried out on semantic segmentation task. We employ the proposed PyraBiNet as our backbone architecture. To ensure a uniform evaluation metric, we strictly adhere to the training configurations set by PVT [34], utilizing a Semantic FPN [16] as our segmentation head. Our PyraBiNet is pre-trained on the ImageNet dataset [4]. The pre-training steps and parameters of our model are the same as the PVT. To entail a fair comparison, we keep the same data augmentation and training settings as the other vision transformers as far as possible. The competitors are all competitive vision transformers, including ResNet18 [10], PVT [34], BiSeNetv2 [40], PoolFormer [42]. PyraBiNet achieved state-of-the-art results on the ADE20K and DOS dataset, in Table 1 and Table 2.

As shown in Table 1, with the exception of BiSeNetV2, all models employ Semantic FPN as their segmentation head. In the nearly equivalent parameter range of 10M-20M, our model achieves the highest mIoU of 37.7. Remarkably,

Table 1. Performance comparisons on the test set of ADE20K. For each method, we report the mean intersection over union (*mIoU*), *Params(M)*, and *GFLOPs*.

Method	Params(M)	GFLOPs	mIoU
ResNet18 [10]	15.5	32.2	32.9
BiSeNetv2 [40]	**14.8**	**12.3**	19.5
PoolFormer [42]	15.7	30.7	37.2
PVT-Tiny [34]	17.0	33.2	35.7
Ours	19.4	37.3	**37.7**

our model outperforms PVT-Tiny by 2.0 points, highlighting the effectiveness of our proposed dual-branch architecture which fuses local and global features. The enhanced global feature extraction of ViT supplemented by the local feature extraction of the CNN increases segmentation precision. Furthermore, compared to the pure CNN-based dual-branch model, BiSeNetv2, our semantic branch possesses a global receptive field, resulting in superior segmentation accuracy in our model. Here, BiSeNetV2 is not pre-trained.

Table 2. Performance comparisons on the test set of DOS. Our model is trained on a single v100 Gpu with 40k iterations, a batchsize of 4, a learning rate of 1e-4, and an input image size of 512*512.

Method	Params(M)	GFLOPs	mIoU
ResNet18 [10]	15.5	31.9	65.2
BiSeNetv2 [40]	**14.8**	**12.0**	67.3
PoolFormer [42]	15.7	30.4	71.0
PVT-Tiny [34]	17.0	32.9	71.3
Ours	19.4	37.0	**72.8**

As illustrated in Table 2, when employing Semantic FPN for semantic segmentation, our model exhibits superior performance on the DOS dataset, achieving a maximum mIoU of 72.8. This score exceeds that of ResNet18 by 7.6 points and PVT-Tiny by 1.5 points, thereby further corroborating the efficacy of our proposed method of combining transformers and CNNs.

4.2   Ablation Study

We carry out ablation studies to validate the effectiveness of the feature fusion module. We compare our PDF-CBAM with several widely used methods, such as 'SUM': element-wise addition, 'MUL': element-wise multiplication, and 'Cat+1 × 1conv': concatenation followed by a 1 × 1 convolution. In addition, '+Stage4' refers to the incorporation of our detail module in the fourth stage of PVT. The results of these experiments can be seen in Table 3. The findings indicate that our mixed-attention feature fusion strategy outperforms simple addition, multiplication, or fusion through 1 × 1 convolution. This superiority

Table 3. Different designs of the feature fusion module to fuse the information from global features and local detail features. Δ denotes mIoU Variation. Ablations were tested on ADE20K.

Method	$Params(M)$	$GFLOPs$	$mIoU$	Δ
SUM	**19.36**	**37.31**	37.2	−0.5
MUL	**19.36**	**37.31**	37.0	−0.7
Cat+1 × 1conv	19.61	37.79	37.5	−0.2
+Stage4	24.79	38.69	37.3	−0.4
PDF-CBAM	19.38	37.32	**37.7**	-

can be attributed to the differing levels of global features extracted by transformers and local features extracted by CNNs, where the application of mixed attention enhances the model's capability to screen features. Furthermore, we discovered that introducing the detail module into the fourth stage of PyraBiNet does not enhance model performance, but instead causes a 0.4 drop in mIoU. This decline is due to the requirement for the extraction of spatial detail information: network depth should be relatively shallow, feature map size large, and a sufficient number of network channels should be available. In the fourth stage of our model, the feature map resolution is excessively small, leading to weakened ability to extract detailed information. This reduction in extraction ability could even introduce noise, resulting in performance degradation.

As shown in Fig. 4, we provide qualitative segmentation results on ADE20K and DOS datasets. The image on the left is the original image, and the image on the right is the semantic segmentation result. As can be observed from the Fig. 4, PyraBiNet demonstrates accurate segmentation of the edges of deformable objects, primarily attributed to the role played by our Detail Feature Block (DF). The DF module enhances the fine details of localized regions, thereby being particularly suited for fine-grained image segmentation tasks. Consequently, this leads to a more precise segmentation of the edges of deformable objects by PyraBiNet.

Fig. 4. Qualitative results of semantic segmentation on ADE20K and DOS datasets.

5   Conclusion

PyraBiNet is a groundbreaking dual-branch architecture adept at navigating the challenges inherent in lightweight semantic segmentation. By strategically integrating the global feature extraction capabilities of PVT with the meticulous local detail extraction of BiSeNet, we realized an efficient feature extraction process. Additionally, our innovative Parallel Dual-Feature Convolutional Block Attention Module facilitated optimal feature fusion while the Detail Feature Block enabled refined feature enhancement. PyraBiNet's superior performance compared to the existing state-of-the-art methods. With its effective and efficient performance, PyraBiNet proves to be an invaluable asset in the domain of mobile robotics, particularly beneficial for applications such as sweeping robots.

References

1. Asgari Taghanaki, S., Abhishek, K., Cohen, J.P., Cohen-Adad, J., Hamarneh, G.: Deep semantic segmentation of natural and medical images: a review. Artif. Intell. Rev. **54**, 137–178 (2021)
2. Chu, X., et al.: Twins: Revisiting the design of spatial attention in vision transformers. Adv. Neural. Inf. Process. Syst. **34**, 9355–9366 (2021)
3. Crespo, J., Castillo, J.C., Mozos, O.M., Barber, R.: Semantic information for robot navigation: A survey. Appl. Sci. **10**(2), 497 (2020)
4. Deng, J., Dong, W., Socher, R., Li, L.J., Li, K., Fei-Fei, L.: Imagenet: a large-scale hierarchical image database. In: 2009 IEEE Conference on Computer Vision and Pattern Recognition, pp. 248–255. Ieee (2009)
5. Dosovitskiy, A., et al.: An image is worth 16×16 words: Transformers for image recognition at scale. arXiv preprint arXiv:2010.11929 (2020)
6. Feng, D., et al.: Deep multi-modal object detection and semantic segmentation for autonomous driving: datasets, methods, and challenges. IEEE Trans. Intell. Transp. Syst. **22**(3), 1341–1360 (2020)
7. Gao, L., Nie, D., Li, B., Ren, X.: Doubly-fused vit: Fuse information from vision transformer doubly with local representation. In: Computer Vision-ECCV 2022: 17th European Conference, Tel Aviv, Israel, October 23–27, 2022, Proceedings, Part XXIII, pp. 744–761. Springer (2022). https://doi.org/10.1007/978-3-031-20050-2_43
8. Graham, B., El-Nouby, A., Touvron, H., Stock, P., Joulin, A., Jégou, H., Douze, M.: Levit: a vision transformer in convnet's clothing for faster inference. In: Proceedings of the IEEE/CVF International Conference on Computer Vision, pp. 12259–12269 (2021)
9. Guo, M.H., Lu, C.Z., Hou, Q., Liu, Z., Cheng, M.M., Hu, S.M.: Segnext: rethinking convolutional attention design for semantic segmentation. arXiv preprint arXiv:2209.08575 (2022)
10. He, K., Zhang, X., Ren, S., Sun, J.: Deep residual learning for image recognition. In: Proceedings of the IEEE Conference on Computer Vision and Pattern Recognition, pp. 770–778 (2016)
11. Ho, J., Kalchbrenner, N., Weissenborn, D., Salimans, T.: Axial attention in multi-dimensional transformers. arXiv preprint arXiv:1912.12180 (2019)
12. Howard, A., et al.: Searching for mobilenetv3. In: Proceedings of the IEEE/CVF International Conference on Computer Vision, pp. 1314–1324 (2019)

13. Howard, A.G., et al.: Mobilenets: efficient convolutional neural networks for mobile vision applications. arXiv preprint arXiv:1704.04861 (2017)
14. Khan, S., Naseer, M., Hayat, M., Zamir, S.W., Khan, F.S., Shah, M.: Transformers in vision: a survey. ACM Comput. Surv. (CSUR) **54**(10s), 1–41 (2022)
15. Kim, W., Seok, J.: Indoor semantic segmentation for robot navigating on mobile. In: 2018 Tenth International Conference on Ubiquitous and Future Networks (ICUFN), pp. 22–25. IEEE (2018)
16. Kirillov, A., Girshick, R., He, K., Dollár, P.: Panoptic feature pyramid networks. In: Proceedings of the IEEE/CVF Conference on Computer Vision and Pattern Recognition, pp. 6399–6408 (2019)
17. Kohli, P., Ladický, L., Torr, P.H.: Robust higher order potentials for enforcing label consistency. Int. J. Comput. Vision **82**, 302–324 (2009)
18. Ladický, L., Russell, C., Kohli, P., Torr, P.H.: Associative hierarchical crfs for object class image segmentation. In: 2009 IEEE 12th International Conference on Computer Vision, pp. 739–746. IEEE (2009)
19. Liu, Y., et al.: A survey of visual transformers. IEEE Trans. Neural Networks Learn. Syst. (2023)
20. Liu, Z., et al.: Swin transformer: hierarchical vision transformer using shifted windows. In: Proceedings of the IEEE/CVF International Conference on Computer Vision, pp. 10012–10022 (2021)
21. Long, J., Shelhamer, E., Darrell, T.: Fully convolutional networks for semantic segmentation. In: Proceedings of the IEEE Conference on Computer Vision and Pattern Recognition, pp. 3431–3440 (2015)
22. Loshchilov, I., Hutter, F.: Decoupled weight decay regularization. arXiv preprint arXiv:1711.05101 (2017)
23. Mehta, S., Rastegari, M.: Mobilevit: light-weight, general-purpose, and mobile-friendly vision transformer. arXiv preprint arXiv:2110.02178 (2021)
24. Mehta, S., Rastegari, M.: Separable self-attention for mobile vision transformers. arXiv preprint arXiv:2206.02680 (2022)
25. Mo, Y., Wu, Y., Yang, X., Liu, F., Liao, Y.: Review the state-of-the-art technologies of semantic segmentation based on deep learning. Neurocomputing **493**, 626–646 (2022)
26. Pan, H., Hong, Y., Sun, W., Jia, Y.: Deep dual-resolution networks for real-time and accurate semantic segmentation of traffic scenes. IEEE Trans. Intell. Transp. Syst. (2022)
27. Russell, B.C., Torralba, A., Murphy, K.P., Freeman, W.T.: Labelme: a database and web-based tool for image. Int. J. of Comput. Vis. **77**(1) (2008). https://doi.org/10.1007/s11263-007-0090-8
28. Sandler, M., Howard, A., Zhu, M., Zhmoginov, A., Chen, L.C.: Mobilenetv 2: Inverted residuals and linear bottlenecks. In: Proceedings of the IEEE Conference on Computer Vision and Pattern Recognition, pp. 4510–4520 (2018)
29. Shotton, J., Winn, J., Rother, C., Criminisi, A.: Textonboost for image understanding: multi-class object recognition and segmentation by jointly modeling texture, layout, and context. Int. J. Comput. Vision **81**, 2–23 (2009)
30. Touvron, H., Cord, M., Douze, M., Massa, F., Sablayrolles, A., Jégou, H.: Training data-efficient image transformers & distillation through attention. In: International Conference on Machine Learning, pp. 10347–10357. PMLR (2021)
31. Tsai, T.H., Tseng, Y.W.: Bisenet v3: bilateral segmentation network with coordinate attention for real-time semantic segmentation. Neurocomputing **532**, 33–42 (2023)

32. Ulku, I., Akagündüz, E.: A survey on deep learning-based architectures for semantic segmentation on 2d images. Appl. Artif. Intell. **36**(1), 2032924 (2022)

33. Wadekar, S.N., Chaurasia, A.: Mobilevitv3: mobile-friendly vision transformer with simple and effective fusion of local, global and input features. arXiv preprint arXiv:2209.15159 (2022)

34. Wang, W., et al.: Pyramid vision transformer: a versatile backbone for dense prediction without convolutions. In: Proceedings of the IEEE/CVF International Conference on Computer Vision, pp. 568–578 (2021)

35. Woo, S., Park, J., Lee, J.-Y., Kweon, I.S.: CBAM: convolutional block attention module. In: Ferrari, V., Hebert, M., Sminchisescu, C., Weiss, Y. (eds.) ECCV 2018. LNCS, vol. 11211, pp. 3–19. Springer, Cham (2018). https://doi.org/10.1007/978-3-030-01234-2_1

36. Wu, H., Xiao, B., Codella, N., Liu, M., Dai, X., Yuan, L., Zhang, L.: Cvt: introducing convolutions to vision transformers. In: Proceedings of the IEEE/CVF International Conference on Computer Vision, pp. 22–31 (2021)

37. Xu, J., Xiong, Z., Bhattacharyya, S.P.: Pidnet: a real-time semantic segmentation network inspired by pid controllers. In: Proceedings of the IEEE/CVF Conference on Computer Vision and Pattern Recognition, pp. 19529–19539 (2023)

38. Yang, C., et al.: Lite vision transformer with enhanced self-attention. In: Proceedings of the IEEE/CVF Conference on Computer Vision and Pattern Recognition, pp. 11998–12008 (2022)

39. Yao, J., Fidler, S., Urtasun, R.: Describing the scene as a whole: Joint object detection, scene classification and semantic segmentation. In: 2012 IEEE Conference on Computer Vision and Pattern Recognition, pp. 702–709. IEEE (2012)

40. Yu, C., Gao, C., Wang, J., Yu, G., Shen, C., Sang, N.: Bisenet v2: bilateral network with guided aggregation for real-time semantic segmentation. Int. J. Comput. Vision **129**, 3051–3068 (2021)

41. Yu, C., Wang, J., Peng, C., Gao, C., Yu, G., Sang, N.: Bisenet: bilateral segmentation network for real-time semantic segmentation. In: Proceedings of the European Conference on Computer Vision (ECCV), pp. 325–341 (2018)

42. Yu, W., et al.: Metaformer is actually what you need for vision. In: Proceedings of the IEEE/CVF Conference on Computer Vision and Pattern Recognition, pp. 10819–10829 (2022)

43. Yuan, K., Guo, S., Liu, Z., Zhou, A., Yu, F., Wu, W.: Incorporating convolution designs into visual transformers. In: Proceedings of the IEEE/CVF International Conference on Computer Vision, pp. 579–588 (2021)

44. Yuan, L., et al.: Tokens-to-token vit: training vision transformers from scratch on imagenet. In: Proceedings of the IEEE/CVF International Conference on Computer Vision, pp. 558–567 (2021)

45. Zhang, W., et al.: Topformer: token pyramid transformer for mobile semantic segmentation. In: Proceedings of the IEEE/CVF Conference on Computer Vision and Pattern Recognition, pp. 12083–12093 (2022)

46. Zhou, B., Zhao, H., Puig, X., Fidler, S., Barriuso, A., Torralba, A.: Scene parsing through ade20k dataset. In: Proceedings of the IEEE Conference on Computer Vision and Pattern Recognition, pp. 633–641 (2017)

Classification of Hard and Soft Wheat Species Using Hyperspectral Imaging and Machine Learning Models

Nitin Tyagi[1]([✉]) [iD], Balasubramanian Raman[1] [iD], and Neerja Garg[2] [iD]

[1] Computer Science and Engineering Department, Indian Institute of Technology, Roorkee, India
{nitin_t,bala}@cs.iitr.ac.in
[2] CSIR-Central Scientific Instruments Organisation, Chandigarh, India
neerjamittal@csio.res.in

Abstract. Ensuring the identification and authenticity of wheat seeds are critical tasks in the food grain industry. In this work, twenty wheat varieties were collected from three different locations in India. The near-infrared (NIR) hyperspectral imaging technique (spectral range 900–1700 nm) was employed in conjunction with machine learning models to discriminate twenty different wheat varieties into two classes: hard wheat and soft wheat. The data images were taken from both sides of the seed (ventral and dorsal side). The dataset includes images of 20,160 seeds. Five different machine learning models were used for classification: Support Vector Machine (SVM), Linear Discriminant Analysis (LDA), Naive Bayes (NB), K-Nearest Neighbor (KNN), and Random Forest (RF). The models were trained using the mean spectral values extracted from the hyperspectral images. Five preprocessing techniques pretreated the mean spectral values of the hyperspectral image: Standard Normal Variate (SNV), Multiplicative Scatter Correction (MSC), Savitzky Golay Smoothing (SG), Savitzky Golay First Derivative (SG-1), and Savitzky Golay Second Derivative (SG-2). The model's performance was evaluated for both raw and preprocessed data. The Support Vector Machine exhibited exceptional performance, attaining an astonishing accuracy rate of 95.01% for amalgamated data (encompassing both ventral and dorsal side data), 95.05% for exclusively ventral side data, and an impressive 95.37% for exclusively dorsal side data.

Keywords: Hyperspectral Imaging (HSI) · Preprocessing · Ventral side (crease up) · Dorsal side (crease down)

1 Introduction

Around the globe, wheat is a cereal grain extensively cultivated and utilized as a food source. India is one of the world's largest wheat producers and is

Supported by the Ministry of Education (MoE) INDIA with reference grant number: OH-3123200428.

ranked second after China. Wheat is classified into two types, hard wheat and soft wheat, with a large number of varieties falling into each category. The production of pasta, noodles, and semolina involves hard wheat, whereas soft wheat is employed to make cakes, pastries, and biscuits that need a finer texture [10]. The identification of the hardness and softness of wheat seeds is a crucial task. Traditionally, it is done manually by professionals based on their experience. Chemically, the hardness and softness of wheat are determined based on its protein and gluten content [11]. More protein and gluten content signifies a hard wheat seed and vice versa. The Kjeldahl method [13] and HPLC (High-Performance Liquid Chromatography) [14] measured protein and gluten content [8]. All these traditional methods are destructive, time-consuming, and require specialized equipment and personnel. Thus, non-destructive techniques are necessary to differentiate between hard and soft wheat varieties [20].

Hyperspectral imaging is a non-destructive technique that effectively captures images of an object in numerous narrow adjacent wavelength bands [7]. This technique provides valuable information about the object's chemical and physical properties and is widely used in various research alongside computer vision. Chaudhary et al. [5] classified eight Canadian wheat classes using HSI of bulk wheat samples. The highest accuracy of 99.1% was obtained by LDA using the top 90 features obtained from wavelet texture analysis. In a study by Bao et al. [3], hyperspectral imaging and chemometric techniques were employed to classify five distinct wheat varieties. The classification task involved the utilization of three classifiers: SVM, LDA, and Extreme Learning Machin (ELM) [22]. Remarkably, the ELM exhibited the highest accuracy, achieving an impressive 91.3% classification accuracy. Sabanci et al. [17] classified wheat seeds using an artificial neural network with a multilayer perceptron and achieved an overall classification accuracy of 99.92%. The majority of studies on wheat seeds have primarily concentrated on the classification of wheat varieties rather than differentiating between hard and soft wheat types. Furthermore, these studies were limited to a few varieties. To our knowledge, no published research has utilized hyperspectral imaging and machine learning techniques to categorize wheat seeds into hard and soft types.

Paper Contribution: The primary work performed in this research is: (1) Wheat sample data were collected from three different states in India, and comprehensive data were acquired by capturing images of both sides of the seed (ventral and dorsal side) utilizing a Hyperspectral Imaging (HSI) system. This HSI system encompasses the wavelength spectrum ranging from 900 nm to 1700 nm. (2) The mean spectral values were computed for each seed, considering both the ventral and dorsal sides, across 147 wavelengths. Additionally, five preprocessing techniques (SNV, MSC, SG, SG-1, SG-2) were implemented to preprocess the mean spectral data. (3) Five machine learning models (SVM, LDA, NB, KNN, RF) were developed, and their efficacy was evaluated and compared in the context of raw and preprocessed data. These models were individually trained for mixed data (ventral and dorsal side data), exclusively for ventral side data, and solely for dorsal side data. A comprehensive assessment of their performance was conducted to determine their comparative capabilities.

The subsequent sections of the paper are framed as follows: Sect. 2 provides an elaborate account of the data collection process and image acquisition techniques. Section 3 presents an extensive overview of the proposed approach. Moving forward, Sect. 4 showcases the results and initiates a detailed discussion. Finally, Sect. 5 encapsulates the study's conclusion.

2   Materials and Methods

2.1   Data Collection

For the current investigation, twenty (20) distinct wheat varieties from the year 2022 were meticulously gathered. These varieties were classified into two categories, namely hard wheat and soft wheat [2]. The wheat samples were acquired from certified wheat grower's institutes in India, ensuring the reliability and authenticity of the collected data. The names of the institutes are Rajasthan Agricultural Research Institute (RARI), DurgaPur Rajasthan, ICAR-Indian Agricultural Research Institute, Indore, Madhya Pradesh, and ICAR-Indian Institute of Wheat and Barley Research, Karnal Haryana. A comprehensive description of the dataset is provided in Table 1. To prepare the dataset, 1008 seeds of each wheat variety were collected manually, with foreign matter and damaged seeds separated and removed. The dataset contains a total of 20,160 seeds (1008 seeds per variety × 20 varieties). The collected sample data was carefully sealed in hermetically sealed polythene bags. The seeds were taken out from the refrigerator and kept at $(25 \pm 1)\,^\circ C$ before 24 h of capturing the image [18].

2.2   Hyperspectral Image Acquisition

The images of the sample seeds were obtained utilizing a line-scan HSI system that operated in reflectance mode [15]. The HSI system included a spectrograph with a wavelength range of 900–1700 nm, an InGaAs camera positioned at a height of 300 mm from the target area, a linear translation stage, a stepping motor to drive the translation stage, and a group of four 35-W halogen bulbs fixed 250 mm above the target area. All components were enclosed in a black metal box and connected to a laptop, as shown in Fig. 1(d). Before capturing the hyperspectral images, the imaging system was turned on for 45 min to ensure thermal and temporal stability, which helped to mitigate any errors that may result from the baseline shift. To obtain clear and undistorted images of the wheat seeds, the scanning speed was set at 14.22 mm/s, while the frame rate was adjusted to 41 Hz. In the beginning, 1008 seeds were initially selected randomly through visual inspection for each variety from the sample seed data. Subsequently, a random subset of seventy-two (72) seeds were manually chosen and arranged in a 12 × 6 matrix on an aluminum plate coated with non-reflective black matte paint for the purpose of capturing images as shown in Fig. 1(c). The images were taken from both sides of the seed [19]. First, the image was captured from the front side (ventral side), and after that, the plate containing the seeds was rotated manually by putting another plate on it; after that, images were taken from the back side (dorsal side). A total of 28 scans (14 plates × 2 sides data) were performed to capture the images. For each variety, images of 2016 seeds (72 seeds per plate × 14 plates × 2 sides data) were captured. The digital intensity values of the raw hyperspectral image were calibrated into reflectance intensity values using the white reference and dark reference using Eq. 1. To acquire the dark reference, the camera lens was diligently obscured with a cap, while a white fluorilon tile was diligently employed as the white reference.

$$Img_c = \frac{Img_{raw} - Img_{dark}}{Img_{white} - Img_{dark}} \times 100 \tag{1}$$

Where Img_c represent the corrected reflectance intensity value, Img_{raw} denotes intensity value of raw hyperspectral image, Img_{dark} denotes intensity value of dark reference image, Img_{white} denotes the intensity value of white reference image. The data was stored in a hypercube containing 168 spectral bands. Nevertheless, the initial fourteen and concluding seven spectral bands were eliminated in this study due to image noise. Consequently, a comprehensive set of 147 spectral bands, spanning from 955.62 nm to 1,688.87 nm, were exclusively employed for the purposes of this investigation.

Table 1. Dataset description

Species	Varieties Name	Location
Soft wheat	RAJ MOLIYA, RAJ 1482, RAJ 3777, RAJ 4037, RAJ 4079, RAJ 4083, RAJ 4120, RAJ 4238, RAJ 3077, RAJ 3765	Rajasthan
Hard wheat	HD 4728, HI 8777, HI 8759, HI 8737, HI 8663, HI 8713, HI 8823	Madhya Pradesh
	DDW 47, PDW 291, A-9-30-1	Haryana

3 Proposed Methodology

3.1 Image Preprocessing and Mean Spectra Extraction

The proposed method consisted of several preprocessing steps to clean the hyperspectral images before extracting the mean spectral values. Hyperspectral images are commonly affected by the dead pixel intensity values that appear in the image due to the poor functioning of the detector during image scanning. The unwanted dead pixel values were eliminated using a median filter of dimension 5×5. The dead pixel was filled with the calculated median value of the surrounding pixels. From the total of 168 spectral bands, a specific spectral band with a wavelength of 1127.65 nm (50th band) was meticulously chosen through visual examination as shown in Fig. 1(g). This selection was based on the optimal contrast observed between the foreground and background regions of the image, ultimately serving the purpose of generating the binary mask. A binary thresholding technique was implemented for the segmentation. A threshold value of 0.19 was selected after many experiments and preliminary tests. A binary mask was created from the chosen spectral band, where any reflectance intensity value greater than 0.19 was assigned the value 1 (white pixel), and values less than or equal to 0.19 were assigned the value 0 (black pixel). Furthermore, in order to extract the region of interest, a meticulous procedure was implemented. The binary mask obtained previously was effectively applied by multiplying it with all 147 spectral bands, ensuring that the desired region was accurately isolated and highlighted for further analysis. The mean spectral values were calculated by taking the average of all the pixel intensities lying in the region of interest. The NIR spectra can be impacted by unwanted interferences, which can be mitigated by utilizing a range of spectral preprocessing methods [12]. This study used six preprocessing methods, including SNV, MSC, SG, SG-1, and SG-2, to preprocess the spectral data. SNV normalizes spectra affected by baseline and path length variations. MSC addresses fluctuations in spectral data arising from variations in sample preparation and measurement conditions. SG is used for smoothing and noise reduction in spectral data by applying a convolution operation with a moving window and polynomial fitting to the data [21]. In addition, the mean spectral values were rigorously utilized as inputs for the machine learning classifiers. This study employed a comprehensive set of five distinct machine-learning models to carry out the classification task, which will be further elaborated in the subsequent section.

3.2 Model Development

In this research, five machine learning classifiers were proficiently utilized for classification purposes. The classifiers, namely SVM, LDA, NB, KNN, and RF, were meticulously employed [6]. To achieve optimal performance, their hyperparameters were rigorously optimized through the grid search approach, employing a five-fold cross-validation approach. In SVM, the penalty parameter (C), gamma, and the radial basis function (RBF) are three critical hyperparameters that were considered [16]. The formula for RBF kernel is given in Eq. 2:

$$K(\mathbf{p}, \mathbf{q}) = \exp\left(-\gamma |\mathbf{p} - \mathbf{q}|^2\right) \tag{2}$$

where, $\mathbf{p}$ and $\mathbf{q}$ are two vectors, $|\cdot|$ represents the Euclidean distance between the vectors, and γ is a hyperparameter that controls the width of the kernel.

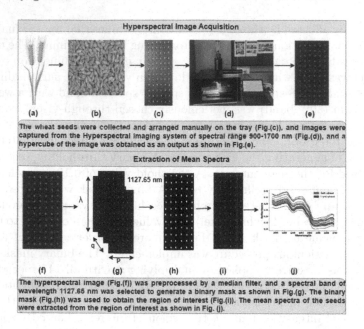

Fig. 1. An overview of image acquisition and proposed methodology.

LDA can be represented as a function $y = \boldsymbol{w}^T\boldsymbol{x}$, where $\boldsymbol{x}$ is an observation and $\boldsymbol{w}$ is the weight vector that best separates the different classes in the dataset. In the Naive Bayes algorithm, the class with the utmost probability was deemed as the optimal prediction for the given item and calculated using Eq. 3 [4].

$$P(S_i|\boldsymbol{y}) = \frac{P(S_i)P(\boldsymbol{y}|S_i)}{P(\boldsymbol{y})} \tag{3}$$

where $P(S_i)$ denotes the preceding probability of class S_i, $P(\boldsymbol{y}|S_i)$ represents the conditional probability of occurring $\boldsymbol{y}$ given that S_i is already happened, and $P(\boldsymbol{y})$ is the marginal probability of observing $\boldsymbol{y}$. In KNN algorithm, the Euclidean distance between the new sample data points and all the training samples was computed utilizing Eq. 4 [23].

$$d(\mathbf{m}, \mathbf{k}) = \sqrt{\sum_{i=1}^{n}(m_i - k_i)^2} \tag{4}$$

where n is the number of features or dimensions in the data points, m_i and k_i are the ith features of the points $\mathbf{m}$ and $\mathbf{k}$, respectively. The random forest algorithm makes predictions based on decision trees [1].

The proposed work used Eq. 5–8 to calculate performance metrics such as accuracy, precision, recall, and F1-score, based on the true positive (TP), true negative (TN), false positive (FP), and false negative (FN) values obtained from the confusion matrix [9].

$$accuracy = \frac{TP + TN}{TP + TN + FP + FN} \qquad (5)$$

$$precision = \frac{TP}{TP + FP} \qquad (6)$$

$$recall = \frac{TP}{TP + FN} \qquad (7)$$

$$F1 - score = 2 \times \frac{precision \times recall}{precision + recall} \qquad (8)$$

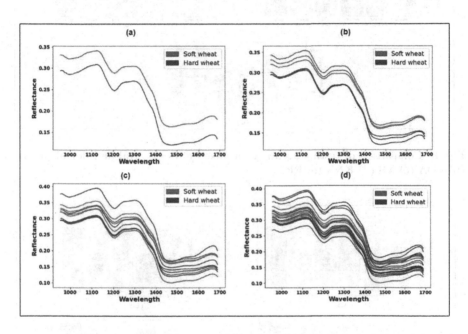

Fig. 2. Comparison of mean spectra of hard and soft wheat varieties (a) mean spectra of two varieties, one from each hard and soft variety (b) mean spectra of six varieties, three from each hard and soft variety (c) mean spectra of ten varieties, five from each hard and soft variety (d) mean spectra of twenty varieties, ten from each hard and soft variety.

4   Results and Discussion

Figure 2 depicts a comparison between the mean spectra of hard and soft wheat covering the wavelength from 955.62 to 1,688.87 nm. These reflectance curves of the wheat sample provide insight into the seed's chemical composition. The absorption peaks observed at 980 nm, 1200 nm, and 1450 nm correspond to the

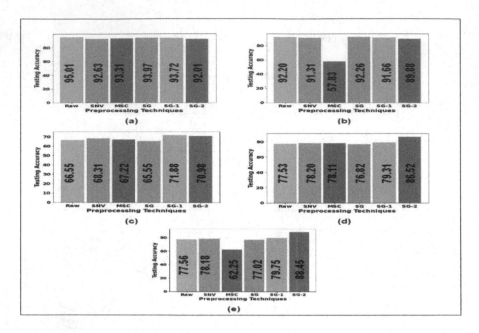

Fig. 3. Comparison of testing accuracy on mixed data (Ventral and dorsal)(a) SVM (b) LDA (c) NB (d) KNN (e) RF

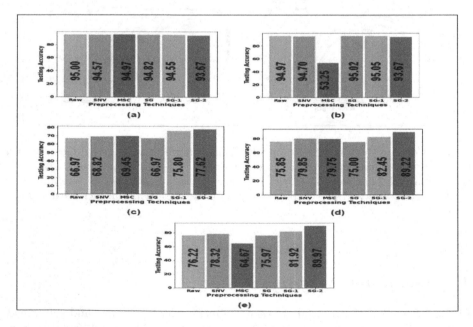

Fig. 4. Comparison of testing accuracy on ventral side wheat data (a) SVM (b) LDA (c) NB (d) KNN (e) RF

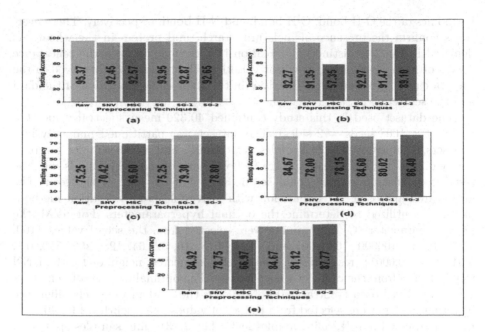

Fig. 5. Comparison of testing accuracy on dorsal side wheat data (a) SVM (b) LDA (c) NB (d) KNN (e) RF

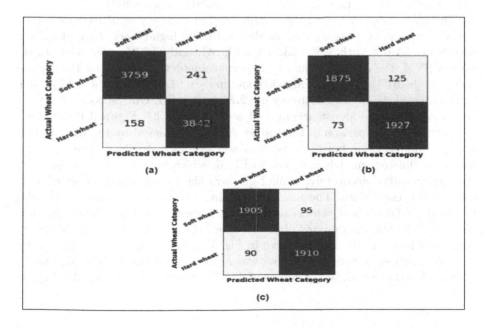

Fig. 6. (a) Confusion matrix of SVM for mixed data (combining ventral and dorsal side) (b) Confusion matrix of LDA combined with SG-1 for ventral side data (c) Confusion matrix of SVM for dorsal side data.

harmonics of the O-H bond, C-H bond, and N-H bond, respectively. These over-tones confirm the presence of moisture, starch, and protein in the wheat. As depicted in Fig. 2, a distinct demarcation is noticeable between the reflectance curves of hard wheat and soft wheat varieties. The fluctuations observed in the spectra can be attributed to disparities in the chemical composition of the data and the scattering of NIR radiation.

The dataset used in this study contained 40,320 mean spectral values (20 varieties × 1008 seeds × 2 sides). The dataset was partitioned manually into two sets, comprising 32,320 samples for training and 8,000 samples for test-ing. We separated the mean spectra of both sides of individual seeds into the test set so that our model could see no values of any side of the seed. The grid search approach, in conjunction with a five-fold cross-validation method-ology, was utilized to determine the optimal hyperparameters. For SVM, the penalty parameter (C) and gamma were selected from the set of values {100, 1000, 10000, 100000, 1000000} and {1, 2, 4, 8, 16, 32, 64, 128, 256, 512, 0.1, 0.01, 0.001, 0.0001}, respectively. The number of nearest neighbor for the KNN were selected from the range of values 2 to 20. The optimal number of neighbors was 7 for mixed data (ventral and dorsal). For the random forest classifier, the hyperparameters were selected from the set of values: max_depth = {30, 40, 50}, max_features = {'auto'}, min_samples_leaf = {1, 2, 3}, min_samples_split = {5, 6, 7}, and n_estimators = {100, 200, 300, 400, 500}. The optimal hyperpa-rameters for the mixed data were {max_depth = 40, max_features = 'auto', min_samples_leaf = 1, min_samples_split = 6, n_estimators = 400}.

The SVM model achieved the highest testing accuracy of 95.01% on unpro-cessed data and surpassed other models with the highest accuracy of 92.63% with SNV, 93.31% with MSC, 93.97% with SG, and 93.72% with SG-1. In the case of SVM, the optimal values of hyperparameters for achieving the highest accuracy (95.01%) were (C = 1000000, gamma = 0.01, kernel = 'rbf'). Moreover, LDA achieved the highest accuracy of 92.26% with SG. On the contrary, Naive Bayes exhibited the lowest accuracy of 65.55% with SG. Figure 3 illustrates a comprehensive comparison of the testing accuracy achieved by the five machine learning models using both unprocessed and preprocessed data. The confusion matrix of different models is shown in Fig. 6, which indicates that wheat vari-eties are classified accurately. Table 2 displays the classification report of SVM using unprocessed data. The machine learning models were also trained using two kinds of data: wheat ventral side data and wheat dorsal side data. The com-parison of the testing accuracy obtained by all the machine learning models for ventral and dorsal side data is shown in Fig. 4 and 5. In the case of ventral side data, the highest accuracy of 95.05% was reported by LDA with SG-1, and for dorsal side data, the highest accuracy of 95.37% was reported by SVM using raw data.

Table 2. Classification report of SVM using mixed data (ventral and dorsal)

Species	Precision	Recall	F1-Score	Support
Soft wheat	95.97	93.97	94.96	4000
Hard wheat	94.10	96.05	95.06	4000

5   Conclusion and Future Work

Our research aimed to create a classification model for differentiating hard and soft wheat varieties based on their chemical properties, utilizing hyperspectral imaging and machine learning models. We evaluated the performance of five different classifiers (SVM, LDA, NB, KNN, and RF) in conjunction with five pre-processing techniques (SNV, MSC, SG, SG-1, SG-2) and compared their results. Our study found that the SVM model outperformed the other models with a high accuracy rate of 95.01% using raw data. This demonstrates the potential of combining HSI and machine learning models for accurate and efficient wheat classification. In the future, we will increase the number of varieties and apply a deep-learning model for the classification.

Acknowledgement. The research work received support from the Ministry of Education (MoE), INDIA under the reference grant number OH-3123200428. Furthermore, one of the authors, Balasubramanian Raman, expressed gratitude for the financial assistance provided by the SERB MATRICS project under file no. MTR/2022/000187.

References

1. Abu Alfeilat, H.A., et al.: Effects of distance measure choice on k-nearest neighbor classifier performance: a review. Big Data **7**(4), 221–248 (2019)
2. Allahverdiyev, T.I., Talai, J.M., Huseynova, I.M., Aliyev, J.A.: Effect of drought stress on some physiological parameters, yield, yield components of durum (Triticum durum desf.) and bread (Triticum aestivum L.) wheat genotypes. Ekin J. Crop Breed. Genet. **1**(1), 50–62 (2015)
3. Bao, Y., Mi, C., Wu, N., Liu, F., He, Y.: Rapid classification of wheat grain varieties using hyperspectral imaging and chemometrics. Appl. Sci. **9**(19), 4119 (2019)
4. Berrar, D.: Bayes' theorem and Naive Bayes classifier. In: Encyclopedia of Bioinformatics and Computational Biology: ABC of Bioinformatics, vol. 403, p. 412 (2018)
5. Choudhary, R., Mahesh, S., Paliwal, J., Jayas, D.: Identification of wheat classes using wavelet features from near infrared hyperspectral images of bulk samples. Biosys. Eng. **102**(2), 115–127 (2009)
6. Fabiyi, S.D., et al.: Comparative study of PCA and LDA for rice seeds quality inspection. In: 2019 IEEE AFRICON, pp. 1–4. IEEE (2019)
7. Feng, L., Zhu, S., Liu, F., He, Y., Bao, Y., Zhang, C.: Hyperspectral imaging for seed quality and safety inspection: a review. Plant Meth. **15**(1), 1–25 (2019)

8. Hacini, N., Djelloul, R., Hadef, A., Samson, M.F., Desclaux, D.: Comparative characterization of grain protein content and composition by chromatography-based separation methods (SE-HPLC and RP-HPLC) of ten wheat varieties grown in different agro-ecological zones of Algeria. Separations **9**(12), 443 (2022)

9. Hossin, M., Sulaiman, M.N.: A review on evaluation metrics for data classification evaluations. Int. J. Data Min. Knowl. Manage. Process **5**(2), 1 (2015)

10. Issarny, C., Cao, W., Falk, D., Seetharaman, K., Bock, J.E.: Exploring functionality of hard and soft wheat flour blends for improved end-use quality prediction. Cereal Chem. **94**(4), 723–732 (2017)

11. Katyal, M., Singh, N., Chopra, N., Kaur, A.: Hard, medium-hard and extraordinarily soft wheat varieties: comparison and relationship between various starch properties. Int. J. Biol. Macromol. **123**, 1143–1149 (2019)

12. Khatri, A., Agrawal, S., Chatterjee, J.M.: Wheat seed classification: utilizing ensemble machine learning approach. Sci. Program. **2022**, 1–9 (2022)

13. Kirk, P.L.: Kjeldahl method for total nitrogen. Anal. Chem. **22**(2), 354–358 (1950)

14. Lozano-Sánchez, J., Borrás-Linares, I., Sass-Kiss, A., Segura-Carretero, A.: Chromatographic technique: high-performance liquid chromatography (HPLC). In: Modern Techniques for Food Authentication, pp. 459–526. Elsevier (2018)

15. Lu, Y., Saeys, W., Kim, M., Peng, Y., Lu, R.: Hyperspectral imaging technology for quality and safety evaluation of horticultural products: a review and celebration of the past 20-year progress. Postharvest Biol. Technol. **170**, 111318 (2020)

16. Qiu, Z., Chen, J., Zhao, Y., Zhu, S., He, Y., Zhang, C.: Variety identification of single rice seed using hyperspectral imaging combined with convolutional neural network. Appl. Sci. **8**(2), 212 (2018)

17. Sabanci, K., Kayabasi, A., Toktas, A.: Computer vision-based method for classification of wheat grains using artificial neural network. J. Sci. Food Agric. **97**(8), 2588–2593 (2017)

18. Sharma, A., Singh, T., Garg, N.: Combining near-infrared hyperspectral imaging and ANN for varietal classification of wheat seeds. In: 2022 Third International Conference on Intelligent Computing Instrumentation and Control Technologies (ICICICT), pp. 1103–1108. IEEE (2022)

19. Singh, T., Garg, N.M., Iyengar, S.R.: Nondestructive identification of barley seeds variety using near-infrared hyperspectral imaging coupled with convolutional neural network. J. Food Process Eng. **44**(10), e13821 (2021)

20. Sricharoonratana, M., Thompson, A.K., Teerachaichayut, S.: Use of near infrared hyperspectral imaging as a nondestructive method of determining and classifying shelf life of cakes. LWT **136**, 110369 (2021)

21. Tyagi, N., Raman, B., Garg, N.M.: Varietal classification of wheat seeds using hyperspectral imaging technique and machine learning models. In: Gupta, D., Bhurchandi, K., Murala, S., Raman, B., Kumar, S. (eds.) Computer Vision and Image Processing, CVIP 2022. Communications in Computer and Information Science, vol. 1777, pp. 253–266. Springer, Cham (2023). https://doi.org/10.1007/978-3-031-31417-9_20

22. Unlersen, M.F., et al.: CNN-SVM hybrid model for varietal classification of wheat based on bulk samples. Eur. Food Res. Technol. **248**(8), 2043–2052 (2022)

23. Zhang, L., et al.: Identification of seed maize fields with high spatial resolution and multiple spectral remote sensing using random forest classifier. Remote Sens. **12**(3), 362 (2020)

Mitigation of Voltage Violation for Battery Fast Charging Based on Data-Driven Optimization

Zheng Xiong, Biao Luo$^{(\boxtimes)}$, and Bingchuan Wang

School of Automation, Central South University, Changsha 410083, China
{xiongzheng,bingcwang}@csu.edu.cn, biao.luo@hotmail.com

Abstract. Fast charging of lithium-ion batteries is a pivotal technology for diminishing charging duration and augmenting user convenience. Nevertheless, with the escalation in battery power, the battery voltage experiences a swift upsurge during fast charging. When the battery voltage surpasses the voltage threshold defined by operational limits, irreversible damage to the battery becomes inevitable. Indeed, the process of fast charging inherently represents a multi-objective optimization challenge, wherein the reduction of charging duration and the prevention of battery voltage violations stand as two opposing objectives. In this work, we introduce a multi-objective reinforcement learning algorithm aimed at devising fast-charging strategies that effectively balance the trade-off between charging duration and battery voltage violation. To begin with, we establish the Doyle-Fuller-Newman (DFN) model for lithium-ion batteries, upon which the framework for the fast-charging process is constructed. Next, the fast charging process is mathematically framed as a Markov decision process (MDP). Subsequently, we present a multi-objective reinforcement learning algorithm tailored to address the MDP problem. Lastly, the effectiveness of the devised algorithm is validated through a series of simulation experiments. The simulation results demonstrate that the introduced algorithm adeptly generates efficient fast-charging strategies for lithium-ion batteries based on the specified preferences.

Keywords: Voltage violation · Battery fast-charge · Reinforcement learning

1 Introduction

In recent years, the proliferation of electric equipment has witnessed a consistent rise, owing to its environmental advantages, heightened efficiency, and convenience. As the most crucial role for energy storage, various types of batteries have been widely used in electric equipment [1]. The lithium-ion battery is one of the most popular batteries due to its high energy density, long cycle life, and low self-discharge rate [2]. However, the charging duration of lithium-ion batteries is

© The Author(s), under exclusive license to Springer Nature Singapore Pte Ltd. 2024
B. Luo et al. (Eds.): ICONIP 2023, CCIS 1968, pp. 577–589, 2024.
https://doi.org/10.1007/978-981-99-8181-6_44

usually long, which is a major obstacle to the development of electric equipment. Therefore, fast charging of lithium-ion batteries is a key technology to reduce charging duration and amplify user convenience [3].

Nowadays, the most common fast-charging method is the constant current constant voltage (CC-CV) charging method [4–6]. The CC-CV charging method is a two-stage charging method, which is divided into the constant current charging stage and the constant voltage charging stage. In the constant current charging stage, the battery is charged with a constant current. And the battery is charged with a constant voltage in the other stage. The CC-CV charging method is simple and easy to implement, but there are potential drawbacks associated with the CC-CV charging method: the uncertainty of charging duration; the degradation of battery capacity; the waste of energy efficiency [7]. Furthermore, a fixed charging strategy fails to accommodate the diverse preferences within the user community. Thus, a novel fast-charging strategy is in urgent need to be developed to overcome the above drawbacks of the CC-CV charging method.

In the past few years, researchers have proposed numerous fast-charging strategies for lithium-ion batteries [8–14]. These strategies can be divided into two categories: model-based strategies; data-driven strategies. Constructing the mathematical battery model with prior knowledge is the most common method to develop model-based strategies. In [8], an early-prediction model and Bayesian optimization algorithm are proposed to predict the battery degradation and optimize the charging strategy, which can alleviate range anxiety and reduce the battery degradation. Zheng et al. [9] proposed an aggregation charging model for a large number of electric vehicles (EVs), and a charging strategy was developed to reduce the power fluctuation level caused by EV charging. In [10], a model-based, health-aware fast-charging strategy for the current control of lithium-ion batteries is proposed to prevent the onset of non-linear aging and to prolong the cycle life of the battery. Model-based methods can effectively improve charging efficiency and reduce battery degradation, but model construction is intricate and time-consuming. Furthermore, model-based techniques often rely on the premise of an accurate battery model, a presumption that may not consistently hold true in practical scenarios.

Data-driven methods are based on the data collected from the real fast-charging process of the battery. In [11], a data-driven multi-agent reinforcement learning algorithm is proposed to learn the charging strategy under different transportation network typologies and fast-charging station planning schemes. Park et al. [12] proposed an optimal-charging procedure based on deep reinforcement learning to minimize the time charging without damaging the cells. Data-driven approaches prove adept at diminishing the intricacies associated with model creation, rendering them better suited for real-world fast-charging scenarios replete with a multitude of uncertain variables.

Current fast-charging strategies primarily concentrate on minimizing charging duration, often disregarding potential battery voltage breaches. As the battery power increases, the battery voltage rises rapidly during fast charging. Once the battery voltage surpasses the maximum allowable operating voltage, the battery may incur unintended damage. Moreover, the permissible operational voltage range varies in accordance with the prevailing operational conditions. Sustaining a high voltage level raises the potential for battery damage, which is incongruent with the preferences of users seeking a prolonged battery lifespan. Conversely, a low voltage prolongs the charging duration, a drawback for users prioritizing quick charging. Consequently, the fast-charging procedure evolves into a multi-objective optimization problem, considering the competing objectives of charging duration and battery voltage violation.

In this work, we propose a multi-objective reinforcement learning algorithm for the fast charging of lithium-ion batteries considering the trade-off between charging duration and battery voltage violation. The main contributions of this work are summarized as follows:

- The lithium-ion battery Doyle-Fuller-Newman (DFN) model is established, and the fast-charge process is formulated into a Markov Decision Process (MDP).
- Aim at the failure of chosen tasks caused by battery limitation, the random task selection strategy is proposed to solve the problem of wrong prediction.
- The multi-objective reinforcement learning algorithm is proposed to address the MDP problem. This algorithm adeptly balances the trade-off between charging duration and battery voltage violation, thereby offering effective solutions.

2 Lithium-Ion Battery Model

2.1 DFN Model

In this section, the mathematical battery model is established based on the porous electrode theory. The porous electrode theory is based on the assumption that the solid-phase diffusion is the rate-limiting step in the charge and discharge process, in which Li-ions are intercalated in spherical particles in the positive and negative electrodes [15]. The Doyle-Fuller-Newman (DFN) model, as the most popular lithium-ion model, is considered [16]. During charging, the Li-ions de-intercalate from a solid phase in the porous positive electrode, represented by spherical particles, where they intercalate in the electrode. And the reverse process occurs during the discharging period. Figure 1 shows the charging/discharging process and the computation domain of the lithium-ion battery model. There are three regions in the computation domain: the positive electrode, the separator, and the negative electrode. The positive electrode and the negative electrode are separated inside of the battery by the separator and connected by the current collector from the outside. In this model, five partial differential equations (PDEs) are used to describe the conservation of the lithium-ions in the solid phase and the electrolyte phase and the conservation of the electrons in the solid phase.

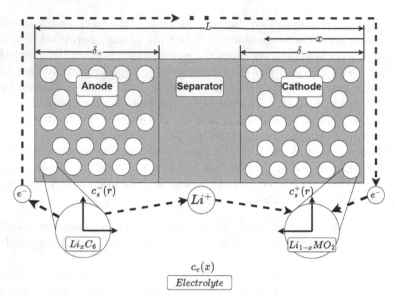

Fig. 1. The DFN model of the lithium-ion battery.

Solid-Phase Diffusion. The diffusion of the lithium-ion in the solid phase of both the positive and negative electrodes is derived from Fick's second law, which is given by:

$$\frac{\partial c_s}{\partial t} = \frac{1}{r^2}\frac{\partial}{\partial r}(r^2 D_s \frac{\partial c_s}{\partial r}), \tag{1}$$

with the boundary conditions:

$$\left.\frac{\partial c_s}{\partial r}\right|_{r=0} = 0, \tag{2}$$

$$-D_s \left.\frac{\partial c_s}{\partial r}\right|_{r=R_s} = \frac{j_{Li}}{a_s F}. \tag{3}$$

where c_s is the concentration of lithium-ion in the solid phase, D_s is the diffusion coefficient of lithium-ion in the solid phase, r is the radial coordinate, R_s is the radius of the spherical particle, j_{Li} is the current density, a_s is the surface area of the spherical particle, and F is the Faraday constant.

Electrolyte-Phase Diffusion. The diffusion of the lithium-ion in the electrolyte phase is similar to the solid phase diffusion, which is given by:

$$\varepsilon_e \frac{\partial c_e}{\partial t} = \frac{\partial}{\partial x}(D_e^{eff}\frac{\partial c_e}{\partial x}) + \frac{1 - t_+^0}{F}j_{Li}, \tag{4}$$

with the boundary conditions:

$$D_e^{eff} \left.\frac{\partial c_e}{\partial x}\right|_{x=0} = D_e^{eff} \left.\frac{\partial c_e}{\partial x}\right|_{x=L} = 0, \tag{5}$$

where ε_e is the porosity of the electrolyte phase, c_e is the concentration of lithium-ion in the electrolyte phase, D_e^{eff} is the effective diffusion coefficient of lithium-ion in the electrolyte phase, x is the coordinate in the direction of the thickness of the electrode, t_+^0 is the transference number of lithium-ion in the electrolyte phase, and L is the thickness of the battery.

Potential of the Solid Phase. The potential of the solid phase ϕ_s is described by Ohm's law, which is given by:

$$\frac{\partial}{\partial x}\left(\sigma^{eff}\frac{\partial \phi_s}{\partial x}\right) = j_{Li}, \tag{6}$$

with the boundary conditions:

$$-\sigma^{eff}\frac{\partial \phi_s}{\partial x}\bigg|_{x=0} = \frac{i_{app}}{A_{surf}}, \tag{7}$$

$$\sigma^{eff}\frac{\partial \phi_s}{\partial x}\bigg|_{x=\delta_-} = 0 \tag{8}$$

for the negative electrode and

$$\sigma^{eff}\frac{\partial \phi_s}{\partial x}\bigg|_{x=L-\delta_+} = 0, \tag{9}$$

$$-\sigma^{eff}\frac{\partial \phi_s}{\partial x}\bigg|_{x=L} = \frac{i_{app}}{A_{surf}} \tag{10}$$

for the positive electrode, where σ^{eff} is the effective conductivity of the solid phase, i_{app} is the applied current ($i_{app} < 0$ indicates charging), A_{surf} is the surface area of the electrode, δ_- is the thickness of the negative electrode current collector, and δ_+ is the thickness of the positive electrode current collector.

Potential of the Electrolyte Phase. The potential of the electrolyte phase ϕ_e is similar to the solid phase potential, which is given by:

$$\frac{\partial}{\partial x}\left(k^{eff}\frac{\partial \phi_e}{\partial x} + k_D^{eff}\frac{\partial \ln c_e}{\partial x}\right) = -j_{Li}, \tag{11}$$

with boundary conditions:

$$k^{eff}\frac{\partial \phi_e}{\partial x}\bigg|_{x=0} = k^{eff}\frac{\partial \phi_e}{\partial x}\bigg|_{x=L} = 0, \tag{12}$$

where k^{eff} is the effective ionic conductivity of the electrolyte phase, and k_D^{eff} is the effective diffusion coefficient of lithium-ion in the electrolyte phase, which is described as:

$$k_D^{eff} = \frac{2RTk^{eff}}{F}(t_+^0 - 1)\left(1 + \frac{\partial \ln f_\pm}{\partial \ln c_e}\right), \tag{13}$$

where R is the universal gas constant, T is the absolute temperature, and $f_{\pm}$ is the activity coefficient of the lithium-ion in the electrolyte phase. The battery voltage, represented as V_{cell}, is established as the potential difference between the current collectors, which is

$$V_{\text{cell}} = \phi_s\Big|_{x=L} - \phi_s\Big|_{x=0}. \tag{14}$$

To accurately track the progress of battery charging, the state of charge (SoC) in this work is defined as the ratio of the lithium-ion concentrations in the negative electrode to the maximum concentrations, which is given by:

$$SoC(t) = \frac{\int_0^{\delta_-} c_s(x,t)dx/\delta_- - c_{s,min}^-}{c_{s,max}^- - c_{s,min}^-}, \tag{15}$$

where $c_{s,min}^-$ is the minimum concentration of lithium-ion in the negative electrode, and $c_{s,max}^-$ is the maximum concentration. Generally speaking, when regarding the DFN model as an opaque system, the applied current i_{app} serves as its input, while the resultant battery voltage V_{cell} and state of charge (SoC) SoC_t constitute its output.

3   Multi-objective Fast-Charging Strategy

In this section, we articulate the fast-charging procedure within the framework of a Markov decision process (MDP). Subsequently, we put forth the multi-objective reinforcement learning algorithm.

3.1   Markov Decision Process Formulation

The MDP is usually defined as a tuple $<S, A, P, R, \gamma>$, where S is the state space, A is the action space, P is the state transition probability, R is the reward function, and γ is the discount factor. Each component is amply defined as follows.

State Space. The state space S is defined as the set of all possible states of the battery, which is given by:

$$s_t = \{SoC_t, V_t\} \in S \tag{16}$$

where s_t is the state of the battery at time step t, which contains the current SoC SoC_t and the current battery voltage V_t.

Action Space. The action space A is defined as the set of all possible actions of the battery, which is given by:

$$a_t = \{i_{\text{app},t}\} \in A \tag{17}$$

where a_t is the action of the battery at time step t, which is the current applied current $i_{\text{app},t}$.

State Transition Probability. The state transition probability P is defined as the probability of the battery transiting from state s_t to state s_{t+1} after taking action a_t, which is given by:

$$P(s_{t+1}|s_t, a_t) \tag{18}$$

This probability is determined by the constructed DFN model, which is described in Sect. 2.1.

Reward Function. The reward $r(s_t, a_t)$ is defined as the reward of the battery after taking action a_t at state s_t, which comprises two distinct components:

$$r(s_t, a_t) = \{r_s(s_t, a_t), r_v(s_t, a_t)\} \in R \tag{19}$$

where $r_s(s_t, a_t)$ is the speed reward quantifying the fast-charging objective, and $r_v(s_t, a_t)$ is the voltage reward quantifying the voltage violation objective, which is separately given by:

$$r_s(s_t, a_t) = \alpha(100(SoC_{t+1} - SoC_t))^2 \tag{20}$$

$$r_v(s_t, a_t) = \beta \frac{V_{max} - V_t}{V_{max} - V_{min}} \tag{21}$$

where α and β are the coefficients of the two objectives, V_{max} and V_{min} are the upper and lower battery cut-off voltage, respectively, and which are set as $4.2V$ and $2.5V$.

3.2 Multi-objective Reinforcement Learning Algorithm

After formulating the fast-charge process as an MDP, the multi-objective reinforcement learning algorithm is proposed to address the MDP problem. The multi-objective reinforcement learning algorithm is based on the prediction-guided multi-objective reinforcement learning algorithm proposed in [17].

The fundamental purpose of fast-charging strategies is to charge the battery considering the trade-off between the charging duration and the battery voltage violation. This intricate task embodies a multi-objective optimization predicament. Thus, given a policy π_θ and a weight vector w, the multi-objective reinforcement learning algorithm is used to find the optimal policy π^* to maximize the weighted-sum reward, which is given by:

$$J(\theta, w) = \sum_{i=1}^{N} w_i J_i^{\pi_\theta} \tag{22}$$

where N is the number of objectives, w_i is the weight of the i_{th} objective, and $J_i^{\pi_\theta}$ is the total reward of the i_{th} objective, which is given by:

$$J_i^{\pi_\theta} = E_{\pi_\theta}[\sum_{t=0}^{T} r_i(s_t, a_t)] \tag{23}$$

To update the policy, the gradient of the weighted-sum reward is calculated, which is given by:

$$\nabla_\theta J(\theta, w) = \sum_{i=1}^{N} w_i \nabla_\theta J_i^{\pi_\theta}$$

$$= E[\sum_{t=0}^{T} w^T A^\pi(s_t, a_t) \nabla_\theta \log \pi_\theta(a_t|s_t)] \qquad (24)$$

$$= E[\sum_{t=0}^{T} A_w^\pi(s_t, a_t) \nabla_\theta \log \pi_\theta(a_t|s_t)]$$

where $A^\pi(s_t, a_t)$ is the vectorized advantage function, and $A_w^\pi(s_t, a_t)$ is the weighted-sum advantage function. In this paper, the Proximal Policy Optimization (PPO) [18] is applied to update the policy, and the Generalized Advantage Estimation (GAE) [19] is used to estimate the advantage function and the target values.

Algorithm 1. Multi-objective reinforcement learning algorithm

---

Input: The number of generations G, the number of tasks n, the weight of the spacing metric α, task iterations m_t, warm-up iterations m_w.

Initialize population $\mathcal{P}$, external pareto archive EP, and RL history record R.

Generate task set $\mathcal{T} = (\pi_i, w_i)_{i=1}^n$ by random initialized policies and evenly distributed weight vectors.

for task $(\pi_i, w_i) \in \mathcal{T}$ **do**

    Run multi-objective RL algorithm with task (π_i, w_i) for m_w iterations.

    Evaluate the trained task (π_i, w_i).

    Collect the final policies and the corresponding objectives in the offspring population $\mathcal{P}'$.

end for

Set the Parato archive EP as $\mathcal{P}'$.

for generation $g = 1, 2, ..G$ **do**

    Randomly select the tasks $\mathcal{T} = \{(\pi_i, w_i)\}_{i=1}^n$ from EP.

    for task $i = 1, 2, ..., n$ **do**

        Run multi-objective RL algorithm with task (π_i, w_i) for m_t iterations.

        Evaluate the trained task (π_i, w_i).

        Collect the final policies and the corresponding objectives in the offspring population $\mathcal{P}'$.

    end for

    Update the Parato archive EP with $\mathcal{P}'$.

end for

Output: The Pareto front EP.

Within the framework of the DFN battery model, an episode concludes either upon the battery voltage attaining the cut-off voltage or upon the completion of the charging task. Additionally, the maximum duration of an episode is stipulated as 200. Hence, it is plausible that the policy improvement prediction model proposed in [17] might not be well-suited for the DFN battery model, primarily due to the presence of discontinuous policy enhancements, as depicted in Fig. 2. To address the issue of inaccurate predictions, the agent adopts a strategy wherein it randomly opts for the next-generation task from the Pareto front of the ongoing generation. Although the random strategy may decrease the efficiency of the algorithm, it can prevent the next generation from facing a shortage of tasks caused by selecting unreachable prediction tasks.

In each generation, once the tasks are selected by the random strategy, the agent will train the policy network and the value network with the selected tasks to seek the maximum weighted-sum metric, i.e., $\mathcal{H}(EP^*) + \alpha_s \mathcal{S}(EP^*)$ of the current Pareto archive EP, where EP^* is the updated Pareto archive after the training process, $\mathcal{H}$ and $\mathcal{S}$ are the hyper-volume and spacing metric, respectively, and α_s is the weight of the spacing metric ($\alpha_s < 0$ for minimizing the sparsity metric). Thus, the optimization process can be formulated as follows:

$$
\max_{\mathcal{T}=\{(\pi_i, w_i)\}_{i=1}^n} \mathcal{Q}(EP, \mathcal{T}) = \mathcal{H}(EP^*) + \alpha_s \mathcal{S}(EP^*)
$$
$$
with \quad EP^* = Pareto(EP \cup \mathcal{P}')
$$
(25)

where $\mathcal{T}$ is the set of tasks, n is the number of tasks, $\mathcal{P}'$ is the collected offspring population containing the objectives and the corresponding policies, and $Pareto$ is the function computing the Pareto front. The whole process of the multi-objective reinforcement learning algorithm is shown in Algorithm 1.

Table 1. The parameters of RL agent

Parameters	Value
Discount factor γ	0.995
Learning rate	0.00005
Entropy coefficient	0
GAE coefficient λ	0.95
PPO clip parameter ϵ	0.2
Batch size	64
Number of epochs	10
Number of tasks n	6
Number of generations G	400
Task iterations m_t	10
Warm-up iterations m_w	20
Weight of spacing metric α_s	−1
Coefficients of reward α, β	2, 0.6

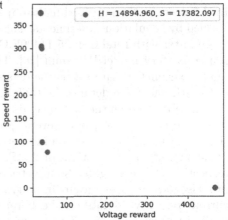

Fig. 2. The Pareto archive obtained by training the randomly selected tasks in the warm-up stage.

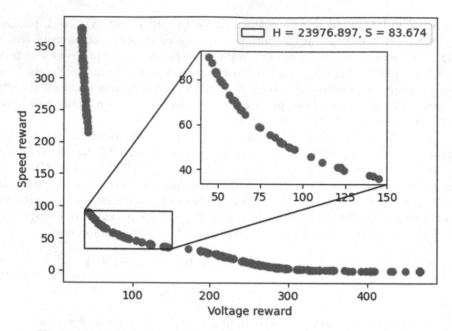

Fig. 3. The Pareto front of the final generation. (Color figure online)

4   Simulation Results

In this section, the proposed multi-objective reinforcement learning algorithm is verified by simulation experiments. The simulation experiments are conducted on a computer with Intel Core i5-11400F CPU and 16 GB RAM, based on an open-source battery model PyBamm [20]. The RL environmental parameters (DFN model parameters) are set as the default in the PyBamm, and the parameters of the RL agent are detailed in Table 1.

The objectives of the first generation Pareto archive are randomly distributed in the objective space, only several points are located in the Pareto front, as shown in Fig. 2. The hyper-volume metric $\mathcal{H}(EP)$ and the spacing metric $\mathcal{S}(EP)$ are shown at the upper right of the Fig. 2, which are 14894.960 and 17382.097, respectively. The weighted-sum metric $\mathcal{Q}(EP, \mathcal{T}) = -2486.137$. As the training process goes on, the Pareto front gradually becomes clear by expanding the current objective points on the Pareto front, as shown in Fig. 3. The hyper-volume metric $\mathcal{H}(EP)$ and the spacing metric $\mathcal{S}(EP)$ are shown at the upper right of Fig. 3, which are 23976.897 and 83.674, respectively. The weighted-sum metric $\mathcal{Q}(EP, \mathcal{T})$ exhibits an increase, reaching a value of 23893.223. It's clear that the Pareto front is divided into three parts. The upper blue points in the figure are the strategies strongly focusing on the charging duration, which leads to the violation of the battery voltage. The lower blue points in the figure are the strategies strongly focusing on the battery voltage violation, which causes the SoC to be unable to reach 80%. Meanwhile, the intermediate green points

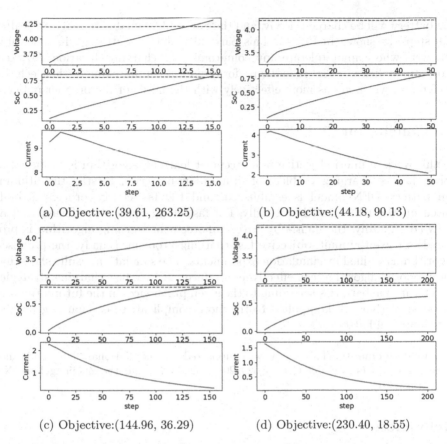

(a) Objective:(39.61, 263.25)

(b) Objective:(44.18, 90.13)

(c) Objective:(144.96, 36.29)

(d) Objective:(230.40, 18.55)

Fig. 4. The charging strategy performance of the several chosen objective points. (Color figure online)

in the figure represent viable strategies that emphasize the delicate equilibrium between charging duration and battery voltage. These points correspond to the authentic Pareto front necessitated by the rapid charging procedure.

For a more precise statement, several results of policies with different preferences, which are the points in Fig. 3, are demonstrated in the following Fig. 4. The results are ordered by the preference of the objectives. The red dashed lines depicted in the illustration correspond to the upper cut-off voltage, whereas the green dashed lines signify the 80% state of charge (SoC) threshold. It's obvious that the voltage violation is mitigated and the charging duration is prolonged as the first entry of the objective vector (voltage reward) increases. As a conflicting objective, the second entry of the objective vector (speed reward) follows the opposite trend of the voltage reward. As shown in Fig. 4a, when over-considering the speed, the battery voltage will surpass the cut-off voltage. Voltage violations pose security risks, rendering them unsuitable for applications requiring a high degree of reliability. On the contrary, when voltage violation is over-considered, the battery will be charged slowly, and the SoC will be unable to reach 80% after 200 steps, as shown in Fig. 4d. This particular charging strategy is unsuitable for users who cannot tolerate any compromise on charging duration. Furthermore, the policies in Fig. 4b and c perform more acceptably due to their eclectic preference, which aligns more effectively with the needs of the broader public.

5   Conclusion

In this work, a multi-objective reinforcement learning algorithm is proposed to solve the fast-charging problem of lithium-ion batteries. Firstly, the lithium-ion battery DFN model is established, and the fast-charge process is built based on the DFN model. Secondly, the fast-charge process is formulated as an MDP. Thirdly, the multi-objective reinforcement learning algorithm is proposed to solve the multi-objective fast-charging problem. Finally, the proposed algorithm is verified by simulation experiments. The simulation results show that the proposed algorithm can effectively produce effective fast-charging strategies for lithium-ion batteries according to the given preference. In the future, the proposed algorithm will be applied to the more complicated fast-charging process of lithium-ion batteries.

Acknowledgements. This study was supported by the National Natural Science Foundation of China under Grants 62022094, 62373375, and the Zhejiang Lab (No. 2021NB0AB01).

References

1. Divya, K.C., Østergaard, J.: Battery energy storage technology for power systems–an overview. Electr. Power Syst. Res. **79**(4), 511–520 (2009)
2. Teki, R., et al.: Nanostructured silicon anodes for lithium ion rechargeable batteries. Small **5**(20), 2236–2242 (2009)
3. Azadfar, E., Sreeram, V., Harries, D.: The investigation of the major factors influencing plug-in electric vehicle driving patterns and charging behaviour. Renew. Sustain. Energy Rev. **42**, 1065–1076 (2015)
4. Cope, R.C., Podrazhansky, Y.: The art of battery charging. In: Proceedings of the Conference on Fourteenth Annual Battery Conference on Applications and Advances (Cat. No. 99TH8371), January 1999
5. Jiang, L., et al.: Optimization of multi-stage constant current charging pattern based on Taguchi method for li-ion battery. Appl. Energy **259**, 114–148 (2020)
6. Yin, Y., Hu, Y., Choe, S.-Y., Cho, H., Joe, W.T.: New fast charging method of lithium-ion batteries based on a reduced order electrochemical model considering side reaction. J. Power Sources **423**, 367–379 (2019)
7. Ruan, H., Barreras, J.V., Engstrom, T., Merla, Y., Millar, R., Wu, B.: Lithium-ion battery lifetime extension: a review of derating methods. J. Power Sources **563**, 232805 (2023)
8. Attia, P.M., et al.: Closed-loop optimization of fast-charging protocols for batteries with machine learning. Nature **578**(7795), 397–402 (2020)

9. Zheng, J., Wang, X., Men, K., Zhu, C., Zhu, S.: Aggregation model-based optimization for electric vehicle charging strategy. IEEE Trans. Smart Grid 4(2), 1058–1066 (2013)
10. Wassiliadis, N., Kriegler, J., Gamra, K.A., Lienkamp, M.: Model-based health-aware fast charging to mitigate the risk of lithium plating and prolong the cycle life of lithium-ion batteries in electric vehicles. J. Power Sources 561, 232586 (2023)
11. Tao, Y., Qiu, J., Lai, S., Sun, X., Zhao, J.: A data-driven agent-based planning strategy of fast-charging stations for electric vehicles. IEEE Trans. Sustain. Energy 14(3), 1357–1369 (2023)
12. Park, S., et al.: A deep reinforcement learning framework for fast charging of Li-ion batteries. IEEE Trans. Transp. Electr. 8(2), 2770–2784 (2022)
13. Chen, K.-H., et al.: Efficient fast-charging of lithium-ion batteries enabled by laser-patterned three-dimensional graphite anode architectures. J. Power Sources 471, 228475 (2020)
14. Wei, Z., Yang, X., Li, Y., He, H., Li, W., Sauer, D.U.: Machine learning-based fast charging of lithium-ion battery by perceiving and regulating internal microscopic states. Energy Storage Mater. 56, 62–75 (2023)
15. Parhizi, M., Pathak, M., Ostanek, J.K., Jain, A.: An iterative analytical model for aging analysis of Li-ion cells. J. Power Sources 517, 230667 (2022)
16. Doyle, M., Fuller, T.F., Newman, J.: Modeling of galvanostatic charge and discharge of the lithium/polymer/insertion cell. J. Electrochem. Soc. (US) 140, 1526 1993
17. Xu, J., Tian, Y., Ma, P., Rus, D., Sueda, S., Matusik, W.: Prediction-guided multi-objective reinforcement learning for continuous robot control. In: Proceedings of the 37th International Conference on Machine Learning (2020)
18. Schulman, J., Wolski, F., Dhariwal, P., Radford, A., Klimov, O.: Proximal policy optimization algorithms. CoRR, abs/1707.06347 (2017)
19. Schulman, J., Moritz, P., Levine, S., Jordan, M., Abbeel, P.: High-dimensional continuous control using generalized advantage estimation (2018)
20. Sulzer, V., Marquis, S.G., Timms, R., Robinson, M., Chapman, S.J.: Python battery mathematical modelling (PyBaMM). J. Open Res. Softw. 9(1), 14 (2021)

Author Index

Printed in the United States
by Baker & Taylor Publisher Services